Lecture Notes in Computer Science 7962

Commenced Publication in 1973
Founding and Former Series Editors:
Gerhard Goos, Juris Hartmanis, and Jan van Leeuwen

Matti Järvisalo   Allen Van Gelder (Eds.)

Theory and Applications of Satisfiability Testing – SAT 2013

16th International Conference
Helsinki, Finland, July 8-12, 2013
Proceedings

 Springer

Volume Editors

Matti Järvisalo
University of Helsinki
HIIT and Department of Computer Science
Gustaf Hällströmin katu 2b
00014 Helsinki, Finland
E-mail: matti.jarvisalo@cs.helsinki.fi

Allen Van Gelder
University of California at Santa Cruz
Computer Science Department
1156 High Street, SOE-3
Santa Cruz, CA 95064, USA
E-mail: avg@cs.ucsc.edu

ISSN 0302-9743   e-ISSN 1611-3349
ISBN 978-3-642-39070-8   e-ISBN 978-3-642-39071-5
DOI 10.1007/978-3-642-39071-5
Springer Heidelberg Dordrecht London New York

Library of Congress Control Number: 2013941358

CR Subject Classification (1998): F.3.1, F.3, F.1, F.4.1, F.2, I.2, B.7, G.1.6

LNCS Sublibrary: SL 1 – Theoretical Computer Science and General Issues

Typesetting: Camera-ready by author, data conversion by Scientific Publishing Services, Chennai, India

Printed on acid-free paper

Springer is part of Springer Science+Business Media (www.springer.com)

Preface

This volume contains the papers presented at the 16th International Conference on Theory and Applications of Satisfiability Testing (SAT 2013) held during July 8–12, 2013, in Helsinki, Finland. SAT 2013 was organized by University of Helsinki in collaboration with Helsinki Institute for Information Technology (HIIT) and the SAT Association.

The International Conference on Theory and Applications of Satisfiability Testing (SAT) is the primary annual meeting for researchers focusing on the theory and applications of the propositional satisfiability problem, broadly construed: besides plain propositional satisfiability, it includes Boolean optimization (including MaxSAT and Pseudo-Boolean (PB) constraints), Quantified Boolean Formulas (QBF), Satisfiability Modulo Theories (SMT), and Constraint Programming (CP) for problems with clear connections to propositional reasoning. Many hard combinatorial problems can be tackled using SAT-based techniques, including problems that arise in formal verification, artificial intelligence, operations research, biology, cryptology, data mining, machine learning, mathematics, etc. Indeed, the theoretical and practical advances in SAT research over the past 20 years have contributed to making SAT technology an indispensable tool in various domains.

SAT 2013 welcomed scientific contributions addressing different aspects of SAT, including (but not restricted to) theoretical advances (including exact algorithms, proof complexity, and other complexity issues), practical search algorithms, knowledge compilation, implementation-level details of SAT solvers and SAT-based systems, problem encodings and reformulations, applications, as well as case studies and reports on insightful findings based on rigorous experimentation.

A total of 71 papers were submitted to SAT 2013, distributed into 50 regular papers, 15 short papers, and six tool papers. Four regular paper submissions were found by the Program Committee to be out of scope for the conference (based on guidelines in the call for papers), and were returned without review. The 67 remaining paper submissions were assigned for review to at least four Program Committee members and their selected external reviewers. Continuing the procedure initiated in SAT 2012, the review process included an author-response period, during which the authors of submitted papers were given the opportunity to respond to the initial reviews for their submissions. For reaching final decisions, a Program Committee discussion period followed the author-response period. This year, external reviewers used by the Program Committee were also invited to participate directly in the discussions for the papers they reviewed. In the end, the Program Committee decided to accept 21 regular papers, five short papers, and five tool papers.

In addition to presentations on the accepted papers, the scientific program of SAT 2013 included three invited talks:

- Albert Atserias (Technical University of Catalonia, Spain):
  The Proof-Search Problem between Bounded-Width Resolution and Bounded-Degree Semi-Algebraic Proofs;
- Edmund M. Clarke (Carnegie Mellon University, USA):
  Turing's Computable Real Numbers and Why They Are Still Important Today;
- Peter Stuckey (NICTA and University of Melbourne, Australia):
  There Are No CNF Problems.

For SAT 2013, open calls for workshops and competitions were issued. As a result, SAT 2013 hosted various affiliated events, including three workshops during July 8–9:

- 11th International Workshop on Satisfiability Modulo Theories (SMT 2013)
  Organizers: Roberto Bruttomesso and Alberto Griggio;
- Fourth International Workshop on Pragmatics of SAT (PoS 2013)
  Organizers: Daniel Le Berre and Allen Van Gelder;
- First International Workshop on Quantified Boolean Formulas (QBF 2013)
  Organizers: Florian Lonsing and Martina Seidl;

and five competitions and system evaluations:

- Configurable SAT Solver Challenge 2013
  Organizers: Frank Hutter, Adrian Balint, Sam Bayless, Holger Hoos, Kevin Leyton-Brown;
- Max-SAT Evaluation 2013
  Organizers: Josep Argelich, Chu-Min Li, Felip Manyà, Jordi Planes;
- SAT Competition 2013
  Organizers: Adrian Balint, Anton Belov, Marijn Heule, Matti Järvisalo;
- SMT-EVAL 2013
  Organizers: David Cok, Aaron Stump, Tjark Weber;
- QBF Gallery 2013
  Organizers: Florian Lonsing, Martina Seidl, Allen Van Gelder.

Olivier Roussel kindly agreed to act as SAT 2013 Competitions Chair, taking the main responsibility for the decisions on competition proposals.

We would like to thank all the people who contributed to making SAT 2013 a success. We thank the SAT Association Chair Armin Biere, Vice Chair John Franco, and Treasurer Hans Kleine Büning for their help and advice in organizational matters. We thank the Steering Committee for selecting Helsinki as the venue for SAT 2013. We thank the local organization team for their efforts with practical aspects of local organization. We thank all authors who submitted their work to SAT 2013. We thank the members of the SAT 2013 Program Committee and the external reviewers for their hard work on guaranteeing the scientific quality of accepted papers. A special thanks goes to the SAT 2013 invited speakers for accepting our invitations. We also thank the organizers of the affiliated workshops and competitions and system evaluations. The EasyChair conference system provided invaluable assistance in coordinating the submission and review process, as well as in the assembly of these proceedings.

Finally, we gratefully thank University of Helsinki, Federation of Finnish Learned Societies (Tieteellisten seurain valtuuskunta), SAT Association, *AI Journal*, IBM Research, Intel, and Microsoft Research for financial and organizational support for SAT 2013.

May 2013   Matti Järvisalo
                                                             Allen Van Gelder

Conference Organization

Program Committee Chairs

Matti Järvisalo   University of Helsinki, Finland
Allen Van Gelder   University of California at Santa Cruz, USA

Competitions Chair

Olivier Roussel   Artois University, France

Local Chair

Matti Järvisalo   University of Helsinki, Finland

Steering Committee

Roberto Sebastiani   University of Trento, Italy
Alessandro Cimatti   FBK-Irst, Italy
Karem Sakallah   University of Michigan, USA
Laurent Simon   University of Paris 11, France
Stefan Szeider   Vienna Institute of Technology, Austria
Ofer Strichman   Technion, Israel
Oliver Kullmann   Swansea University, UK

Program Committee

Gilles Audemard   Artois University, France
Fahiem Bacchus   University of Toronto, Canada
Armin Biere   Johannes Kepler University, Austria
María Luisa Bonet   Technical University of Catalonia, Spain
Lucas Bordeaux   Microsoft Research, UK
Uwe Bubeck   University of Paderborn, Germany
Samuel Buss   University of California at San Diego, USA
Nadia Creignou   Aix-Marseille University, France
Leonardo de Moura   Microsoft Research, USA
John Franco   University of Cincinnati, USA
Enrico Giunchiglia   University of Genoa, Italy
Ziyad Hanna   Jasper Automation, USA
Marijn Heule   The University of Texas at Austin, USA
Holger H. Hoos   University of British Columbia, Canada
Jinbo Huang   NICTA, Australia
Tommi Junttila   Aalto University, Finland
Matti Järvisalo   University of Helsinki, Finland
Arist Kojevnikov   Intel, Germany

Daniel Kröning	Oxford University, UK
Oliver Kullmann	Swansea University, UK
Daniel Le Berre	Artois University, France
Florian Lonsing	Vienna University of Technology, Austria
Inês Lynce	Technical University of Lisbon, Portugal
Joao Marques-Silva	University College Dublin, Ireland
Alexander Nadel	Intel, Israel
Jakob Nordström	KTH Royal Institute of Technology, Sweden
Albert Oliveras	Technical University of Catalonia, Spain
Ramamohan Paturi	University of California at San Diego, USA
Jussi Rintanen	Aalto University, Finland
Olivier Roussel	Artois University, France
Ashish Sabharwal	IBM Research, USA
Lakhdar Sais	Artois University, France
Roberto Sebastiani	University of Trento, Italy
Bart Selman	Cornell University, USA
Peter Stuckey	NICTA and University of Melbourne, Australia
Stefan Szeider	Vienna University of Technology, Austria
Naoyuki Tamura	Kobe University, Japan
Allen Van Gelder	University of California at Santa Cruz, USA
Toby Walsh	NICTA and University of New South Wales, Australia

Additional Reviewers

Abío, Ignasi	Fröhlich, Andreas
Alban, Grastien	Grastien, Alban
Ansótegui, Carlos	Griggio, Alberto
Balint, Adrian	Gwynne, Matthew
Banbara, Mutsunori	Heras, Federico
Bayless, Sam	Hoessen, Benoît
Belov, Anton	Ignatiev, Alexey
Beyersdorff, Olaf	Itsykson, Dmitry
Bruttomesso, Roberto	Jabbour, Saïd
Cao, Weiwei	Janota, Mikoláš
Chu, Geoffrey	Kleine Büning, Hans
Codish, Michael	Kovásznai, Gergeley
Dantchev, Stefan	Lagniez, Jean-Marie
David, Cristina	Lauria, Massimo
Dilkina, Bistra	Le Bras, Ronan
Egly, Uwe	Lettmann, Theodor
Ermon, Stefano	Lewis, Matt
Feydy, Thibaut	Li, Chu-Min
Forejt, Vojtěch	Manquinho, Vasco
Franzén, Anders	Manyà, Felip

Maratea, Marco
Martins, Ruben
Meier, Arne
Mikša, Mladen
Morgado, Antonio
Nabeshima, Hidetomo
Navas, Jorge A.
Niemetz, Aina
Nikolić, Mladen
Novikov, Yakov
Olivetti, Nicola
Ordyniak, Sebastian
Piette, Cédric
Preiner, Mathias
Previti, Alessandro
Rollini, Simone Fulvio
Roussel, Stephanie
Ryvchin, Vadim
Schaafsma, Bas
Scheder, Dominik
Schmidt, Johannes

Schneider, Marius
Schneider, Stefan
Schrammel, Peter
Seidl, Martina
Simon, Laurent
Slivovsky, Friedrich
Soh, Takehide
Tack, Guido
Tautschnig, Michael
Tomasi, Silvia
Tonetta, Stefano
Vinyals, Marc
Vizel, Yakir
Widl, Magdalena
Wieringa, Siert
Wintersteiger, Christoph M.
Wolfovitz, Guy
Xu, Lin
Xue, Yexiang
Yue, Weiya

Table of Contents

Parallel Solving

Maximum Satisfiability

Encodings and Applications

Beyond SAT

Solver Techniques and Algorithms

Clique-Width and SAT

Propositional Proof Complexity II

Parameterized Complexity

Tool Papers

The Proof-Search Problem
between Bounded-Width Resolution
and Bounded-Degree Semi-algebraic Proofs*

Albert Atserias

Universitat Politècnica de Catalunya
Barcelona, Spain

Abstract. In recent years there has been some progress in our understanding of the proof-search problem for very low-depth proof systems,
e.g. proof systems that manipulate formulas of very low complexity such
as clauses (i.e. resolution), DNF-formulas (i.e. R(k) systems), or polynomial inequalities (i.e. semi-algebraic proof systems). In this talk I
will overview this progress. I will start with bounded-width resolution,
whose specialized proof-search algorithm is as easy as uninteresting, but
whose proof-search problem is unintentionally solved by certain versions
of conflict-driven clause-learning algorithms with restarts. I will continue
with R(k) systems, whose proof-search problem turned out to hide the
complexity of certain two-player games of interest in the area of systems
synthesis and verification. And I will close with bounded-degree semi-
algebraic proof systems, whose proof-search problem turned out to hide
the complexity of systems of linear equations over finite fields, among
other problems.

1 Introduction

Let P be a propositional proof system, which we think of, abstractly, as a
polynomial-time verifiable relation between *tautologies* and *proofs*, or dually,
between *contradictions* and *refutations* [21]. The proof-search problem for P
asks, for a given tautology as input, to find one of its P-proofs. However, since
we cannot expect all tautologies to have polynomial-size P-proofs (as this would
imply NP = co-NP), we will feel satisfied if we are able to find P-proofs that are
not too far from optimal. More formally, a proof system P is called *automatizable
in time t* if there exists an algorithm that, when it is given a tautology as input,
finds one of its P-proofs in time $t(s)$, where s is the size of its smallest P-proof.
Note that we do not insist that the found proof is the shortest possible [17].

The question whether there is an interesting proof system that is automatizable in polynomial time is open. The admittedly vague term *interesting* should
mean that the proof system is powerful enough to admit *some* short proofs.
For (a non-)example, the proof system whose proof for a given tautology is
its full truth-table is *not* interesting for it does not have short proofs at all.

---

* Research partially supported by project TIN2010-20967-C04-05 (TASSAT).

M. Järvisalo and A. Van Gelder (Eds.): SAT 2013, LNCS 7962, pp. 1–17, 2013.

This makes it trivially automatizable in polynomial time, but for a silly reason. For contrast, interesting proof systems in this sense do include propositional resolution, for example, whose reasoning power is able to produce short proofs of non-trivial tautologies arising in multiple application contexts. For example, resolution admits polynomial-size proofs of the least-number principle (every finite linear order has a least element) [40], which underlies many inductive proofs.

The purpose of this paper is to discuss the status of the proof-search problem for inference-based proof systems that work with formulas of very low complexity. These include resolution or DNF-resolution, which work with clauses and DNF-formulas, respectively, and semi-algebraic proofs, which work with polynomial inequalities over the reals. We also take the opportunity to discuss the connection to some of the lift-and-project methods in mathematical programming.

2 Inference-Based Proof Systems

Most classical proof systems are inference-based: starting with a set of given *hypotheses*, some *conclusions* are produced syntactically by means of one or more inference rules, which are then added to the set of hypotheses to proceed. In producing *proofs* for a tautology, an inference-based proof system starts with the empty set of hypotheses and the goal is to produce the tautology. Of course this will mean that the set of inference rules includes some *axioms*, i.e. inference rules that can be fired without any hypotheses. In producing *refutations* for a contradiction, an inference-based proof system starts with the given contradiction and the goal is to produce some blatant inconsistency.

All typical inference-based proof systems manipulate some particular type of formulas, be them clauses, DNF or CNF-formulas, propositional formulas of some higher but fixed depth of alternations between disjunctions and conjunctions, general propositional formulas, polynomial equations over some ring, polynomial inequalities over some ordered ring, disjunctions of those, decision trees branching on variables or more complicated formulas, binary decision diagrams of various sorts, Boolean circuits, etc. The inference rules are typically some more or less obvious, non-interesting, and polynomially checkable ways of producing some logical consequence of the hypotheses. In this sense, what makes an inference-based proof system more or less powerful is the expressive power of the type of formulas it manipulates.

2.1 Systems that Manipulate Propositional Formulas

In resolution, the formulas are *clauses*, disjunctions of variables or negated variables, and the only inference rule is the *resolution rule*:

$$\frac{A \vee x \quad B \vee \neg x}{A \vee B},$$

where A and B are clauses and x is a variable. We will see this proof system as a special case of a proof system that manipulates arbitrary propositional formulas and that has the following inference rules:

$$\frac{}{A \vee \overline{A}} \qquad \frac{A}{A \vee B} \qquad \frac{A \vee C \quad B \vee D}{A \vee B \vee (C \wedge D)} \qquad \frac{A \vee C \quad B \vee \overline{C}}{A \vee B},$$

where A, B, C and D denote propositional formulas in *negation normal form* (i.e. all its negations appear in front of variables), and a bar on top of a formula denotes its *dual* (i.e. $\overline{A \vee B} = \overline{A} \wedge \overline{B}$, $\overline{A \wedge B} = \overline{A} \vee \overline{B}$, $\overline{x} = \neg x$, and $\overline{\neg x} = x$). The four rules above are called *axiom, weakening, introduction of conjunction,* and *cut.* Besides these rules, the proof system is allowed to produce *structural manipulations*, which means that it is allowed to rewrite a propositional formula into an equivalent one that is obtained by repeated applications of the straight-forward rules of commutativity, associativity, and idempotency of disjunctions and conjunctions. We refer to this proof system as F, for *Frege system* [21].

The proof system F is *implicationally complete*, which means that if A is a logical consequence of $A_1, \ldots, A_m$, then there is an F-proof that takes $A_1, \ldots, A_m$ as hypotheses and produces A as conclusion. By the classical results of Cook and Reckhow [21], the reasoning power of F is hence equivalent to any other *Frege proof system*, i.e. any Hilbert-style textbook proof system for propositional logic, and also equivalent to the propositional sequent calculus. By this we mean that every proof in any one of these proof systems can be converted to an F-proof in polynomial time on the size of the proof, and conversely. Here, the size of a proof is the sum of the sizes of the formulas that make it (this includes all the hypotheses, and of course the conclusion). When such efficient conversions from P-proofs into P'-proofs are possible we say that P' *polynomially simulates* P.

As said, resolution can be seen as the special case of this proof system in which the only allowed formulas are clauses and the only allowed rule is cut. When the only allowed formulas are k-DNF-formulas, i.e. disjunctions of conjunctions of up to k literals, the corresponding restriction has been named $R(k)$ or k-DNF-resolution [29]. It is not hard to see that $R(1)$ is equivalent to resolution. When the only allowed formulas are arbitrary DNF-formulas, the proof system is called DNF-resolution.

2.2   Tree-Like, Dag-Like, and Bounded-Width Proofs

An essential feature of inference-based proofs as defined up to now is that, as soon as a conclusion is derived, it can be used multiple times as a hypothesis at no additional cost. On the other hand, it is obvious that every multiple use of a derived hypothesis could be replaced by multiple proofs of that hypothesis, from which it looks like the feature is not that essential after all. However, the point is that in doing this conversion, the proof-size could get exponentially bigger because at every re-derivation we could be doubling the size of the proof up to that point. In the following, we say that a proof in an inference-based proof system is in *tree form*, or *tree-like*, if every derived formula is used at most once as the hypothesis of an inference. Sometimes we use the term *dag form*, or *dag-like*, to emphasize the fact that a certain proof is not in tree form.

The intuition that the dag form of proofs is an essential feature that could lead to exponential savings is indeed correct, but only for proof systems that work with formulas of very low complexity. As will appear clear soon in this section, dag-like proofs can usually be converted efficiently into tree-like proofs whose lines are disjunctions of formulas of the starting dag-like proof or their negations. In particular, this means that in any Frege system such as F, the tree-like and dag-like versions polynomially simulate each other [30]. On the other hand, for proof systems such as resolution or R(k) with constant k, it is known that dag-like proofs could be exponentially shorter [15], [11], [25].

Since for resolution and R(k) tree-like proofs are much less powerful than their dag-like versions, an obvious question arises: why do we even consider the tree-likeness restriction at all? The answer is to be found on the fact that tree-like proofs appear naturally as the result of backtracking procedures. For example, the straightforward backtracking procedure to verify that a given set of clauses is contradictory by branching on the truth values of unset variables, and by pruning each branch as soon as some clause is falsified by the assignment of that branch, corresponds to a tree-like refutation in resolution: turn the recursion tree upside-down, label each leaf by one of the falsified clauses, and label the internal nodes of the tree by a resolution inference on the branched variable [6].

At this point we can ask for the proof system that corresponds to backtracking procedures that branch on the truth value of more complicated formulas and that stop as soon as the assigned truth values incurs into a blatant contradiction with the semantics of the connectives (for example by assigning $A \wedge B$ to true but A to false), or with the given clauses. The correspondence with natural tree-like proofs persists. For example, if the branching formulas are conjunctions of up to k literals, what we get is equivalent to tree-like R(k) [25].

Interestingly, tree-like R(k)-proofs appear naturally in a different context. Suppose $C_1, \ldots, C_m, C_{m+1}, \ldots, C_t$ is a resolution proof of C_t from $C_1, \ldots, C_m$ in which every clause has at most k literals; in that case we say that the resolution proof has *width* k. Since the resolution rule is sound, the last inference-step in this proof is indeed a tautology of the form $C_{\ell(t)} \wedge C_{r(t)} \rightarrow C_t$, or equivalently $\overline{C_{\ell(t)}} \vee \overline{C_{r(t)}} \vee C_t$, where $0 < \ell(t) < r(t) < t$. This tautology depends on no more than $3k$ variables and is a k-DNF, and hence has a tree-like R(k)-proof of size $2^{O(k)}$, and indeed size $O(k)$ because it has very special form. Of course, this is also the case for any inference in the proof. Now, starting at the tautology that corresponds to the inference that derives C_t, and cutting it with the tautologies that correspond to the inferences that derive $C_{\ell(t)}$ and $C_{r(t)}$, we get a k-DNF of the form $\overline{C_{\ell(\ell(t))}} \vee \overline{C_{r(\ell(t))}} \vee \overline{C_{\ell(r(t))}} \vee \overline{C_{r(r(t))}} \vee C_t$. Repeating for every inference in the proof we get $\overline{C_1} \vee \cdots \vee \overline{C_m} \vee C_t$, from which C_t follows by m cuts with the m initial clauses $C_1, \ldots, C_m$. Observe that the result is a tree-like R(k)-proof whose size is a factor $O(k)$ bigger than the original proof (and note also that this argument works equally well to polynomially simulate dag-like F-proofs by tree-like F-proofs [30]).

As we will see later on, the width of a resolution proof as defined in the beginning of the previous paragraph is a very important parameter for the understanding of resolution. For this reason, let us write R_k for the restriction of resolution in which all clauses have at most k literals. Note that R_k is obviously a restriction of $R(1)$, and from the above, it can also be thought as a restriction of tree-like $R(k)$. However, let us also note that, for $k < n$, the restriction R_k need not be complete on sets of clauses with n variables. Certainly R_k with $k < n$ cannot derive any clause with $k + 1$ literals, or cannot even start if the initial set of clauses contains one with $k + 1$ literals, but even explicit n-variable contradictory sets of 3-clauses are known for which all resolution refutations must use a clause with $\Omega(n)$ literals [12].

2.3   Systems that Manipulate Polynomial Inequalities

If we represent *true* by 1 and *false* by 0, propositional clauses are obviously represented by linear inequalities over the reals. For example, the clause $x \vee \overline{y} \vee z$ is represented by the linear inequality $x + (1 - y) + z \geq 1$, which may be rewritten as $x - y + z \geq 0$. In this sense, resolution may be seen as a proof system that manipulates linear inequalities of special form, over the reals. There are several ways in which this can be generalized to arbitrary linear inequalities. In the cutting planes proof system [18], seen as a proof system for refuting sets of propositional clauses, the hypotheses are represented by linear inequalities of special form as above, the inequalities $x_i \geq 0$ and $1 - x_i \geq 0$ are added to the set of hypotheses, and arbitrary inequalities with integer coefficients may be inferred by means of *positive linear combinations* and *integer rounding*. Although the published work on the cutting planes proof system is very extensive, in this paper we want to focus on a more general family of proof systems that manipulates inequalities over the reals that we call semi-algebraic proof systems.

In the most general semi-algebraic proof system the primary objects are arbitrary polynomial inequalities over the reals. These are inequalities of the form $P \geq 0$, where P is a multi-variate polynomial in the ring of polynomials $\mathbb{R}[x_1, \ldots, x_n]$. The proof system has the following simple rules of inference:

$$\frac{P \geq 0 \quad Q \geq 0}{c \cdot P + d \cdot Q \geq 0} \qquad \frac{P \geq 0 \quad Q \geq 0}{P \cdot Q \geq 0} \qquad \frac{}{P^2 \geq 0}$$

where P and Q are polynomials, and c and d are positive real constants. These rules are called *positive linear combination*, *multiplication rule*, and *positivity of squares*, respectively. Of course a representation issue arises here as the coefficients of the polynomials, as well as the multipliers c and d, could be arbitrary reals. Whenever this issue is important (e.g. when we consider the proof-search problem for such proofs) we will restrict the valid proofs to those that involve rational coefficients that are represented in binary. For the cases of interest, this will not be a severe restriction, as we will see.

Obviously the rules above are sound: if $(x_1, \ldots, x_n) \in \mathbb{R}^n$ satisfies the hypotheses of a rule, then it must also satisfy the conclusion. Moreover, a deep result in real algebraic geometry known as Stengle's Positivstellensatz [41] implies that the rules make up a proof system that is *refutationally complete* for systems of arbitrary polynomial inequalities. More precisely, if $P_1, \ldots, P_m$ are polynomials in $\mathbb{R}[x_1, \ldots, x_n]$ such that the system $P_1 \geq 0, \ldots, P_m \geq 0$ is unfeasible over $\mathbb{R}^n$, then there is a proof of $-1 \geq 0$ from the hypotheses $P_1 \geq 0, \ldots, P_m \geq 0$ (see [35], [13]). We should also point out that this proof system is *not* implicationally complete for arbitrary polynomial inequalities (see [35], [13] again). As will be evident in the forthcoming, it is often convenient to restrict the degree of all the polynomials appearing in the proof to some bound k. The semi-algebraic proof system restricted to using polynomials of degree at most k will be called $S^+(k)$. The version without positivity of squares is called $S(k)$. Let us note that very closely related proof system were called $LS^k_{+,*}$ and LS^k_* in [27].

Just as clauses can be represented by linear inequalities, there is an obvious way of representing k-DNF-formulas as polynomial inequalities of degree k through *sums of extended monomials*, i.e. inequalities of the form

$$\sum_{t=1}^{m} \prod_{i \in I_t} x_i \prod_{i \in J_t} (1 - x_i) \geq 1$$

where $|I_t \cup J_t| \leq k$ for every $t \in \{1, \ldots, m\}$. Moreover, it is a rather pleasant fact that, under this translation plus some additional axioms stating that $0 \leq x_i \leq 1$ and $x_i^2 = x_i$, the system $R(k)$ is polynomially simulated by $S(2k)$ (note $2k$ vs. k). More precisely, if $A_1, \ldots, A_m$ and A are k-DNF-formulas and there is an $R(k)$-proof of A from $A_1, \ldots, A_m$ of size s, then there is a semi-algebraic proof (of the translation) of A from (the translations of) $A_1, \ldots, A_m$ and the additional axioms $x_i \geq 0$, $1 - x_i \geq 0$, $x_i - x_i^2 \geq 0$ and $x_i^2 - x_i \geq 0$, all whose polynomials have rational coefficients, degree at most $2k$, and the total size of the proof is polynomial in s. The proof of this is not completely trivial, so we give a sketch.

We start by noting that there is a small degree-$2k$ proof of $A + B \geq 1$ from $A + \prod_{i \in I} x_i \prod_{j \in J} (1 - x_j) \geq 1$ and $B + \sum_{i \in I} (1 - x_i) + \sum_{j \in J} x_j \geq 1$ for every two sums of degree-k extended monomials A and B, and $|I \cup J| \leq k$. First observe that if M is a degree-k extended monomial then, in the presence of the four axioms stating $0 \leq x_i \leq 1$ and $x_i^2 = x_i$, there are small degree-$2k$ proofs of $0 \leq M \leq 1$ and $M^2 = M$. In what follows, write $M(I, J)$ for the extended monomial $\prod_{i \in I} x_i \prod_{j \in J} (1 - x_j)$. Now take the second hypothesis $B + \sum_{i \in I} (1 - x_i) + \sum_{j \in J} x_j \geq 1$ and, iteratively for each $i \in I$, multiply by $x_i \geq 0$ and then eliminate $(1 - x_i) x_i$ using $x_i^2 = x_i$. Continuing, iteratively for each $j \in J$, multiply the result by $1 - x_j \geq 0$ and eliminate $x_j (1 - x_j) \prod_{i \in I} x_i$ using $x_j^2 = x_j$ and hence $x_j^2 \prod_{i \in I} x_i = x_j \prod_{i \in I} x_i$. The result is $B \cdot M(I, J) \geq M(I, J)$. Add this to the first hypothesis to get $A + B \cdot M(I, J) \geq 1$. Now, using the fact that B is a sum of extended monomials, derive $B \geq 0$. Derive also $1 - M(I, J) \geq 0$, and multiply together to get $B - B \cdot M(I, J) \geq 0$. Adding this to the above we get $A + B \geq 1$.

The derivation above allows the simulation of cuts except that we also need contraction of repeated terms. In other words, we need small degree-$2k$ proof of $A + Q \geq 1$ from $A + 2Q \geq 1$ for every sum of degree-k extended monomials A and every degree-k extended monomial Q. Proceed as follows: Multiply $A + 2Q \geq 1$ by $1 - Q \geq 0$ to get $A + 3Q - 1 - AQ - 2Q^2 \geq 0$. Then use the fact that A

is a sum of degree-k extended monomials to get $A \geq 0$ and hence $AQ \geq 0$ by multiplication, and add it to the previous inequality to get $A + 3Q - 1 - 2Q^2 \geq 0$. Using $Q^2 = Q$ we get $A + Q \geq 1$.

We leave the simulation of axioms, weakenings, and introductions of conjunctions as exercises.

2.4   Connection with Lift-and-Project Methods

In linear programming we are given a collection of linear inequalities $L_1 \geq 0, \ldots, L_m \geq 0$ that define a polyhedron over $\mathbb{R}^n$ and we are asked to optimize a linear function L over the polyhedron. Proving that the optimum is at least some bound c is of course an instance of the general problem of the previous section: prove that $L \geq 0$ follows from given assumptions $L_1 \geq 0, \ldots, L_m \geq 0$, over $\mathbb{R}^n$. However, $L_1, \ldots, L_m$ and L are all linear, and in this case the fundamental duality theorem for linear programming implies that, whenever the implication holds, there is a *linear programming proof*, i.e. one that derives the conclusion as a positive linear combination of the hypotheses and the trivial inequality $1 \geq 0$. Moreover, any polynomial-time algorithm for linear programming can be used to find the proof (by solving the dual).

All this is very good but not directly suited to an arbitrary combinatorial problem in which the implications that matter are over a discrete domain, such as $\{0,1\}^n$, instead of $\mathbb{R}^n$ or $[0,1]^n$. Of course, the domain $\{0,1\}^n$ can be enforced by adding the quadratic constraints $x_i^2 - x_i \geq 0$ and $x_i - x_i^2 \geq 0$, but now, if we want to make use of these constraints, we are forced to go beyond positive linear combinations and use some multiplications or squares. The lift-and-project method of Lovász and Schrijver [33] allows these rules but only in the following limited forms:

$$\frac{P \geq 0 \quad Q \geq 0}{c \cdot P + d \cdot Q \geq 0} \qquad \frac{L \geq 0}{L \cdot x_i \geq 0} \qquad \frac{L \geq 0}{L \cdot (1 - x_i) \geq 0} \qquad \frac{}{L^2 \geq 0}$$

where P and Q are polynomials, L is linear, and c and d are positive real constants. The second and third rules are called *lifting rules*. Besides these rules, the axioms $x_i \geq 0$, $1 - x_i \geq 0$, $x_i - x_i^2 \geq 0$ and $x_i^2 - x_i \geq 0$ are always present (note also that by adding the first two axioms we get $1 \geq 0$).

The proof system introduced by Lovász and Schrijver is called LS^+ in the literature. The version in which positivity of squares is not allowed is called LS. Note that LS and LS^+ are restrictions of $S(2)$ and $S^+(2)$, respectively. It is also known that LS polynomially simulates resolution [38]. The restrictions of LS and LS^+ to *lifting rank* less than k are denoted by LS_k and LS_k^+, respectively. Here, the *lifting rank* of a proof is the maximum number of applications of the lifting rules in a path from the hypotheses to the conclusion. Let us note that for $k < n$, the restrictions LS_k and LS_k^+ are not complete over $\{0,1\}^n$; in other words, there exist linear inequalities $L_1, \ldots, L_m$ and L with n variables such that $L \geq 0$ follows from $L_1 \geq 0, \ldots, L_m \geq 0$ over $\{0,1\}^n$, but LS_k and LS_k^+ are not able to prove $L \geq 0$ from $L_1 \geq 0, \ldots, L_m \geq 0$. On the other hand, Lovász and Schrijver argued that LS_n, and hence LS_n^+, is complete for deriving linear inequalities over $\{0,1\}^n$.

The name "lift-and-project" comes from the idea that the linear inequalities that define the initial polyhedron are lifted to linear inequalities over $\mathbb{R}^{n^2}$ (by thinking of each product $x_i x_j$ as a new variable), and projected back to $\mathbb{R}^n$ through linear combinations (by cancelling all products $x_i x_j$) before a new lifting is allowed.

A different lift-and-project method was suggested by Sherali and Adams [39]. Chronologically, this came before Lovász and Schrijver, but for the purposes of exposition it makes more sense to reverse the order. In the method of Sherali and Adams, the liftings are more powerful, but the way they are combined together is more restricted. Precisely, instead of lifting linear inequalities by multiplication by one literal, we allow lifting of arbitrary polynomials:

$$\frac{P \geq 0}{P \cdot x_i \geq 0} \qquad \frac{P \geq 0}{P \cdot (1 - x_i) \geq 0}$$

where P is an arbitrary polynomial. However, the proofs must have a very special form: they start at the given inequalities $L_1 \geq 0, \ldots, L_m \geq 0$, $x_i \geq 0$, $1 - x_i \geq 0$, $x_i - x_i^2 \geq 0$ and $x_i^2 - x_i \geq 0$, perform a few liftings, and combine them by positive linear combinations (with no further liftings). Thus, all liftings come before all positive linear combinations, and positivity of squares is not allowed. This rather special form will look more natural if we think of the Sherali-Adams method as making a single lift-and-project round, instead of making multiple rounds as in LS, but using dimension n^k for some $k \geq 2$ in the middle stage, instead of dimension n^2 as in LS. The restriction of the Sherali-Adams proof system to polynomials that do not exceed degree k is called SA_k. We call SA_k^+ the natural extension in which, besides the initial inequalities, arbitrary squares are also allowed, but again all restricted to degree at most k.

It is not too hard to see that every proof in LS_k or LS_k^+ can be converted, in polynomial time, into a proof in SA_k or SA_k^+ by *moving the liftings up* towards the hypotheses. Note also that SA_k is a restriction of $\mathrm{S}(k)$ but not a restriction of $\mathrm{S}(k-1)$. Compare this with the fact that, since LS_k is a restriction of LS which in turn is a restriction of $\mathrm{S}(2)$, each LS_k is a restriction of $\mathrm{S}(2)$. As for LS_k and LS_k^+, the restrictions SA_k and SA_k^+ are not complete over $\{0,1\}^n$ when $k < n$, but Sherali and Adams proved that SA_n, and hence SA_n^+, is complete for deriving linear inequalities over $\{0,1\}^n$. Of course nothing prevents us from considering a proof system that allows *multiple rounds* of SA_k as a generalization of LS. This would keep it a subsystem of $\mathrm{S}(k)$ and, indeed, if the number of rounds is unbounded, it would make it equivalent for systems of inequalities that include $x_i \geq 0$ and $1 - x_i \geq 0$. The multiple-round version of SA_k was called LS^k in [27].

One last interesting thing to notice is that SA_k polynomially simulates R_k for sets of clauses. This follows from three facts: 1) that every k-clause of the form $\bigvee_{i \in I} \overline{x_i} \vee \bigvee_{i \in J} x_i$ may be represented by a degree-k inequality of the form $0 \geq M(I, J)$, where $M(I, J)$ is the shorthand notation for extended monomials used earlier, 2) that this representation may be obtained from the given form $\sum_{i \in I}(1 - x_i) + \sum_{i \in J} x_i \geq 1$ of a clause by at most k liftings (ignoring terms that are 0 modulo $x_i^2 = x_i$), and 3) that, in this representation, any width-k resolution step may be simulated by addition with a valid inequality of the form $M(I \cup \{i\}, J) + M(I', J' \cup \{i\}) \geq M(I \cup I', J \cup J')$, which has an SA_k-proof itself.

Let us note that Sherali and Adams did not phrase their lift-and-project method in terms of inference rules. Also, they did not consider anything like SA_k^+, which is very closely related to the method of Lasserre. See [32] for a comparison of the three methods.

3   The Proof-Search Problem

After this long introduction, we move now to the proof-search problem for the proof systems introduced in Section 2. We start by stating some positive results and their consequences, then we discuss negative (i.e. conditional hardness) results, and we close with some observations concerning the cases in-between.

3.1   Width-Related Algorithms

The *width* of a clause is defined as the number of literals it has. In the following, a *k-clause* is one of width at most k. A generous bound on the number of k-clauses on a set of n variables is $(2n + 1)^k / k! \leq 2(n + 1)^k$. In particular this means that if a contradictory set of clauses has a resolution refutation of width k, then it also has one of size $O(k(n + 1)^k)$, where n is the number of variables. It also means that if such a refutation exists, then one can be found in time $n^{O(k)}$ by repeatedly resolving upon known clauses provided the result is an as yet unknown k-clause. This solves the proof-search problem for R_k in time $n^{O(k)}$, where n is the number of variables of the given set of clauses.

As we just noticed, small width refutations entail short refutations. One of the fundamental facts about resolution is that a partial converse is also true: building on the work of Clegg, Edmonds, and Impagliazzo [19] and Beame and Pitassi [8], Ben-Sasson and Wigderson [12] proved if a contradictory set of clauses has a resolution refutation of size s, then it also has a resolution refutation of width $O(\sqrt{n \log s} + w)$, where n is again the number of variables, and w is the width of the widest clause in the given set of clauses. To appreciate the depth of this result let us look at the case of polynomial s and constant w. In that case the width becomes $O(\sqrt{n \log n})$ which is very significantly smaller than the maximum possible width n. It is also known that this trade-off is worst-case optimal (up to logarithmic factors): there exist n-variable sets of 3-clauses that have polynomial-size resolution refutations but that do not have resolution refutations of width $o(\sqrt{n})$ (see [16]).

Among other applications, the fundamental size-width tradeoff result for resolution can be used to argue that, for contradictory sets of w-clauses with constant w, resolution is automatizable in non-trivial time. Consider the algorithm that solves the proof-search problem for R_k in time $n^{O(k)}$ and run it on increasing values of k until the empty clause is found. By the size-width tradeoff, k will not exceed $O(\sqrt{n \log s})$ where s is the size of the shortest resolution refutation (recall that we are assuming that w is a constant, but we could even afford $w = O(\sqrt{n \log n})$ for this to be true). Hence the algorithm runs in time $n^{O(\sqrt{n \log s})}$. Note how this is a non-trivial time-bound: if s is polynomial, the running time is of subexponential type $2^{O(n^{0.51})}$.

The width-based algorithm from the preceding paragraph is not terribly satisfying in that it completely ignores any structure that the input set of clauses could have, and blindly derives all possible clauses (in increasing order of width). In contrast, practically-used resolution-based algorithms exploit very fine-tuned heuristics that *learn* strategically chosen clauses with the hope of deriving the empty clause earlier or pruning the search-space to a point where exhaustive search for a satisfying assignment becomes successful [34]. Of course one could always run the width-based algorithm in parallel to a fine-tuned heuristic-based algorithm in order to guarantee the worst-case bound from the first with the practical features of the second. But, somewhat surprisingly, it turned out that the architecture of most practically-used algorithms does not require this. The relatively recent result from [3] shows that if a standard *conflict-driven clause-learning algorithm* (CDCL algorithm) is given the opportunity to restart and branch on randomly chosen literals often enough, then the resulting algorithm is guaranteed to have high probability of finding a refutation after no more than $n^{O(k)}$ iterations, if a resolution refutation of width k exists. Interestingly, the validity of this result is quite robust to the actual tuning of the underlying CDCL algorithm. We refer the reader to the reference [3] for details.

3.2   Degree-Related Algorithms

The original motivation for the lift-and-project methods from Section 2.4 was to devise a method by which an initial polytope P over $[0,1]^n$ could be *tightened* into better and better approximations $P \supseteq P_1 \supseteq P_2 \supseteq \cdots \supseteq P_n = P^*$, where P^* denotes the convex hull of the 0-1 points in P. Both SA and LS achieve this by letting P_k be the polytope defined by the inequalities that have an SA_k or LS_k-proof from the inequalities that define the initial polytope. Moreover, and this is the main point of the methods, in both cases there is an algorithm running in time $n^{O(k)}$ to optimize any given linear objective function over the polytope P_k (see [39], [33]).

For SA_k even more is true. Not only it is possible to optimize linear functions over P_k, but even SA_k-proofs of optimality can be found. More precisely, there exists an algorithm that, given linear functions $L_1, \ldots, L_m$ and L, finds an SA_k-proof of $L \geq 0$ from $L_1 \geq 0, \ldots, L_m \geq 0$, if there is one, and does so in time $n^{O(k)}$, where n is the number of variables. One way to see this is by first observing that, during the phase of liftings in an SA_k-proof, all we are doing is multiplying the given inequalities and the axioms $x_i^2 - x_i \geq 0$ by an extended monomial of the form $\prod_{i \in I} x_i \prod_{j \in J}(1 - x_j)$ with $|I \cup J| \leq k$, of which there are no more than $(2n)^k$. The second observation is that what is left to do in the phase of positive linear-combinations is a linear programming problem over $\mathbb{R}^{(n+1)^{k+1}}$ (by interpreting each monomial of degree at most $k + 1$ as a new variable). Thus, any algorithm solving linear programming in polynomial time will give an algorithm to solve the proof-search problem for SA_k in time $n^{O(k)}$. Observe that, by a simple binary search argument, this is a stronger claim than the ability to optimize over the polytope P_k.

For the lift-and-project method LS_k of Lovász and Schrijver only the weaker claim about optimization is known. The difficulty in providing explicit LS_k-proofs of optimality is that the optimization algorithm works by providing a polynomial-time separation oracle for the polytope P_{i+1} given a separation oracle for P_i, and using this recursion to apply the ellipsoid method on P_k. An intriguing observation is that if we are happy with an SA_k-proof of optimality, then we can still get it time $n^{O(k)}$. The reason for this is that, as mentioned in Section 2.4, there is a polynomial translation of LS_k-proofs into SA_k-proofs. Thus, the algorithm from the previous paragraph applies. This also shows that the optimization problem for LS_k can also be solved without resorting to the ellipsoid method.

For SA_k^+ and LS_k^+ similar statements are true by using polynomial-time algorithms for semi-definite programming in one case, and the ellipsoid method in the other. In both cases, the key observation is that the sums of squares of linear forms are in one-to-one correspondance with the positive semi-definite quadratic forms. For these reasons, SA_k^+ and LS_k^+ are called the *semi-definite versions* of SA_k and LS_k. The catchy acronym SoS (for sum-of-squares) is also used for certain versions of SA_k^+ (see [35], [5]).

3.3 Reductions from Tree-Form to Bounded Width or Degree

The version of the Ben-Sasson-Wigderson size-width tradeoff for tree-like resolution is this: if a set of clauses has a tree-like resolution refutation of size s, then it also has a resolution refutation of width $O(\log s + w)$, where w is the width of the largest clause in the given set of clauses. In particular, this means that for constant w, by running the proof-search algorithm for R_k with increasing values of k until we find the empty clause, we succeed in time $n^{O(\log s)}$, where s is the size of the shortest tree-like refutation and n is the number of variables. Note however that the proof is not necessarily tree-like. In other words, the algorithm runs within a non-trivial time-bound that depends on the size of the shortest tree-like refutation, but the obtained proof is in a different proof system.

For tree-like LS and LS$^+$ what happens is closer to what happens for dag-like resolution. The analogue size-rank tradeoff for tree-like LS and LS$^+$ was shown by Pitassi and Segerlind [36]: if a system of linear inequalities with n variables has a tree-like LS-refutation of size s, then it also has an LS_k-refutation with $k = O(\sqrt{n \log s})$, and the same for LS$^+$. Again this gives an algorithm that, given a system of linear inequalities that is contradictory over $\{0, 1\}^n$, finds an LS-refutation in time $n^{O(\sqrt{n \log s})}$, where s is the size of the shortest tree-like LS-refutation, and n is the number of variables. However, the obtained proof is not necessarily in tree form.

When this happens, namely that there is an algorithm that given a tautology finds one of its P'-proofs in time $t(s)$ where s is the size of the smallest P-proof, we say that P is *weakly automatizable* (in terms of P') in time t. For later use, let us point out that it is not hard to see that P is weak automatizable in polynomial time if and only if there is a proof system P' that is automatizable in polynomial time and that polynomially simulates P [37].

Using the fact that $s \geq n$ because the hypotheses are counted in the size of any proof, what the first paragraph of this section says is that tree-like resolution is weakly automatizable in quasi-polynomial time of the type $s^{O(\log s)}$. For the sake of completeness, let us also mention that for tree-like resolution, a direct (non-weak) proof-search algorithm that runs in quasi-polynomial time $s^{O(\log s)}$ is known [8]. The latter has the added advantage that it works for arbitrary sets of clauses and not only those of limited width.

3.4   Hardness Results

The weak automatizability of a proof or refutation system is closely related to the concept of *feasible interpolation* [31], [37]. In short, the interpolation problem for a refutation system P is the following: given a P-refutation of a conjunction $A_0 \wedge A_1$, where A_0 and A_1 are formulas on disjoint sets of variables, output $b \in \{0,1\}$ such that the formula A_b is contradictory by itself. Under a very mild closure condition on the set of P-refutations, the connection is that if P is weakly automatizable in polynomial time, then the interpolation problem for P can also be solved in polynomial time. The mild closure condition, called *natural* in [7], is that if a contradictory formula A has a P-refutation of size at most s, then the result of assigning any truth value to any one of the variables of A also has a P-refutation of size at most s. This is true of virtually any refutation system one can think of (but see [7] where it is pointed out that this is not so clear for proofs produced by CDCL algorithms).

To see the connection pointed out above argue as follows: Let P' be the refutation system that is automatizable in polynomial time and that polynomially simulates P. Given a P-refutation of $A_0 \wedge A_1$ as input, first we run the proof-search algorithm for P' on input A_0 until either it finds a P'-refutation or it runs for more than $t(p(s))$ steps, where s is the size of the given P-refutation of $A_0 \wedge A_1$, and t and p are, respectively, the polynomials that bound the running time of the proof-search algorithm for P', and the size of the P'-refutations as a function of the size of the P-refutations. In the first case we output 0. In the second case we know that A_1 cannot be satisfiable and it is safe to output 1 (otherwise, by the mild closure condition, plugging one of its satisfying assignment into the P-refutation of $A_0 \wedge A_1$ would give a size-s P-refutation of A_0). See [37], [2] for more on this.

Several interesting proof systems have feasible interpolation, which means that their interpolation problem can be solved in polynomial time. These include resolution, cutting planes, and LS [31], [38]. On the other hand, if we want to show that a proof system P is not automatizable in polynomial time, it suffices to argue that it does not have feasible interpolation. Typically this is done by reducing a (conjecturally) hard problem to the interpolation problem for P. More precisely, starting at a problem for which distinguishing the YES-instances from the NO-instances requires more than polynomial time, we want to find a polynomial-time translation from instances into P-refutations of certain formulas of the form $A_0 \wedge A_1$, in such a way that YES-instances give a satisfiable A_0 and NO-instances give a satisfiable A_1.

This strategy for arguing the failure of feasible interpolation can be made to work for several proof systems. For example, in the same paper where the concept of automatization was defined, Bonet, Pitassi and Raz followed this strategy to prove that no Frege system P has feasible interpolation unless factoring Blum integers can be solved in polynomial time. Intuitively, the formula $A_0 \wedge A_1$ states that a *hard bit* of the given Blum integer is both 0 and 1 at the same time, and a short P-proof is given for the impossibility of this fact. This established that no Frege system is automatizable or weakly automatizable in polynomial time under a reasonable cryptographic conjecture. Of course, this applies as well to the Frege system F from Section 2.1. A few years later the argument was refined to prove a weaker negative result for Frege-systems working with formulas of fixed (but large) AND/OR alternation depth [14]. It was shown that for every large enough depth d, such systems do not have feasible interpolation unless factoring Blum integers can be solved in subexponential time $2^{n^{1/d^{O(1)}}}$.

For resolution and other *low-depth* proof systems such as $R(k)$ or $S(k)$ for constant k, LS, or DNF-resolution, the situation is less clear. Part of the difficulty is that resolution and LS *do* have feasible interpolation and therefore the type of arguments above cannot be made to work. Thus, if we want to make progress in our understanding of the automatizability of resolution we need to focus on the proof-search problem itself. That is what Alekhnovich and Razborov did, i.e. they reduced a conjecturally hard problem to the problem of distinguishing formulas with small resolution refutations from formulas that do not have much larger resolution refutations [1]. This way they proved that resolution and tree-like resolution are not automatizable in polynomial time unless W[P] is tractable. Without entering the details of W[P], let us mention that the intractability of W[P] is quite likely as otherwise it would mean that the k-clique problem on graphs with n vertices can be solved by a probabilistic algorithm in time $f(k) \cdot p(n)$ for some fixed computable function f independent of n and some fixed polynomial p independent of k (see [22]).

3.5   Games and Propositional Proofs

The hardness result of Alekhnovich and Razborov says nothing about the possibility that resolution could be automatizable in quasi-polynomial time. Indeed, as mentioned in Section 3.3, tree-like resolution *is* automatizable in quasi-polynomial time and this is not incompatible with the result of Alekhnovich and Razborov. Also, it says nothing about the possibility that resolution or tree-like resolution could be weakly automatizable in polynomial time. In particular, it says nothing about the possibility that any of $R(k)$ or $S(k)$ for $k \geq 2$, LS or cutting planes, or DNF-resolution could be automatizable in polynomial time. All these are important proof systems for which their automatizability could be an important breakthrough. In view of this, since the proof-search problem is about distinguishing formulas with small proofs from those without, it makes sense to continue the search for large families of formulas that admit small proofs in these systems. With this in mind, a line of recent work has uncovered an interesting connection between some classical problems in game theory and some of these low-depth proof systems. We discuss this in the rest of this section.

In a mean-payoff game (MPG) two players take rounds at extending a path on a finite weighted directed graph. The game ends as soon as the path intersects itself, forming a cycle. The first player wins if the average weight of the cycle is positive. Otherwise the second player wins. Classically, the game is played for infinitely many rounds, but for our purposes this finite version suffices [23], [43]. The problem of mean-payoff games asks, for a given game graph with weights written in binary, whether the first player has a winning strategy. The exact complexity of this problem is unknown, but is known to lie in NP ∩ co-NP (see [43]). The connection with low-depth proof systems was found by Atserias and Maneva who showed that every MPG can be converted in polynomial time into a set of clauses that is either satisfiable, in which case the first player has a winning strategy, or has a polynomial-size DNF-refutation, in which case the second player has a winning strategy [4]. In particular, this shows that if DNF-resolution were automatizable or even weakly automatizable in polynomial time, then MPGs would be solvable in polynomial time.

Shortly after this was shown, Huang and Pitassi improved this to the simple stochastic games (SSG) of Condon [20], a class of games to which MPGs reduce in polynomial time. Indeed, their proof showed more since they reduced SSGs to the interpolation problem for DNF-resolution. In the conference version of their paper [28], the reduction was stated to produce depth-3 refutations instead of DNF-resolution-refutations, but it was later pointed out that the refutations are indeed in DNF-resolution. We close this section by describing the latest development in this line of research which takes us to a third type of games.

In a parity game (PG) again two players take rounds at extending a path on a finite directed graph, this time unweighted. The game ends as soon as the path intersects itself, forming a cycle. The first player wins if the least numbered vertex in the cycle is odd. Otherwise the second player wins. As with MPGs, classically the game is played indefinitely, but for us the finite version will be enough. The problem of parity games asks for the winner of a given PG. Parity games have their origins in automata theory where they are used to give combinatorial semantics to the modal μ-calculus, among other things [24]. Again the complexity of the problem of PGs is unknown, but it is known to reduce to MPGs and in particular belongs to NP ∩ co-NP (see [43]).

An interesting recent discovery of Beckmann, Pudlák and Thapen [9] is that the problem of parity games (PG) reduces to the interpolation problem for $R(k)$ for a fixed constant $k \geq 2$. In particular, by known results relating interpolation of $R(k)$ with weak automatizability of resolution (see [2]), this means that if resolution were weakly automatizable, then PGs would be solvable in polynomial time. This last possibility is not fully unlikely, even conjectured by some, but at least it shows that the proof-search problem for resolution must be at least as hard as PGs, a notorious 20 year-old unsolved problem. It also reinforces the claim that resolution, and R(k) with constant k, are "interesting" from the point of view of the proof-search problem in the sense it was meant in the introduction.

3.6   Semi-algebraic Proofs and Linear Equations Mod 2

Note that the results mentioned in the last paragraph of the previous section also apply to the semi-algebraic systems LS and S(k) for $k \geq 2$. This is because the results were referring to weak automatizability of resolution, and the systems LS and S(2) polynomially simulate resolution. We want to finish this paper by pointing out what we believe is an important characteristic that distinguishes these semi-algebraic systems from resolution and even DNF-resolution.

Hand-in-hand with the pigeonhole principle, unsolvable systems of linear equations over the 2-element field make one of the classical sources of hardness for resolution-based proof systems, and even bounded-depth Frege systems; the celebrated Tseitin formulas illustrate the point [42], [12], [10]. On the other hand, it was shown by Grigoriev, Hirsch, and Pasechnik [27] that the Tseitin formulas are not hard for a system very related to S(k), for some constant k. Even more, a careful look at their proof shows that any unsolvable system of linear equations mod 2 in which each equation has at most three non-zero coefficients, when appropriately encoded as a set of clauses, has polynomial-size refutations in S(5) by simulating Gaussian elimination. This should be put in contrast with the results of Grigoriev [26] that imply that any SA_k^+-refutation of the Tseitin formulas requires $k = \Omega(n)$, where n is the number of variables.

What these observations say is that the proof-search problem for S(k) for constant $k \geq 5$ hides the complexity of systems of linear equations over the 2-element field. Of course this is not a computationally hard problem since Gaussian elimination solves it in polynomial time. But the point is that if S(k) is to have an efficiently solvable proof-search algorithm, this algorithm will need to be at least as clever as it takes to solve systems of linear equations. In particular, it also says that the distance between SA_k and S(k) is much bigger than it could look from the definitions, and that the methods for solving the proof-search problem for SA_k are probably completely irrelevant to S(k). We would love to be wrong on this and be shown that a clever application of the ellipsoid algorithm, say, is able to lift the proof-search algorithm for SA_k to a linear optimization algorithm for (some interesting version of) S(k).

Acknowledgments. We thank the comments of Allen Van Gelder and an anonymous referee on the preliminary draft of this paper.

References

1. Alekhnovich, M., Razborov, A.A.: Resolution Is Not Automatizable Unless W[P] Is Tractable. SIAM J. Comput. 38(4), 1347–1363 (2008)
2. Atserias, A., Bonet, M.L.: On the Automatizability of Resolution and Related Propositional Proof Systems. Information and Computation 189(2), 182–201 (2004)

3. Atserias, A., Fichte, J.K., Thurley, M.: Clause-Learning Algorithms with Many Restarts and Bounded-Width Resolution. Journal of Artificial Intelligence Research 40, 353–373 (2011)
4. Atserias, A., Maneva, E.: Mean-payoff games and propositional proofs. Information and Computation 209(4), 664–691 (2011)
5. Barak, B., Brandão, F., Harrow, A., Kelner, J., Zhou, Y.: Hypercontractivity, Sum-of-Squares Proofs, and their Applications. In: Proc. of 44th ACM Symposium on Theory of Computing (STOC), pp. 307–326 (2012)
6. Beame, P., Karp, R., Pitassi, T., Saks, M.: The efficiency of resolution and Davis-Putnam procedures. SIAM J. Comput. 31(4), 1048–1075 (2002)
7. Beame, P., Kautz, H., Sabharwal, A.: Towards Understanding and Harnessing the Potential of Clause Learning. Journal of Artificial Intelligence Research 22, 319–351 (2004)
8. Beame, P., Pitassi, T.: Simplified and improved Resolution lower bounds. In: Proc. of the 27th IEEE Foundations of Computer Science (FOCS), pp. 274–282 (1996)
9. Beckmann, A., Pudlák, P., Thapen, N.: Parity games and propositional proofs (April 2013) (in preparation)
10. Ben-Sasson, E.: Hard examples for the bounded depth Frege proof system. Computational Complexity 11, 109–136 (2002)
11. Ben-Sasson, E., Impagliazzo, R., Wigderson, A.: Near Optimal Separation of Tree-Like and General Resolution. Combinatorica 24(4), 585–604 (2003)
12. Ben-Sasson, E., Wigderson, A.: Short proofs are narrow–resolution made simple. Journal of the ACM 48(2), 149–169 (2001)
13. Bochnak, J., Coste, M., Roy, M.-F.: Real algebraic geometry. Springer (1999)
14. Bonet, M.L., Domingo, C., Gavaldà, R., Maciel, A., Pitassi, T.: Non-automatizability of bounded-depth Frege proofs. Computational Complexity 13, 47–68 (2004)
15. Bonet, M.L., Esteban, J.L., Galesi, N., Johansen, J.: On the Relative Complexity of Resolution Refinements and Cutting Planes Proof Systems. SIAM J. Comput. 30(5), 1462–1484 (2000)
16. Bonet, M.L., Galesi, N.: Optimality of Size-Width Tradeoffs for Resolution. Computational Complexity 10(4), 261–276 (2001)
17. Bonet, M.L., Pitassi, T., Raz, R.: On Interpolation and Automatization for Frege Systems. SIAM J. Comput. 29(6), 1939–1967 (2000)
18. Chvátal, V.: Edmonds polytopes and a hierarchy of combinatorial problems. Discrete Mathematics 4(4), 305–337 (1973)
19. Clegg, M., Edmonds, J., Impagliazzo, R.: Using the Groebner basis algorithm to find proofs of unsatisfiability. In: Proc. of the 28th Annual ACM Symposium on Theory of Computing (STOC), pp. 174–183 (1996)
20. Condon, A.: The Complexity of Stochastic Games. Information and Computation 96, 203–224 (1992)
21. Cook, S.A., Reckhow, R.A.: The Relative Efficiency of Propositional Proof Systems. J. Symbolic Logic 44(1), 36–50 (1979)
22. Downey, R.G., Fellows, M.R.: Parameterized Complexity. Monographs in Computer Science. Springer (1999)
23. Ehrenfeucht, A., Mycielsky, J.: Positional strategies for mean payoff games. International Journal of Game Theory 8(2), 109–113 (1979)
24. Emerson, E.A., Jutla, C.S.: Tree Automata, Mu-Calculus and Determinacy. In: Proc. of 32nd IEEE Foundations of Computer Science (FOCS), pp. 368–377 (1991)
25. Esteban, J.L., Galesi, N., Messner, J.: On the Complexity of Resolution with Bounded Conjunctions. Theoretical Computer Science 321(2-3), 347–370 (2004)

26. Grigoriev, D.: Linear Lower Bound on Degrees of Positivstellensatz Calculus Proofs for the Parity. Theor. Comput. Sci. 259, 613–622 (2001)
27. Grigoriev, D., Hirsch, E.A., Pasechnik, D.V.: Complexity of semi-algebraic proofs. Moscow Mathematical Journal 2(4), 647–679 (2002)
28. Huang, L., Pitassi, T.: Automatizability and Simple Stochastic Games. In: Aceto, L., Henzinger, M., Sgall, J. (eds.) ICALP 2011, Part I. LNCS, vol. 6755, pp. 605–617. Springer, Heidelberg (2011)
29. Krajíček, J.: On the weak pigeonhole principle. Fundamenta Mathematicae 170(1-3), 123–140 (2001)
30. Krajíček, J.: Lower Bounds to the Size of Constant-Depth Propositional Proofs. J. of Symbolic Logic 59(1), 73–86 (1994)
31. Krajíček, J.: Interpolation theorems, lower bounds for proof systems, and independence results for bounded arithmetic. J. of Symbolic Logic 62(2), 457–486 (1997)
32. Laurent, M.: A comparison of the Sherali-Adams, Lovász-Schrijver and Lasserre relaxations for 0-1 programming. Mathematics of Operations Research 28, 470–496 (2001)
33. Lovász, L., Schrijver, A.: Cones of matrices and set-functions and 0-1 optimization. SIAM J. Optimization 1, 166–190 (1991)
34. Marques-Silva, J., Lynce, I., Malik, S.: Conflict-Driven Clause Learning SAT Solvers. In: Biere, A., Heule, M., van Maaren, H., Walsh, T. (eds.) Handbook of Satisfiability. IOS Press (2009)
35. Parrilo, P.A.: Structured Semidefinite Programs and Semialgebraic Geometry Methods in Robustness and Optimization. Ph.D. Thesis, California Institute of Technology, Pasadena, CA (2000)
36. Pitassi, T., Segerlind, N.: Exponential Lower Bounds and Integrality Gaps for Tree-Like Lovász-Schrijver Procedures. SIAM J. Comput. 41(1), 128–159 (2012)
37. Pudlák, P.: On reducibility and symmetry of disjoint NP-pairs. Theor. Comput. Science 295, 323–339 (2003)
38. Pudlák, P.: On the complexity of propositional calculus. In: Sets and Proofs, Invited papers from Logic Colloquium 1997, pp. 197–218. Cambridge Univ. Press (1999)
39. Sherali, H.D., Adams, W.P.: A hierarchy of relaxations and convex hull characterizations for mixed-integer zero-one programming problems. Discrete Applied Mathematics 52(1), 83–106 (1994)
40. Stålmarck, G.: Short resolution proofs for a sequence of tricky formulas. Acta Informatica 33(3), 277–280 (1996)
41. Stengle, G.: A Nullstellensatz and a Positivstellensatz in Semialgebraic Geometry. Mathematische Annalen 207(2), 87–97 (1974)
42. Urquhart, A.: Hard examples for resolution. Journal of the ACM 34(1), 209–219 (1987)
43. Zwick, U., Paterson, M.S.: The complexity of mean payoff games on graphs. Theoretical Computer Science 158, 343–359 (1996)

Turing's Computable Real Numbers and Why They Are Still Important Today

Edmund M. Clarke

Carnegie Mellon University, Pittsburgh, PA, 15213 USA

Abstract. Although every undergraduate in computer science learns about Turing Machines, it is not well known that they were originally proposed as a means of characterizing computable real numbers. For a long time, formal verification paid little attention to computational applications that involve the manipulation of continuous quantities, even though such applications are ubiquitous. In recent years, however, there has been great interest in safety-critical hybrid systems involving both discrete and continuous behaviors, including autonomous automotive and aerospace applications, medical devices of various sorts, control programs for electric power plants, and so on. As a result, the formal analysis of numerical computation can no longer be ignored. In this talk, we focus on one of the most successful verification techniques, temporal logic model checking. Current industrial model checkers do not scale to handle realistic hybrid systems. We believe that the key to handling more complex systems is to make better use of the theory of the computable reals, and computable analysis more generally. We argue that new formal methods for hybrid systems should combine existing discrete methods in model checking with new algorithms based on computable analysis. In particular we discuss a model checker we are currently developing along these lines.

M. Järvisalo and A. Van Gelder (Eds.): SAT 2013, LNCS 7962, p. 18, 2013.
© Springer-Verlag Berlin Heidelberg 2013

There Are No CNF Problems

Peter J. Stuckey

National ICT Australia, Victoria Laboratory
Department of Computing and Information Systems,
University of Melbourne, Australia
pstuckey@unimelb.edu.au

Abstract. SAT technology has improved rapidly in recent years, to the point now where it can solve CNF problems of immense size. But solving CNF problems ignores one important fact: *there are NO problems that are originally CNF*. All the CNF that SAT solvers tackle is the result of modelling some real world problem, and mapping the high-level constraints and decisions modelling the problem into clauses on binary variables. But by throwing away the high level view of the problem SAT solving may have lost a lot of important insight into how the problem is best solved. In this talk I will hope to persuade you that by keeping the original high level model of the problem one can realise immense benefits in solving hard real world problems.

1 Introduction

SAT technology has improved markedly in the last 12 years, to the point where it can solve CNF problems of immense size. Hence it is now a generic tool for tackling many combinatorial satisfaction and optimization problems. But solving CNF problems ignores one important fact: *there are NO problems that are originally CNF*. The CNF that SAT solvers tackle is the result of modelling a real world problem, mapping the decisions of the problem into binary decisions, and the complex constraints of the problem into possibly very large sets of clauses. By throwing away this high level view of the problem SAT solving has been able to concentrate on solving a very tightly defined problem, and doing it very effectively, but it has also lost a lot of knowledge about the problem.

Constraint programming (CP) [1,2] is a very flexible approach to modelling and solving combinatorial problems, which makes use of the high level structure of the problem. Modern CP solvers make use of SAT solving technology to solve most effectively, but without throwing away the structure. In this talk I hope to convince you that no one should solve a CNF problem *per se*. By considering the high level structure one can reason about the high level structure of a problem to

- improve the mapping of the complex constraints to clauses [3]; or
- during runtime lazily create a SAT model of the problem we are solving [4]; or
- during runtime choose how we wish to map complex constraints to clauses, and indeed if we should [5].

M. Järvisalo and A. Van Gelder (Eds.): SAT 2013, LNCS 7962, pp. 19–21, 2013.

In this manner we can gain the benefits of both the complex reasoning possible on the high level model, and the SAT style reasoning on the low level model.

Constraint programming with learning, or *lazy clause generation* (LCG) [4], had proved remarkably successful in tackling hard combinatorial optimization problems. It defines the state of the art complete method in many well studied scheduling problems, such as resource constraint project scheduling (RCPSP) [6], and its variations RCPSP with generalized precedences [7] and RCPSP with discounted cashflows [8]. LCG has led to substantial benefits in real life packing problems, such as carpet cutting [9]. LCG solvers have dominated the Mini-Zinc challenge competition www.minizinc.org since 2010, although they are not eligible for prizes, illustrating the approach is applicable over a wide range of problem classes.

Constraint programming with learning is in some sense a special case of SAT modulo theories [10]. CP concentrates on individual propagators, with a rich language of communication at the SAT level, between each propagator, while SMT concentrates on putting all the constraints of one form in a theory to reason about them conjunctively. CP concentrates on complex propagators like alldifferent that reflect some substructure (assignment subproblem) of the overall problem, rather than theories for classes of constraints which have some tractable reasoning.

MiniZinc [11] is an emerging standard for constraint programming modelling which captures the model at a high level, but can automatically map it to low level constraints. The language is currently supported by constraint programming solvers such as Gecode [12] and Google's OR-tools [13], mixed integer programming solvers such as SCIP [14], and SMT solvers such as fzn2smt [15]. MiniZinc would appear to provide an excellent starting point for high-level modelling that can then be mapped to CNF. Indeed I hope to persuade you that SAT solvers should be lifted to solve problems described at the high level in a language like MiniZinc.

Constraint programming researchers have learnt a lot from the SAT community, and I hope that I can demonstrate that SAT may gain from some insights from the CP community.

References

1. Hentenryck, P.V.: Constraint Satisfaction in Logic Programming. MIT Press (1989)
2. Marriott, K., Stuckey, P.: Programming with Constraints: an Introduction. MIT Press (1998)
3. Metodi, A., Codish, M., Stuckey, P.J.: Boolean equi-propagation for concise and efficient sat encodings of combinatorial problems. Journal of Artificial Intelligence Research 46, 303–341 (2013), http://www.jair.org/papers/paper3809.html
4. Ohrimenko, O., Stuckey, P., Codish, M.: Propagation via lazy clause generation. Constraints 14(3), 357–391 (2009)
5. Abío, I., Stuckey, P.J.: Conflict directed lazy decomposition. In: Milano, M. (ed.) CP 2012. LNCS, vol. 7514, pp. 70–85. Springer, Heidelberg (2012)
6. Schutt, A., Feydy, T., Stuckey, P., Wallace, M.: Explaining the cumulative propagator. Constraints 16(3), 250–282 (2011)

7. Schutt, A., Feydy, T., Stuckey, P., Wallace, M.: Solving RCPSP/max by lazy clause generation. Journal of Scheduling (2012) (online first: August 2012), http://dx.doi.org/10.1007/s10951-012-0285-x

8. Schutt, A., Chu, G., Stuckey, P.J., Wallace, M.G.: Maximising the net present value for resource-constrained project scheduling. In: Beldiceanu, N., Jussien, N., Pinson, É. (eds.) CPAIOR 2012. LNCS, vol. 7298, pp. 362–378. Springer, Heidelberg (2012)

9. Schutt, A., Stuckey, P.J., Verden, A.R.: Optimal carpet cutting. In: Lee, J. (ed.) CP 2011. LNCS, vol. 6876, pp. 69–84. Springer, Heidelberg (2011)

10. Nieuwenhuis, R., Oliveras, A., Tinelli, C.: Solving SAT and SAT Modulo Theories: From an abstract Davis–Putnam–Logemann–Loveland procedure to DPLL(T). J. ACM 53(6), 937–977 (2006)

11. Nethercote, N., Stuckey, P.J., Becket, R., Brand, S., Duck, G.J., Tack, G.: MiniZinc: Towards a standard CP modelling language. In: Bessière, C. (ed.) CP 2007. LNCS, vol. 4741, pp. 529–543. Springer, Heidelberg (2007)

12. Schulte, C., Lagerkvist, M., Tack, G.: Gecode, http://www.gecode.org/

13. Perron, L.: OR-tools: Operations research tools developed at Google, https://code.google.com/p/or-tools/

14. SCIP: Solving constraint integer programs, http://scip.zib.de/scip.shtml

15. Bofill, M., Palahí, M., Suy, J., Villaret, M.: Solving constraint satisfaction problems with SAT modulo theories. Constraints 17(3), 273–303 (2012)

Soundness of Inprocessing
in Clause Sharing SAT Solvers*

Norbert Manthey, Tobias Philipp, and Christoph Wernhard

Knowledge Representation and Reasoning Group
Technische Universität Dresden

Abstract. We present a formalism that models the computation of
clause sharing portfolio solvers with inprocessing. The soundness of these
solvers is not a straightforward property since shared clauses can make a
formula unsatisfiable. Therefore, we develop characterizations of simplification techniques and suggest various settings how clause sharing and
inprocessing can be combined. Our formalization models most of the recent implemented portfolio systems and we indicate possibilities to improve these. A particular improvement is a novel way to combine clause
addition techniques – like blocked clause addition – with clause deletion
techniques – like blocked clause elimination or variable elimination.

1 Introduction

The satisfiability problem (SAT) is one of the most prominent problems in
theoretical computer science and has many applications in verification, planning, model checking [7] or scheduling [14]. Modern SAT solvers employ many
advanced techniques like *clause learning* [33], *non-chronological backtracking*,
restarts [13], *clause removal* [11,4] and advanced *decision heuristics* [35,4,3],
making SAT very attractive for problem-solving. Among all these additional
techniques, clause learning is the most powerful one, both from a theoretical and
an empirical point of view [36,5,27]. Formula simplification techniques like *variable elimination* [10,40]and *bounded variable addition* [32] are applied to further
improve the efficiency. In recent SAT solvers, (e.g. LINGELING [6]), simplifications are also applied during search as *inprocessing* techniques, on all clauses at
hand, including learned clauses [26,31].

The parallel architecture of today's computers is utilized in portfolio solvers
like PPFOLIO [37], PFOLIOUZK [42], MANYSAT [18] and PLINGELING [6]. In
the portfolio approach, different sequential solvers work on the same formula.
For instance, PFOLIOUZK consists of several different conflict-driven systematic
solvers, look-ahead SAT solvers and incomplete SAT solvers. Each of them is
allowed to apply arbitrary simplification techniques. In contrast to PPFOLIO,
MANYSAT's portfolio consists only of systematic CDCL SAT solvers, but ex-
changes learned clauses. Thus, MANYSAT can solve instances faster than the

---

* The second author was supported by the European Master's Program in Computational Logic (EMCL).

M. Järvisalo and A. Van Gelder (Eds.): SAT 2013, LNCS 7962, pp. 22–39, 2013.

best sequential solver of its portfolio. However, no simplification techniques are applied during search. PLINGELING combines the two worlds: The solver applies a restricted form of inprocessing and shares unit clauses. Proving soundness of such clause sharing portfolio solvers where simplification techniques are applied during search is not trivial.

To the best of the authors' knowledge, there does not exists a soundness proof of any parallel clause sharing SAT solver with inprocessing. The authors are not aware of a formalization of parallel SAT solvers, which can be used to verify when clause sharing is sound. The research focus so far has been mainly on clause sharing strategies [17,15,16,1], and on utilizing computing resources [22,23].

In this paper, we bridge the gap between simplification techniques and clause sharing SAT solvers. A formal *sharing model* is introduced, which models the computation of clause sharing portfolio solvers with inprocessing. This model allows to reason about the behavior of portfolio solvers. Additionally, it allows to explain how portfolio solvers are working in a compact and easy way. Abstracting from the specific solver implementation is important since modern systems are complex and are written in programming languages with side effects. Equipped with this sharing model, we can show that, assumed the base solving engine is correct, inprocessing does not harm portfolio solvers on satisfiable instances. In general, inprocessing has to be restricted to obtain a valid system. In particular, we will see the soundness of a combination of restricted inprocessing rules and clause sharing as present in the PLINGELING system. We will also see the soundness of specific combinations with further inprocessing techniques, for instance that adding blocked clauses is possible in only a single solver incarnation of the portfolio. In addition to soundness proofs, several examples demonstrate that certain combinations of techniques cannot be applied in the presence of clause sharing. The result presented here can be easily lifted to parallel SAT solvers based on the guiding path approach [43], since these solvers only restrict the way how the search space is traversed but work on the same input formula.

The paper is structured as follows: Followed by a specification of the used formal notation and an outline of the necessary preliminaries of modern SAT solving in Section 2, the new sharing model is introduced in Section 3, variations of it are discussed, and the soundness of several combinations of inprocessing techniques and clause sharing is proven. Section 4 concludes the paper.

2   Preliminaries

We introduce the used notion of *Satisfiability Testing* (SAT), briefly discuss recent solving approaches that exploit parallel computing resources and finally give some insights in modern formula simplification techniques.

2.1   Notation and Basic Concepts

We assume a fixed set $\mathcal{V}$ of Boolean variables, or briefly just *variables*. A *literal* is a variable v (*positive literal*) or a negated variable $\overline{v}$ (*negative literal*).

We overload the overbar notation: The *complement* $\overline{x}$ of a positive (negative, resp.) literal x is the negative (positive, resp.) literal with the same variable as x. A *clause* is a finite set of literals, a *formula* is a finite set of clauses. We write a clause $\{x_1, \ldots, x_n\}$ also as disjunction $x_1 \vee \ldots \vee x_n$, and a formula $\{C_1, \ldots, C_n\}$ also as conjunction $C_1 \wedge \ldots \wedge C_n$. In this notation, the empty clause is written as $\bot$ (falsum) and the the empty formula as $\top$ (verum). Occasionally, we write formulas with further standard connectives such as $\leftrightarrow$. This can be understood as meta level notation for equivalent formulas in conjunctive normal form.

An *interpretation* is a mapping from the set $\mathcal{V}$ of all Boolean variables to the set $\{\top, \bot\}$ of truth values. The satisfaction relation $\models$ can then be defined inductively as follows: If x is a positive (negative, resp.) literal with variable v, then $I \models x$ holds if and only if $I(v) = \top$ ($I(v) = \bot$, resp.). If C is a clause, then $I \models C$ if and only if there is a literal $l \in C$ such that $I \models l$. If F is a formula, then $I \models F$ if and only if for all clauses $C \in F$ it holds that $I \models C$. A *general formula* is a literal, a clause or a formula. If, for an interpretation I and general formula F it holds that $I \models F$, we say that I is a *model* of F, or that I *satisfies* F. If there exists an interpretation that satisfies F, then F is called *satisfiable*, otherwise *unsatisfiable*. Two general formulas are *equisatisfiable* if and only if either both of them are satisfiable or both are unsatisfiable. A general formula F *entails* a general formula G, written $F \models G$, if and only if for all interpretations I such that $I \models F$ it holds that $I \models G$. This definition of entailment has two nice properties that simplify the proofs in this paper: transitivity and monotonicity. The drawback is that some formula simplification rules must be treated in an extra step. Two general formulas F and G are *equivalent*, written $F \equiv G$, if and only if $F \models G$ and $G \models F$.

A clause that contains exactly a single literal is called a *unit clause*. A clause that contains a literal and its complement is called a *tautology*. The set of all variables occurring in a formula F (in positive or negative literals) is denoted by $\mathsf{vars}(F)$. The set of all literals in the clauses in F by $\mathsf{lits}(F)$. A literal x such that $x \in \mathsf{lits}(F)$ and $\overline{x} \notin \mathsf{lits}(F)$ is called *pure* in F. If F is a formula and x is a literal, then the formula consisting of all clauses in F that contain x is denoted by F_x. If v is a variable and $C = v \vee C'$ as well as $D = \overline{v} \vee D'$ are clauses, then the clause $C' \vee D'$ is called the *resolvent of C and D upon v*. For two formulas F, G and variable v, the set of all resolvents of a clause in F with a clause in G upon v is denoted by $F \otimes_v G$. The formula F after substituting all occurrences of the variable v with the variable w is denoted by $F[v \mapsto w]$.

2.2   Parallel SAT Solving

The first parallel SAT solvers have been based on the DPLL procedure and divided the search space among the parallel resources [8]. Later on, parallel architecture has been exploited in different ways. A survey on parallel SAT solving is given in [21,34]. Parallel SAT solvers can be classified into competitive approaches and cooperative approaches. The first class contains the *portfolio* approach, where several solver incarnations try to solve the same formula (e.g. [18,6,1]). The latter class contains the *search space partitioning* methods (e.g. [19]). Here, the search space of the input formula is divided into several sub spaces, represented by several modified formulas. Finally, there exist hybrid approaches, which combine competitive and cooperative strategies (e.g. [22,23]). Guiding path solvers [43] are hard to categorize: on the one hand, this approach partitions the search space. On the other hand, these systems do not alter the input formula, but control the partitioning based on the current interpretation of the specific solvers. From a modeling point of view, guiding path solvers could be added to search space partitioning. Since our sharing model is based on the formula that each solver uses, portfolio and guiding path solvers are quite similar.

Since clause learning is essential for sequential solvers, this technique has been lifted to parallel solvers as well: Clauses are shared among solver incarnations [18,17], even if different subformulas are solved [24,29]. However, most of these solver incarnations do not use clause simplification techniques for *inprocessing*. During inprocessing clause simplification techniques are interleaved with search and are also applied to learned and shared clauses. To the best of the authors' knowledge, all sharing implementations do not check the validity of the received clause, but the implementation assumes that using received clauses is sound. Obvious reasons for not performing such checks are the computational costs of testing properties like $F \models C$ or testing equisatisfiability of F and $F \wedge C$ with respect to a clause C received by a solver whose working formula is F.

In the following, we focus on the portfolio approach, and discuss some systems in detail. A nice property of this approach is that the search can be stopped, as soon as one solver incarnation found the solution for the input formula. This property motivates combining special solvers for formulas from special categories to a single portfolio solver, which then solves a given formula in the time required by the fastest of the sequential component solvers. Portfolio solvers in recent competitions like PPFOLIO [37] and PFOLIOUZK [42] simply execute several powerful SAT solvers in parallel, even scheduled on a sequential machine. This easy approach scales with the number of available solvers. If the number of available parallel computing resources exceeds the number of solvers, parallel solvers will be added to the portfolio.

More sophisticated portfolio approaches use a single solving engine and execute multiple incarnations in parallel (e.g. MANYSAT [18] or PENELOPE [2]). An advantage of single engine portfolio solvers is that learned clauses can be shared among the incarnations. A soundness proof is obvious: since all incarnations are solving the same formula, and the solver always preserve the equivalence of the formula, all learned clauses are a consequence of the input formula. Knowledge sharing enables the portfolio solver to solve a formula even faster than the best sequential incarnation, because the search of this best incarnation is enhanced with more clauses that cut off search space. Another difference to these single engine portfolio solvers has been introduced with the PLINGELING system [6]. This solver executes LINGELING as the solving engine, which applies simplification technique during search, and also shared learned clauses.

2.3   Inprocessing

Simplifying the formula before giving it to the SAT solver has become a crucial part of the solving chain. Several techniques, such as bounded variable elimination [40,10], blocked clause elimination [25] and addition [28,26], equivalent literal elimination [12], probing [30], and automated re-encoding by using extended resolution [32,41] have been proposed for simplification. These techniques have been originally suggested with preprocessing, application before search, in mind. However, with minor modifications to treat learned clauses correctly, these techniques can also be used during search as *inprocessing*. Successful SAT solvers that participated in recent competitions, such as PRECOSAT [6], CRYPTOMINISAT [38] or LINGELING [6], utilize this. Many simplification techniques result in formulas that have a certain semantic relationship to the input formula, defined as follows:

Definition 1 (Unsat-Preserving Consequence). *A formula F' is called an* unsat-preserving consequence *of a formula F if and only if*

1. *$F \models F'$, and*
2. *If F is unsatisfiable, then F' is unsatisfiable.*

Note that if F' is an unsat-preserving consequence of F, it immediately follows that F and F' are equisatisfiable. If F' is an unsat-preserving consequence of F, then F' is also a consequence of F in the standard sense. That is, any model of F is also a model of F'. A special case of unsat-preserving consequence applies if $F \equiv F'$. As examples with respect to simplifications that involve addition and removal of clauses consider that if C is a clause entailed by F, then $F \wedge C$ is an unsat-preserving consequence of F; and if $F = F' \wedge C$ and F' is equisatisfiable with F, then F' is an unsat-preserving consequence of F.

In [26] a formal framework that can be used to prove the correctness of existing and new inprocessing rules has been presented. From the properties of simplification techniques discussed there we will drop the introduced redundancy property, because this property is related to very specific simplification procedures that go beyond the scope of this paper. The particular simplifications that we will consider can be described as follows, where we adopt the convention that F denotes the input formula and F' the output formula:

Variable Elimination. (VE) [40,10] eliminates a variable v by resolution: $F' = (F \cup F_v \otimes_v F_{\overline{v}}) \setminus (F_v \cup F_{\overline{v}})$. Note that $v \notin \mathsf{vars}(F')$. It follows from properties of propositional resolution that F' is an unsat-preserving consequence of F.

Equivalence Elimination. (EE) [12] replaces all occurrences of a variable v with the variable w, if the input F entails the equivalence $w \leftrightarrow v$, that is, $F' = F[v \mapsto w]$. Since $F \equiv F \wedge (w \leftrightarrow v) \equiv F[v \mapsto w] \wedge (w \leftrightarrow v) \equiv F' \wedge (w \leftrightarrow v)$, it follows that $F \models F'$. Moreover, this technique preserves satisfiability [20] and thus F' is an unsat-preserving consequence of F.

Blocked Clause Elimination. (BCE) [25] removes a *blocked clause* C from the input formula F. Then $F' = F \setminus \{C\}$. A clause C is called *blocked*, if it contains a literal x such that for all clauses $D \in F_{\overline{x}}$ the resolvent of C and D upon the variable of x is a tautology. Obviously, $F \models F'$ and since equisatisfiability is preserved by this technique, F' is an unsat-preserving consequence of F [25].

Extended Resolution. (ER) [41] adds a definition of a fresh variable v to the formula: $F' = F \land (\overline{v} \lor x \lor y) \land (\overline{x} \lor v) \land (\overline{y} \lor v)$, where the two literals x and y occur already in the formula, that is, $x, y \in \text{lits}(F)$. As presented in [41], this addition preserves satisfiability of the formula, that is, F' and F are equisatisfiable. However, $F \not\models F'$, because of the fresh variable, and therefore F' is not an unsat-preserving consequence of F.

Blocked Clause Addition. (BCA) [28,26] is the dual technique of BCE: BCA adds a clause C, such that this clause is blocked in the new formula: $F' = F \land C$. Note that the formula F' is not an unsat-preserving consequence of F, as can be seen in the following small counterexample: Let $F = (x \lor y)$. Then the clause $(\overline{x} \lor \overline{y})$ is blocked with the blocking literal $\overline{y}$. However, $F \not\models (\overline{x} \lor \overline{y})$.

Bounded Variable Addition. (BVA) [32] can be understood as adding a partial definition of a fresh variable v to the formula: First, a fresh variable is introduced like in extended resolution, resulting in the intermediate formula $G = F \land (v \lor \overline{x} \lor \overline{y}) \land (\overline{v} \lor x) \land (\overline{v} \lor y)$, where $x, y \in \text{lits}(F)$. The formulas F and G are equisatisfiable. Next, all clauses $C, D \in (G \setminus \{(\overline{v} \lor x), (\overline{v} \lor y)\})$ which have a common subclause E such that $C = (x \lor E)$ and $D = (y \lor E)$ are replaced by the new clause $(v \lor E)$. The resulting formula H is equivalent to G. Finally, the formula F', the result of BVA, is obtained from H by removing the clause $(v \lor \overline{x} \lor \overline{y})$. As presented in [32], BVA preserves satisfiability, that is, F' and F are equisatisfiable, because the old clauses can be restored by resolution. However, $F \not\models F'$ and thus F' is not an unsat-preserving consequence of F.

3   Sharing Clauses in Portfolio Solvers

We model the computation of clause-sharing solver portfolios by means of state transition systems as follows: A state of computation of a single sequential solver is a formula. A state of computation of a portfolio of n sequential solvers $Solver_1, \ldots, Solver_n$ is an n-tuple $(F_1, \ldots, F_n)$ of formulas, where F_i is the state of the $Solver_i$. Thus, a state of a portfolio is a snapshot of the states of its member solvers. A *portfolio system with input formula F_0 and multiplicity n* is a state transition system whose set of states is $\{(F_1, \ldots, F_n) \mid F_1, \ldots, F_n \text{ are formulas}\} \cup \{\text{SAT}, \text{UNSAT}\}$, whose initial state $\text{init}(n, F_0)$ is the n-tuple $(F_0, \ldots, F_0)$, and whose set of terminal states is $\{\text{SAT}, \text{UNSAT}\}$. For a transition relation $\leadsto$ of a portfolio system we define $\leadsto^*$ as the reflexive transitive closure of $\leadsto$, we define $x \leadsto^0 x$, and for all natural numbers $n > 0$ we define: $x \leadsto^n y$ if and only if $x \leadsto^{n-1} z \leadsto y$. In the course of the paper, we will investigate the soundness of portfolio systems with different transition relations corresponding to different allowed inprocessing methods. Soundness of a SAT solver, and thus also of a portfolio system, is a combination of two properties: refutational soundness and soundness with respect to satisfiability. More precisely, a portfolio system with input formula F_0, multiplicity n and transition

SAT-rule: $(F_1, \ldots, F_i, \ldots, F_n) \leadsto_{\mathsf{SAT}} \mathsf{SAT}$
iff F_i is satisfiable.

UNSAT-rule: $(F_1, \ldots, F_i, \ldots, F_n) \leadsto_{\mathsf{UNSAT}} \mathsf{UNSAT}$
iff F_i is unsatisfiable.

CM-rule: $(F_1, \ldots, F_{i-1}, F_i, F_{i+1}, \ldots, F_n) \leadsto_{\mathsf{CM}} (F_1, \ldots F_{i-1}, F_i', F_{i+1}, \ldots, F_n)$
iff $F_i \equiv F_i'$.

CS-rule: $(F_1, \ldots, F_{i-1}, F_i, F_{i+1}, \ldots, F_n) \leadsto_{\mathsf{CS}} (F_1, \ldots, F_{i-1}, F_i \wedge C, F_{i+1}, \ldots, F_n)$
iff $C \in F_j$ for some $j \in \{1, \ldots, i-1, i+1, \ldots, n\}$.

UI-rule: $(F_1, \ldots, F_{i-1}, F_i, F_{i+1}, \ldots, F_n) \leadsto_{\mathsf{UI}} (F_1, \ldots, F_{i-1}, F_i', F_{i+1}, \ldots, F_n)$
iff F_i and F_i' are equisatisfiable.

ER-rule: $(F_1, \ldots, F_{i-1}, F_i, F_{i+1}, \ldots, F_n) \leadsto_{\mathsf{ER}} (F_1, \ldots, F_{i-1}, F_i', F_{i+1}, \ldots, F_n)$
iff F_i' is obtained by extended resolution from F_i.

BVA-rule: $(F_1, \ldots, F_{i-1}, F_i, F_{i+1}, \ldots, F_n) \leadsto_{\mathsf{BVA}} (F_1, \ldots, F_{i-1}, F_i', F_{i+1}, \ldots, F_n)$
iff F_i' is obtained by bounded variable addition from F_i.

RI-rule: $(F_1, \ldots, F_{i-1}, F_i, F_{i+1}, \ldots, F_n) \leadsto_{\mathsf{RI}} (F_1, \ldots, F_{i-1}, F_i', F_{i+1}, \ldots, F_n)$
iff F_i' is an unsat-preserving consequence of F_i.

ADD-rule: $(F_1, \ldots, F_n) \leadsto_{\mathsf{ADD}} (F_1 \wedge C, F_2, \ldots, F_n)$
iff $\mathsf{vars}(C) \subseteq \mathsf{vars}(F_0)$, and
the formulas $F_1 \wedge C$ and F_1 are equisatisfiable.

DEL-rule: $(F_1, \ldots, F_{i-1}, F_i, F_{i+1}, \ldots, F_n) \leadsto_{\mathsf{DEL}} (F_1, \ldots, F_{i-1}, F_i', F_{i+1}, \ldots, F_n)$
iff $i > 1$ and F_i is an unsat-preserving consequence of F_i'.

Fig. 1. Transition relations used to characterize clause sharing models by means of portfolio systems with input formula F_0 and multiplicity n. These definitions apply to all formulas $F_1, \ldots, F_n, F_1', \ldots, F_n'$, clauses C and $i \in \{1, \ldots, n\}$. Fresh variables introduced by extended resolution (ER-rule) and by bounded variable addition (BVA-rule) have to be *globally* fresh, that is, they are not allowed to occur in $F_1, \ldots, F_n$.

relation $\leadsto$ is called *sound* if and only if $\mathsf{init}(n, F_0) \leadsto^* \mathsf{UNSAT}$ implies that F is unsatisfiable and $\mathsf{init}(n, F_0) \leadsto^* \mathsf{SAT}$ implies that F is satisfiable.

We investigate four different portfolio systems whose transition relation is composed of the relations presented in Fig. 1, which we also call *rules*. The particular systems that we are going to investigate are $\mathsf{SysA}, \mathsf{SysB}, \mathsf{SysC}$ and SysD, characterized by the following respective transition relations:

$$\leadsto_{\mathsf{SysA}} := \leadsto_{\mathsf{SAT}} \cup \leadsto_{\mathsf{UNSAT}} \cup \leadsto_{\mathsf{CM}} \cup \leadsto_{\mathsf{CS}}.$$
$$\leadsto_{\mathsf{SysB}} := \leadsto_{\mathsf{SysA}} \cup \leadsto_{\mathsf{UI}}.$$
$$\leadsto_{\mathsf{SysC}} := \leadsto_{\mathsf{SysA}} \cup \leadsto_{\mathsf{RI}} \cup \leadsto_{\mathsf{ER}} \cup \leadsto_{\mathsf{BVA}}.$$
$$\leadsto_{\mathsf{SysD}} := \leadsto_{\mathsf{SysA}} \cup \leadsto_{\mathsf{ADD}} \cup \leadsto_{\mathsf{DEL}} \cup \leadsto_{\mathsf{ER}} \cup \leadsto_{\mathsf{BVA}}.$$

▶ SysA is a basic portfolio system that models clause sharing portfolios where inprocessing rules preserve equivalence: The first termination rule is the SAT-rule that terminates the computation with the answer SAT, if one formula

F_i in the state is satisfiable. Likewise, the second termination rules is the UNSAT-rule that leads to the answer UNSAT, if one formula F_i in the state is unsatisfiable. Modern complete solvers are modifying the formula by adding and removing learned clauses. Such clause management is modelled by the CM-rule that rewrites a formula F_i of a solver incarnation into an equivalent formula F_i'. Finally, clause sharing is captured by the CS-rule that adds a clause C from the formula F_j to the formula F_i, where $i \neq j$. As we will see later, solvers like MANYSAT [18] or PENELOPE [2] can be modeled as instances of SysA.

▶ SysB extends the system SysA by the UI-rule that performs inprocessing, just constrained to preserve equisatisfiability: The UI-rule replaces a formula F_i with another formula F_i', if they are equisatisfiable. We will later see that the general UI-rule does not harmonize with clause sharing, leading to refutational unsoundness.

▶ SysC extends the system SysA by the RI-, ER- and BVA-rule. The RI-rule replaces a formula F_i with a formula F_i', if the F_i' is an unsat-preserving consequence of F_i. Extended resolution and bounded variable addition can be performed on a formula F_i by the ER- and BVA-rule. We will later see that the formalism SysC is sound, and that PLINGELING [6] is an instance of this formalism.

▶ SysD extends the system SysA by the ER- and BVA-rule and two new rules that allow to perform clause addition techniques on a single designated solver whose state is represented without loss of generality by F_1, and only clause deletion techniques on the remaining solvers, represented by $F_2, \ldots, F_n$ to obtain a sound system: The ADD-rule adds a clause C to the formula F_1, if F_1 and $F_1 \wedge C$ are equisatisfiable and $\mathsf{vars}(C) \subseteq \mathsf{vars}(F_0)$. On the other hand, the clause deletion is captured by the DEL-rule, which like the RI-rule, except that the DEL-rule can only modify $F_2, \ldots, F_n$.

3.1 Sharing Model SysA – Where Equivalence Is Preserved

Let us start with the system SysA that models clause sharing portfolios where inprocessing preserves equivalence. The rules of SysA guarantee that all formulas $F_i \in (F_1, \ldots, F_n)$ of the current state are equivalent to the input formula, as expressed by the following lemma:

Lemma 2 (Key Invariant of SysA). *Let $n \geq 1$, let $F_0, F_1, \ldots, F_n$ be formulas, and let $m \geq 0$. If $\mathsf{init}(n, F_0) \leadsto_{\mathsf{SysA}}^m (F_1, \ldots, F_n)$, then for all $i \in \{1, \ldots, n\}$ it holds that $F_i \equiv F_0$.*

Proof. The CM-rule preserves equivalence and therefore the CS-rule only adds clauses that are entailed by the corresponding formula. The claim then holds since equivalence is transitive. □

The following theorem is a straightforward consequence of the above lemma:

Theorem 3 (Soundness of SysA). SysA *is sound.*

Proof. If the input formula F_0 is satisfiable (unsatisfiable, resp.), the UNSAT-rule (SAT-rule, resp.) is never applicable, since by Lemma 2 all formulas in states reached by SysA are equivalent to F_0.   □

The portfolio solvers MANYSAT and PENELOPE are built upon MINISAT [18,2]. PENELOPE differs from MANYSAT in the clause import and export strategy and in the applied heuristics. Since the learned clauses in MINISAT can be obtained by resolution [39], MINISAT preserves the equivalence of formulas. Before searching for a model of the input formula F_0, the formula is subjected to a preprocessor in MANYSAT and PENELOPE. The portfolio systems then work on the result of the preprocessor. Consequently, SysA models the behavior of MANYSAT and PENELOPE with respect to the preprocessed formula as input.

3.2   Sharing Model SysB – Inprocessing without Limits

The portfolio transition system SysB models clause sharing portfolios where inprocessing can be applied constrained only by equisatisfiability. Many inprocessing techniques like variable elimination (VE), equivalence elimination (EE) and blocked clause addition (BCA) do preserve equisatisfiability, but not equivalence. Combining clause sharing and the general UI-rule inprocessing is problematic:

Example 4. Consider the two formulas x and $\overline{x}$, where x is a variable. Both formulas x and $\overline{x}$ are satisfiable and equisatisfiable. Since $x \wedge \overline{x}$ is unsatisfiable, the execution $\mathsf{init}(2, x) = (x, x) \leadsto_{\mathsf{UI}} (x, \overline{x}) \leadsto_{\mathsf{CS}} (x \wedge \overline{x}, \overline{x}) \leadsto_{\mathsf{UNSAT}} \mathsf{UNSAT}$ of a parallel SAT solver with transition relation $\leadsto_{\mathsf{SysB}}$ produces an incorrect result.

Thus, the system SysB is not refutationally sound. This small example shows that general inprocessing is incompatible with clause sharing, since the answer UNSAT can be incorrect. The phenomenon that the addition of a clause makes a formula unsatisfiable is the reason why clause sharing is a non-trivial problem. On the other hand, the system SysB is sound with respect to satisfiability. SAT answers obtained with SysB are still correct, as stated in the Theorem 5 below, preceded by a lemma that shows the underlying invariant of SysB transitions:

Lemma 5 (Key Invariant of SysB). *Let $n \geq 1$, let $F_0, F_1, \ldots, F_n$ be formulas, and let $m \geq 0$. If $\mathsf{init}(n, F_0) \leadsto_{\mathsf{SysB}}^m (F_1, \ldots, F_n)$ and for some $i \in \{1, \ldots, n\}$ it holds that if F_i is satisfiable, then F_0 is satisfiable.*

Proof. By induction on m: For the base case $m = 0$, the statement is trivially true since every component formula in $\mathsf{init}(n, F_0)$ is identical to F_0. For the induction step, assume that the claim holds for the state $(F_1, \ldots, F_n)$ and that $(F_1, \ldots, F_n) \leadsto_R (F_1, \ldots, F_{i-1}, F_i', F_{i+1}, \ldots, F_n)$ for some $1 \leq i \leq n$ and some rule R in SysB. The induction step follows from the fact that if F_i' is satisfiable, then F_i is satisfiable, which holds for each non-terminating rule of SysB:

– CM-rule: Since $F_i' \equiv F_i$.
– UI-rule: Since F_i' and F_i are equisatisfiable.

– CS-rule: In this case $F_i' = F_i \wedge C$ for some clause C. Thus, it holds that $F_i' \models F_i$ which implies that if F_i' is satisfiable, then F_i is satisfiable.   □

Theorem 6 (Soundness of SysB w.r.t Satisfiability). *Let $n \geq 1$ and let F_0 be a formula. If $\mathsf{init}(n, F_0) \leadsto_{\mathsf{SysB}}^m SAT$, then F_0 is satisfiable.*

Proof. Immediate from Lemma 5 and the definition of the SAT-rule.   □

3.3   Sharing Model SysC – With Clause Deletion Techniques

System SysC models clause sharing portfolios where clause deletion techniques are used. In contrast to model SysB, the result of a computation in SysC is always sound. The parallel SAT solver PLINGELING is an instance of SysC. The portfolio of PLINGELING consists of differently configured incarnations of the sequential solver LINGELING. The following inprocessing rules are used in PLINGELING [6]: Variable elimination (VE) [10], equivalence elimination (EE) [12], and blocked clause elimination (BCE) [25]. We do not discuss the other used techniques that preserve equivalence like hyper binary resolution since they are modelled by the CM-rule. Since these techniques transform a formula F into unsat-preserving consequence of F, these techniques can be modelled by the RI-rule.

   In the following, we need to trace the clauses introduced by extended resolution or bounded variable addition. For a particular sequence of transition steps $S_0 \leadsto S_1 \leadsto \ldots \leadsto S_m$, we let D_m denote the set of all definition clauses that were introduced by the ER- and BVA-rule in the sequence. Note that an implementation does not need to construct the set D_m, but we will use this set in the proofs to show that if $\mathsf{init}(n, F_0) \leadsto^m (F_1, \ldots, F_n)$, then F_0 and F_i are always equisatisfiable for $i \in \{0, \ldots, n\}$. On the other hand, an implementation has to guarantee that the variables introduced by extended resolution or bounded variable addition are *globally* fresh, that is, fresh throughout all involved sequential solvers. Incorrect UNSAT answers can then not be obtained with SysC.

   Theorem 8 below states the soundness of SysC. This theorem is based again on a key invariant stated in Lemma 7 as claim (iii). The lemma shows two further invariants of SysC as claims (i) and (ii), which involve the D_m. In the context of this paper these are just applied to prove invariant (iii).

Lemma 7 (Key Invariants of SysC). *Let $n \geq 1$, let $F_0, F_1, \ldots, F_n$ be formulas, and let $m \geq 0$. Assume $\mathsf{init}(n, F_0) \leadsto_{\mathsf{SysC}}^m (F_1, \ldots, F_n)$ with a transition sequence that has the formula D_m as the set of clauses introduced by the ER- and BVA-rule. Then the following properties hold:*

(i) $F_0 \wedge D_m \models F_1 \wedge \ldots \wedge F_n$.
(ii) F_0 *and* $F_0 \wedge D_m$ *are equisatisfiable.*
(iii)   *For all $i \in \{1, \ldots, n\}$ it holds that F_i and F_0 are equisatisfiable.*

Proof. We show the statement by induction on the number m of transition steps. For the base case $m = 0$, the claims (i)–(iii) are easy to see, since $D_0 = \top$ and $\mathsf{init}(n, F_0)$ is the n-tuple $(F_0, \ldots, F_0)$. For the induction step, assume that the claim holds for the state $(F_1, \ldots, F_n)$ and that $(F_1, \ldots, F_n) \leadsto_{\mathsf{R}}$

$(F_1, \ldots, F_{i-1}, F'_i, F_{i+1}, \ldots, F_n)$ for some rule R in SysC. We distinguish according to the applied rule R:

- CM-rule: Then $F'_i \equiv F_i$ and $D_{m+1} = D_m$. (i) holds since the substitution of a subformula by an equivalent formula preserves equivalence. (ii) holds since $D_{m+1} = D_m$. (iii) holds since F'_i and F_i are equivalent and equisatisfiable.
- RI-rule: Then $F_i \models F'_i$, F_i and F'_i are equisatisfiable and $D_{m+1} = D_m$. (i) holds since the entailment relation is transitive. (ii) holds since $D_{m+1} = D_m$. (iii) holds by the definition of the RI-rule.
- ER-rule: Then $F'_i = F_i \wedge (\overline{v} \vee x \vee y) \wedge (\overline{x} \vee v) \wedge (\overline{y} \vee v)$ and $D_{m+1} = D_m \wedge (\overline{v} \vee x \vee y) \wedge (\overline{x} \vee v \wedge (\overline{y} \vee v)$, where v is a fresh variable and $x, y \in \mathsf{lits}(F_i)$. Then (i) holds since the formula $v \leftrightarrow x \vee y$ is added on the left hand side and the right hand side of the entailment. (ii) holds since v is a fresh variable and (iii) holds since extended resolution preserves satisfiability.
- BVA-rule: Bounded variable addition consists of three steps: First, a fresh variable v is defined by $(v \vee \overline{x} \vee \overline{y}) \wedge (\overline{v} \vee x) \wedge (\overline{v} \vee y)$ like in the ER-rule, then the formula is rewritten into an equivalent formula, as in the CM-rule. Finally, the clause $(v \vee \overline{x} \vee \overline{y})$ is deleted, which can be modeled by the RI-rule.
- CS-rule: (i) is clear since $F_1 \wedge \ldots \wedge F_{i-1} \wedge F_i \wedge F_{i+1} \wedge \ldots \wedge F_n \equiv F_1 \wedge \ldots \wedge F_{i-1} \wedge F_i \wedge C \wedge F_{i+1} \wedge \ldots \wedge F_n$, which follows since $C \in F_j$ for some $j \in \{1, \ldots, i-1, i+1, \ldots, n\}$. (ii) holds since $D_{m+1} = D_m$. (iii) holds since the equisatisfiability of F_0 and $F'_i = F_i \wedge C$ can be proven as follows: In case F_0 is unsatisfiable, by the induction assumption F_i is also unsatisfiable, and thus $F_i \wedge C = F'_i$ is also unsatisfiable. Assume now the other case, that F_0 is satisfiable. By statement (ii) of the induction assumption it follows that $F_0 \wedge D_m$ is satisfiable, and thus by statement (i) of the induction assumption that also $F_1 \wedge \ldots \wedge F_n$ is satisfiable. Since $i, j \in \{1, \ldots, n\}$ and $F_j \models C$ it follows that also $F_i \wedge C = F'_i$ is satisfiable. □

Theorem 8 (Soundness of SysC). SysC *is sound.*

Proof. Let $n \geq 0$ and let $F_0, F_1, \ldots, F_n$ be formulas such that $\mathsf{init}(n, F_0) \rightsquigarrow^*_{\mathsf{SysC}}$ $(F_1, \ldots, F_n) \rightsquigarrow_{\mathsf{SysC}} \mathsf{UNSAT}$ (SAT, resp.). From the definition of the UNSAT-rule (SAT-rule, resp.) it follows that there exists an $i \in \{1, \ldots, n\}$ s.t. the formula F_i is unsatisfiable (satisfiable, resp.). By Lemma 7.iii it follows that F_0 is unsatisfiable (satisfiable, resp.). □

The above theorem shows in particular that combining clause sharing and applying inprocessing rules as done in PLINGELING is sound.

3.4 Sharing Model SysD – With Clause Addition Techniques

Motivated by the blocked clause addition technique, we will now study how to combine clause addition with clause deletion techniques and develop the sharing model SysD. To the best of our knowledge, blocked clause addition is not applied in clause sharing portfolios. The following examples illustrate two main problems that occur: First, if we allow two sequential solver to perform clause addition techniques like blocked clause addition, UNSAT-answers may be incorrect.

Example 9. Consider the satisfiable formula $F_0 = (x \vee \bar{y}) \wedge (\bar{x} \vee y) \wedge (x \vee z) \wedge (y \vee \bar{z})$ from [26]. Note that the clause $(\bar{x} \vee \bar{z})$ is blocked w.r.t. $\bar{z}$ and the clause $(\bar{y} \vee z)$ is blocked w.r.t. z. By using BCA, $\mathsf{init}(2, F_0) \leadsto_{\mathsf{UI}} (F_0 \wedge (\bar{x} \vee \bar{z}), F_0) \leadsto_{\mathsf{UI}} (F_0 \wedge (\bar{x} \vee \bar{z}), F_0 \wedge (\bar{y} \vee z)) \leadsto_{\mathsf{CS}} (F_0 \wedge (\bar{x} \vee \bar{z}) \wedge (\bar{y} \vee z), F_0 \wedge (\bar{y} \vee z)) \leadsto_{\mathsf{UNSAT}}$ UNSAT, but this answer is incorrect.

This incorrect answer is the reason, why we restrict clause addition techniques such that they are permitted only in one designated solver, the first component of the portfolio. The second problem is combining clause deletion techniques and clause addition techniques in one solver: it may lead to incorrect UNSAT-answers:

Example 10. Consider the satisfiable formula $F_0 = F' \wedge (\bar{x} \vee \bar{z})$ from [26] where $F' = (x \vee \bar{y}) \wedge (\bar{x} \vee y) \wedge (x \vee z) \wedge (y \vee \bar{z})$. Then, $(\bar{x} \vee \bar{z})$ is blocked and after removing this clause, the clause $(\bar{y} \vee z)$ can be again added by blocked clause addition. Then, $\mathsf{init}(2, F_0) \leadsto_{\mathsf{UI}} (F', F' \wedge (\bar{x} \vee \bar{z})) \leadsto_{\mathsf{UI}} (F' \wedge (\bar{y} \vee z), F' \wedge (\bar{x} \vee \bar{z})) \leadsto_{\mathsf{CS}} (F' \wedge (\bar{y} \vee z) \wedge (\bar{x} \vee \bar{z}), F' \wedge (\bar{x} \vee \bar{z})) \leadsto_{\mathsf{UNSAT}}$ UNSAT, but this answer is incorrect.

For this reason, the presented system in general applies clause deletion techniques in all solvers *except* the designated solver in which clause addition is permitted. With these examples in mind, we can show the following property in SysD:

Lemma 11 (Key Invariants of SysD). *Let $n \geq 1$, let $F_0, F_1, \ldots, F_n$ be formulas, and let $m \geq 0$. Assume $\mathsf{init}(n, F_0) \leadsto_{\mathsf{SysD}}^m (F_1, \ldots, F_n)$ with a transition sequence that has the formula D_m as the set of clauses introduced by the ER- and BVA-rule. Then the following properties hold:*

(i) $F_1 \wedge D_m \models F_2 \wedge \ldots \wedge F_n$.

(ii) F_1 *and* $F_1 \wedge D_m$ *are equisatisfiable.*

(iii) *for all $i \in \{1, \ldots, n\}$ it holds that F_i and F_0 are equisatisfiable.*

Proof. As before, we show the statement by induction on the number m of transition steps. For the base case $m = 0$, the claims (i)–(iii) are easy to see, since $D_0 = \top$ and $\mathsf{init}(n, F_0)$ is the n-tuple $(F_0, \ldots, F_0)$. For the induction step, assume that the claim holds for the state $(F_1, \ldots, F_n)$ and that $(F_1, \ldots, F_n) \leadsto_{\mathsf{R}} (F_1, \ldots, F_{i-1}, F_i', F_{i+1}, \ldots, F_n)$ for some rule R in SysD. We distinguish according to the applied rule R:

- CM-rule: Then $F_i' \equiv F_i$ and $D_m = D_{m+1}$. Claims (i) and (ii) follow since replacement of a subformula with an equivalent formula preserves equivalence, claim (iii) follows since equivalence implies equisatisfiability.
- ADD-rule: Then $F_1' = F_1 \wedge C$ and F_1' is equisatisfiable to F_1. Since the entailment relation is monotone, (i) holds. (ii) follows since by the definition of the ADD-rule $\mathsf{vars}(C) \subseteq \mathsf{vars}(F_0)$ and D_m only contains clauses introduced by extended resolution or bounded variable addition. (iii) is an immediate consequence of the definition of the ADD-rule.
- DEL-rule. Then $F_i \models F_i'$ for $i \in \{2, \ldots, n\}$. Consequently, $F_1 \wedge D_m \models F_2 \wedge \ldots F_{i-1} \wedge F_i \wedge F_{i+1} \wedge \ldots \wedge F_n \models F_2 \wedge \ldots F_{i-1} \wedge F_i' \wedge F_{i+1} \wedge \ldots \wedge F_n$ and thus (i) holds. Since $i \neq 1$, (ii) holds and (iii) is a consequence of the requirement that F_i and F_i' are equisatisfiable implied by the unsat-preserving consequence precondition of the rule.

- ER-rule: Then $F_i' = F_i \wedge (\overline{v} \vee x \vee y) \wedge (\overline{x} \vee v) \wedge (\overline{y} \vee v)$, where v is a globally fresh variable and $x, y \in \mathsf{lits}(F_i)$. (iii) holds as extended resolution preserves satisfiability. For showing (i) and (ii) we consider the following cases:
  - $i = 1$: Then, (i) is an immediate consequence of the entailment relation being monotone. (ii) holds since a globally fresh variable was defined.
  - $i > 1$: Then (i) holds as the added clause of F_i is also contained in D_{m+1}. (ii) holds as $i \neq 1$ and a globally fresh variable was defined.
- BVA-rule: Then $F_i' = F_i \wedge C$ and $D_{m+1} = D_m \wedge C'$ where $C = (\overline{v} \vee x) \wedge (\overline{v} \vee y)$ and $C' = C \wedge (\overline{v} \vee x \vee y)$. Similar to the ER-rule, (i) holds. Since $F_i' \wedge D_{m+1} \equiv F_i \wedge D_{m+1}$ and F_i and F_i' are equisatisfiable, (ii) holds. (iii) holds since bounded variable addition preserves equisatisfiability.
- CS-rule: Then $F_i' = F_i \wedge C$ where $C \in F_j$ for some $j \in \{1, \ldots, i - 1, i + 1, \ldots, n\}$. We consider the following cases:
  - $j = 1$ and $i > 1$: Then F_1 exports a clause. (i) holds as the entailment relation is monotone. (ii) holds since $i > 1$. (iii) is shown as follows: If F_i is unsatisfiable, then $F_i \wedge C$ is unsatisfiable and by the induction assumption, F_0 is unsatisfiable. Otherwise, if F_i is satisfiable, it follows by induction assumption that F_0 is satisfiable and then that F_1 is satisfiable. Then $F_1 \wedge D_m$ is satisfiable (since D_m only contains definitions) and hence by (i) $F_1 \wedge D_m \wedge F_2 \wedge \ldots \wedge F_n$ is satisfiable. We can conclude that $F_i \wedge C$ is satisfiable as $C \in F_1$.
  - $i = 1$ and $j > 1$: Then F_1 imports the clause C from F_j. (i) holds since the entailment relation is monotone. (ii) is proven as follows: Suppose F_1 is satisfiable. Then $F_1 \wedge D_m$ is satisfiable by induction assumption and since (ii) holds we know that $F_1 \wedge D_m \models C$. Consequently $F_1 \wedge C \wedge D_m$ is satisfiable. Otherwise, if F_1 is unsatisfiable, $F_1 \wedge D_m$ is unsatisfiable by induction assumption and consequently $F_1 \wedge C \wedge D_m$ is unsatisfiable. For showing (iii), suppose that F_1 is unsatisfiable, then $F_1 \wedge C$ is unsatisfiable. Otherwise, let F_1 be satisfiable, then $F_1 \wedge D_m$ is satisfiable by induction assumption and consequently $F_1 \wedge D_m \wedge F_2 \wedge \ldots F_n$ is satisfiable by (i). Therefore, $F_1 \wedge C$ is satisfiable since $C \in F_j$.
  - $j \neq 1$ and $i \neq 1$: This can be done similar to the proof of Prop. 7.   □

Theorem 12 (Soundness of SysD). SysD *is sound.*

Proof. Analogously to the proof of Theorem 8, but based on Lemma 11.iii.   □

Theorem 12 justifies that we can use clause addition procedures in a single solver and clause deletion procedures in the remaining solvers. The authors are not aware of a parallel SAT solver that uses this combination of techniques.

Example 10 shows that mixing deletion and addition procedures in a single sequential solver does not work. Moreover, using clause addition techniques in more than a single solver incarnation is problematic. If solver S_1 adds a clause C by *technique A* and solver S_2 adds clause D by *technique B*, then S_1 might not be able to apply B anymore since applying B might not preserve equisatisfiability. As already pointed out in [26], the application of simplification rules must take the learned clauses into account. Accordingly, the solver incarnations that apply clause addition techniques must take all clauses of the other solvers into account. Hence, if several solvers in a portfolio apply clause addition techniques, the implementation has to guarantee that the learned clause database is synchronized before. Otherwise, learned clauses of other solvers could be received after the clause addition and then the soundness of the overall system can break. We require that the ADD-rule only adds clauses that have variables occurring in the original formula. The reason is illustrated in the following:

Example 13. Consider the satisfiable formula $F_0 = (x \wedge y)$. Suppose that the ADD-rule does not require that the added clause C only contains variables that occur in F_0. Then $\text{init}(2, F_0) \rightsquigarrow_{\mathsf{BVA}} (F_0, F_0 \wedge (v \vee \overline{x} \vee \overline{y}) \wedge (\overline{v} \vee x) \wedge (\overline{v} \vee y)) \rightsquigarrow_{\mathsf{CS}}$ $(F_0 \wedge (\overline{v} \vee x), F_0 \wedge (v \vee \overline{x} \vee \overline{y}) \wedge (\overline{v} \vee x) \wedge (\overline{v} \vee y)) \rightsquigarrow_{\mathsf{ADD}} (F_0 \wedge (\overline{v} \vee x) \wedge \overline{v}, F_0 \wedge (v \vee \overline{x} \vee \overline{y}) \wedge (\overline{v} \vee x) \wedge (\overline{v} \vee y)) \rightsquigarrow_{\mathsf{CS}} (F_0 \wedge (\overline{v} \vee x) \wedge \overline{v}, F_0 \wedge (v \vee \overline{x} \vee \overline{y}) \wedge (\overline{v} \vee x) \wedge (\overline{v} \vee y) \wedge \overline{v}) \rightsquigarrow_{\mathsf{UNSAT}} \mathsf{UNSAT}$, but this answer is incorrect.

Restricting the variables of the added clause is not a big restriction. Basically, instead of adding a clause $(v \vee C)$, where v is defined by as $(x \vee y)$, one can add the clause $(x \vee y \vee C)$ by the ADD-rule. Afterwards, the clause $(x \vee C)$ can be obtained by resolution. This technique only works, if the definition is completely present at the solver. An alternative to this syntactical restriction of the added clause is the requirement that F_1 and $F_1 \wedge C \wedge D_m$ are equisatisfiable.

3.5 Guiding Path Solvers

Rooted in the DPLL algorithm [9], a *guiding path* [43] is the current partial interpretation of a (sequential) SAT solver. In the parallel setting, assume the sequence of decision literals D_i of the solver S_i to be $D_i = (l_1, \ldots, l_n)$. Now, if another solver incarnation S_j should be added to the parallel solver, this solver would simply create another sequence $D_j = D_i[l_k \mapsto \overline{l_k}]$, for exactly one $k \leq n$. Thus, the new sequence D_j differs exactly in the polarity of one literal of the sequence D_i. Now, the two solvers proceed with their search until one of the two either finds a solution or backtracks to the sequence $D = (l_1, \ldots, l_{k-1})$. If the latter case is reached, the whole sub search space can be closed, and thus also the other solver can be terminated and given a new sequence. Note, none of the steps in the guiding path approach touched the underlying formula F.

Therefore, our presented sharing model is fully applicable to guiding path solvers, because it models only the formulas but not the search process of the single solver incarnations.

4 Conclusion

In SAT solving, it is desirable to explore advanced techniques in combination with each other. However, soundness of such combinations is not always easy to see. Combinations that are sound and useful, but involve constraints that are too complicated to be easily captured by intuition might be missed with experimental verification approaches. Here we have seen an approach to overcome this situation for the case of parallel SAT solvers that perform clause sharing in combination with a variety of inprocessing techniques. The behavior of portfolio solvers is represented formally as a state transition system. Different sets of transition rules allow to represent different combinations of inprocessing simplifications and ways of clause sharing among the component solvers running in parallel. The three currently most successful clause sharing portfolio solving systems can be modeled as such transition systems, allowing to prove their soundness. Moreover, the soundness of further, not yet implemented, combinations of particular ways of clause sharing with particular inprocessing techniques has been shown.

The considered simplifications include variable elimination, equivalence elimination, blocked clause elimination, extended resolution, blocked clause addition and bounded variable addition. For many of these, the preservation of a semantic relation, which we call *unsat-preserving consequence*, has been identified as the crucial property behind soundness. We have inspected four transition systems, SysA, SysB, SysC, and SysD, characterized by different sets of inprocessing rules. The four models allow unrestricted clause sharing among the component solvers running in parallel. SysA models solvers that employ only equivalence preserving techniques. With SysB, we considered parallel solvers that allow unrestricted satisfiability preserving simplifications. Such solvers would be sound with respect to satisfiability, but not refutationally sound, and thus hardly of practical use. SysC covers parallel solvers that utilize inprocessing to *weaken* formulas, for example by removing clauses, and extended resolution. SysD shows the soundness of clause sharing with combinations of clause addition techniques, and clause deletion techniques. To the best of the authors' knowledge such parallel solvers have not yet been considered in the literature. The following table gives a summary of the results proven in this paper, and the currently implemented systems to which they apply.

Presented sharing models with properties and covered systems

Model	Covered inprocessing techniques	Soundness	Systems
SysA	preserve equivalence	✓	MANYSAT, PENELOPE
SysB	preserve equisatisfiability	no	–
SysC	clause deletion techniques	✓	PLINGELING
SysD	clause addition and deletion techniques	✓	open

The presented modeling also applies to parallel SAT solvers that follow the guiding path approach. As future work, we plan to extend it to parallel SAT solvers that rely on iterative partitioning [22,23], where clause sharing has also been introduced [24,29]. With the characteristics of SysD, another open question arises: can modern portfolio solvers be improved by adding a single distinguished solver incarnation that performs clause addition techniques?

Acknowledgments. The authors would like to thank Armin Biere for insights into the architecture of PLINGELING.

References

1. Audemard, G., Hoessen, B., Jabbour, S., Lagniez, J.-M., Piette, C.: Revisiting clause exchange in parallel SAT solving. In: Cimatti, A., Sebastiani, R. (eds.) SAT 2012. LNCS, vol. 7317, pp. 200–213. Springer, Heidelberg (2012)
2. Audemard, G., Hoessen, B., Jabbour, S., Lagniez, J.M., Piette, C.: Penelope, a parallel clause-freezer solver. In: SAT Challenge 2012; Solver and Benchmark Descriptions, pp. 43–44 (2012)
3. Audemard, G., Lagniez, J.-M., Mazure, B., Saïs, L.: On freezing and reactivating learnt clauses. In: Sakallah, K.A., Simon, L. (eds.) SAT 2011. LNCS, vol. 6695, pp. 188–200. Springer, Heidelberg (2011)
4. Audemard, G., Simon, L.: Predicting learnt clauses quality in modern SAT solvers. In: Proc. 21st Int. Joint Conf. on Artifical Intelligence (IJCAI 2009), pp. 399–404. Morgan Kaufmann (2009)
5. Beame, P., Kautz, H., Sabharwal, A.: Towards understanding and harnessing the potential of clause learning. Journal of Artificial Intelligence Research 22(1), 319–351 (2004)
6. Biere, A.: Lingeling, Plingeling, PicoSAT and PrecoSAT at SAT Race 2010. FMV Report Series Technical Report 10/1, Johannes Kepler University, Linz, Austria (2010)
7. Biere, A., Cimatti, A., Clarke, E.M., Fujita, M., Zhu, Y.: Symbolic model checking using SAT procedures instead of BDDs. In: Proc. 36th Annual ACM/IEEE Design Automation Conf (DAC), pp. 317–320. ACM (1999)
8. Böhm, M., Speckenmeyer, E.: A fast parallel SAT-solver – efficient workload balancing. Annals of Mathematics and Artificial Intelligence 17, 381–400 (1996), (Based on a technical report published already in 1994)
9. Davis, M., Logemann, G., Loveland, D.: A machine program for theorem-proving. CACM 5(7), 394–397 (1962)
10. Eén, N., Biere, A.: Effective preprocessing in SAT through variable and clause elimination. In: Bacchus, F., Walsh, T. (eds.) SAT 2005. LNCS, vol. 3569, pp. 61–75. Springer, Heidelberg (2005)
11. Eén, N., Sörensson, N.: An extensible SAT-solver. In: Giunchiglia, E., Tacchella, A. (eds.) SAT 2003. LNCS, vol. 2919, pp. 502–518. Springer, Heidelberg (2004)
12. Gelder, A.V.: Toward leaner binary-clause reasoning in a satisfiability solver. Annals of Mathematics and Artificial Intelligence 43(1), 239–253 (2005)
13. Gomes, C.P., Selman, B., Crato, N., Kautz, H.: Heavy-tailed phenomena in satisfiability and constraint satisfaction problems. Journal of Automated Reasoning 24(1-2), 67–100 (2000)
14. Großmann, P., Hölldobler, S., Manthey, N., Nachtigall, K., Opitz, J., Steinke, P.: Solving periodic event scheduling problems with SAT. In: Jiang, H., Ding, W., Ali, M., Wu, X. (eds.) IEA/AIE 2012. LNCS, vol. 7345, pp. 166–175. Springer, Heidelberg (2012)
15. Guo, L., Hamadi, Y., Jabbour, S., Sais, L.: Diversification and intensification in parallel SAT solving. In: Cohen, D. (ed.) CP 2010. LNCS, vol. 6308, pp. 252–265. Springer, Heidelberg (2010)

16. Hamadi, Y., Jabbour, S., Piette, C., Sais, L.: Deterministic parallel DPLL. JSAT 7(4), 127–132 (2011)
17. Hamadi, Y., Jabbour, S., Sais, L.: Control-based clause sharing in parallel SAT solving. In: Proc. 21st Int. Joint Conf. on Artificial Intelligence (IJCAI 2009), pp. 499–504 (2009)
18. Hamadi, Y., Jabbour, S., Sais, L.: ManySAT: a parallel SAT solver. JSAT 6(4), 245–262 (2009)
19. Heule, M.J.H., Kullmann, O., Wieringa, S., Biere, A.: Cube and conquer: Guiding CDCL SAT solvers by lookaheads. In: Eder, K., Lourenço, J., Shehory, O. (eds.) HVC 2011. LNCS, vol. 7261, pp. 50–65. Springer, Heidelberg (2012)
20. Heule, M.J.H., Järvisalo, M., Biere, A.: Efficient CNF simplification based on binary implication graphs. In: Sakallah, K.A., Simon, L. (eds.) SAT 2011. LNCS, vol. 6695, pp. 201–215. Springer, Heidelberg (2011)
21. Hölldobler, S., Manthey, N., Nguyen, V., Stecklina, J., Steinke, P.: A short overview on modern parallel SAT-solvers. In: Proc. Int. Conf. on Advanced Computer Science and Information Systems (ICACSIS 2011), pp. 201–206. IEEE (2011)
22. Hyvärinen, A.E.J., Junttila, T., Niemelä, I.: Partitioning SAT instances for distributed solving. In: Fermüller, C.G., Voronkov, A. (eds.) LPAR-17. LNCS, vol. 6397, pp. 372–386. Springer, Heidelberg (2010)
23. Hyvärinen, A.E.J., Manthey, N.: Designing scalable parallel SAT solvers. In: Cimatti, A., Sebastiani, R. (eds.) SAT 2012. LNCS, vol. 7317, pp. 214–227. Springer, Heidelberg (2012)
24. Hyvärinen, A.E.J., Junttila, T., Niemelä, I.: Grid-based SAT solving with iterative partitioning and clause learning. In: Lee, J. (ed.) CP 2011. LNCS, vol. 6876, pp. 385–399. Springer, Heidelberg (2011)
25. Järvisalo, M., Biere, A., Heule, M.J.H.: Blocked clause elimination. In: Esparza, J., Majumdar, R. (eds.) TACAS 2010. LNCS, vol. 6015, pp. 129–144. Springer, Heidelberg (2010)
26. Järvisalo, M., Heule, M.J.H., Biere, A.: Inprocessing rules. In: Gramlich, B., Miller, D., Sattler, U. (eds.) IJCAR 2012. LNCS, vol. 7364, pp. 355–370. Springer, Heidelberg (2012)
27. Katebi, H., Sakallah, K.A., Marques-Silva, J.P.: Empirical study of the anatomy of modern SAT solvers. In: Sakallah, K.A., Simon, L. (eds.) SAT 2011. LNCS, vol. 6695, pp. 343–356. Springer, Heidelberg (2011)
28. Kullmann, O.: On a generalization of extended resolution. Discrete Applied Mathematics 96- 97(1), 149–176 (1999)
29. Lanti, D., Manthey, N.: Sharing information in parallel search with search space partitioning. In: Learning and Intelligent Optimization – 7th Int. Conf. (LION 7) (to appear, 2013)
30. Lynce, I., Marques-Silva, J.P.: Probing-based preprocessing techniques for propositional satisfiability. In: Proc. 15th IEEE Int. Conf. on Tools with Artificial Intelligence (ICTAI 2003), pp. 105–110. IEEE (2003)
31. Manthey, N.: Coprocessor 2.0 – A flexible CNF simplifier. In: Cimatti, A., Sebastiani, R. (eds.) SAT 2012. LNCS, vol. 7317, pp. 436–441. Springer, Heidelberg (2012)
32. Manthey, N., Heule, M.J.H., Biere, A.: Automated reencoding of Boolean formulas. In: Hardware and Software: Verification and Testing – 8th Int. Haifa Verification Conf. (HVC 2012). LNCS, Springer (to appear, 2013)
33. Marques-Silva, J.P., Sakallah, K.A.: Grasp: A search algorithm for propositional satisfiability. IEEE Transactions on Computers 48(5), 506–521 (1999)

34. Martins, R., Manquinho, V., Lynce, I.: An overview of parallel SAT solving. Constraints 17(3), 304–347 (2012)
35. Moskewicz, M.W., Madigan, C.F., Zhao, Y., Zhang, L., Malik, S.: Chaff: Engineering an efficient SAT solver. In: Proc. 38th Annual ACM/IEEE Design Automation Conf. (DAC), pp. 530–535. ACM (2001)
36. Pipatsrisawat, K., Darwiche, A.: On the power of clause-learning SAT solvers as resolution engines. Artificial Intelligence 175(2), 512–525 (2011)
37. Roussel, O.: Description of ppfolio 2012. In: Proc. SAT Challenge 2012; Solver and Benchmark Descriptions, p. 46. Univ. of Helsinki (2012),
http://hdl.handle.net/10138/34218
38. Soos, M.: Cryptominisat 2.5.0. In: SAT Race Competitive Event Booklet (July 2010), http://baldur.iti.uka.de/sat-race-2010/descriptions/solver_13.pdf (retrieved February 11, 2013)
39. Sörensson, N., Biere, A.: Minimizing learned clauses. In: Kullmann, O. (ed.) SAT 2009. LNCS, vol. 5584, pp. 237–243. Springer, Heidelberg (2009)
40. Subbarayan, S., Pradhan, D.K.: NiVER: Non-increasing variable elimination resolution for preprocessing SAT instances. In: Hoos, H.H., Mitchell, D.G. (eds.) SAT 2004. LNCS, vol. 3542, pp. 276–291. Springer, Heidelberg (2005)
41. Tseitin, G.S.: On the complexity of derivations in the propositional calculus. Studies in Mathematics and Mathematical Logic Part II, 115–125 (1968)
42. Wotzlaw, A., van der Grinten, A., Speckenmeyer, E., Porschen, S.: pfoliouzk: Solver description. In: Proc. SAT Challenge 2012; Solver and Benchmark Descriptions, p. 45. Univ. of Helsinki (2012), http://hdl.handle.net/10138/34218
43. Zhang, H., Bonacina, M.P., Hsiang, J.: Psato: a distributed propositional prover and its application to quasigroup problems. Journal of Symbolic Computation 21(4), 543–560 (1996)

Exponential Separations in a Hierarchy of Clause Learning Proof Systems

Jan Johannsen

Institut für Informatik
Ludwig-Maximilians-Universität München
jan.johannsen@ifi.lmu.de

Abstract. Resolution trees with lemmas (RTL) are a resolution-based propositional proof system that is related to the DPLL algorithm with clause learning. Its fragments RTL(k) are related to clause learning algorithms where the width of learned clauses is bounded by k.

For every k up to $O(\log n)$, an exponential separation between the proof systems RTL(k) and RTL($k+1$) is shown.

1 Introduction

Many of the most efficient contemporary SAT solvers belong to the class of conflict-driven clause learning (CDCL) solvers. Historically, these solvers developed as extensions of the basic backtracking procedure known as the DPLL algorithm [9, 8], even though their most recent versions use more general forms of backtracking.

This recursive DPLL procedure is called for a formula F in conjunctive normal form and a partial assignment α (which is empty in the outermost call). If α satisfies F, then it is returned, and if α causes a conflict, i.e., falsifies a clause in F, then the call fails. Otherwise a variable x not set by α is chosen, and the procedure is called recursively twice, with α extended by $x := 1$ and by $x := 0$. If one of the recursive calls returns a satisfying assignment, then it is returned, otherwise the call fails.

The first generations of CDCL solvers employed several refinements and extensions of the basic DPLL algorithm, including *clause learning* [14], *non-chronological backtracking* [1] and *restarts* [10]. Crucial for their success is the technique of *clause learning* [14]: when the procedure finds a conflict, a sub-assignment α' of the current assignment α is computed such that α' suffices to cause this conflict. This sub-assignment α', the reason for the conflict, can then be stored in form of a new clause added to the formula, viz. the unique largest clause $C_{\alpha'}$ falsified by α'. This way, when in a later branch of the search another partial assignment extending α' occurs, earlier backtracking is possible since then the added clause $C_{\alpha'}$ causes a conflict.

When clause learning is implemented, a strategy is needed to decide which learnable clauses to keep in memory, because learning too many clauses leads to excessive memory consumption. Early learning strategies were such that the

M. Järvisalo and A. Van Gelder (Eds.): SAT 2013, LNCS 7962, pp. 40–51, 2013.

width, i.e., the number of literals, of learned clauses was restricted (see e.g. [14, Sec. 3.2]). Experience has shown that such learning strategies are not very helpful, i.e., learning only short clauses does not significantly improve the performance of a DPLL algorithm for hard formulas. This experience is supported by several lower bound theorems.

Contemporary CDCL solvers use more general forms of backtracking, which are not represented by the recursive DPLL algorithm scheme above, since it may be the case that branches in the search tree pruned in backtracking contain satisfying assignments. Therefore, in the following we speak about DPLL algorithms with clause learning instead of CDCL algorithms, to make it clear that our results apply to these earlier class of algorithms, where it is enforced that no satisfying branches are pruned. It remains to be investigated whether the results carry over to more contemporary CDCL algorithms.

The first lower bound for width-restricted clause learning was shown [6] for the well-known pigeonhole principle clauses PHP_n. These formulas require time $2^{\Omega(n \log n)}$ to solve when learning clauses of width up to $n/2$ only, whereas they can be solved in time $2^{O(n)}$ when learning arbitrary clauses. Another lower bound was shown [13] for a a set of clauses Ord_n based on the principle that every finite ordering has a maximal element. These formulas can be solved in polynomial time when learning arbitrary clauses, but require exponential time to solve when learning clauses of size up to $n/4$ only. This lower bound was generalized [4] to a lower bound exponential in w for all formulas for which a lower bound w on the width of resolution refutations holds.

All these lower bounds are shown by proving the same lower bounds on the length of refutations in a certain resolution based propositional proof system RTL. The relationship of this proof system to the DPLL algorithm with clause learning was established in several earlier works [6, 12]. The learned clauses correspond to so-called *lemmas* in the proof systems, so the mentioned lower bounds were shown for restricted version RTL(k) of RTL which allows only lemmas of width k, for the respective values of k.

In this work, we show that the restricted systems RTL(k) form a strict hierarchy: for every k, we prove an exponential separation of RTL($k + 1$) from RTL(k). In other words, increasing the width of lemmas that can be used by one can give an exponential speed-up.

2 Preliminaries

A *literal* is a variable x or a negated variable $\bar{x}$. A *clause* is a disjunction $C = a_1 \vee \ldots \vee a_k$ of literals a_i. The *width* of C is k, the number of literals in C. We identify a clause with the set of literals occurring in it, even though for clarity we still write it as a disjunction.

A formula in *conjunctive normal form* (CNF) is a conjunction $F = C_1 \wedge \ldots \wedge C_m$ of clauses, it is usually identified with the set of clauses $\{C_1, \ldots, C_m\}$. A formula F in CNF is in k-CNF if every clause C in F is of width $w(C) \leq k$.

We consider resolution-based refutation systems for formulas in CNF, which are strongly related to DPLL algorithms. These proof systems have as their only inference rule the *resolution rule*, which allows to infer the clause $C \lor D$ from the two clauses $C \lor x$ and $D \lor \bar{x}$, provided that the variable x does not occur in either C or D, pictorially:

$$\frac{C \lor x \qquad D \lor \bar{x}}{C \lor D}$$

We say that the variable x is *eliminated* in this inference.

A more general form of resolution inference is *w-resolution*, which allows to perform the inference even if the eliminated variable does not occur in one (or both) of the premises. More precisely, let C and D be clauses such C does not contain $\bar{x}$ and D does not contain x, then the w-resolution inference eliminating x allows to infer the clause $(C \setminus \{x\}) \cup (D \setminus \{\bar{x}\})$ from these.

The w-resolution rule can be simulated by the usual resolution rule together with the rule of *weakening* – which allows to conclude from a clause C any super-clause $D \supseteq C$ – as follows: infer $C \lor x$ and $D \lor \bar{x}$ by (possibly empty) weakenings, then apply resolution.

An *ordered binary tree* is a rooted tree in which every inner node has two children, a distinguished *left* and *right* child. The post-ordering $\prec$ of an ordered binary tree is the order in which its nodes are visited by a post-order traversal, i.e., $u \prec v$ holds for nodes u, v if u is a descendant of v, or if there is a common ancestor w of u and v such that u is a descendant of the left child of w and v is a descendant of the right child of w.

An RTL-derivation of a clause C from a CNF-formula F is an ordered binary tree, in which every node v is labeled with a clause C_v such that:

1. The root is labeled with C.
2. If v is an inner node with children u_1, u_2, then C_v follows from C_{u_1} and C_{u_2} by the resolution rule.
3. A leaf v is labeled by a clause D in F, or by a clause C labeling some node $u \prec v$. In the latter case we call C a *lemma*.

An RTL-derivation is an RTL(k)-derivation if every lemma C is of width $w(C) \leq k$. An RTL-refutation of F is an RTL-derivation of the empty clause from F.

A tree-like resolution derivation is an RTL-derivation that does not use any lemmas. An RTL-derivation is called *regular* if on every path, no variable is eliminated twice. This condition is inessential for tree-like resolution since minimal tree-like refutations are always regular [15]. It is not known whether RTL-refutations can be simulated by regular RTL-refutations without increasing the size super-polynomially.

Let V be a set of variables. A *restriction* ρ of V is a partial assignment $V \to \{0, 1\}$. A restriction ρ is extended to literals by setting

$$\rho(\bar{x}) := \begin{cases} 1 & \text{if } \rho(x) = 0 \\ 0 & \text{if } \rho(x) = 1 \end{cases}$$

For a clause C in variables V, we define

$$C{\restriction}\rho := \begin{cases} 1 & \text{if } \rho(a) = 1 \text{ for some } a \in C \\ \bigvee_{a \in C,\, \rho(a) \neq 0} a & \text{otherwise,} \end{cases}$$

where the empty disjunction is identified with the constant 0. For a CNF-formula F over V, we define

$$F{\restriction}\rho := \begin{cases} 0 & \text{if } C{\restriction}\rho = 0 \text{ for some } C \in F \\ \bigwedge_{C \in F,\, C{\restriction}\rho \neq 1} C{\restriction}\rho & \text{otherwise,} \end{cases}$$

where the empty conjunction is identified with 1.

Proposition 1. *Let R be a tree-like resolution derivation of C from F of size s, and ρ a restriction. Then there is a tree-like resolution derivation R' of $C{\restriction}\rho$ from $F{\restriction}\rho$ of size at most s.*

In particular, if $C{\restriction}\rho = 0$ then R' is a tree-like resolution refutation of $F{\restriction}\rho$. As usual, we denote the derivation R' by $R{\restriction}\rho$.

Tree-like resolution exactly corresponds to the DPLL algorithm by the following well-known correspondence: the search tree produced by the run of a DPLL algorithm on an unsatisfiable formula F forms a tree-like resolution refutation of F, and from a given tree-like regular resolution refutation of F one can construct a run of a DPLL algorithm showing the unsatisfiability of F that produces essentially the given search tree.

Buss et al. [6] define a variant WRTI of RTL which exactly corresponds to a general formulation of the DPLL algorithm with clause learning. Proofs in WRTI are regular resolution trees with lemmas using the w-resolution rule, but in which a clause can only be used as a lemma if it was derived by *input resolution*. An input resolution derivation is one in which in every inference step, one of the children is a leaf, i.e., labeled by a clause from the input formula or a lemma derived earlier.

The size of a refutation of an unsatisfiable formula F in WRTI has been shown [6] to be polynomially related to the run-time of a schematic algorithm DLL-L-UP on F. This schema DLL-L-UP subsumes many clause learning strategies commonly used in practice [6]. It follows from these results that if an unsatisfiable formula F can be solved by a DPLL algorithm with clause learning in time t, then it has an WRTI-refutation, and hence an RTL-refutation of size polynomial in t. Moreover, if the algorithm learns only clauses of width at most k, then the refutation is in RTL(k).

3   The Result

Our main result is an exponential separation between the systems RTL with lemmas restricted to be of width k, for every k:

Theorem 2. *For every k, there is a family of formulas $F_n^{(k)}$ such that*

- *$F_n^{(k)}$ have RTL($k+1$)-refutations of polynomial size $n^{O(1)}$.*
- *$F_n^{(k)}$ requires RTL(k)-refutations of exponential size $2^{\Omega(n/\log n)}$,*

This even holds for $k = k(n)$ depending on n when $k(n) = O(\log n)$.

The lower bound also holds for a stronger system that also includes a weakening rule, the proof requires little to no modification. Therefore, it also applies for the systems with the w-resolution rule of Buss et al. [6].

On the other hand, the upper bound is shown for the weaker system with only the usual resolution rule, and the refutations given are regular.

4   Graph Pebbling

Let $G = (V, E)$ be a *pointed dag*, i.e., a directed acyclic graph having exactly one sink t, such that every vertex has either in-degree 0 or 2, and let $S, T \subseteq V$. The pebble game on (G, S, T) is played by placing pebbles onto the vertices of G according to the rules below until a pebble is placed onto a vertex in T. Formally, a pebbling of (G, S, T) is a sequence $C_0, C_1, \ldots, C_\ell$ of subsets $C_i \subseteq V$, where C_i should be pictured as the set of vertices carrying pebbles at time i, with $C_0 = \emptyset$ and $C_\ell \cap T \neq \emptyset$ such that for all $i < \ell$ one of the following properties holds:

1. $C_{i+1} = C_i \cup \{u\}$ for some $u \in S$, i.e., a pebble can be put onto a vertex in S.
2. $C_{i+1} = C_i \cup \{u\}$ for some u such that all immediate predecessors of u are in C_i, i.e., if all predecessors of u are pebbled, then u can be pebbled.
3. $C_{i+1} \subset C_i$, i.e., pebbles can be removed from vertices.

By (2), a source vertex can be pebbled at any time, so we can always assume that S contains all sources of G.

The complexity of a pebbling is $\max_{i \leq \ell} |C_i|$, i.e., the maximal number of pebbles used. The *pebbling number* $\mathrm{peb}(G, S, T)$ is the minimal complexity of a pebbling of (G, S, T). The pebbling number $\mathrm{peb}(G)$ of G is $\mathrm{peb}(G, \emptyset, \{t\})$.

We shall need the following well-known property of the pebbling number [3].

Lemma 3. *For every pointed dag $G = (V, E)$, disjoint subsets $S, T \subseteq V$ and $v \in V \setminus S \cup T$, we have $\mathrm{peb}(G, S, T) \leq \max(\mathrm{peb}(G, S \cup \{v\}, T), \mathrm{peb}(G, S, T \cup \{v\})) + 1$.*

Graphs with a maximally large pebbling number were constructed by Celoni et al. [7]:

Theorem 4. *There are pointed graphs G_n with n vertices such that $\mathrm{peb}(G_n) \geq \Omega(n/\log n)$.*

5   Pebbling Formulas

For a pointed dag $G = (V, E)$, the pebbling formula $\mathrm{Peb}(G)$ is the unsatisfiable formula in variables x_v for $v \in V$ consisting of the following clauses:

- x_s for every source s
- $\bar{x}_u \vee \bar{x}_v \vee x_w$ for every inner vertex w with predecessors u and v
- $\bar{x}_t$ for the sink t

The formula $\mathrm{Peb}(G)$ has a short tree-like resolution refutation of linear size, since it is a Horn formula. Ben-Sasson et al. [3] construct harder to refute formulas from them by replacing every variable x with the disjunction of two new variables $x_1 \vee x_2$. For the resulting formulas $\mathrm{Peb}^2(G)$ they show a lower bound for tree-like resolution that is exponential in the pebbling number $\mathrm{peb}(G)$.

6   Generalized Xorification

A different way to make a boolean formula harder is *xorification*, i.e., replacing every variable by the XOR of two or more variables. This technique has been used in proof complexity so far mainly for space lower bounds [2, 5]. It also has been applied in circuit complexity, e.g. to obtain cubic lower bounds on formula size[1] [11].

The formulas that witness the separations in Theorem 2 are obtained by xorification from the pebbling formulas $\mathrm{Peb}(G)$. In the lower bound argument, restrictions will be applied to these formulas, and in order to understand and analyze the restricted formulas, we introduce a generalized form of xorification.

We generalize xorification in two ways: first, some variables are replaced by the XOR of k variables, whereas some other variables are replaced by the negation of the XOR of k variables. Second, some designated variables are not replaced at all, but remain a single variable or its negation. Thus, for every variable two bits β_0 and β_1 specify how it occurs in the xorification: β_0 controls whether it is replaced by an XOR or not, and β_1 specifies whether it is negated or not. This is made precise in the following definition:

Let F be a formula in variables from a set V. Recall that $\neg x$ is equivalent to $x \oplus 1$. For $k \in \mathbb{N}$ and a function $\beta : V \to \{0,1\}^2$, where we denote the components of β by $\beta(x) = (\beta_0(x), \beta_1(x))$, the generalized xorification $X(F, k, \beta)$ is defined by:

- $X(x, k, \beta) = x_1 \oplus \ldots \oplus x_k \oplus \beta_1(x)$ for a variable $x \in V$ with $\beta_0(x) = 0$.
- $X(x, k, \beta) = x_1 \oplus \beta_1(x)$ for a variable $x \in V$ with $\beta_0(x) = 1$.
- $X(\bar{x}, k, \beta) = X(x, k, \beta) \oplus 1$ for a negated variable $x \in V$.
- $X(C, k, \beta) = \bigvee_{a \in C} X(a, k, \beta)$ expanded into CNF, for a clause C.
- $X(F, k, \beta) = \bigwedge_{C \in F} X(C, k, \beta)$ for a CNF formula F.

For the pebbling formulas $\mathrm{Peb}(G)$, we use the abbreviation $\mathrm{Peb}_\beta^{\oplus k}(G)$ for $X(\mathrm{Peb}(G), k, \beta)$, and we omit the lower index if β is the constant function $\beta \equiv (0,0)$. More generally, for a clause C we write $C^{\oplus k}$ for $X(C, k, \beta)$ when $\beta \equiv (0,0)$. Also we abbreviate $\beta(x_v)$ by $\beta(v)$, i.e., we identify the vertices of G with the variables of $\mathrm{Peb}(G)$.

---

[1] I am grateful to Ryan Williams for providing this reference on
   `cstheory.stackexchange.com`

We picture the variables of $\mathrm{Peb}^{\oplus k}(G)$ as a rectangular matrix, with a column for every vertex v of G and a row for every index $1 \leq i \leq k$.

The following lower bound for tree-like resolution is a generalization of the result of Ben-Sasson et al. [3], the proof is an adaptation[2] of their proof.

Theorem 5. *For every pointed dag $G = (V, E)$ and every $\beta : V \to \{0,1\}^2$, tree-like resolution refutations of $\mathrm{Peb}_\beta^{\oplus 2}(G)$ require size $2^{\Omega(\mathrm{peb}(G)-b)}$, where b is the number of $v \in V$ with $\beta_0(v) = 1$.*

Proof. Let R be a tree-like resolution refutation of $\mathrm{Peb}_\beta^{\oplus 2}(G)$, we show that $|R| \geq 2^{\mathrm{peb}(G)-b-2} - 1$.

To that end, we define a sequence $C_0, C_1, \ldots, C_h$ of clauses in R, with $C_0 = 0$ and C_{i+1} one of the predecessors of C_i for every $i < h$, and C_h a leaf, i.e., an axiom from $\mathrm{Peb}_\beta^{\oplus 2}(G)$, together with an increasing sequence of restrictions $\rho_0 \subseteq \rho_1 \subseteq \ldots \subseteq \rho_h$ such that $C_i\lceil \rho_i = 0$ for every $i \leq h$, and sets $S_0, S_1, \ldots, S_h$ and $T_0, T_1, \ldots, T_h$ with $S_i \cap T_i = \emptyset$.

We let S_0 be the set of sources in G and $T_0 = \{t\}$ where t is the sink of G, and $\rho_0 = \emptyset$. Now assume C_i, ρ_i, S_i and T_i are defined, and assume that C_i is derived from $D_i \vee x$ and $D_i' \vee \bar{x}$, where x is a variable $x_{v,\epsilon}$ for $v \in V$ and $\epsilon \in \{1,2\}$. Let $\bar{\epsilon} := 3 - \epsilon$ so that $x_{v,\bar{\epsilon}}$ is the other variable in column v.

We define $C_{i+1}, \rho_{i+1}, S_{i+1}$ and T_{i+1} by distinguishing cases, where in each case, ρ_{i+1} is obtained from ρ_i by specifying the value for the variable $x_{v,\epsilon}$.

- Case 1a: $v \in T_i$, and $\beta_0(v) = 1$ or $x_{v,\bar{\epsilon}} \notin \mathrm{dom}\,\rho_i$.
  Set $\rho_{i+1}(x_{v,\epsilon}) = \beta_1(v)$, $S_{i+1} = S_i$ and $T_{i+1} = T_i$.
- Case 1b: $v \in T_i$ and $x_{v,\bar{\epsilon}} \in \mathrm{dom}\,\rho_i$.
  Set $\rho_{i+1}(x_{v,\epsilon}) = \rho_i(x_{v,\bar{\epsilon}}) \oplus \beta_1(v)$, $S_{i+1} = S_i$ and $T_{i+1} = T_i$.
- Case 2a: $v \in S_i$, and $\beta_0(v) = 1$ or $x_{v,\bar{\epsilon}} \notin \mathrm{dom}\,\rho_i$.
  Set $\rho_{i+1}(x_{v,\epsilon}) = \beta_1(v) \oplus 1$, $S_{i+1} = S_i$ and $T_{i+1} = T_i$.
- Case 2b: $v \in S_i$ and $x_{v,\bar{\epsilon}} \in \mathrm{dom}\,\rho_i$.
  Set $\rho_{i+1}(x_{v,\epsilon}) = \rho_i(x_{v,\bar{\epsilon}}) \oplus \beta_1(v) \oplus 1$, $S_{i+1} = S_i$ and $T_{i+1} = T_i$.
- Case 3: $v \notin S_i \cup T_i$ and $\mathrm{peb}(G, S_i, T_i \cup \{v\}) = \mathrm{peb}(G, S_i, T_i)$.
  Set $\rho_{i+1}(x_{v,\epsilon}) = \beta_1(v)$, $S_{i+1} = S_i$ and $T_{i+1} = T_i \cup \{v\}$.
- Case 4a: $v \notin S_i \cup T_i$ and $\mathrm{peb}(G, S_i, T_i \cup \{v\}) < \mathrm{peb}(G, S_i, T_i)$ and $\beta_0(v) = 1$.
  Set $\rho_{i+1}(x_{v,\epsilon}) = \beta_1(v) \oplus 1$, $S_{i+1} = S_i \cup \{v\}$ and $T_{i+1} = T_i$.

In all these cases 1a - 4a, define C_{i+1} to be the parent clause of C_i that is falsified by ρ_{i+1}.

- Case 4b: $v \notin S_i \cup T_i$ and $\mathrm{peb}(G, S_i, T_i \cup \{v\}) < \mathrm{peb}(G, S_i, T_i)$ and $\beta_0(v) = 0$.
  Choose C_{i+1} as that parent clause of C_i s.t. the subtree rooted at C_{i+1} is the smaller among the two, and set a value of $\rho_{i+1}(x_{v,\epsilon})$ such that C_{i+1} is falsified by ρ_{i+1}. Moreover, let $S_{i+1} = S_i \cup \{v\}$ and $T_{i+1} = T_i$.

---

[2] Urquhart [16] claims that a lower bound for tree-like resolution refutations of $\mathrm{Peb}^{\oplus 2}(G)$ can be obtained by imitating the proof of Ben-Sasson et al. [3] "almost word for word". We found however that it requires some subtle modifications even for the non-generalized case.

In the following, we denote $X(x_v, 2, \beta)$ by $x_v^\oplus$, i.e., $x_v^\oplus = x_{v,1} \oplus x_{v,2} \oplus \beta_1(v)$ if $\beta_0(v) = 0$ and $x_v^\oplus = x_{v,1} \oplus \beta_1(v)$ if $\beta_0(v) = 1$.

Claim. If $\rho_i(x_v^\oplus) = 0$, then $v \in T_i$.

If the assumption of the claim holds, then in the case $\beta_0(v) = 1$, the value of $x_{v,1}$ must have been set in case 1a or in case 3. In either case $v \in T_i$.

In the case $\beta_0(v) = 0$, the variable among $x_{v,\epsilon}$ and $x_{v,\bar\epsilon}$ whose value was set later, must have been set by Case 1b, and hence $v \in T_i$.

Claim. If $\rho_i(x_v^\oplus) = 1$, then $v \in S_i$.

The proof is similar to that of the previous claim.

It follows that C_h is not a clause from a target axiom $X(\bar{x}_t, k, \beta)$. If this were the case, then $\rho_h(x_t^\oplus) = 1$, and hence $t \in S_h$ by the claim above, whereas we have $t \in T_0 \subseteq T_h$ and $S_h \cap T_h = \emptyset$. By analogous reasoning, C_h cannot be a clause from a source axiom $X(x_s, k, \beta)$ for a source s of G.

For $i \le h$, let b_i be the number of $v \in V$ with $\beta_0(v) = 1$ such that $x_{v,1} \in \operatorname{dom} \rho_i$.

Claim. For every $i \le h$, the size s_i of the subtree of R rooted at C_i is at least
$$s_i \ge 2^{\operatorname{peb}(G, S_i, T_i) - b_h + b_i - 2} - 1.$$

The claim is proven by induction on i, downward from h to 0.

By the considerations above, C_h must be a clause from $X(\bar{x}_u \vee \bar{x}_v \vee x_w, 2, \beta)$ for some $w \in V$ with predecessors u and v. Therefore $\rho_h(x_u^\oplus) = \rho_h(x_v^\oplus) = 1$, so $u, v \in S_h$ by the claim above, and $\rho_h(x_w^\oplus) = 0$, and thus $w \in T_h$. We get $\operatorname{peb}(G, S_h, T_h) = 3$, and hence $s_h = 1 = 2^{\operatorname{peb}(G, S_h, T_h) - b_h + b_h - 2} - 1$, which shows the induction base for $i = h$.

Assume the claimed lower bound holds for s_{i+1}. Since s_i is the size of the tree rooted at C_i, which contains the subtree rooted at C_{i+1} of size s_{i+1}, we obviously have $s_i \ge s_{i+1}$.

If C_{i+1} is defined by one of the cases 1a through 3, then $\operatorname{peb}(G, S_{i+1}, T_{i+1}) = \operatorname{peb}(G, S_i, T_i)$ and $b_{i+1} \ge b_i$, and thus
$$s_i \ge s_{i+1} \ge 2^{\operatorname{peb}(G, S_{i+1}, T_{i+1}) - b_h + b_{i+1} - 2} - 1 \ge 2^{\operatorname{peb}(G, S_i, T_i) - b_h + b_i - 2} - 1 ,$$

which shows the claim for s_i.

If C_{i+1} was defined using case 4a, then we have
$$\operatorname{peb}(G, S_{i+1}, T_{i+1}) \ge \operatorname{peb}(G, S_i, T_i) - 1$$

by Lemma 3, and $b_{i+1} = b_i + 1$, thus we get
$$s_i \ge s_{i+1} \ge 2^{\operatorname{peb}(G, S_{i+1}, T_{i+1}) - b_h + b_{i+1} - 2} - 1 \ge 2^{\operatorname{peb}(G, S_i, T_i) - 1 - b_h + b_i + 1 - 2} - 1 ,$$

which shows the claim for s_i.

If C_{i+1} was defined using case 4b, then we have
$$\operatorname{peb}(G, S_{i+1}, T_{i+1}) \ge \operatorname{peb}(G, S_i, T_i) - 1$$

by Lemma 3 again, and $b_{i+1} = b_i$, therefore we obtain

$$s_i \geq 2s_{i+1} + 1 \geq 2^{\mathrm{peb}(G,S_{i+1},T_{i+1})-b_h+b_{i+1}-2} - 1 \geq 2^{\mathrm{peb}(G,S_i,T_i)-b_h+b_i-2} - 1 \, ,$$

which shows the claim for s_i.

The theorem follows, since we have $|R| = s_0$, and $b_0 = 0$, and $b_h \leq b$, and $\mathrm{peb}(G, S_0, T_0) = \mathrm{peb}(G)$. $\square$

7   The Lower Bound

We will now prove one half of our main result, a lower bound on the size of RTL(k)-refutation of the $(k+1)$-fold xorification of the pebbling formulas.

Theorem 6. *For a pointed dag G, every RTL(k)-refutation of $\mathrm{Peb}^{\oplus(k+1)}(G)$ requires size $2^{\Omega(\mathrm{peb}(G))}$.*

Proof. Let R be an RTL(k)-refutation of $F := \mathrm{Peb}^{\oplus(k+1)}(G)$. Note that every clause in F is of width at least $k+1$. Let C be the first clause in R with $w(C) \leq k$, so that C could possibly be used as a lemma. Then the subtree R_C of R rooted at C is a tree-like resolution derivation of C from F.

Let ρ be the smallest restriction with $C\lceil\rho = 0$, and note that $|\rho| \leq k$. Recall that we picture the variables of $\mathrm{Peb}^{\oplus k}(G)$ as arranged in a rectangular matrix, with a column for every vertex v of G and a row for every index $1 \leq i \leq k$. There are two cases: either the variables set by ρ are all in the same column, or ρ sets variables from at least two different columns.

In the latter case, there are at most $k - 1$ rows set in every column, thus for each column v there are two rows $i(v)$ and $i'(v)$ such that $x_{v,i(v)}$ and $x_{v,i'(v)}$ are not set by ρ. In this case, we can set all but these two rows in every column, i.e., extend ρ to a restriction ρ^* by setting $\rho^*(x_{v,j}) = 0$ for every variable $x_{v,j} \notin \mathrm{dom}\,\rho$ with $j \notin \{i(v), i'(v)\}$. Define $\beta_0(v) = 0$ for every $v \in V$, and $\beta_1(v) := \bigoplus_{j \notin \{i(v),i'(v)\}} \rho^*(x_{v,j})$.

In the first case, let v be the column containing all variables set by ρ. If there are fewer than k variables set, then we can proceed as in the other case. Otherwise, there is one row i such that $x_{v,j}$ is set by ρ for all $j \neq i$. In this case, we set all but two rows in every other column, and in column v only one variable remains. Thus we pick a row i' with $i' \neq i$ arbitrarily, and extend ρ to a restriction ρ^* by setting $\rho^*(x_{u,j}) = 0$ for every column $u \neq v$ and row $j \notin \{i, i'\}$. Set $\beta_0(v) = 1$ and $\beta_1(v) = \bigoplus_{j \neq i} \rho(x_{v,j})$ for the vertex v, and for all other vertices $u \neq v$, set $\beta_0(u) = 0$ and $\beta_1(u) = 0$.

In both cases, for the so defined function β we have $F\lceil\rho^* \equiv \mathrm{Peb}_\beta^{\oplus 2}(G)$ after a renaming of the variables that changes only the numbering of the rows.

Thus in both cases $R_C\lceil\rho^*$ is a tree-like resolution refutation of $\mathrm{Peb}_\beta^{\oplus 2}(G)$, and the number b of $v \in V$ with $\beta_0(v) = 1$ is at most 1, therefore $|R_C| \geq 2^{\Omega(\mathrm{peb}(G))}$ by Theorem 5. The size lower bound for R follows. $\square$

8 The Upper Bound

We now prove the remaining half of our result, the upper bound.

Theorem 7. *For every pointed dag G with n vertices, the formulas $\mathrm{Peb}^{\oplus k}(G)$ have regular $\mathrm{RTL}(k)$-refutations of size $O(2^{3k}n)$.*

Proof. Fix a topological ordering $\prec$ of G, and let S be the set of sources of G. We first show the following claim:

Claim. Let $w \in V$ with predecessors u and v, where $u \prec v$. For every clause C in $x_w^{\oplus k}$, there is a tree-like regular resolution derivation of C from $x_u^{\oplus k}$ and $x_v^{\oplus k}$ and $(\bar{x}_u \vee \bar{x}_v \vee x_w)^{\oplus k}$ of size $O(2^{2k})$. Moreover, in this derivation only the variables from columns u and v are eliminated, and on every path from a leaf labeled with a clause from $x_v^{\oplus k}$ to C, only the variables from column v are eliminated.

Proof. Take a regular tree-like resolution refutation R_v of $x_v^{\oplus k}$ and $\bar{x}_v^{\oplus k}$, of size $O(2^k)$. Add the clause C to every clause in R_v except the leaves from $x_v^{\oplus k}$. This yields a derivation R'_v of C from $x_v^{\oplus k}$ and $\bar{x}_v^{\oplus k} \vee C$.

Now take a regular tree-like resolution refutation R_u of $x_u^{\oplus k}$ and $\bar{x}_u^{\oplus k}$, of size $O(2^k)$. For every clause C' in $\bar{x}_v^{\oplus k} \vee C$, take a copy of R_u, and add C' to every clause in it except the leaves from $x_u^{\oplus k}$. Replace the leaf in R'_v labeled C' by the result. This gives the desired derivation and thus proves the claim.   □

To prove the theorem, we construct a sequence $R_1, \ldots, R_\ell$ of *partial* resolution trees with lemmas, in which some leaves are labeled by clauses that are not axioms or lemmas derived earlier, these are called the *open leaves*. In addition, we define a sequence $U_1, \ldots, U_\ell$ of subsets of $V \setminus S$, such that the following invariants hold:

- The open leaves in R_i are all among the clauses from $x_{u,1} \oplus \ldots \oplus x_{u,k}$ for an $u \in U_i$.
- On the path from an open leaf with a clause from $x_{u,1} \oplus \ldots \oplus x_{u,k}$ to the root, all variables resolved are from a column $v \in V$ with $u \preceq v$.

Let R_1 be a tree-like regular resolution refutation of the clauses from $\bar{x}_t^{\oplus k}$, which are axioms of $\mathrm{Peb}^{\oplus k}(G)$, and those from $x_t^{\oplus k}$, which are the open leaves of R_1, and let $U_1 := \{t\}$. Obviously the invariants hold, and the size of R_1 is 2^k.

Assume we have constructed R_i, we show how to construct R_{i+1}. Let v be the maximal element of U_i w.r.t. the ordering $\prec$, and let u_1 and u_2 be its predecessors. For each clause C from $x_v^{\oplus k}$, replace its first occurrence in R_i by the derivation R_C of C from $x_{u_1}^{\oplus k}$ and $x_{u_2}^{\oplus k}$ given by the claim above. The other occurrences of C will then become lemmas.

Let the result be R_{i+1}, then the open leaves of R_{i+1} are those of R_i without the clauses from $x_v^{\oplus k}$, plus those leaves of R_C labeled by clauses from $x_{u_1}^{\oplus k}$ and $x_{u_2}^{\oplus k}$, except when u_1 or u_2 are sources. Thus if we define $U_{i+1} := (U_i \setminus \{v\}) \cup (\{u_1, u_2\} \setminus S)$, then the first invariant holds.

Since in a path from an open leaf C of R_i to the root, only variables from columns w with $v \preceq w$ are eliminated, and in R_C only variables from columns u_1 and u_2, the second invariant as well as regularity of R_{i+1} hold.

For each of the 2^{k-1} clauses in $x_v^{\oplus k}$, we have added one derivation of size 2^{2k}, hence the size of R_{i+1} is $|R_{i+1}| \leq |R_i| + 2^{3k}$.

The process terminates after at most n iterations, since $\max_\prec U_i$ strictly decreases in every step. Since in R_ℓ, there are no open leaves left, it is a regular RTL-refutation. Since every lemma used is a clause from $x_v^{\oplus k}$ for some $v \in V$, they are of size k, hence R_ℓ is a regular RTL(k)-refutation of $\mathrm{Peb}^{\oplus k}(G)$ of size $2^{3k} \cdot n$.   □

9   Wrapping Up

Finally, we put everything together to prove the main theorem.

Proof (of Theorem 2). Let $F_n^{(k)}$ be the formula $\mathrm{Peb}^{\oplus(k+1)}(G_n)$, where G_n are the graphs given by Theorem 4 with n vertices and pebbling number $\mathrm{peb}(G_n) = \Omega(n/\log n)$. For $k = O(\log n)$, these formulas are of polynomial size $n^{O(1)}$.

By Theorem 6, the formulas $F_n^{(k)}$ require RTL(k)-refutations of exponential size $2^{\Omega(n/\log n)}$, and by Theorem 7, they have regular RTL(k)-refutations of polynomial size $n^{O(1)}$.   □

Note that for k larger than $O(\log n)$, the formulas $\mathrm{Peb}^{\oplus(k+1)}(G_n)$ are themselves of super-polynomial size in n, and therefore have no proofs of size polynomial in the size of the underlying graph G_n.

10   Conclusion

We have shown that for resolution trees with lemmas – a resolution-based propositional proof system that forms the basis of a family of proof systems capturing the complexity of clause-learning algorithms – an increase of one in the width of clauses that may be used as lemmas can lead to an exponential speed-up.

The lower bounds hold for the strongest form of these proof systems with no regularity restrictions, and even with the weakening rule. The upper bounds, on the other hand, are given for a rather weak variant, the given refutations are regular and do not use any weakenings.

Unfortunately, we cannot immediately conclude an exponential speed-up of the DPLL algorithm with clause learning with learned clauses of width $k + 1$ over the version with learned clauses of width k from this. In order for that conclusion to hold, the upper bound would have to be given for a still weaker variant of the system, in which only lemmas derived by input resolution can be used, i.e. a restricted version of WRTI without use of the w-resolution rule and with lemmas restricted to width $k + 1$.

References

[1] Bayardo Jr., R.J., Schrag, R.C.: Using CSP look-back techniques to solve real-world SAT instances. In: Proc. 14th Natl. Conference on Artificial Intelligence, pp. 203–208 (1997)

[2] Ben-Sasson, E.: Size-space tradeoffs for resolution. SIAM Journal on Computing 38(6), 2511–2525 (2009)

[3] Ben-Sasson, E., Impagliazzo, R., Wigderson, A.: Near-optimal separation of general and tree-like resolution. Combinatorica 24(4), 585–604 (2004)

[4] Ben-Sasson, E., Johannsen, J.: Lower bounds for width-restricted clause learning on small width formulas. In: Strichman, O., Szeider, S. (eds.) SAT 2010. LNCS, vol. 6175, pp. 16–29. Springer, Heidelberg (2010)

[5] Ben-Sasson, E., Nordström, J.: Understanding space in proof complexity: Separations and trade-offs via substitutions. In: Chazelle, B. (ed.) Innovations in Computer Science, ICS 2011, pp. 401–416. Tsinghua University Press (2011)

[6] Buss, S.R., Hoffmann, J., Johannsen, J.: Resolution trees with lemmas: Resolution refinements that characterize DLL algorithms with clause learning. Logical Methods in Computer Science 4(4) (2008)

[7] Celoni, J., Paul, W., Tarjan, R.: Space bounds for a game on graphs. Mathematical Systems Theory 10, 239–251 (1977)

[8] Davis, M., Logemann, G., Loveland, D.W.: A machine program for theorem-proving. Communications of the ACM 5(7), 394–397 (1962)

[9] Davis, M., Putnam, H.: A computing procedure for quantification theory. Journal of the Association for Computing Machinery 7(3), 201–215 (1960)

[10] Gomes, C.P., Selman, B., Crato, N.: Heavy-tailed distributions in combinatorial search. In: Smolka, G. (ed.) CP 1997. LNCS, vol. 1330, pp. 121–135. Springer, Heidelberg (1997)

[11] Håstad, J.: The shrinkage exponent of De Morgan formulas is 2. SIAM Journal on Computing 27(1), 48–64 (1998)

[12] Hertel, P., Bacchus, F., Pitassi, T., van Gelder, A.: Clause learning can effectively p-simulate general propositional resolution. In: Fox, D., Gomes, C.P. (eds.) Proceedings of the 23rd AAAI Conference on Artificial Intelligence, AAAI 2008, pp. 283–290. AAAI Press (2008)

[13] Johannsen, J.: An exponential lower bound for width-restricted clause learning. In: Kullmann, O. (ed.) SAT 2009. LNCS, vol. 5584, pp. 128–140. Springer, Heidelberg (2009)

[14] Marques-Silva, J.P., Sakallah, K.A.: GRASP - a new search algorithm for satisfiability. In: Proc. IEEE/ACM International Conference on Computer Aided Design, ICCAD, pp. 220–227 (1996)

[15] Tseitin, G.S.: On the complexity of derivation in propositional calculus. Studies in Constructive Mathematics and Mathematical Logic, Part 2, 115–125 (1968)

[16] Urquhart, A.: A near-optimal separation of regular and general resolution. SIAM Journal on Computing 40(1), 107–121 (2011)

On the Resolution Complexity
of Graph Non-isomorphism

Jacobo Torán

Institut für Theoretische Informatik
Universität Ulm
Oberer Eselsberg,
D-89069 Ulm, Germany
jacobo.toran@uni-ulm.de

Abstract. For a pair of given graphs we encode the isomorphism principle in the natural way as a CNF formula of polynomial size in the number
of vertices, which is satisfiable if and only if the graphs are isomorphic.
Using the CFI graphs from [12], we can transform any undirected graph
G into a pair of non-isomorphic graphs. We prove that the resolution
width of any refutation of the formula stating that these graphs are isomorphic has a lower bound related to the expansion properties of G.
Using this fact, we provide an explicit family of non-isomorphic graph
pairs for which any resolution refutation requires an exponential number
of clauses in the size of the initial formula. These graphs pairs are colored
with color multiplicity bounded by 4. In contrast we show that when the
color classes are restricted to have size 3 or less, the non-isomorphism
formulas have tree-like resolution refutations of polynomial size.

1 Introduction

Resolution is one of the most popular and best studied proof systems for propositional logic. Since the first exponential lower bound for the size of resolution
refutations proven by Haken [17] for the family of formulas encoding the pigeonhole principle, many other combinatorial principles have been shown to have
exponential lower bounds [26,13,9,7,8]. With the recent development of modern
SAT-solvers based on DPLL algorithms and the fact that the resolution principle lies in the core of such algorithms, resolution lower bounds have gained in
importance because they also provide lower bounds for the running time of the
SAT-solvers. We study here the complexity of testing graph non-isomorphism
using resolution. The graph isomorphism problem, GI, asks whether there is a
bijection between the nodes of two given graphs preserving the adjacency relationship. The problem has been extensively studied in the past (see e.g [21])
because its intrinsic importance and also because it is one of the few problems in
NP that is not known to be solvable in polynomial time but also is not expected
to be NP-complete.

The impressive improvement of the performance of SAT-solvers based on
DPLL algorithms in the last years has motivated a new way for dealing with NP

M. Järvisalo and A. Van Gelder (Eds.): SAT 2013, LNCS 7962, pp. 52–66, 2013.

problems. For many practical applications, these problems are reduced to formulas than are then tested for satisfiability using the SAT-solvers. This method works well in practice for several problems, although strong resolution lower bounds for random instances of some NP-complete problems are known [7,8]. It is natural to ask how well this approach works for problems in NP that are not believed to be complete in the class, like graph isomorphism. We study here the size of resolution refutations for formulas encoding graph isomorphism in the natural way. Given two graphs $G_1 = (V_1, E_1)$ and $G_2 = (V_2, E_2)$, with n nodes each, the formula $F(G_1, G_2)$ over the set of variables $\{x_{i,j} \mid i, j \in [n]\}$ is satisfiable if and only if there if an isomorphism between G_1 and G_2. Each satisfying assignments of the formula encodes an isomorphisms. In such an assignment the variable $x_{i,j}$ receives value 1 if and only if the encoded isomorphism maps vertex $v_i \in V_1$ to $v_j \in V_2$.

Definition 1.1. *For a pair of graphs G_1, G_2 with n vertices each, $F(G_1, G_2)$ is the conjunction of the following sets of clauses:*

> *Type 1 clauses: for every $i \in [n]$ the clause $(x_{i,1} \vee x_{i,2} \vee \cdots \vee x_{i,n})$ indicating that vertex $v_i \in V_1$ is mapped to some vertex in V_2.*
> *Type 2 clauses: for every $i, j, k \in [n]$ with $i \neq j$ the clause $(\overline{x_{i,k}} \vee \overline{x_{j,k}})$ indicating that not two different vertices are mapped to the same one.*
> *Type 3 clauses:, for every $i, j, k, l \in [n]$ $i < j$ and $k \neq l$ with $(v_i, v_j) \in E_1 \leftrightarrow (v_k, v_l) \notin E_2$, the clause $(\overline{x_{i,j}} \vee \overline{x_{i,k}})$ expressing the adjacency relation (an edge cannot be mapped to a non edge and vice-versa).*

Formula $F(G_1, G_2)$ has n^2 variables and $O(n^4)$ clauses. The clauses of Types 2 and 3 have width 2, while the clauses of Type 1 have width n.

It is not hard to find pairs of non-isomorphic graphs whose formulas require exponential size resolution refutations. For example if graph G_1 consists of $n+1$ isolated vertices and G_2 n isolated vertices (no edges), then $F(G_1, G_2)$ is exactly $PHP(n+1, n)$, the formula encoding the pigeonhole principle with $n+1$ pigeons and n holes. It is well known that this formula requires exponential size resolution refutations [17]. More elaborate examples can be constructed for example encoding PHP in pairs of connected graphs with the same number of vertices. In order to find more interesting examples and to investigate whether the apparent inability of resolution for dealing with GI comes only from the difficulty to count, we consider here graphs with colored vertices and bounded color multiplicities (there is a bound on the number of vertices of each color). In an isomorphism between colored graphs, colors must be preserved. A vertex coloring is reflected very naturally in the clauses of Type 1, since for a vertex i in the first graph we only have to include the variables $x_{i,j}$ for the vertices j in the second graph with the same color as i. If the maximum color multiplicity is bounded by k, the clauses of Type 1 are reduced to have at most k literals. This restricts the isomorphism search space. This also prevents from encoding the pigeonhole principle in the formula. Also in this case we can ignore all the variables $x_{i,j}$ when i and j have different colors. This means that the corresponding isomorphism formulas have at most kn variables.

When the input graphs G_1 and G_2 are colored, we will also denote by $F(G_1, G_2)$ the formula defined as above, but with the Type 1 clauses restricted according to the colors.

For any constant k, it is known that GI for graphs with color multiplicity bounded by k can be solved in polynomial time, and even using more restricted resources [6,15,5]. In contrast to this fact, we show in this paper than in the case of resolution, there is a big difference between color classes of size 3 and larger classes. When the maximum color multiplicity is 3, the non-isomorphism formulas have polynomial size resolution refutations, (even tree-like refutations). On the other hand we prove an exponential lower bound for the resolution refutation of certain pairs of non-isomorphic graphs with color classes of size 4 or larger. The gap in the complexity of resolution depending on the size of the color classes coincides with the gap in the number of variables required for graph identification [19].

For our lower bound we consider the CFI graphs used by Cai, Fürer and Immerman in [12] to prove the impossibility of Weisfeiler-Lehmann based algorithms for solving GI. In this important paper the authors gave a method to transform any graph G with n vertices and maximum degree d into a pair of non-isomorphic graphs of size $nd2^d$ based on G. We show here that any resolution refutation of the related isomorphism formulas must have exponential size in $\frac{ex(G)}{d}$, were $ex(G)$ is the expansion of the graph G (Definition 4.5). The exponential lower bound follows by considering constant degree graphs with linear expansion. The lower bound holds even for pairs of colored graphs of degree at most 3 and color classes of size at most 4.

The idea behind the proof of the resolution lower bound resembles that from Urquhart [26] for proving resolution lower bounds for Tseitin formulas [25]. We profit however from several newer results that help us to simplify the proof. Especially, we make use of the relationship between resolution size and width (the maximum number of literals in a clause in the refutation) proven by Ben-Sasson and Wigderson in [11], which imply size lower bounds by proving bounds on the width.

2   Preliminaries

We deal with Boolean formulas in conjunctive normal form, CNF. A CNF formula F on the set of variables V is a conjunction of clauses $C_1, \ldots, C_m$. Each clause is a disjunction of literals. A literal is either a variable or a negated variable from V. A (partial) assignment α is a (partial) mapping from V in $\{0, 1\}$. For a clause C and an assignment α, we denote by $C|_\alpha$ the result of applying α to C. This is 1 if α assigns value 1 some literal in C, or the result of deleting the literals in C being assigned to 0 otherwise. For a CNF formula F, $F|_\alpha$ is the conjunction of the clauses $C|_\alpha$ for every C in F.

2.1   Resolution

The concept of *resolution* was introduced by Robinson in [22]. Resolution is a refutation proof system for propositional formulas in conjunctive normal form. The only inference rule in this proof system is the resolution rule:

$$\frac{C \vee x \qquad D \vee \bar{x}}{C \vee D}.$$

Resolving variable x from clauses $C \vee x$ and $D \vee \bar{x}$ we get the *resolvent* clause $C \vee D$. A resolution refutation of a CNF formula F is a sequence of clauses $C_1, \ldots, C_s$ where each C_i is either a clause from F or is inferred from earlier clauses by the resolution rule, and C_s is the empty clause.

A resolution refutation can be pictured as a directed acyclic graph in which the clauses are the vertices and there are edges from the clauses to their resolvents. The restriction of resolution in which the underlying graph is a tree is called tree-like resolution.

Definition 2.1. *The size of a resolution refutation is the number of clauses it contains. For an unsatisfiable formula F, $size(Res(F))$ denotes the minimal size of a resolution refutation of F.*

We denote the size of the smallest tree-like refutation for F, by $size(TRes(F))$. Families of unsatisfiable formulas exist, for which there is an exponential separation between the size of tree-like resolution refutation and that of resolution refutations without restrictions [10]. It is well known that the size of a tree-like resolution refutations for an unsatisfiable formula corresponds to the running time of a DPLL algorithm on the formula (see e.g. [23]).

Definition 2.2. *[11] The width of a clause is the number of literals appearing in it. For a set of clauses C (C can be for example a formula in CNF or a resolution refutation) the width of C, denoted by $width(C)$, is the maximal width of a clause in the set C.*

The width needed for the resolution of an unsatisfiable CNF formula F, denoted by $width(Res(F))$, is the minimal width needed in a resolution of F, that is, the minimum of $width(\pi)$ over all resolution refutations π of F.

For proving the lower bound on the resolution size we will use the relationship between width and size of a refutation introduced by Ben-Sasson and Widgerson in [11]. This approach allows to reduce the problem of giving lower bounds on the size of a refutation to that of giving lower bounds on the width.

Theorem 2.1. *[11] For an unsatisfiable formula F in CNF with n variables*

$$size(Res(F)) = exp(\Omega(\frac{[width(Res(F)) - width(F)]^2}{n})).$$

2.2 Graph Isomorphism

The graphs considered in this paper will be undirected simple graphs, usually denoted by $G = (V, E)$, where V is the vertex set and $E \subseteq \binom{V}{2}$. We say two graphs G_1 and G_2 are *isomorphic* if there is a bijection $\varphi : V_1 \longrightarrow V_2$ such that $(u, v) \in E_1$ iff $(\varphi(u), \varphi(v)) \in E_2$. We write $G_1 \cong G_2$ and call φ an isomorphism. An *automorphism* of a graph G is an isomorphism from G to G. Automorphisms are permutations on the set V, and the set of automorphisms $\mathrm{Aut}(X)$ forms a group under permutation composition. We say that a set of vertices $V' \subseteq V$ is set-wise stabilized by an automomphism φ, if $V' = \varphi(V')$.

We will deal with graphs with colored vertices. A coloring with k colors is a function $f : V \to \{1, \dots k\}$. In an isomorphism between colored graphs, the colors have to be preserved. This restricts the search space when looking for isomorphisms. For a color c, the color class corresponding to c is the set of vertices with this color in V. The set of color classes defines a partition of the graph vertices. A *refinement* of a given set of color classes, is a refinement of this partition, that is, every color class in the refinement is a subset of the original partition. When a pair of graphs is given as input for the isomorphism problem, there are two color classes of each color, one in each graph. It should be clear from the context, from which one of the classes we are talking about in the text.

3 Polynomial Size Tree-Like Resolution Refutations for Color Multiplicity 2 and 3

For the case of graphs with color classes of size 2, all the clauses in the non-isomorphism formulas have width 2. It is well known that an unsatisfiable set of clauses of width at most 2 has polynomial size tree-like resolution refutations. This simple observation provides an alternative proof for the fact that GI for graphs with color multiplicities at most 2 is in P. We extend this observation to the case of graphs with color classes at most 3.

Theorem 3.1. *Let G_1 and G_2 be two colored non-isomorphic graphs with color classes of size at most 3. Then $F(G_1, G_2)$ has tree-like resolution refutations of polynomial size.*

Proof. We can suppose that the subgraphs induced by every pair of color classes in G_1 and G_2 are isomorphic, because otherwise, there would be an unsatisfiable subformula of constant size in $F(G_1, G_2)$, having a constant size tree-like resolution refutation. For some pairs of color classes, the subgraphs S_1 and S_2, induced by the vertices with these colors in G_1 and G_2 can restrict the set of possible isomorphisms between the subgraphs, thus implying a further refinement in the vertex partition defined by the colors.

For example in Figure 1 every possible isomorphism between S_1 and S_2 must map 1 to b or c, 2 to b or c, and 3 to a. This implies that every possible isomorphism between the subgraphs must map 4 to f, 5 to d or e and 6 to d or e. Observe that by considering the subgraphs induced just by the white color classes, no further refinement would have followed.

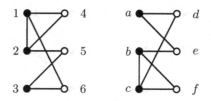

Fig. 1. The subgraphs S_1 and S_2 induced by a pair of color classes (black,white)

One can use this refinement in the set of possible isomorphisms between S_1 and S_2 to derive a refinement of the clauses of Type 1 by a constant size resolution refutation from the (constant size) initial set of clauses in $F(S_1, S_2)$. In our example these new clauses would be: $(x_{1,b}, x_{1,c}), (x_{2,b}, x_{2,c}), x_{3,a}, x_{4,e}, (x_{5,d}, x_{5,f})$ and $(x_{6,d}, x_{6,f})$. A way to see that in case of a refinement in the color classes it is always possible to obtain the reduced clauses, is by noticing that if there is no isomorphism between the subgraphs mapping 1 to a, for example, then the subformula obtained by setting $x_{1,a}$ to 1 in $F(S_1, S_2)$ is unsatisfiable and therefore it has a constant size tree-like refutation $\mathcal{R}$. By the standard trick of using the structure of the refutation $\mathcal{R}$, but starting with the clauses in $F(S_1, S_2)$ (instead of $F(S_1, S_2)|_{x_{1,a}=1}$), one derives the literal $\overline{x_{1,a}}$ (or maybe the empty clause in case $F(S_1, S_2)$ was unsatisfiable). By resolving these literals (unitary clauses) with the corresponding clauses of Type 1, one obtains the desired refined clauses.

Because of these observations, we can suppose that in G_1 and G_2 the partition on the vertex set defined by the color classes, cannot be further refined by considering the subgraphs induced by pairs of color classes. Considering this, by inspecting the few possible cases of edge connections between two color classes, it can be seen that for every pair of colors, say black and white, there are two possible situations in the subgraphs S_1 and S_2 induced by these color classes in G_1 and G_2, (this fact has been previously observed [19,20]). Either:

1. For every possible bijective mapping of the black vertices, there is a *unique* extension to the white vertices that is an isomorphism from S_1 to S_2, or
2. For every possible bijective mapping of the black vertices, every possible bijective extension to the white vertices is an isomorphism from S_1 to S_2.

(This property does not hold when the color classes can have size larger than 3.) We show in Figure 2 the possible edge connections between two color classes when they do not imply a refinement. The first and last situations belong to Case 2, while the second and third belong to Case 1. The situations involving color classes of size smaller than 3 are not included in the figure but are also easy to check.

Translating this to resolution, this property intuitively means that in Case 1, an assignment for a possible mapping of one of the color classes fixes an assignment of the variables for another color, and so on, until (in the case of non-isomorphic graphs) a contradiction is found.

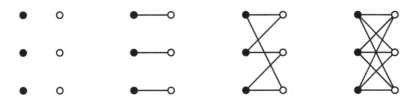

Fig. 2. The possible (non refining) edge connections between two color classes of size 3

Suppose that black and white are two color classes with edge connections as in Case 1, and suppose we had three unitary clauses, specifying a mapping of the black vertices (like for example $x_{1,a}$, $x_{2,b}$, and $x_{3,c}$). By a unit-resolution refutation of constant size, resolving first these clauses with clauses of Type 3 and then using the obtained resolvents together with the clauses of Type 1 for the white vertices, the unitary clauses specifying the corresponding mapping of the white vertices can be obtained. Observe that one would also obtain (by unit-resolution) the unit clauses for the white vertices if instead of the unit clauses for the black vertices one would have started with a partial assignment α defining a mapping of the black vertices (like $x_{1,a} = 1$, $x_{2,b} = 1$, and $x_{3,c} = 1$) and considering the formula $F(G_1, G_2)|_\alpha$.

We can now define a new graph $\mathcal{C}$ in which there is a vertex for each color class in G_1, and there is an edge between two color classes if and only if the edge connections between the vertices of the corresponding classes in G_1 or G_2 are as in Case 1.

If G_1 and G_2 are not isomorphic, then there must be a set of color classes so that the subgraphs induced by these classes in G_1 and G_2 are non-isomorphic. These color classes define a connected component in $\mathcal{C}$. Moreover, if for every pair of color classes the corresponding induced subgraphs in G_1 and G_2 are isomorphic, then there has to be a cycle in $\mathcal{C}$ so that the graphs induced by the colors in this cycle are non isomorphic. Otherwise, it would be possible to extend the isomorphism between two color classes to an isomorphism between G_1 and G_2.

Let black be any color class in such a cycle. By the above observations, from any of the possible partial assignments α of the variables corresponding to a bijective mapping of the black color class, the clauses corresponding to the unique possible mapping of a neighboring color class in the cycle can be derived (by a constant unit-resolution refutation) and so on, coming back to black. A partial isomorphism different from the initial one is then forced on this class, thus forcing a contradiction.

Since the number of colors in the cycle is bounded by the number of vertices, any partial assignment α defining a bijective mapping of the black vertices, defines a polynomial size tree-like (and unit-resolution) refutation of $F(G_1, G_2)|_\alpha$. There are only 6 possible such bijective mappings α. By using again the trick repeating these refutations separately on $F(G_1, G_2)$, one obtains a tree-like derivation of an unsatisfiable set of clauses involving only variables $x_{i,j}$ with i and j in the black classes. This set has constant size, and from it, the empty clause can be derived. ∎

4   The CFI Graphs and Their Formulas

We define now the graphs that will be used for the resolution lower bounds. These graphs were considered in [12] to prove lower bounds for the Weisfeiler-Lehman method in isomorphism testing. In [24] a generalization of these graphs was used in order to show that GI is hard for the complexity class DET.

Definition 4.1. *For $k \geq 2$ the graph $X_k = (V_k, E_k)$ is defined as follows:*
$V_k = A_k \cup B_k \cup M_k$ where $A_k = \{a_i \mid i \in [k]\}$, $B_k = \{b_i \mid i \in [k]\}$ and $M_k = \{m_S \mid S \subseteq [k], |S| \text{ even}\}$. The graph is bipartite, the set of edges connect a and b vertices with m vertices $E_k = \{(m_S, a_i) \mid i \in S\} \cup \{(m_S, b_i) \mid i \notin S\}$.

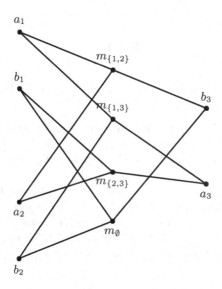

Fig. 3. The graph X_3

Graph X_k consists of $2^{k-1} + 2k$ vertices and $k2^{k-1}$ edges. Let us give some intuition on the definition. Suppose that for each i we color the vertex set $\{a_i, b_i\}$ with color i so that any automorphism of X_k must set-wise stabilize these vertex sets. An automorphism in the colored graph, might map some a_i vertices to the corresponding b_i vertex, while fixing the rest of the a and b vertices. As stated in the following Lemma from [12], describing the set of automorphism in X_k, the graph is constructed in such a way, that their automorphisms correspond to the situations in which the number of $i \in [n]$ with vertex a_i being mapped to b_i is even.

Lemma 4.1. *[12] There are exactly 2^{k-1} automorphisms in X_k stabilizing the sets $\{a_i, b_i\}, i \in [k]$. Each such automorphism is determined by interchanging a_i and b_i for each i in some subset $S \subseteq [k]$ of even cardinality.*

More intuitively, the construction can be understood with the simplification of X_k given in Figure 4. Here we have one v_i vertex for each pair a_i, b_i, and these are connected to a single m vertex for all the vertices in M_k. If we assign $\{0, 1\}$ values to the v_i vertices, the previous lemma just says that the sum of these values has to be even.

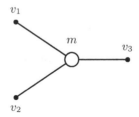

Fig. 4.

We now transform a graph G into a graph $X(G)$ by substituting its vertices by the gadgets of Definition 4.1.

Definition 4.2. *Let $G = (V, E)$ be a connected graph with minimum degree at least 2. We transform G in a new graph $X(G)$ in which every vertex v of degree d in G is substituted by a copy $X(v)$ of the gadget X_d, and these are connected in the following way:*

To each edge $e = (u, v)$ having v as endpoint we associate two vertices $\{a_e^v, b_e^v\}$ in $X(v)$ and two vertices $\{a_e^u, b_e^u\}$ in $X(u)$. We then join with an edge the a_e vertices in $X(v)$ and $X(u)$ and the b_e vertices in $X(v)$ and $X(u)$. This means that every edge in G is transformed into two edges in $X(G)$. $X(G)$ can be intuitively understood as the result of going back from the graph in Figure 4, to the one in Figure 3, for every vertex.

If G has maximum degree d then $X(G)$ has at most $|V| d 2^d$ vertices and $2|E| + |V| d 2^{d-1}$ edges. It should be clear that the set of automorphisms of $X(G)$ stabilizing the pairs $\{a_e^v, b_e^v\}$ have to be edge respecting in the sense of the following definition.

Definition 4.3. *A permutation φ acting on the set $\{a_e^v, b_e^v \mid e$ is an incident edge with vertex v in $G\}$ is called edge respecting if it stabilizes all the pairs $\{a_e^v, b_e^v\}$ and has the property that for every edge $e = (u, v)$ in G, $\varphi(a_e^u) = a_e^u$ if and only if $\varphi(a_e^v) = a_e^v$.*

The following observation is a direct consequence of Lemma 4.1.

Observation 4.1 *There is a 1-1 correspondence between the set of edge respecting permutations φ acting on the set $\{a_e^v, b_e^v \mid e$ is an incident edge with vertex v in $G\}$ and with the property that for every vertex v, φ interchanges the vertices a_e^v and b_e^v for an even number of edges e incident with v, and the set of automorphism in $Aut(X(G))$ stabilizing the sets $\{a_e^v, b_e^v\}$.*

For $E' \subseteq E$, let $\tilde{X}(G, E')$ be a copy of $X(G)$ but in which all the edges $e = (u, v) \in E'$ are twisted, that is a_e^u is connected to b_e^v and b_e^u is connected to a_e^v. The next lemma shows that depending on the number of twisted edges in $\tilde{X}(G, E')$ we can only have two possible isomorphism classes.

Lemma 4.2. *[12] Let $G = (V, E)$ be a connected graph with minimal degree at least 2 and let $E' \subseteq E$ with $\|E'\| = t$. If t is even then $\tilde{X}(G, E')$ is isomorphic to $X(G)$, and if t is odd, then $\tilde{X}(G, E')$ is isomorphic to $\tilde{X}(G, \{e\})$, for any edge $e \in E$. Moreover, $X(G)$ and $\tilde{X}(G, \{e\})$ are non-isomorphic.*

We will say that an edge (u, v) in G is *straight*, if the corresponding edge in $\tilde{X}(G, E')$ has not been twisted. For simplicity, we will denote by $\tilde{X}(G)$ the graph $\tilde{X}(G, \{e\})$ for some fixed $e \in E$. Since all the graphs defined in this way are isomorphic, for our purposes it does not matter which of these graphs we are considering. Analogously we will refer to the formula $F(X(G), \tilde{X}(G))$, considering that $\tilde{X}(G)$ is a fixed graph.

We extend Definition 4.3 to the set of bijections between the vertices of $X(G)$ and $\tilde{X}(G)$.

Definition 4.4. *A bijection φ between the sets $\{a_e^v, b_e^v \mid a_e^v, b_e^v \in V(X(G))\}$ and $\{a_e^v, b_e^v \mid a_e^v, b_e^v \in V(\tilde{X}(G))\}$ is called edge respecting if for every vertex v and incident edge e, $\{\varphi(a_e^v), \varphi(b_e^v)\} = \{a_e^v, b_e^v\}$ and fulfills the following property:*
For every edge $e = (u, v)$ in G, if e is straight then $\varphi(a_e^u) = a_e^u$ if and only if $\varphi(a_e^v) = a_e^v$, and if e is twisted then $\varphi(a_e^u) = a_e^u$ if and only if $\varphi(a_e^v) = b_e^v$.

For graph G, the variables in the formula $F(X(G), \tilde{X}(G))$ are of the form $x_{i,j}$ representing the mapping of vertex v_i in $X(G)$ to vertex v_j in $\tilde{X}(G)$. For clarity we will divide the set of x variables in two kinds during the exposition:

The y variables correspond to the endpoints of the original edges in G, (that have been doubled in $X(G)$). For a vertex v in G and an edge e incident with v, the vertices corresponding to v and e in $X(G)$ are $\{a_e^v, b_e^v\}$. $y_{a_e^v, b_e^v}$, for example is the variable representing the mapping from a_e^v in $X(G)$ to b_e^v in $\tilde{X}(G)$. For simplicity, when the edge is clear from the context, we will sometimes denote this variable by $y_{a,b}^v$. Also we will consider the graphs $X(G)$ and $\tilde{X}(G)$ to be colored so that for a vertex v in G and an edge e incident with v, $\{a_e^v, b_e^v\}$ are the only two vertices having the color (v, e). Since we are only interested in color preserving isomorphisms, the clause of Type 1 for a vertex a_e^v is $(y_{a_e, a_e}^v \vee y_{a_e, b_e}^v)$ and has width 2 (analogous for the vertex b_e^v). This only restricts the number of possible isomorphisms and makes it easier to refute the formula.

The z variables correspond to the m vertices in the $X(v)$ gadgets. For a vertex v of degree d in G, there are 2^{d-1} vertices m_S^v, in $X(G)$ and in $\tilde{X}(G)$. The variables $z_{S,S'}^v$ are the ones representing the mappings between these vertices. Analogously as in the case of the y variables, for a vertex v we will consider that the vertices m_S^v, in $X(G)$ and $\tilde{X}(G)$ are the only ones colored with color v. This implies that the clauses of Type 1 for a vertex m_S^v have width 2^{d-1}.

For a vertex v in G we denote by $F(X(v))$ the set of initial clauses in $F(X(G), \tilde{X}(G))$ containing some variable y^v or z^v (observe that for two vertices u and v, $F(X(u))$ and $F(X(v))$ might not be disjoint. Analogously, for a set of vertices $C \subseteq V$ we denote by $F(X(C))$ the union of the clauses $F(X(v))$ for $v \in C$.

By Lemma 4.2, for any graph G, the formula $F(X(G), \tilde{X}(G))$ is unsatisfiable. However, as stated in the next lemma, for any vertex from G there are assignments satisfying simultaneously all the formula clauses except some of the clauses in $F(X(v))$.

Lemma 4.3. *For any graph $G = (V, E)$ and for every $v \in V$ there is an assignment satisfying all the clauses in $F(X(G), \tilde{X}(G))$ except two clauses in $F(X(v))$.*

Proof. Consider first the easy case in which $\tilde{X}(G)$ is the version of $X(G)$ in which exactly one edge $e = (u, v)$ is twisted, for some neighbor u of v in G. The assignment $x_{i,j} = 1$ if and only if $j = i$ satisfies all the clauses in $F(X(G), \tilde{X}(G))$ except the two Type 3 clauses $(\overline{y_{a_e^u, a_e^u}} \vee \overline{y_{a_e^v, a_e^v}})$ and $(\overline{y_{b_e^u, b_e^u}} \vee \overline{y_{b_e^v, b_e^v}})$. In the general case, by Lemma 4.2, $\tilde{X}(G)$ is isomorphic to the copy of $X(G)$ with only twisted edge $e = (u, v)$. Let φ be an isomorphism between these graphs. The assignment $x_{i,j} = 1$ if and only if $j = \varphi(i)$ satisfies all the clauses in $F(X(G), \tilde{X}(G))$ except the two Type 3 clauses $(\overline{y_{a_e^u, \varphi(a_e^u)}} \vee \overline{y_{a_e^v, \varphi(a_e^v)}})$ and $(\overline{y_{b_e^u, \varphi(b_e^u)}} \vee \overline{y_{b_e^v, \varphi(b_e^v)}})$. ∎

We will show in the next section that the size of the resolution refutations of $F(X(G), \tilde{X}(G))$ for any graph G are related to the expansion of G.

Definition 4.5. *Let $G = (V, E)$ be an undirected graph with $|V| = n$. The expansion of G, $ex(G)$ is defined as:*

$$ex(G) = \min k : \exists S \subseteq V, |S| \in \left[\frac{n}{3}, \frac{2n}{3}\right], |\{(x, y) \in E : x \in S, y \notin S\}| = k.$$

Intuitively this represents the minimum number of edges that have to be cut in order to separate a big component of G from the rest.

5 Resolution Lower Bounds for Color Multiplicity Larger than 3

We show next that for certain pairs of non-isomorphic graphs G_1, G_2, the size of any resolution refutation of $F(G_1, G_2)$ is exponential in n, the number of vertices. The proof follows the ideas introduced in [9] and [11] for proving resolution lower bounds. We will prove that for any connected graph G of minimum degree at least 2, the width of any refutation of $F(X(G)\tilde{X}(G))$ is at least the expansion of G. The lower bound on the size follows by considering a graph G with large expansion and applying Theorem 2.1.

Theorem 5.1. *Let $G = (V, E)$ be a connected graph with maximum degree d and minimum degree at least 2. Any resolution refutation of the colored version of $F(X(G), \tilde{X}(G))$ requires width at least $\frac{ex(G)}{d}$.*

Proof. Let $\mathcal{R}$ be a resolution refutation for $F(X(G), \tilde{X}(G))$. For a vertex v in G, let $G - \{v\}$ be the subgraph of G induced by the set of vertices $V \setminus \{v\}$. An assignment α of all the variables in $F(X(G), \tilde{X}(G))$ is called *v-critical*, if it satisfies all the clauses in $F(X(G), \tilde{X}(G))$ except maybe some clauses in $X(v)$. Observe that by Lemmas 4.2 and 4.3, that that there are v-critical assignments for every vertex v, and that if α is v-critical, then the number of vertices a_e^v being mapped to b_e^v for some edge e incident with v, is odd, while for every other vertex $u \neq v$ the number of such vertices is even.

We define the *significance* of a clause C in $\mathcal{R}$, abbreviated by $\sigma(C)$, as the number of vertices v such that there is a v-critical assignment, that falsifies C. It should be clear, that the initial clauses in $F(X(G), \tilde{X}(G))$ have significance 1 or 0. The empty clause, at the end of the resolution refutation $\mathcal{R}$, has significance n. Moreover, when K is the resolvent of two clauses K_1, K_2, having significance s_1 and s_2, then the significance from K is at most $s_1 + s_2$, since every assignment that falsifies K, falsifies also K_1 or K_2. From this follows, that there must be a clause C in $\mathcal{R}$ with significance $s \in [\frac{n}{3}, \frac{2n}{3}]$. (One can choose the first clause C in $\mathcal{R}$ with $\sigma(C) \geq \frac{n}{3}$.) Let V' be the set of vertices v for which there exists some v-critical assignment α, falsifying C. $|V'| = s$. For every vertex $w \in V \setminus V'$, it holds that *all* w-critical assignments satisfy clause C. Since $s \in [\frac{n}{3}, \frac{2n}{3}]$, there are at least $ex(G)$ edges joining a vertex in V' with a vertex in $V \setminus V'$. Let $e = (v, w)$ be such an edge and let d be the degree of v. We modify α in a few positions, so that it mutates to a w-critical assignment α_e: we toggle the values of the variables related to the end points of e, $y_{a,a}^v, y_{a,b}^v, y_{b,a}^v, y_{b,b}^v$, as well as toggling the values from $y_{a,a}^w, y_{a,b}^w, y_{b,a}^w, y_{b,b}^w$. Moreover, we set in α_e the values of the z variables from vertex v so that the assignment restricted to $X(v)$ defines a partial isomorphism. This is always possible since the number of vertices a^v being mapped to b^v for the edges incident with v, by α_e is even. All the other values in α are not changed in α_e. Because of this, α_e is w-critical. As a consequence of the modification, α_e satisfies the clause C. This implies that at least one of the changed variables must occur in C. Observe that for two edges $e = (v, w), e' = (v', w')$ with $v, v' \in V'$ and $w, w' \in V \setminus V'$ if $v \neq v'$ then the sets of changed variables in α_e and $\alpha_{e'}$ are disjoint. If $v = v'$ then α_e and $\alpha_{e'}$ can coincide in the values of some of the changed z variables. But v has degree at most d. This implies that for every set of d edges $e = (v, w)$ with $v \in V'$ and $w \in V \setminus V'$ a different variable must occur in C and therefore $\text{width}(C) \geq \frac{ex(G)}{d}$. ∎

The lower bound follows:

Corollary 5.1. *There exists a family of graphs $\mathcal{G}$ such that for any n, $G_n \in \mathcal{G}$ has n vertices and the resolution refutation of the formula $F(X(G_n), \tilde{X}(G_n))$ expressing that the graphs $X(G_n)$ and $\tilde{X}(G_n)$ are non-isomorphic, requires size $\exp(\Omega(n))$. $X(G_n)$ and $\tilde{X}(G_n)$ are colored graphs with color multiplicity at most 4.*

Proof. It is known that there are constructive families $\mathcal{G}$ of graphs of degree 3 and with an expansion that is linear in the number of vertices (see e.g. [1]). For a graph $G_n \in \mathcal{G}$ with n vertices, the graph $X(G_n)$ has $O(n)$ vertices, and color multiplicity at most 4. The formula $F(X(G_n), \tilde{X}(G_n))$ contains $O(n)$ variables and $O(n^2)$

clauses. Observe that the number of variables is linear in n because the size of the color classes is bounded. The width of these clauses is at most 4. By the above result, the width of any resolution refutation of the formula is $\Omega(n)$. By Theorem 2.1, the size of any resolution refutation of $F(X(G_n), \tilde{X}(G_n))$ is $\exp(\Omega(n))$. ∎

6   Discussion

We have shown that the natural encoding of the isomorphism problem in CNF formulas requires exponential size resolution refutations for a certain family of colored graphs. These graphs have colored classes of size 4 and maximum degree 3. In contrast, when the size of the color classes is bounded by 3, the formulas have polynomial size tree-like resolution refutations. The formulas used for the lower bound are based on the CFI graphs from [12]. In these pairs of graphs, every vertex of a certain color has the same degree, the same number of neighbors of another color or the same distance to any color. Therefore, the difficulty of the resolution system in performing counting (as shown for example in the resolution lower bounds for the pigeon hole principle), is not the reason for the large refutations, since counting does not help in this context. As shown in [12], the non-isomorphic graphs we use, are indistinguishable using even inductive logic with counting. The lower bound can be explained as an "encoding" of the Tseitin tautologies (for which resolution lower bounds are known), into graph isomorphism instances. I believe that this new connection between Tseitin tautologies and isomorphism might help to solve some open question in the area of proof complexity. An example of this might be the proof of exponential lower bounds for Tseitin tautologies in stronger systems, like the cutting plane proof system. Such a result is only known for the case in which a parameter called the degree of falsity is bounded [16,18]. Knowledge on graph isomorphism problem might help to attack the question from another perspective.

Although the main interest for the results has a theoretical motivation, the isomorphism formulas discussed here could be used as benchmarks for testing sat-solvers. To my knowledge this has only been done for formulas encoding sub-graph isomorphism [3,4]. A way to do this, for example, would be to consider a (regular) graph and color its vertices with color classes of a bounded size. Considering then a random permutation of the vertices, one obtains an isomorphic copy of the graph. If the size of the color classes is at most 3, we know by Theorem 3.1 that there is a variable ordering under which the running time of a DPLL algorithm testing isomorphism is polynomial. For color classes of size larger than 3 we only have non trivial resolution upper bounds (that might guide the sat-solvers) for the case of the CFI graph pairs. Because of the connection between such isomorphism formulas and the Tseitin tautologies, the results from [2] relating the width of a resolution refutation for a Tseitin formula with a structural parameter (branch-width) of the underlying graph, can also be applied to the isomorphism formulas. This provides a way to design example instances for isomorphism formulas with bounded resolution width.

Acknowledgement. The author would like to thank Nicola Galesi for interesting discussions related to this paper and the anonymous reviewers for many useful suggestions.

References

1. Ajtai, M.: Recursive construction for 3-regular expanders. Combinatorica 14(4), 379–416 (1994)
2. Alekhnovich, M., Razborov, A.A.: Satisfiability, branch-width and Tseitin tautologies. Computational Complexity 20(4), 649–678 (2011)
3. Anton, C., Neal, C.: Notes on generating satisfiable SAT instances using random subgraph isomorphism. In: Farzindar, A., Kešelj, V. (eds.) Canadian AI 2010. LNCS, vol. 6085, pp. 315–318. Springer, Heidelberg (2010)
4. Anton, C.: An improved satisfiable SAT generator based on random subgraph isomorphism. In: Butz, C., Lingras, P. (eds.) Canadian AI 2011. LNCS, vol. 6657, pp. 44–49. Springer, Heidelberg (2011)
5. Arvind, V., Kurur, P.P., Vijayaraghavan, T.C.: Bounded color multiplicity graph isomorphism is in the #L Hierarchy. In: Proceedings of the 20th Conference on Computational Complexity, pp. 13–27 (2005)
6. Babai, L.: Monte Carlo algorithms for Graph Isomorphism testing. Tech. Rep. 79-10, Dép. Math. et Stat., Univ. de Montréal (1979)
7. Beame, P., Culberson, J.C., Mitchell, D.G., Moore, C.: The resolution complexity of random graph k-colorability. Discrete Applied Mathematics 153(1-3), 25–47 (2005)
8. Beame, P., Impagliazzo, R., Sabharwal, A.: The resolution complexity of independent sets and vertex covers in random graphs. Computational Complexity 16(3), 245–297 (2007)
9. Beame, P., Pitassi, T.: Simplified and improved resolution lower bounds. In: 37th Annual IEEE Symposium on Foundations of Computer Science, pp. 274–282 (1996)
10. Ben-Sasson, E., Impagliazzo, R., Wigderson, A.: Near-optimal separation of treelike and general resolution. Combinatorica 24(4), 585–603 (2004)
11. Ben-Sasson, E., Wigderson, A.: Short proofs are narrow – resolution made simple. Journal of the ACM 48(2), 149–169 (2001)
12. Cai, J., Fürer, M., Immerman, N.: An optimal lower bound on the number of variables for graph identifications. Combinatorica 12(4), 389–410 (1992)
13. Chvátal, V., Szemerédi, E.: Many hard examples for resolution. Journal of the ACM 35, 759–768 (1988)
14. Davis, M., Logemann, G., Loveland, D.: A machine program for theorem proving. Communications of the ACM 5, 394–397 (1962)
15. Furst, M., Hopcroft, J., Luks, E.: Polynomial time algorithms for permutation groups. In: Proc. 21st IEEE Symp. on Foundations of Computer Science, pp. 36–41 (1980)
16. Goerdt, A.: The cutting plane proof system with bounded degree of falsity. In: Kleine Büning, H., Jäger, G., Börger, E., Richter, M.M. (eds.) CSL 1991. LNCS, vol. 626, pp. 119–133. Springer, Heidelberg (1992)
17. Haken, A.: The intractability of resolution. Theoretical Computer Science 39(2-3), 297–308 (1985)
18. Hirsch, E.A., Kojevnikov, A., Kulikov, A.S., Nikolenko, S.I.: Complexity of semi-algebraic proofs with restricted degree of falsity. Journal on Satisfiability, Boolean Modeling and Computation 6, 53–69 (2008)

19. Immerman, N., Lander, E.: Describing graphs: a first-order approach to graph canonization. In: Selman, A.L. (ed.) Complexity Theory Retrospective, pp. 59–81. Springer (1990)

20. Jenner, B., Köbler, J., McKenzie, P., Torán, J.: Completeness results for graph isomorphism. J. Comput. Syst. Sci. 66(3), 549–566 (2003)

21. Köbler, J., Schöning, U., Torán, J.: The Graph Isomorphism problem: Its structural complexity. Birkhauser (1993)

22. Robinson, J.A.: A machine oriented logic based on the resolution principle. Journal of the ACM 12(1), 23–41 (1965)

23. Schöning, U., Torán, J.: Das Erfüllbarkeitsproblem SAT - Algorithmen und Analysen, Lehmann (2012)

24. Torán, J.: On the hardness of Graph Isomorphism. SIAM Journal on Computing 33(5), 1093–1108 (2004)

25. Tseitin, G.S.: On the complexity of derivation in propositional calculus. In: Studies in Constructive Mathematics and Mathematical Logic, Part 2, pp. 115–125. Consultants Bureau (1968)

26. Urquhart, A.: Hard examples for resolution. Journal of the ACM 34, 209–219 (1987)

On Propositional QBF Expansions and Q-Resolution

Mikoláš Janota[1] and Joao Marques-Silva[2]

[1] IST/INESC-ID, Lisbon, Portugal
[2] University College Dublin, Ireland

Abstract. Over the years, proof systems for propositional satisfiability (SAT) have been extensively studied. Recently, proof systems for quantified Boolean formulas (QBFs) have also been gaining attention. Q-resolution is a calculus enabling producing proofs from DPLL-based QBF solvers. While DPLL has become a dominating technique for SAT, QBF has been tackled by other complementary and competitive approaches. One of these approaches is based on expanding variables until the formula contains only one type of quantifier; upon which a SAT solver is invoked. This approach motivates the theoretical analysis carried out in this paper. We focus on a two phase proof system, which expands the formula in the first phase and applies propositional resolution in the second. Fragments of this proof system are defined and compared to Q-resolution.

This paper follows the line of research on proof systems for propositional and quantified Boolean formulas (QBFs). This research is motivated by complexity theory and more recently by the objective to develop and certify QBF solvers [11,18,8,14]. Proof systems for QBF come in different styles and flavors. Krajíček and Pudlák propose a Genzen-style calculus *KP* for QBF [18]. Büning et al. propose a refutation calculus *Q-resolution* [8], an extension of propositional resolution. Giunchiglia et al. extend the work of Büning et al. into *term resolution* for proofs of true formulas [14] . Certain separation results were shown between KP and Q-resolution recently by Egly [12].

While many QBF solvers are based on the DPLL procedure [21,9,23,20,13], other solvers tackle the given formula by *expanding* out quantifiers until a single quantifier type is left. At that point, this formula is handed to a SAT solver [1,4,19,15]. Experimental results show that expansion-based QBF solvers can outperform DPLL-based solvers on a number of families of practical instances. Also, expansion can be used in QBF preprocessing [6,5].

This practical importance of expansion motivates the study carried out in this paper. We define a proof system ∀Exp+Res, which eliminates universal quantification from the given *false* formula and then applies propositional resolution to refute the remainder.

We show that ∀Exp+Res can p-simulate *tree* Q-resolution refutations. Conversely, we show that Q-resolution can p-simulate ∀Exp+Res refutations under certain restrictions on the propositional resolution part of the proofs.

1 Preliminaries

A *literal* is a Boolean variable or its negation. The literal complementary to a literal l is denoted as $\bar{l}$, i.e. $\bar{x} = \neg x$, $\overline{\neg x} = x$. A *clause* is a disjunction of zero or more noncomplementary literals. A formula in *conjunctive normal form* (CNF) is a conjunction of

M. Järvisalo and A. Van Gelder (Eds.): SAT 2013, LNCS 7962, pp. 67–82, 2013.

clauses. Whenever convenient, a clause is treated as a set of literals and a CNF formula as a set of sets of literals. For a literal $l = x$ or $l = \bar{x}$, we write $\mathsf{var}(l)$ for x. For a clause C, we write $\mathsf{var}(C)$ to denote $\{\mathsf{var}(l) \mid l \in C\}$ and for a CNF ψ, $\mathsf{var}(C)$ denotes $\{l \mid l \in \mathsf{var}(C), C \in \psi\}$

Substitutions are denoted as $x_1/\psi_1, \ldots, x_n/\psi_n$, with $x_i \neq x_j$ for $i \neq j$. The set of variables $x_1, \ldots, x_n$ is called the *domain* of the substitution. An application of a substitution is denoted as $\phi[x_1/\psi_1, \ldots, x_n/\psi_n]$ meaning that variables x_i are simultaneously substituted with corresponding ψ_i in ϕ. A substitution is called an *assignment* iff each ψ_i is one of the constants 0, 1. An assignment is called *total*, or *complete*, for a set of variables $\mathcal{X}$ if each $x \in X$ is in the domain of the assignment. For substitutions $\tau_1 = x_1/\psi_1, \ldots, x_n/\psi_n$ and $\tau_2 = y_1/\xi_1, \ldots, y_m/\xi_m$ with distinct domains we write $\tau_1 \cup \tau_2$ for the substitution $x_1/\psi_1, \ldots, x_n/\psi_n, y_1/\xi_1, \ldots, y_m/\xi_m$.

Quantified Boolean Formulas (QBFs) [7] are an extension of propositional logic with quantifiers with the standard semantics that $\forall x. \Psi$ is satisfied by the same truth assignments as $\Psi[x/0] \wedge \Psi[x/1]$ and $\exists x. \Psi$ as $\Psi[x/0] \vee \Psi[x/1]$. Unless specified otherwise, we assume that QBFs are in *closed prenex* form with a CNF *matrix*, i.e. $Q_1 X_1 \ldots Q_k X_k. \phi$, where X_i are pairwise disjoint sets of variables; $Q_i \in \{\exists, \forall\}$ and $Q_i \neq Q_{i+1}$. The formula ϕ is in CNF and is defined only on variables $X_1 \cup \ldots \cup X_k$. The propositional part ϕ is called the *matrix* and the rest the *prefix*. If a variable x is in the set X_i, we say that x is at *level* i and write $\mathsf{lv}(x) = i$; we write $\mathsf{lv}(l)$ for $\mathsf{lv}(\mathsf{var}(l))$. A closed QBF is *false* (resp. *true*), iff it is semantically equivalent to the constant 0 (resp. 1).

For a clause C, a universal literal $l \in C$ is *blocked* by an existential literal $k \in C$ iff $\mathsf{lv}(l) < \mathsf{lv}(k)$. $\forall$-*reduction* is the operation of removing from a clause C all universal literals that are *not* blocked by some literal. For two $\forall$-reduced clauses $x \vee C_1$ and $\bar{x} \vee C_2$, where x is an existential variable, a *Q-resolvent* [8] is obtained in two steps. (1) Compute $C_u = C_1 \cup C_2 \setminus \{x, \bar{x}\}$. If C_u contains complementary literals, the $Q-$ *resolvent* is undefined. (2) $\forall$-reduce C_u. For a QBF $\mathcal{P}.\phi$, a A *Q-resolution proof* of a clause C is a sequence of clauses $C_1, \ldots, C_n$ where $C_n = C$ and any C_i in the sequence is part of the given matrix ϕ or it is a Q-resolvent for some pair of the preceding clauses. A Q-resolution proof is called a *refutation* iff C is the empty clause, denoted $\bot$.

In this paper Q-resolution proofs treated as connected directed acyclic graphs so that the each clause in the proof corresponds to some node p_n labeled with that clause. We assume that the input clauses are already $\forall$-reduced. Q-resolution steps are depicted as on the right. Note that $\forall$-reduction corresponds to a separate node. A proof system P_1 *p-simulates* a proof system P_2 iff any proof in P_2 of a formula Φ can be translated into a proof in P_1 of Φ in polynomial time (c.f. [11,22]).

2   Expansions

Modern SAT solvers can be easily used in a black box setting which suggests a straightforward approach to solving QBF by expanding variables until only one type of quantifier is left; at that point a SAT solver can be invoked. Here we are assuming the mainstream type of a SAT solver that accepts formula in CNF and produces resolution proofs for unsatisfiable inputs.

Existential quantification can be expanded by the equivalence $\exists x. \Phi = \Phi[x/0] \vee \Phi[x/1]$ and universal quantification by the equivalence $\forall x. \Phi = \Phi[x/0] \wedge \Phi[x/1]$. These equivalences reveal two main obstacles to developing a calculus using both expansion and plain resolution (besides the exponential growth). The first obstacle is that the result of an expansion is not in prenex form; this can be overcome by prenexing the expansion. The second obstacle is that the result of expanding the existential quantifier does not yield CNF. Hence, in this paper we focus only on expansion of the universal quantifier. We show that this limitation still leads to a refutation complete calculus with many interesting properties.

Expansion of universal quantifiers enables decreasing the number of quantifiers and maintain prenex normal form at the cost of introducing fresh variables. For instance, expanding $\exists x \forall y \exists z. \phi$ yields $\exists x. (\exists z. \phi[y/0]) \wedge (\exists z. \phi[y/1])$. To get back to prenex form, we add two fresh copies of z, one for the sub-QBF where $y = 0$ and one for the sub-QBF where $y = 1$, thus obtaining $\exists x z^0 z^1. \phi[y/0, z/z^0] \wedge \phi[y/1, z/z^1]$.

A significant drawback of expansion is that the formula grows in size exponentially. This effect can be mitigated by observing that only *partial expansions* may be sufficient to show unsatisfiability. For instance, for the formula $\forall y \exists x. (y \vee x) \wedge (y \vee \bar{x})$ it is sufficient to consider an expansion with $y/0$ to show the formula false. Another source of rapid growth lies in the number of the formula's quantification levels. Expanding y in $\exists x \forall y \exists z \forall u \exists w. \phi$ yields $\exists x. (\exists z \forall u \exists w. \phi[y/0]) \wedge (\exists z \forall u \exists w. \phi[y/1])$. We could again prenex all variables but since we are aiming at eventually expanding *all* universal variables, we can expand more carefully by prenexing only z: $\exists x z^0 z^1. \forall u \exists w. \phi[y/0, z/z^0] \wedge \forall u \exists w. \phi[y/1, z/z^1]$. Such expansion gives us a finer control over the expansion process (see [15, Sec. 3.1] for more detailed discussion). If for instance now we wish to expand u as 1 in the first sub-formula and 0 in the second sub-formula we obtain the following:

$$\exists x z^0 z^1 w^{01} w^{10}. \phi[y/0, z/z^0, u/1, w/w^{01}] \wedge \phi[y/1, z/z^1, u/0, w/w^{10}]$$

Consider a general QBF $\Phi = \forall \mathcal{U}_1 \exists \mathcal{E}_2 \ldots \forall \mathcal{U}_{2N-1} \exists \mathcal{E}_{2N}. \phi$ (WLOG we start with a universal quantifier to simplify notation). For succinctness reasons, from now on Φ refers to this formula.

An expansion consists of expanding variables $\mathcal{U}_1$ with some values and introducing fresh variables for $\mathcal{E}_2$ variables yielding a sub-QBF for each considered assignment to the $\mathcal{U}_1$ variables. These sub-QBFs are recursively expanded in an analogous fashion. Note that if we expanded from the highest quantification level (innermost level), we would lose the structural information, which is enabling the above-mentioned finer expansion steps. The following definitions formalize this process.

Definition 1 ($\forall$-expansion tree). *A $\forall$-expansion tree is a rooted tree $\mathcal{T}$ such that each path $p_0 \xrightarrow{\tau_1} p_1 \ldots \xrightarrow{\tau_N} p_N$ in $\mathcal{T}$ from the root p_0 to some leaf p_N has exactly N edges and each edge $p_{i-1} \xrightarrow{\tau_i} p_i$ is labeled with a total assignment τ_i to the variables $\mathcal{U}_{2i-1}$, for $i \in 1..N$. Each path in $\mathcal{T}$ is uniquely determined by its labeling.*

Convention Since paths from the root in an $\forall$-expansion tree are uniquely determined by the labeling of the edges, i.e. assignments, we treat paths and the union of the appropriate assignments interchangeably.

Definition 2 ($\forall$-expansion). *Let $\mathcal{T}$ be a $\forall$-expansion tree. For a root-to-leaf path P in $\mathcal{T}$ and a clause C, the following rules define $\forall$-expansion of C by P, $\forall$-expansion of ϕ*

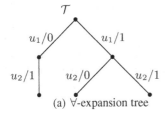

(a) ∀-expansion tree

$\Psi = \forall u_1 \exists e_1 \forall u_2 \exists e_2.\,\psi$	$\mathscr{E}(\mathcal{T},\Psi)$
$u_1 \vee e_1 \vee \bar{u}_2 \vee e_2$	$e_1^{u_1/0} \vee e_2^{u_1/0,u_2/1}$
$u_1 \vee e_1 \vee \bar{u}_2 \vee \bar{e}_2$	$e_1^{u_1/0} \vee \bar{e}_2^{u_1/0,u_2/1}$
$u_1 \vee \bar{e}_1$	$\bar{e}_1^{u_1/0}$
$\bar{u}_1 \vee \bar{e}_1 \vee \bar{u}_2 \vee \bar{e}_2$	$\bar{e}_1^{u_1/1} \vee \bar{e}_2^{u_1/1,u_2/1}$
$\bar{u}_1 \vee \bar{e}_1 \vee \bar{u}_2 \vee e_2$	$\bar{e}_1^{u_1/1} \vee e_2^{u_1/1,u_2/1}$
$\bar{u}_1 \vee e_1 \vee u_2 \vee \bar{e}_2$	$e_1^{u_1/1} \vee \bar{e}_2^{u_1/1,u_2/0}$
$\bar{u}_1 \vee e_1 \vee u_2 \vee e_2$	$e_1^{u_1/1} \vee e_2^{u_1/1,u_2/0}$

(b) ∀-expansion

Fig. 1. Example expansion tree and its application

by P, and ∀-expansion of Φ by $\mathcal{T}$. These expansions are denoted as $\mathscr{E}(P,C)$, $\mathscr{E}(P,\psi)$, and $\mathscr{E}(\mathcal{T},\Phi)$, respectively.

1. For each path P_k in $\mathcal{T}$ from the root, labeled by assignments $\tau_1,\dots,\tau_k$, and an existential variable x with $\mathsf{lv}(x) = 2k$ define a fresh variable $x^{\tau_1,\dots,\tau_k}$.
2. For each path P in $\mathcal{T}$ from the root to some leaf labeled by $\tau_1,\dots,\tau_N$, and a clause $C \in \phi$ define $\mathscr{E}(P,C)$ as $C[\tau_1 \cup \dots \tau_N \cup \tau_R]$ where

$$\tau_R = \{x/x^{\tau_1,\dots,\tau_k} \mid 1 \le k \le N, x \text{ an existential variable s.t. } \mathsf{lv}(x) = 2k\}$$

3. For each path P in $\mathcal{T}$ from the root to some leaf define $\mathscr{E}(P,\phi)$ as a union of $\mathscr{E}(P,C)$ for $C \in \phi$.
4. Define $\mathscr{E}(\mathcal{T},\Phi)$ as the union of all $\mathscr{E}(P,\phi)$ for each root-to-leaf path P in $\mathcal{T}$.

Example 1. Figure 1(a) shows an example of a ∀-expansion tree and Figure 1(b) shows a ∀-expansion of some formula Ψ based on this tree. The expansion considers both values of u_1 but only the value 1 is considered for u_2 when $u_1 = 0$. The tree has 3 leafs so the formula could potentially grow 3 times. But because the formula is very simple, for each clause C there is only a single path P from the root to some leaf for which $\mathscr{E}(P,C) \ne 1$. Hence, the expansion has the same size as the original formula. Note that there are as many copies of e_2 as there are leafs in the expansion tree ($e_2^{u_1/0,u_2/1}$, $e_2^{u_1/1,u_2/0}$, $e_2^{u_1/1,u_2/1}$) but only two copies of e_1 ($e_1^{u_1/0}$, $e_1^{u_1/1}$).

Definition 3 (∀Exp+Res). ∀Exp+Res refutation *for* Φ is a pair $(\mathcal{T},\pi)$ where $\mathcal{T}$ is a ∀-expansion tree for Φ and π is a resolution refutation for $\mathscr{E}(\mathcal{T},\Phi)$. A size of $(\mathcal{T},\pi)$, denoted $|(\mathcal{T},\pi)|$, is the sum of the numbers of nodes in $\mathcal{T}$ and π.

Note that for a ∀-expansion $\mathcal{T}$ the size of $\mathscr{E}(\mathcal{T},\Phi)$ is bounded by the number of leafs of $\mathcal{T}$ times the size of the matrix ϕ. Therefore a ∀Exp+Res refutation can be validated in polynomial time.

Theorem 1. *A formula Φ is false iff there exists a ∀Exp+Res refutation for Φ.*

Proof. If Φ is false, consider $\mathcal{T}_{\text{full}}$ capturing a full expansion of all of the quantifiers. More precisely, each node p_i of $\mathcal{T}_{\text{full}}$ at depth i (with the root being at depth 0) has $2^{|\mathcal{U}_{2i+1}|}$ children, each corresponding to a total assignment to variables $\mathcal{U}_{2i+1}$. Since this expansion mirrors semantics of QBF, $\mathscr{E}(\mathcal{T}_{\text{full}}, \Phi)$ is false iff Φ is false.

Throughout the $\forall$-expansion process, (sub-)QBFs $\forall \mathcal{U}. \Psi$ are replaced with the conjuncts $\Xi = \bigwedge_{\tau \in \omega} \Psi[\tau]$ for some ω, a set of total assignments to $\mathcal{U}$. Since Ξ is equivalent to $\forall \mathcal{U}. \Psi$ when ω is the set of *all* assignments, it is weaker if ω is a set of only some total assignments, i.e. $(\forall \mathcal{U}. \Psi) \to \Xi$. Consequently $\Phi \to \mathscr{E}(\mathcal{T}, \Phi)$ for any $\forall$-expansion tree $\mathcal{T}$. Therefore, if $\mathscr{E}(\mathcal{T}, \Phi)$ is false, then Φ is false. $\square$

3 Simulating Tree Q-Resolution by ∀Exp+Res

Consider a tree Q-resolution refutation π of Φ. Our objective is to construct a $\forall$Exp+Res refutation $(\mathcal{T}, \pi')$ based on π. We should stress that DPLL-based solvers enable producing non-tree Q-resolution proofs due to learning [23]. Hence, this proof is *not* a proof of the fact $\forall$Exp+Res can simulate DPLL-based solving in general.

We will construct $\mathcal{T}$ and π' so that π' will share its basic structure with π but with universal variables removed and existential variables renamed (according to the definition of $\mathscr{E}$). We observe that if π consists of a single node $\bot$, $\mathcal{T}$ and π' are easily constructed by setting $\mathcal{T}$ to the empty tree and setting π'. Therefore, from now on, we assume that all leafs of π are labeled with nonempty clauses. For the sake of succinctness, in this section, π always refers to the given Q-resolution proof that we wish to translate to a $\forall$Exp+Res refutation.

We first observe that if two clauses $x \vee C_1$ and $\bar{x} \vee C_2$ are resolved in π, the $\forall$-expansion tree being constructed must ensure that x is substituted by the same fresh variable x' in both clauses so that the same resolution step can be carried out in π' on variable x'. The literals x and $\bar{x}$ can appear inside the Q-resolution tree π only if they were introduced by some of its leafs. Consequently, the corresponding leafs of the resolution tree π' must contain the same copy of x. This observation motivates the construction. In the first phase of the construction, we identify sets of leafs of π where a certain existential variable must be substituted by the same fresh copy. In the second phase we construct a $\forall$-expansion tree $\mathcal{T}$ that will respect the sets identified in the first phase. The $\forall$-expansion tree $\mathcal{T}$ will provide us with the leafs of π'.

Consider a resolution step in π on some variable x corresponding to nodes p_1 and p_2 with the resolvent (parent) node r. Let C_1, C_2, and C_r be the clauses labeling p_1, p_2, and r, respectively. Hence, $C_r = C_1 \cup C_2 \setminus \{x, \bar{x}\}$ (recall that $\forall$-reduction is modeled as a separate step). Let D be the set of universal literals $l \in C_1 \cup C_2$ such that $\mathsf{lv}(l) < \mathsf{lv}(x)$. Let S be the set of leafs p of π such that there is a path from either p_1 or p_2 to p for which all clauses on the path contain the variable x (including the clause labeling p). Record the quadruple (r, x, D, S). In the following text we write $\mathcal{Q}_\pi$ to denote the set of quadruples generated for each resolution step in π.

Consider any two leafs p_1, p_2 of π s.t. $p_1, p_2 \in S$ for some $(r, x, D, S) \in \mathcal{Q}_\pi$. Once we ensure that x is replaced with the same fresh copy in the clauses labeling p_1 and p_2, the plain resolution refutation π' is easy to construct.

Proposition 1. *Let $\mathcal{T}$ be a $\forall$-expansion tree of Φ and let M be a total mapping from the leafs of π to paths of $\mathcal{T}$. If the following conditions $\mathscr{C}_1$–$\mathscr{C}_3$ hold for $\mathcal{T}$ and M, then there is a resolution refutation π' of $\mathscr{E}(\mathcal{T}, \Phi)$ linear in size of π.*

($\mathscr{C}_1$) *If p is a leaf of π, then $M(p)$ is a path from the root to some leaf in $\mathcal{T}$.*

($\mathscr{C}_2$) *If p is a leaf of π, labeled by a clause C, and $M(p) = P$, then P assigns to 0 all universal literals of C.*

($\mathscr{C}_3$) *If leafs p_1, p_2 of π appear in the same S for some quadruple $(r, x, D, S) \in \mathcal{Q}_\pi$, $M(p_1) = P_1$, and $M(p_2) = P_2$, then P_1 and P_2 assign the same value to all universal variables with level $l < \text{lv}(x)$.*

Proof. We construct π' from π in the leaf-to-root direction; during this construction we mark each node of p' in π' as *corresponding* with some node p in π. The construction follows the following rules $\mathscr{R}_l, \mathscr{R}_r, \mathscr{R}_u$.

($\mathscr{R}_l$) For each leaf p in π labeled with C create a leaf $p' \in \pi'$ labeled with $\mathscr{E}(M(p), C)$; mark p and p' as corresponding.

($\mathscr{R}_r$) Let r be a node, with children p_1, p_2 labeled C, C_1, and C_2, respectively, where $C = C_1 \cup C_2 \setminus \{x, \bar{x}\}$. Further, consider the nodes p_1' and p_2' corresponding to p_1 and p_2, respectively, and their respective labels C_1' and C_2'. If there is a literal $x^P \in C_1' \cup C_2'$ for some P, create a node r' in π' and label it with $C' = C_1' \cup C_2' \setminus \{x^P, \bar{x}^P\}$. Mark r and r' as corresponding.

($\mathscr{R}_u$) Let p_u be node in π with a single child r labeled C_u and C_r, respectively, where C_u is a result of $\forall$-reduction of C_r. If p_r corresponds to p_r' mark p_u and p_r' also corresponding.

By induction on resolution depth, we show that the above construction results in a valid resolution tree π'. Additionally we prove, that if p' in π', labeled with a clause C', corresponds to some p in π, labeled with a clause C, then for any existential literal $l \in C$, with $\text{var}(l) = x$ there is one and only one literal $l' \in C'$ s.t. $\text{var}(l') = x^P$, for some P, and, the literals l, l' have the same polarity. Consequently, the root of π' must be labeled with the empty clause.

Rule $\mathscr{R}_l$ is well-defined due to conditions ($\mathscr{C}_1$) and ($\mathscr{C}_2$); it establishes the induction hypothesis due to definition of $\mathscr{E}$. For rule $\mathscr{R}_r$ we first observe that there must be a $x^{P_1} \in C_1' \cup C_2'$, for some P_1, from the induction hypothesis because $x \in C_1 \cup C_2$. WLOG let $x^{P_1} \in C_1'$. From induction hypothesis we also have, $x \in C_1$, $\bar{x} \in C_2$, and $\bar{x}^{P_2} \in C_2'$ for some P_2. Since C_1' and C_2' were obtained by valid resolution steps, there must be a path in π' from some leaf p_{l_1}' to p_1' where all clauses contain the literal x^{P_1}; analogously there a is path in π' from some leaf p_{l_2}' to p_2' where all clauses contain the literal $\bar{x}^{P_2}$. Both paths correspond to some paths from p_{l_1} to p_1 and p_{l_2} to p_2 in π. Hence, $p_{l_1}, p_{l_2} \in S$ for some $(r, x, D, S) \in \mathcal{Q}_\pi$. Due to condition ($\mathscr{C}_3$), the variable x must be substituted with the same copy in the leafs and therefore also $P_1 = P_2$. Because $x^{P_1} \in C_1'$ and $\bar{x}^{P_1} \in C_2'$, the resolution step on C_1' and C_2' is possible. It remains to be shown that the resolution step does not introduce more than one copy of some literal. Assume that there are literals y^{R_1} and y^{R_2} in C_1' and C_2', respectively, where $y \neq x$. From induction hypothesis, $y \in C_1$ and $y \in C_2$. Consequently, there are some leafs p_{l_1}, p_{l_2} of π s.t. y appears in all clauses on the paths from p_{l_1} to p_1 and from p_{l_2} to p_2. Because π is a refutation proof, y gets eventually resolved away. Therefore there is some $(r_y, y, D_y, S_y) \in \mathcal{Q}_\pi$ for which $p_{l_1}, p_{l_2} \in S_y$ and therefore $R_1 = R_2$ from

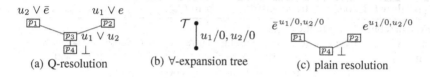

Fig. 2. Examples

condition ($\mathscr{C}_3$). Rule $\mathscr{R}_u$ preserves the induction hypothesis as universal reduction does not modify the set of existential literals. $\qquad\square$

Example 2. Consider $\forall u_1 u_2 \exists e. (u_1 \vee e) \wedge (u_2 \vee \bar{e})$ with the Q-resolution refutation in Figure 2(a), which induces a single quadruple $(p_4, e, \{u_1, u_2\}, \{p_1, p_2\})$. To obtain a $\forall$Exp+Res refutation, generate the single-branch tree $\mathcal{T}$ in Figure 2(b) and mapping M with $M(p_1) = M(p_2) = \{u_1/0, u_2/0\}$ yielding the $\forall$-expansion $e^{u_1/0, u_2/0} \wedge \bar{e}^{u_1/0, u_2/0}$ with the corresponding resolution tree Figure 2(c). Observe that conditions $\mathscr{C}_1$–$\mathscr{C}_3$ from Proposition 1 are fulfilled. Clauses participating in the Q-resolution step are expanded so that e is replaced with the same copy. The universal literals u_1, u_2 are assigned to 0 by the expansion. Consequently, this Q-resolution step can be reproduced in a plain resolution refutation. Note that universal reduction steps are unnecessary in the resolution refutation since expansions remove all universal literals.

3.1   Construction of $\mathcal{T}$ and M

Proposition 1 gives us conditions $\mathscr{C}_1$–$\mathscr{C}_3$ on a $\forall$-expansion tree $\mathcal{T}$ and a mapping M so that any $\mathcal{T}$ and M satisfying these conditions enable us to construct the desired plain-resolution refutation π' for $\mathscr{E}(\mathcal{T}, \Phi)$. This subsection shows that such $\mathcal{T}$ and M can be constructed for any given Q-resolution refutation π.

For a quadruple $q = (r, x, D, S) \in \mathcal{Q}_\pi$ we say that q *is at level* $\mathsf{lv}(x)$ and we say that a leaf p of π *is in* q iff $p \in S$. Recall that the intuition behind a quadruple $(r, x, D, S) \in \mathcal{Q}_\pi$ is that the expanded counterparts of clauses labeling the leafs in S will contain the same fresh copy of x. Further, the assignment used for the expansion must assign to 0 the universal literals in those clauses. This poses the following question: *If some leaf p of π is in two different quadruples $q_1, q_2 \in \mathcal{Q}_\pi$, how do we ensure that the conditions are not conflicting?*

We say that $(r, x, D, S), (r', x', D', S') \in \mathcal{Q}_\pi$ are *connected* iff $S \cap S' \neq \emptyset$. We say that leafs p_1, p_2 of π *share level* k iff there exists a sequence (with possible repetitions) of quadruples $q_1, \ldots, q_n \subseteq \mathcal{Q}_\pi$, s.t. p_1 is in q_1; p_2 is in q_n; each q_i in the sequence has a level $l \geq k$; and each two adjacent quadruples are connected.

Observation 1. *The relation "share level k" is an equivalence relation on the leafs of π. All leafs of π share level 2 (recall that existential variables start at level 2). If two leafs share level k, then they share a level $l \leq k$.*

Let us look more closely at quadruples that share some level k. Recall that the given Φ formula has the prefix $\forall \mathcal{U}_1 \exists \mathcal{E}_2 \ldots \forall \mathcal{U}_{2N-1} \exists \mathcal{E}_{2N}$. Consider two connected quadruples

---

Algorithm 1. Expansion tree construction from $\mathcal{Q}_\pi$

---

1 **Function** Build (k, StopLev, L)
   in : StopLev..base-case level, $k \leq$ StopLev..current level, L..subset of leafs of π
   out: a pair $(\mathcal{T}', M')$, where $\mathcal{T}'$ is an expansion tree for universal variables with
        level $\geq k$, M' is a mapping from leafs in L to root-to-leaf paths in $\mathcal{T}'$

2 **begin**
3   **if** $k =$ StopLev **then**
4       $\mathcal{T}' \leftarrow$ create a tree with a single node, the root r
5       $M' \leftarrow$ map all nodes in L to the empty path starting in r
6   **return** $(\mathcal{T}', M')$
7    $T' \leftarrow$ a tree with the root node r
8    $M' \leftarrow$ empty mapping
9    $\Xi \leftarrow$ partition nodes L by the "share level $k + 1$" relation
10   **foreach** $\rho \in \Xi$ **do**
11       $Q_\rho \leftarrow \{q \in \mathcal{Q}_\pi \mid$ there exists $p \in \rho$ in q, q is at level $> k\}$
12       $D_\rho \leftarrow \{l \mid (p, e, D, S) \in Q_\rho, l \in D, \mathsf{lv}(l) = k\}$
13       $\tau_\rho \leftarrow \{u/0 \mid u \in D_\rho\} \cup \{u/1 \mid \bar{u} \in D_\rho\} \cup \{u/0 \mid u, \bar{u} \notin D_\rho, \mathsf{lv}(u) = k\}$
14       $(\mathcal{T}_\rho, M_\rho) \leftarrow$ Build($k + 2$, StopLev, ρ)
15   add $\mathcal{T}_\rho$ to $\mathcal{T}'$, connect r to the root of $\mathcal{T}_\rho$ with an edge labeled with τ_ρ
16   if M_ρ maps a leaf $p \in L$ to τ, map p to $\tau_\rho \cup \tau$ in M'
17   **return** $(\mathcal{T}', M')$

$(r, x, D, S), (r', x', D', S') \in \mathcal{Q}_\pi$, both at some level $\geq k$, i.e. $\mathsf{lv}(x) \geq k$ and $\mathsf{lv}(x') \geq k$. Our objective is to build such mapping M that for any two $p_1, p_2 \in S$, the paths $M(p_1)$ and $M(p_2)$ share the prefix of length $\mathsf{lv}(x)/2$ corresponding to assignments to variables $\mathcal{U}_1\mathcal{U}_2 \ldots \mathcal{U}_{\mathsf{lv}(x)-1}$; this ensures that x is renamed to the same fresh copy in clauses of the leafs. The same holds for leafs in S'. Since the quadruples are connected, there is some leaf p that belongs to $p \in S \cap S'$. Further, since both x and x' are at a level greater or equal to k, by transitivity, *all* leafs in $S \cup S'$ must be mapped to such paths of the $\forall$-expansion tree $\mathcal{T}$ that they share their prefixes of length $k/2$. This immediately generalizes to sequences of connected quadruples. If two leafs p_1, p_2 of π share level $k = 2l$, then $M(p_1)$ and $M(p_2)$ must have common prefix of length l, corresponding to assignments to variables $\mathcal{U}_1\mathcal{U}_2 \ldots \mathcal{U}_{k-1}$.

This observation motivates Algorithm 1, which is represented as a recursive function. The recursion is initiated by the call Build($1, 2N + 1, L_{\text{all}}$) where L_{all} is the set of leafs of π. After this initial call terminates, any root-to-leaf paths with the same labeling in the returned tree are merged to obtain the required $\mathcal{T}$.

The function returns $\mathcal{T}'$, a subtree of the tree $\mathcal{T}$ being constructed, and a mapping M' that maps the given leafs L to paths of $\mathcal{T}'$. The labeling of root-to-leaf paths in $\mathcal{T}'$ are total assignments to variables $\mathcal{U}_k, \mathcal{U}_{k+2}, \ldots, \mathcal{U}_{2N-1}$, where k is an odd natural number. Hence, for the base case of the recursion, i.e. $k = 2N + 1$, the function creates a single-node tree $\mathcal{T}'$ and maps all given leafs L to an empty path starting and ending in the root of $\mathcal{T}'$.

For the non-base case, the function partitions the given leafs L of π by the "share level $k + 1$" relation. From the conditions on $\mathcal{T}$, clauses labeling leafs that share level $k + 1$ must be expanded such that existential variables with level $> k$ are replaced with the same copies. At the same time, the universal literals in these clauses with level $\leq k$ must be assigned to 0. The algorithm visits each partition ρ of the "share level $k + 1$" partition and collects quadruples $q \in \mathcal{Q}_\pi$ for which there is some leaf $p \in \rho$ in q. Subsequently, it collects all universal literals at level k that appear in these quadruples and computes an assignment τ_ρ which assigns them to 0 and other literals assigns arbitrarily (line 14).

Example 3. Consider the following Q-resolution proof π with the prefix $\forall u_1 \exists e_2 \forall u_3 \exists e_4$.

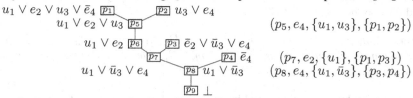

$$(p_5, e_4, \{u_1, u_3\}, \{p_1, p_2\})$$

$$(p_7, e_2, \{u_1\}, \{p_1, p_3\})$$
$$(p_8, e_4, \{u_1, \bar{u}_3\}, \{p_3, p_4\})$$

This yields the quadruples depicted on the right hand side. All leafs share level $1 + 1$ and are put into a single partition $\rho = \{p_1, p_2, p_3, p_4\}$ labeled with $\{u_1/0\}$. Based on sharing of level $3 + 1$, ρ is split into $\{p_1, p_2\}$ and $\{p_3, p_4\}$, labeled $\{u_3/0\}$ and $\{u_3/1\}$, respectively. The resulting mapping is $M(p_1) = M(p_2) = \{u_1/0, u_3/0\}$ and $M(p_3) = M(p_4) = \{u_1/0, u_3/1\}$.

Let us now focus on the correctness of Algorithm 1. The algorithm is terminating because the set of quadruples $\mathcal{Q}_\pi$ is finite. That the algorithm constructs mapping M and the tree $\mathcal{T}$ satisfying the conditions $(\mathscr{C}_1)$–$(\mathscr{C}_3)$ of Proposition 1 hinges on proving that the set of literals D_ρ (line 12) does not contain complementary literals. Consequently, that the assignment τ_ρ (line 14) is indeed an assignment. For now we assume that this holds and show it later in order to first focus on the overall workings of the algorithm.

Since π has no empty clauses in leafs and all input clauses are $\forall$-reduced, every leaf p labeled with some clause C must be in some quadruple in $\mathcal{Q}_\pi$. At each level k, quadruples are partitioned so eventually there will be one and only one path P in $\mathcal{T}$ s.t. $M(p) = P$. Thus satisfying condition $(\mathscr{C}_1)$ of Proposition 1. If C contains some universal literal l with $\mathsf{lv}(l) = k$, l must be blocked by some existential literal $b \in C$ with $\mathsf{lv}(b) > k$. This literal b is eventually resolved away and therefore there must be a quadruple $q_b = (r, \mathsf{var}(b), D_b, S_b) \in \mathcal{Q}_\pi$ s.t. $p \in S_b$. Since b blocks l on a path from p to some child of r, it also holds that $l \in D_b$. Hence $q_b \in Q_\rho$, defined on line 11, and $l \in D_\rho$, defined on line 12. The algorithm places p into a subtree prepended by an edge labeled with τ_ρ, which sets l to 0. Thus satisfying condition $(\mathscr{C}_2)$. Consider two leafs p_1, p_2 of π such that they are in the same quadruple q at some level l. These leafs are connected at level $\leq l$. Hence they will be part of the same partition for levels $k < l$. Therefore, the algorithm puts the leafs in the same subtree while $k < l$ and therefore $M(p_1)$ and $M(p_2)$ assign the same value to all universal variables with level $k < l$ thus satisfying condition $(\mathscr{C}_3)$.

Now it remains to be shown that the set D_ρ constructed on line 12 is not contradictory. This will be shown in Lemma 5. However, before we reach this lemma, a series of auxiliary lemmas need to be derived. Since Q-resolution enables resolving two clauses $C_1 \vee x$ and $C_2 \vee \bar{x}$ only if $C_1 \cup C_2$ does not contain complementary literals, we can make the following observation.

Observation 2. *For any $(r, x, D, S) \in \mathcal{Q}_\pi$, the literals D are noncontradictory.*

Lemma 1. *If any two quadruples $(r_1, x_1, D_1, S_1), (r_2, x_2, D_2, S_2) \in \mathcal{Q}_\pi$ are connected, then r_1 dominates r_2, i.e. r_2 is in a subtree of r_1, or r_2 dominates r_1.*

Proof. Since the quadruples are connected, there is some leaf p_l of π s.t. $p_l \in S_1$ and $p_l \in S_2$. At the same time there is an undirected path from both r_1 and r_2 to p_l. If neither r_1 dominated r_2 nor r_2 dominated r_1 there would be a cycle from root to r_1, p_l, r_2, and back to the root. □

Lemma 2. *Consider any two quadruples $(r_1, x_1, D_1, S_1), (r_2, x_2, D_2, S_2) \in \mathcal{Q}_\pi$ such that r_1 dominates r_2 and r_2 dominates some $p_l \in S_1$. Then all the clauses on the path from r_1 to r_2 except for r_1 contain a literal $b \in \{x_1, \bar{x}_1\}$.*

Proof. Since the leaf p_l is dominated by both r_1 and r_2, there is a path from the root of π going through r_1, r_2, and ending in p_l. Since $p_l \in S_1$, from definition of the quadruples, there is a literal $b \in x_1, \bar{x}_1$ that appears everywhere on the path except for the node r_1. □

The following lemma shows that for any sequence of connected quadruples that are all at some level $l \geq k$, there is a quadruple pertaining to a resolution node r such that r dominates all the other resolution nodes in the sequence, and, all paths from this node to these resolution nodes contain some existential literal b with $\mathsf{lv}(b) \geq k$. Consequently, these literals block all universal literals with level $l < k$ on these paths.

Lemma 3. *Consider a sequence of quadruples $\gamma = q_1, \ldots, q_n$, such that each $q_i \in \mathcal{Q}_\pi$ in the sequence has a level $l \geq k$ and each two adjacent quadruples are connected. Then there is $(r, x, D, S) \in \gamma$ such that for any quadruple $(r_j, x_j, D_j, S_j) \in \gamma$ the node r dominates r_j and all the clauses on the path from r to r_j, except for r, contain some existential literal b with $\mathsf{lv}(b) \geq k$.*

Proof. Proof by induction on the length of prefix of γ. For the base case choose (r, x, D, S) as q_1. For the inductive case consider $i > 1$ and $q' = (r', x', D', S')$ from the induction hypothesis such that q' satisfies the condition for $q_1, \ldots, q_{i-1}$. Since adjacent quadruples are connected, for $q_i = (r_i, x_i, D_i, S_i)$ and $q_{i-1} = (r_{i-1}, x_{i-1}, D_{i-1}, S_{i-1})$ there is a leaf $p_c \in S_{i-1} \cap S_i$. Split on the following cases.

If q_i is equal to any of the q_j for $j < i$, choose (r, x, D, S) to be q'. If r_i dominates r' then invoke Lemma 2 whose preconditions are satisfied because r_i dominates r' and r' dominates p_c, from the induction hypothesis. Hence there is a path from one of the children of r_i to r containing the literal $b \in \{x_i, \bar{x}_i\}$. Note that b does not appear in r_i but does appear in r'. From induction hypothesis, for any r_j, $j < i$ there is a path from a child of r' to r_j where each clause is blocked by some literal with level $l \geq k$. Concatenating the path from r_i to r' with the path r' to r_j satisfies the condition for j. Choose (r, x, D, S) to be q_i.

From Lemma 1, either r_i is dominated by r_{i-1} or r_{i-1} is dominated by r_i. Hence we need to consider only these two remaining cases. If r_{i-1} dominates r_i, then from Lemma 2 there is a $b_{i-1} \in \{x_{i-1}, \bar{x}_{i-1}\}$ that appears on the path from one of the children of r_{i-1} to r_i (inclusively). From induction hypothesis, there is a path from r' to r_{i-1}, excluding r', that contains some existential literals b with $\mathsf{lv}(b) \geq k$. Concatenating this path with the path from r_{i-1} to r_i gives us a path satisfying the required condition for the node r_i. In particular, there is a path from a child of r' to r_i such that each clause on the path contains a some existential literals b with $\mathsf{lv}(b) \geq k$.

If r_{i-1} is dominated by r_i and r_i does not dominate r', then r' must dominate r_i otherwise there would be a cycle from the root to r', r_{i-1}, r_i, and back to root. From induction hypothesis, each clause on the path from r' to r_{i-1} contains some existential literal b with $\mathsf{lv}(b) \geq k$. Since r' dominates r_i, which in turn dominates r_{i-1}, the path from r' to r_i is a prefix of the path from r' to r_{i-1} and therefore also satisfies the required condition. Choose (r, x, D, S) to be q'.   □

Lemma 4. *Consider ρ a subset of leafs of π that is an equivalence class of the share level $k + 1$ relation for some odd number k. Define $Q_\rho^k \subseteq Q_\pi$ as follows.*

$$Q_\rho^k = \{(r, x, D, S) \in Q_\pi \mid p \in \rho, p \in S, \mathsf{lv}(x) > k\}$$

Then for any $q_a, q_b \in Q_\rho^k$ there is a sequence of quadruples $q_1, \ldots, q_m$ where $q_a = q_1$, $q_b = q_m$, each q_i is at a level $> k$ and $q_i \in Q_\rho^k$, and each two adjacent $q_i, q_{i+1} \in Q_\rho^k$ are connected.

Proof. From definition of Q_ρ^k there are leafs $p_a, p_b \in \rho$ s.t. $p_a \in q_a$, $p_b \in q_b$. Since ρ is an equivalence class of *share level $k + 1$* relation, there is a sequence of connected quadruples $s_1, \ldots, s_n$ such that p_a is in s_1 and p_b is in s_n, and each quadruple in the sequence is at a level $> k$. Since for any $s_i = (r_i, x_i, D_i, S_i)$, the set S_i is non-empty, all leafs $p \in S_i$ share level $k + 1$ with p_a and $p \in \rho$. Hence, all the quadruples s_i in the sequence are in Q_ρ^k. Since q_a and s_1 are connected because of p_a and q_b and q_b are connected because of p_b, constructing the sequence $q_a, s_1, \ldots, s_n, q_b$ yields the required sequence.   □

Lemma 5. *Let k, ρ, and Q_ρ^k be defined as in Lemma 4. Define a set of literals D_ρ^k as $D_\rho^k = \{l \mid (r, x, D, S) \in Q_\rho^k, \mathsf{lv}(l) = k, l \in D\}$. The set D_ρ^k does not contain complementary literals.*

Proof. Lemma 4 gives us that Q_ρ^k can be organized into a sequence γ where each two adjacent quadruples are connected and each $q_i \in \gamma$ is at a level $> k$. From Lemma 3 there is a quadruple $(r_d, x_d, D_d, S_d) \in \gamma$ s.t. for any quadruple $(r_j, x_j, D_j, S_j) \in \gamma$ the node r_d dominates r_j and all the clauses on the path from r_d to r_j, except for r_d, contain some existential literal b with $\mathsf{lv}(b) > k$. Hence, no universal literals with level $l \leq k$ can be $\forall$-reduced on a path from r_j to r_d in π. Therefore necessarily, D_d contains all literals D_j. Consequently, $D_\rho^k \subseteq D_d$. From Observation 2, the set D_d is noncontradictory and therefore D_ρ^k is also noncontradictory.   □

This last lemma gives us what we needed to conclude the correctness of Algorithm 1, i.e. that the set of literals D_ρ, constructed on line 12 is not contradictory. Algorithm 1 operates in time polynomial to the size of π because the size of the set $\mathcal{Q}_\pi$ is linear to the size of π and partitioning by "share level $k+1$" relation can be done in polynomial time. This fact, together with Proposition 1 lets us derive the following.

Theorem 2. *For any tree Q-resolution refutation π there exists a $\forall$Exp+Res refutation $(\mathcal{T}, \pi_\mathcal{T})$ s.t. both $\mathcal{T}$ and $\pi_\mathcal{T}$ are polynomial in size of π. This $\forall$Exp+Res refutation can be constructed in time polynomial to π. Hence, $\forall$Exp+Res p-simulates tree Q-resolution.*

4   Simulating Restricted ∀Exp+Res by Q-Resolution

This section shows that a certain *fragment* of $\forall$Exp+Res refutations can be simulated by Q-resolution. This fragment allows expansions of universal quantifiers as before but puts a restriction on the resolution proof of the expansion. In particular, it allows only resolutions that follow the order of the quantifier prefix.

Definition 4 (level-ordered). *Consider a $\forall$Exp+Res refutation $(\mathcal{T}, \pi)$ of Φ. We say that $(\mathcal{T}, \pi)$ is* level-ordered *iff the following holds. Let $x^P \vee C_1$ and $\bar{x}^P \vee C_2$ be some clauses resolved in π, then $\mathsf{lv}(y) \leq \mathsf{lv}(x)$ for any $y^{P_1} \in \mathsf{var}(C_1 \vee C_2)$.*

Lemma 6. *Let $(\mathcal{T}, \pi)$ be a level-ordered $\forall$Exp+Res refutation of Φ. Let C be some clause in π and $x_1^{P_1}, x_2^{P_2} \in \mathsf{var}(C)$. If $\mathsf{lv}(x_1) \leq \mathsf{lv}(x_2)$, then the path P_1 is a prefix of the path P_2.*

Proof. By induction on the number of resolution steps that led to C. The condition is true for the leafs of π from the definition of $\mathscr{E}$. For the induction step consider clauses $C_1 \vee \bar{x}_r^P$ and $C_2 \vee x_r^P$ with the resolvent $C = C_1 \vee C_2$. If C is empty or unit, the condition is trivially satisfied. Let $x_1^{P_1}, x_2^{P_2} \in \mathsf{var}(C)$ with $\mathsf{lv}(x_1) \leq \mathsf{lv}(x_2)$. Because π is level-ordered, $\mathsf{lv}(x_1) \leq \mathsf{lv}(x_r)$ and $\mathsf{lv}(x_2) \leq \mathsf{lv}(x_r)$, from which the induction hypothesis gives that both paths P_1 and P_2 are prefixes of the path P. Since $\mathsf{lv}(x_1) \leq \mathsf{lv}(x_2)$, then $|P_1| \leq |P_2|$ from definition of $\mathscr{E}$. Hence the path P_1 is a prefix of the path P_2.   □

Lemma 7. *Let $(\mathcal{T}, \pi)$ be a level-ordered $\forall$Exp+Res refutation of Φ. Let C be a clause in π and $x^{P_1}, x^{P_2} \in \mathsf{var}(C)$, then $P_1 = P_2$.*

Proof. Immediate consequence of Lemma 6.   □

Theorem 3. *Let $(\mathcal{T}, \pi)$ be a level-ordered $\forall$Exp+Res refutation of Φ. Then a Q-resolution refutation of Φ can be constructed in polynomial time with respect to $|(\mathcal{T}, \pi)|$. Hence, Q-resolution p-simulates level-ordered $\forall$Exp+Res.*

Proof (sketch). The proof is similar to the one of Proposition 1, i.e. we construct a Q-resolution refutation π' based on π and prove its correctness by induction on resolution depth. For each leaf p in π labeled with a clause C, there exists a path P from the root to some leaf in $\mathcal{T}$ and a clause $C' \in \phi$ such that $\mathscr{E}(P, C') = C$. Replace C with C'. Whenever there is a resolution on some variable x^P in π, perform resolution on x in π'. Add $\forall$-reduction steps after each resolution step. Effectively, the Q-resolution refutation will have the same shape as the plain resolution refutation but each variable x^P will be

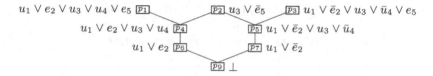

Fig. 3. Nontree Q-resolution example

replaced with the variable x ("removed superscripts"), and, some universal literals will be inserted into the clauses.

The correctness of the resulting π' follows from Lemmas 6 and 7. Lemma 7 guarantees that in the plain resolution refutation there are no clauses containing variables x^{P_1} and x^{P_2} with $P_1 \neq P_2$. Consequently, removing the superscripts does not yield complementary existential literals in clauses of π'.

It remains to be shown that there are no complementary universal literals within clauses of π'. If there's a universal literal $k \in C'$ for some clause $C' \in \pi$, there most be some existential literal $x \in C'$ that blocks it. At the same time there's a corresponding literal $x^P \in C$ for the corresponding clause in π. We observe that P assigns k to 0. For leaf clauses this follows from the definition of $\mathcal{E}$. For resolution steps this follows from the level-orderndess which guarantees that the literal being resolved on blocks all universal literals in the clause. So if there's a resolution on a x^P in π, the clauses involved in the corresponding resolution in π' may contain only universal literals that are assigned to 0 by P and therefore complementary universal literals cannot meet.

5   Examples

This section illustrates some of the practical implications of the results derived so far. Section 3 shows that *tree* Q-resolution refutations can be simulated by ∀Exp+Res refutations. This result points in the direction of formulas where ∀Exp+Res will perform significantly worse than Q-resolution. In particular, this hints that *non-tree* Q-resolution refutations might prove nontrivial to simulate for ∀Exp+Res. The following example illustrates why that is the case.

For the quantifier prefix $\forall u_1 \exists e_2 \forall u_3 u_4 \exists e_5$, Figure 3 shows a simple non-tree Q-resolution proof that demonstrates a drawback of ∀-expansion-based proofs. Assume that clauses on p_1, p_3 are expanded to some clauses C_1', C_3', respectively. The clauses will contain some copies of e_5: $e_5^{P_1} \in C_1'$, $e_5^{P_3} \in C_3'$, let's say. It must be that $P_1(u_1) = P_1(u_3) = P_1(u_4) = 0$ and $P_3(u_1) = P_3(u_3) = P_3(\bar{u}_4) = 0$ Because of the different polarity of literal u_4 in the assignments, $P_1 \neq P_3$. This means that there must be 2 different expansions of clause on p_2. Hence, formulas leading to a high level of sharing in Q-resolution are likely to be easier for DPLL-based solvers than for expansion-based solvers.

Section 4 shows that Q-resolution can simulate ∀Exp+Res refutations where the plain resolution part follows a certain variable order. Again, this points us in the direction of formulas where ∀Exp+Res might perform better than Q-resolution, i.e. formulas with proofs not respecting this order. To support this hypothesis, we construct

$\mathbf{x_i} \vee \mathbf{z} \vee \mathbf{C_i^1}$	$\mathbf{\bar{x}_i} \vee \mathbf{\bar{z}} \vee \mathbf{C_i^2}$	$\mathbf{z/0}$	$\mathbf{z/1}$
$x_1 \vee z \vee \bar{y}_1$	$\bar{x}_1 \vee \bar{z} \vee \bar{y}_1$	$x_1 \vee \bar{y}_1^{z/0}$	$\bar{x}_1 \vee \bar{y}_1^{z/1}$
$x_2 \vee z \vee y_1$	$\bar{x}_2 \vee \bar{z} \vee \bar{y}_1$	$x_2 \vee y_1^{z/0}$	$\bar{x}_2 \vee \bar{y}_1^{z/1}$
$x_3 \vee z \vee \bar{y}_1$	$\bar{x}_3 \vee \bar{z} \vee y_1$	$x_3 \vee \bar{y}_1^{z/0}$	$\bar{x}_3 \vee y_1^{z/1}$
$x_4 \vee z \vee y_1$	$\bar{x}_4 \vee \bar{z} \vee y_1$	$x_4 \vee y_1^{z/0}$	$\bar{x}_4 \vee y_1^{z/1}$

Fig. 4. Example formula for $n = 1$

the following formula[1]. Let $n \in \mathbb{N}^+$ and $H = 2^{2n}$. Consider the set of variables $y_1, \ldots, y_n, x_1, \ldots, x_H, z$ and the prefix $\exists x_1, \ldots, x_H \forall z \exists y_1, \ldots, y_n$. We construct the matrix as follows. For each $i \in 1 \ldots H$ construct two clauses of the form $x_{i+1} \vee z \vee C_i^1$, $\bar{x}_{i+1} \vee \bar{z} \vee C_i^2$, where $\mathrm{var}(C_i^1) = \mathrm{var}(C_i^2) = y_1 \ldots y_n$ and the pair C_i^1, C_i^2 goes over all the possible $2^{2n} = H$ pairs of sets of literals on the pertaining variables. More precisely, Let i_j be the jth bit of i, where $j \in 0..(2N-1)$. Add to C_i^1 the literal $\bar{y}_j$ if $i_j = 0$, where $j \in 0..(N-1)$. Add to C_i^1 the literal y_j if $i_j = 1$, where $j \in 0..(N-1)$. Add to C_i^2 the literal $\bar{y}_j$ if $i_j = 0$, where $j \in N..(2N-1)$. Add to C_i^2 the literal y_j if $i_j = 1$, where $j \in N..(2N-1)$. For the expansion we consider an expansion that includes both possible assignments: $z/0$ and $z/1$. Figure 4 shows the matrix and the expansion for $n = 1$.

While the expansion duplicates the y_i variables, it is easily shown unsatisfiable. Any total assignment to the copies of y_i variables gives a conflict and therefore a SAT solver that assigns these variables first, will need at most $2^{2n} = H$ conflicts to show unsatisfiability.

We show that this formula requires exponential computation by a conflict-driven DPLL QBF solver [23]. (However, this does not mean that there is no polynomial Q-resolution proof.) We first make the following observation.

Lemma 8. *If a CNF ψ is unsatisfiable and $|C| \geq k$ for all $C \in \psi$, then $|\psi| \geq 2^k$.*

Proof. Let $V = \mathrm{var}(\psi)$. Each clause $C \in \psi$ is 0 under $2^{|V|-|C|} \leq 2^{|V|-k}$ assignments to variables V. Since ψ is unsatisfiable, for *each* assignment τ to variables V there is a clause that is 0 under τ. By averaging $|\psi| \geq \frac{2^{|V|}}{2^{|V|-k}} = 2^k$.

A conflict-driven QBF solver first assigns the x_i variables, then z, and then y_i variables. Since long-distance resolution is not invoked in this example, clauses containing z do not give propagation while x_i variables are being assigned. Since the formula is false, after all x_i variables are assigned by some assignment τ_x, the solver eventually finds such value v_z for z that $\phi[\tau_x, z/v_z]$ is unsatisfiable. Once z is assigned a value, either all $x_i \vee z \vee C_i^1$ are satisfied or all $\bar{x}_i \vee \bar{z} \vee C_i^2$ are satisfied. For the solver to backtrack to the level of x_i variables, it must learn a clause containing only x_i variables. From Lemma 8, 2^n clauses must be used in learning this clause since this clause is a result of a resolution tree that forms a refutation proof once all z and x_i variables are removed from it. Consequently, the learned clause containing only x_i variables has at least 2^n variables. This is repeated until the set of learned clauses containing only x_i variables is unsatisfiable. Invoking again Lemma 8 gives that this must be repeated at

---

[1] The formula's generator is found at http://sat.inesc-id.pt/~mikolas/sat13

least $2^{2^n} = 2^{\sqrt{H}}$ times (exponentially more than the expansion approach). We note that QuBE7.2 [13], DepQBF [20], and, non-CEGAR version of GhostQ [17] were able solve this formula only for $n \le 3$. The expansion-based solver RAReQS [15] was able to solve the formula up to $n = 10$ (which has $1,048,587$ variables).

6 Conclusions and Future Work

This paper introduces and studies a proof system ∀Exp+Res aimed at refuting false QBFs based on expansion of universal variables and propositional resolution. Besides preprocessing [6,5] expansion of variables plays an important role in QBF solving. The solvers QUBOS [1], Nenofex [19], Quantor [4] expand universal variables from inner- to outermost levels. However, these expansions are possibly interleaved with operations for removal of existential quantifiers. In future work, we wish to investigate if these interleaved expansions give additional proving power to the solvers. The solver sKizzo [3] expands all universal quantifiers as is done in ∀Exp+Res (even though the process is called Skolemization). sKizzo expands the formula clause by clause, ignoring assignments to universal variables that satisfy the clause. So even though sKizzo does not explicitly avail of partial expansions, trivial parts of the expansion are not generated.

The solver RaReQS [16,15] constructs two types of expansions: one for universal variables and one for existential ones. For false QBFs, universal expansion eventually becomes false. Hence, the workings of RaReQS mimics the ∀Exp+Res in the case of false formulas. It should also be noted that out of the mentioned solvers, only RaReQS constructs *partial* expansions, i.e. both polarities of the expanded variable are considered in the other solvers.

It is the ability of ∀Exp+Res to expand partially that was crucial in showing that ∀Exp+Res can p-simulate *tree* Q-resolution refutations. In the opposite direction, we showed that Q-resolution can polynomially simulate ∀Exp+Res if the plain resolution part follows certain order of variables.

Hence, at this point it remains open how unrestricted ∀Exp+Res compares to unrestricted Q-resolution or possibly *long distance Q-resolution* [23,2,17]. However, Section 5 hints towards formulas that will be easy for one calculus and hard for the other. We conjecture that exponential separations can be shown in both directions. Such separation would be of high practical importance. Firstly, it would explain why expansion-based solvers are better for some classes of instances than DPLL solvers, and the other way around. Secondly, the separation would necessitate QBF certification formats supporting both types of solvers.

Acknowledgments. We would like to thank Uwe Egly and Will Klieber for various helpful conversations on QBFs. We would also like to thank the anonymous reviewers for their stimulating feedback. This work is partially supported by SFI PI grant BEACON (09/IN.1/I2618), FCT grants ATTEST (CMU-PT/ELE/0009/2009) and PO-LARIS (PTDC/EIA-CCO/123051/2010), and multiannual PIDDAC program funds (PEst-OE/EEI/LA0021/2011).

References

1. Ayari, A., Basin, D.: QUBOS: Deciding quantified Boolean logic using propositional satisfiability solvers. In: Aagaard, M.D., O'Leary, J.W. (eds.) FMCAD 2002. LNCS, vol. 2517, pp. 187–201. Springer, Heidelberg (2002)
2. Balabanov, V., Jiang, J.H.R.: Unified QBF certification and its applications. Formal Methods in System Design 41(1), 45–65 (2012)
3. Benedetti, M.: Evaluating QBFs via symbolic Skolemization. In: Baader, F., Voronkov, A. (eds.) LPAR 2004. LNCS (LNAI), vol. 3452, pp. 285–300. Springer, Heidelberg (2005)
4. Biere, A.: Resolve and expand. In: Hoos, H.H., Mitchell, D.G. (eds.) SAT 2004. LNCS, vol. 3542, pp. 59–70. Springer, Heidelberg (2005)
5. Bubeck, U.: Model-based transformations for quantified Boolean formulas. Ph.D. thesis, University of Paderborn (2010)
6. Bubeck, U., Büning, H.K.: Bounded universal expansion for preprocessing QBF. In: Marques-Silva, J., Sakallah, K.A. (eds.) SAT 2007. LNCS, vol. 4501, pp. 244–257. Springer, Heidelberg (2007)
7. Büning, H.K., Bubeck, U.: Theory of quantified boolean formulas. In: Handbook of Satisfiability. IOS Press (2009)
8. Büning, H.K., Karpinski, M., Flögel, A.: Resolution for quantified Boolean formulas. Inf. Comput. 117(1) (1995)
9. Cadoli, M., Schaerf, M., Giovanardi, A., Giovanardi, M.: An algorithm to evaluate quantified Boolean formulae and its experimental evaluation. J. Autom. Reasoning 28(2), 101–142 (2002)
10. Cimatti, A., Sebastiani, R. (eds.): SAT 2012. LNCS, vol. 7317. Springer, Heidelberg (2012)
11. Cook, S.A., Reckhow, R.A.: The relative efficiency of propositional proof systems. J. Symb. Log. 44(1), 36–50 (1979)
12. Egly, U.: On sequent systems and resolution for QBFs. In: Cimatti, Sebastiani (eds.) [10], pp. 100–113
13. Giunchiglia, E., Marin, P., Narizzano, M.: QuBE 7.0 system description. Journal on Satisfiability, Boolean Modeling and Computation 7 (2010)
14. Giunchiglia, E., Narizzano, M., Tacchella, A.: Clause/term resolution and learning in the evaluation of quantified Boolean formulas. Journal of Artificial Intelligence Research 26(1), 371–416 (2006)
15. Janota, M., Klieber, W., Marques-Silva, J., Clarke, E.M.: Solving QBF with counterexample guided refinement. In: Cimatti, Sebastiani (eds.) [10], pp. 114–128
16. Janota, M., Marques-Silva, J.: Abstraction-based algorithm for 2QBF. In: Sakallah, K.A., Simon, L. (eds.) SAT 2011. LNCS, vol. 6695, pp. 230–244. Springer, Heidelberg (2011)
17. Klieber, W., Sapra, S., Gao, S., Clarke, E.: A non-prenex, non-clausal QBF solver with game-state learning. In: Strichman, O., Szeider, S. (eds.) SAT 2010. LNCS, vol. 6175, pp. 128–142. Springer, Heidelberg (2010)
18. Krajíček, J., Pudlák, P.: Quantified propositional calculi and fragments of bounded arithmetic. Mathematical Logic Quarterly 36(1), 29–46 (1990)
19. Lonsing, F., Biere, A.: Nenofex: Expanding NNF for QBF solving. In: Kleine Büning, H., Zhao, X. (eds.) SAT 2008. LNCS, vol. 4996, pp. 196–210. Springer, Heidelberg (2008)
20. Lonsing, F., Biere, A.: DepQBF: A dependency-aware QBF solver. JSAT (2010)
21. Rintanen, J.: Improvements to the evaluation of quantified Boolean formulae. In: Dean, T. (ed.) IJCAI, pp. 1192–1197. Morgan Kaufmann (1999)
22. Urquhart, A.: The complexity of propositional proofs. Bulletin of the EATCS 64 (1998)
23. Zhang, L., Malik, S.: Conflict driven learning in a quantified Boolean satisfiability solver. In: ICCAD (2002)

Recovering and Utilizing Partial Duality in QBF

Alexandra Goultiaeva and Fahiem Bacchus

Department of Computer Science
University of Toronto
{alexia,fbacchus}@cs.toronto.edu

Abstract. Quantified Boolean Formula (QBF) solvers that utilize non-CNF representations are able to reason dually about conflicts and solutions by accessing structural information contained in the non-CNF representation. This structure is not as easily accessed from a CNF representation, hence CNF based solvers are not able to perform the same kind of reasoning. Recent work has shown how this additional structure can be extracted from a non-CNF representation and encoded in a form that can be fed directly to a CNF-based QBF solver without requiring major changes to the solver's architecture. This combines the benefits of specialized CNF-based techniques and dual reasoning.

This approach, however, only works if one has access to a non-CNF representation of the problem, which is often not the case in practice. In this paper we address this problem and show how working only with the CNF encoding we can successfully extract partial structural information in a form that can be soundly given to a CNF-based solver. This yields performance benefits even though the information extracted is incomplete, and allows CNF-based solvers to obtain some of the benefits of dual reasoning in a more general context. To further increase the applicability of our approach we develop a new method for extracting structure from a CNF generated with the commonly used Plaisted-Greenbaum transformation.

1 Introduction

The problem of deciding the truth of a Quantified Boolean Formula (QBF) is PSPACE-Complete. Hence, any problem in PSPACE can be compactly encoded as a QBF decision problem. This includes many problems that would require exponentially sized SAT encodings. QBF's representational power comes from its use of universal and existential quantified variables and from the arbitrary interleaving of these quantifiers. This makes QBF a particularly useful representation for problems involving adversarial situations, incomplete information, and non-deterministic behavior. It also makes the development of efficient QBF solvers an important research goal.

In this paper we present a method for improving the performance of CNF-based QBF solvers. We develop a technique that allows such solvers to better exploit the duality inherent in the QBF formalism. The novelty of our approach is that, unlike prior related efforts, it does not require us to possess additional

M. Järvisalo and A. Van Gelder (Eds.): SAT 2013, LNCS 7962, pp. 83–99, 2013.
© Springer-Verlag Berlin Heidelberg 2013

information about the problem instance beyond what is already contained in its CNF representation. This is important because many QBF problems are presented only in CNF with no easy access to additional information.

Duality has been effectively exploited in prior non-CNF QBF solvers [20,13,10]. This technique involves reasoning about the given QBF formula as well as about its negation. By detecting when a partial truth assignment, π, falsifies the formula's negation we have detected that π satisfies the formula. In QBF solving, as we will explain, it is important to detect both falsifying and satisfying *partial* assignments. Non-CNF representations, e.g., the circuit representation used in [10] provide easy access to the formula and its negation and thus support this kind of dual reasoning. In contrast, although detecting falsifying partial assignments is easy with a CNF representation (the assignment need only falsify a single clause), detecting satisfying partial assignments is not. This has a negative impact on the performance of CNF-based QBF solvers.

It has been shown in prior work that *if we have access to a circuit description of the formula* we can achieve the speedups of dual reasoning within a CNF-based solver without significant changes to the solver's architecture [11]. This work shows that one of the standard features of modern QBF solvers, cube learning, can be exploited to achieve this speedup by simply providing the solver an initial set of input cubes derived from the negation of the formula. Empirically, this yields a significant performance gain in such solvers. Furthermore, since the clause and cube data structures are simpler and highly optimized in CNF solvers, a CNF solver supplied with "input cubes" can perform better than a native non-CNF solver which has to support reasoning on a more complex representation.

The core of our approach is to recognize that when a circuit description of the formula is unavailable we can still extract partial information about the circuit from the CNF encoding. In particular a number of techniques have been developed to extract original structure from CNF [4]. These techniques are usually incomplete and employ heuristic or greedy techniques, since in general it is NP-Hard to optimally extract the original structure from a CNF encoding. Nevertheless, a portion of the original structure can often be extracted.

We show how this partial information can be used generate an incomplete set of initial cubes that we prove can be soundly added to the solver and further exploited with a slight modification the solver. We also show empirically that, although incomplete, these cubes still yield useful performance improvements. Finally, we address for the first time the issue of extracting circuit structure information from the commonly used Plaisted-Greenbaum encoding [17].

2 Background and Definitions

A propositional formula in CNF is a conjunction of **clauses** each of which is a disjunction of literals each of which is a propositional variable or its negation. A formula in DNF is a disjunction of **cubes** each of which is a conjunction of literals. Negating a CNF yields a DNF and vice versa. We assume that all clauses and cubes contain at most one literal for any given variable and view them as sets of literals. CNFs (DNFs) are viewed as sets of clauses (cubes).

An assignment is a set of variable value pairs, (v = TRUE or V = FALSE) represented as a set of literals. It is **complete** if it includes all variables in the problem, otherwise it is **partial**. We use $\phi|_\pi$ to denote the reduction of a formula ϕ by the assignment π. The reduction is computed by replacing variables in ϕ by the values assigned to them in π followed by simplifying. We use "$\equiv$" to denote that two expressions evaluate to the same value if they are closed QBFs, or are equisatisfiable if they have free variables. The empty clause is equivalent to $\bot$ (FALSE) and the empty cube is equivalent to $\top$ (TRUE).

Note that a partial assignment π need only satisfy a single cube for us to detect that it satisfies a DNF, and similarly π need only falsify a single clause for us to detect that it falsifies a CNF. On the other hand detecting that a partial assignment falsifies a DNF or satisfies a CNF requires checking all cubes or all clauses.

Definition 1. *A **QBF formula** $Q.\phi$ contains a **prefix** Q which is sequence of universally ($\forall$) and existentially ($\exists$) quantified variables, and a **matrix** ϕ which is a propositional formula over the variables in Q. The truth of a QBF formula is defined recursively: $\exists x Q.\phi \equiv (Q.\phi|_x) \vee (Q.\phi|_{\neg x})$ and $\forall x Q.\phi \equiv (Q.\phi|_x) \wedge (Q.\phi|_{\neg x})$. We assume that all variables of ϕ are contained in Q. Hence ϕ will eventually be reduced to the constant 1 or 0: $Q.1$ is always TRUE and $Q.0$ is always FALSE.*

A QBF $Q.\phi$ is in CNF if ϕ is in CNF. The Tseitin transformation [19] converts an arbitrary formula into CNF by introducing a new variable to represent each subformula along with clauses to ensure the new variable and its subformula are equivalent. We use $\mathcal{T}(\psi)$ to denote the Tseitin transformation of the formula ψ. To apply this transformation to a QBF we have to insert the newly new variables into the prefix [2]. The new variables are existentially quantified and are placed in the prefix immediately after all of the variables of the sub-formula they encode: given a setting of these variables the sub-formula has a fixed truth value and so does the new auxiliary variable. Thus $\mathcal{T}(Q.\psi)$ will be $Q^+.\mathcal{T}(\psi)$ with Q^+ being Q with the new variables of $\mathcal{T}(\psi)$ added.

Example 1. QBF $\exists ab \forall cd.(a \wedge c) \vee (a \wedge d) \vee b$ can be encoded in CNF as $\exists ab \forall cd \exists fg.$ $(\neg f \vee a) \wedge (\neg f \vee c) \wedge (f \vee \neg a \vee \neg c) \wedge (\neg g \vee a) \wedge (\neg g \vee d) \wedge (g \vee \neg a \vee \neg d) \wedge (f \vee g \vee b).$

The **decision tree** of a QBF is a complete binary tree, where all nodes at level i split on the value of the ith variable of the quantifier prefix. Thus each node represents a partial assignment described by the path to it from the root and leaves correspond to complete assignments. A leaf ℓ is labeled with with the value $\phi|_{\pi(\ell)}$ where $\pi(\ell)$ is the complete assignment corresponding to ℓ. (Hence, each leaf is labeled with TRUE or FALSE). A node that splits on an existential variable is labeled TRUE iff at least one of its children is TRUE, a node that splits on a universal is labeled TRUE iff both of its children are TRUE. It follows that the root is labeled with the truth value of the QBF is true.

A QBF **model** is a subset of the TRUE nodes of the decision tree that includes the root, exactly one child for every existential node and both children for a universal node. Similarly, a **countermodel** is a subset of the FALSE nodes that includes the root, both children of existential nodes and one child of a universal node. See [11] for examples. A true QBF will have one or more models, and a false QBF will have one or more countermodels. Since the root has a unique label a QBF has at least one model or countermodel but never both.

Resolution combines two non-tautological clauses $c_1 = (\alpha \vee x)$ and $c_2 = (\beta \vee \neg x)$ to obtain a new clause $c_3 = (\alpha \vee \beta)$. It has the property that $c_1 \wedge c_2 \equiv c_1 \wedge c_2 \wedge c_3$, so adding the new resolvent to a set of clauses does not affect any models or countermodels that set has under a relevant prefix. **Term resolution** combines two non-tautological cubes $c_1 = (\alpha \wedge x)$ and $c_2 = (\beta \wedge \neg x)$ to obtain a new cube $c_3 = (\alpha \wedge \beta)$. Similarly, the set of models or countermodels of a set of cubes is preserved by adding term resolvants.

A literal u is **tailing** in a clause c if u comes later in the prefix that the other literals in c. Universally quantified tailing literals can be removed from a clause, a procedure called **universal reduction**. If C is a set of clauses, Q a quantifier, and u is a **universal** literal that is tailing in $c \in C$ then it can be shown that $Q.C$ has the same set of models and countermodels as $Q.(C - c) \cup \{(c - u)\}$. **Existential reduction** is the dual case of removing a tailing existential from a cube. Just like universal reduction, existential reduction preserves the models of a set of cubes.

2.1 QDPLL

Algorithm 1. QDPLL

Input: $Q.\phi$ (ϕ in CNF)

1 C = ϕ; U = $\emptyset$;
2 **while** TRUE **do**
3 | c = propagate(C, U);
4 | **if** $c \equiv$ NEXTVAR **then**
5 | | v = select_var();
6 | | assign_var (v);
7 | **else**
8 | | **if** $c \equiv$ COMPLETE **then**
9 | | | c = gatherCube()
10 | | (btlevel, c) = analyze();
11 | | **if** c *is a clause* **then**
12 | | | C = C $\cup$ c;
13 | | **else**
14 | | | U = U $\cup$ c;
15 | | backtrack (btlevel);

Algorithm 1 presents an outline of QDPLL, a DPLL style algorithm for solving QBF [9], [21]. The solver maintains a set of clauses C, initialized to the clauses of the input formula, and a set of cubes U, initialized to the empty set. The sets are maintained so that every model of $Q.\phi$ is also a model of $Q.C$, and every countermodel of $Q.\phi$ is a countermodel of $Q.U$ (note that initially $Q.U \equiv Q.\bot \equiv \bot$).

During propagation C and U are examined to find unit clauses or cubes. Clauses are reduced by falsified literals and universal reduction and when they become unit they imply forced existential literals. (Unit universal clause become empty by universal reduction rather than implying the remaining universal literal). Dually, cubes are reduced by true literals and existential reduction and when they become unit they imply forced universal literals. Propagation might detect an empty clause or cube: such a clause or cube is called conflicting. If propagation finds a conflict it returns it in the variable c.

If propagation completes without finding a conflict, another variable is selected and assigned a value (an unassigned variable with outer most scope must be selected). If no more variables remain, a new conflicting cube is constructed from the current set of assignments. Since $\phi \subseteq C$ and no conflicting clause has been found, all of the input clauses are satisfied by the currently set of assignments π. A cube is constructed by selecting from π a subset τ sufficient to satisfy all clauses in ϕ. This subset includes at least one true literal from every input clause; hence τ can be quite large. The conjunction of literals in τ is the new cube. Note that $\tau \rightarrow \phi$, and ϕ is falsified at every leaf of every countermodel of $Q.\phi$, so τ must also be falsified and the set of countermodels is preserved. τ is also conflicting at this point of the search: every literal in it is TRUE.

Hence, when line 10 is executed either 'propagate' or 'gatherCube' has set c to a conflict. An empty clause indicates that there is no model extending the current node, while an empty cube indicates that there is no countermodel extending the current node. In either case the solver can analyze the contradicted clause (cube) by performing a series of resolution (term resolution) and universal (existential) reduction steps based on previously forced literals. The result is a new clause (cube) that can be added to C (U) whilst preserving models (countermodels), since the new clause, or cube, is a logical entailment of the current C or U, and the solver can backtrack to 'btlevel': the level at which it is once again possible that a model or countermodel exists.

We note that while the clause database is initialized with clauses from the input problem, the cube database is initially empty. Thus, it can do no propagation, and cannot be used to recognize solutions (i.e., the non-existence of counter-models). Also, the cubes τ generated by 'gatherCube' are often very long. Since new cubes are generated by term resolution from these initial long cubes the effectiveness of cube learning in the QDPLL is negatively impacted.

2.2 Dual Propagation

Consider the formula $\exists ab \forall cd.(a \wedge c) \vee (a \wedge d) \vee b$. Recall its Tseitin encoding:

$$(\neg f \vee a) \wedge (\neg f \vee c) \wedge (f \vee \neg a \vee \neg c) \wedge (\neg g \vee a) \wedge (\neg g \vee d) \wedge (g \vee \neg a \vee \neg d) \wedge (f \vee g \vee b)$$

with the prefix $\exists ab \forall cd \exists fg$. The assignment $\{b\}$ is a solution, and is enough to certify that the formula is true. However, a CNF solver working with the Tseitin encoding is unable to detect that. All the variables have to be set before a solution is detected: suppose all variables are set to true. Then the solver chooses a subset of variables that satisfies all clauses. Every literal from $\{a, c, d, f, g\}$ must be included since there is some clause where it is the only true literal. Since the clause $(f \vee g \vee b)$ is already satisfied by that assignment, b does not have to be included. After existential reduction, the cube $(a \wedge c \wedge d)$ is added to the database. As the solver continues its search, it might consider the assignment $\{a, b, c, \neg d, f, \neg g\}$. This generates cube $(a, c, \neg d)$, which can be resolved with the previous one to obtain (a, c). So, the next assignment considered will include $\neg c$, which renders all previously learnt cubes useless. The search might go through every combination of values of c and d before terminating.

Given the right representation, however, this formula can be easily solved with no search. The difficulty is that a CNF-based solver starts out with an empty cube database U. However, recent work has presented a technique for initializing that database, and has argued that this is identical to the technique of dual propagation as employed by non-CNF solvers [11].

The main idea is to convert the formula to DNF and use that to initialize the cube database. The conversion has to introduce new variables to avoid an exponential explosion in the formula's size. Suppose $Q.\phi$ is a non-CNF QBF, and consider its negation $\neg(Q.\phi)$. If we push in the negation through the quantifier prefix, by De Morgan's law, we get $(\neg Q).\neg\phi$, where $(\neg Q)$ is the same prefix as Q except the quantifiers are flipped. Then let $Q^n.\phi^n = \mathcal{T}((\neg Q).\neg\phi)$ be the Tseitin encoding. Note that Q^n is simply $(\neg Q)$ with some new existential variables introduced. Then $Q^n.\phi^n \equiv \neg(Q.\phi)$, and thus $Q.\phi \equiv \neg(Q^n.\phi^n) \equiv (\neg Q^n).\neg\phi^n$. Also, note that the quantifier prefix $(\neg Q^n)$ has the same variables as in Q, with quantifiers negated and then negated back, and some auxiliary variables, which have been existential in Q^n but have since then been negated and became universal variables. So, $(\neg Q^n)$ is the same as Q except a number of auxiliary universal variables have been added. We can show that $Q.\phi \equiv (\neg Q^n).\phi$, since introducing dummy variables to the prefix does not change the value of a QBF. Note that for any such new variable, the two subtrees that are its children in the decision tree are identical. So, $(\neg Q^n).\phi \equiv (\neg Q^n).\neg(\phi^n)$. Note that since ϕ^n is in CNF, pushing in the negation in $\neg(\phi^n)$ yields a DNF.

Example 2. Applied to our example, this gives a DNF $\exists ab \forall cd \forall hi.(h \wedge \neg a) \vee (h \wedge \neg c) \vee (\neg h \wedge a \wedge c) \vee (i \wedge \neg a) \vee (i \wedge \neg d) \vee (\neg i \wedge a \wedge d) \vee (h) \vee (i) \vee (b)$. Note that in this form there is a unit cube (b) which after existential reduction becomes empty. This immediately indicates that the formula is true without any search.

Let $Q^t.\phi^t = \mathcal{T}(Q.\phi)$. Let Q^+ be the quantifier Q^t with the universal auxiliary variables from $(\neg Q^n) - Q$ added. It is sound to use $Q^+.\phi^t$ as the initial formula, and $\neg(\phi^n)$ as the cubes. This result follows from a more general one described below. Applying this method to our example, the cube database will initially contain an empty cube, which would allow the solver to immediately recognize that the formula is true.

Note that $Q^+.\phi^t \equiv Q^+.\phi$, and if $Q^+.\phi$ is false, then a countermodel of $Q.\phi$ extended by the valid settings of all auxiliary variables is a countermodel for both $Q^+.\phi^t$ and $Q^+.\neg(\phi^n)$. As we'll see below, these are sufficient conditions to show that QDPLL would give a correct answer given $Q^+.\phi^t$ as input, and with cube database $\neg\phi^n$.

The conditions that guarantee soundness for seeding the cube database have been formalized [11], and can be reformulated as follows:

Theorem 1. *Let $Q.\phi$ be a QBF, and C and U be sets of clauses and cubes respectively, such that:*

- $Q.C \equiv Q.\phi$
- *If $Q.\phi \equiv \bot$, then at least one countermodel of $Q.C$ is a countermodel of $Q.U$*

Then a QDPLL solver is guaranteed to return a correct value of $Q.\phi$

Proof. A QDPLL solver only returns a value if it obtains an empty clause or an empty cube. The majority of its steps (resolution/term resolution and universal/existential reduction) preserve the models and countermodels of its databases: if the set of clauses(cubes) S' is obtained from S by zero or more of these steps, then the sets of models and countermodels of $Q.S'$ and $Q.S$ are preserved.

The only step that is not model/countermodel-preserving is the extraction of a new cube. If S_u is a set of cubes, and c is a new cube gathered from a set of clauses S_c, then $S_u \cup \{c\}$ might potentially have less countermodels than S_u (but never more). Since c includes a literal from every clause of S_c, we are guaranteed that $c \to S_c$. In particular, this means that c is false at every leaf of every countermodel of S_c. So, any countermodel of S_u which is also a countermodel of S_c is preserved.

Suppose QDPLL starts with C and U as its initial databases, and let C' and U' be its databases at an arbitrary point in the search. We can then say that: any model of $Q.C$ is a model of $Q.C'$; any common countermodel of $Q.U$ and $Q.C$ is also a countermodel of $Q.U'$. So, an empty clause in C' guarantees that $Q.C'$ has no models, and thus neither does $Q.C$, so $Q.\phi \equiv \bot$. An empty cube guarantees that there were no shared countermodels of $Q.U$ and $Q.C$, and thus $Q.\phi \equiv \top$.

The original motivation for the formulation of these rules was the ability to use preprocessing on the two formulas separately. However, as we will show below, we can use it to use incomplete information to seed the cube database.

2.3   CNF Encodings

In general, given a CNF, figuring out which variables are auxiliary (used to represent subformulas) is as hard as solving the problem. However, it is possible to develop incomplete techniques, and/or techniques aimed at particular methodologies for encoding CNF. Most existing techniques are aimed at finding *functional dependencies*, cases a value of a variable is completely specified by some set of clauses as a function over preceding variables.

The simplest technique is to look for patterns commonly used to encode logical gates. For example, AND and OR gates are often encoded as $(y \vee x_1) \wedge (y \vee x_2) \wedge ... \wedge (y \vee x_n) \wedge (\neg y \vee \neg x_1 \vee \neg x_2 \vee ... \vee \neg x_n)$, which encode the equivalence $\neg y \equiv x_1 \wedge x_2 \wedge ... \wedge x_n$. Note that here the literals may be any combination of polarities. In the above example, the gates encoding F and G can be recognized by this method.

This kind of pattern matching can be implemented efficiently. However, there is another problem: the resultant circuit has to be acyclic. Finding an optimal subset of gates can be costly, so often a greedy approach is taken. One application of such gate recognition is the simplification of SAT problems [3].

More advanced technique exist that use propagation and other reasoning techniques to uncover equivalences [15], [12], or to select which definitions would give a maximal circuit [4]. Although our method would benefit from better structure reconstruction, comparison and review of different reconstruction techniques is beyond the scope of this paper.

Some forms of structural reconstruction have also been applied to QBF [13], [16]. Most of the techniques can be generalized from CNF in a straightforward manner, while keeping in mind an additional constraint: that the auxiliary variable must be scoped by the variables it depends in.

For the first part of this paper we are using the greedy pattern matching technique which ensures that the dependent variable is deeper in the quantifier prefix than the variables it depends on.

The *Plaisted-Greenbaum* transformation is an improvement over the Tseitin encoding [17]. It recognizes that if the literal x only appears positively in the rest of the formula, then instead of the equivalence $x \equiv \alpha$, it is sufficient to encode $\neg\alpha \rightarrow \neg x$. Thus the Plaisted-Greenbaum transformation of our example formula would be $\exists ab \forall cd \exists fg.(\neg f \vee a) \wedge (\neg f \vee c) \wedge (\neg g \vee a) \wedge (\neg g \vee d) \wedge (f \vee g \vee b)$.

Obviously, the formulas transformed this way do not have any gate definitions that an algorithm could find. Actually, they have no functional dependencies, since only one polarity of the variable is forced. So, no method mentioned above is directly applicable. We are unaware of any work on reconstructing formulas transformed using the Plaisted-Greenbaum transformation.

3   Utilizing Partial Information

Let $Q.\gamma$ be a QBF formula in CNF, and suppose that the set of clauses $A \subset \gamma$ encodes a definition of a variable x. Let V be a set of variables in A except for x. When we say that a set of clauses encodes a *definition*, we assume the following set of conditions:

- x is an existential variable.
- All variables in V precede x in Q.
- For any assignment to variables in V, π can be extended by exactly one setting of x such that the new assignment satisfies A.

The second condition is specific to QBF, and is based on the fact that in order to preserve the value of the QBF, the auxiliary variables must follow all the variables in the subformula to which they are equivalent. Consider, for example, a formula $\forall a \exists b.(\neg a \wedge \neg b) \vee (a \wedge b)$. This formula is true, since b can always be set to the same value as a. Suppose we create an auxiliary variable c to represent $a \wedge b$. The matrix will then be $((\neg a \wedge \neg b) \vee c) \wedge (c \equiv a \wedge b)$. If we insert c in the wrong place in the prefix, we might get $\exists c \forall a \exists b$. Now, however, the formula becomes false. If c is true, a can be set to false, violating $c \equiv a \wedge b$. If c is false, a can be set to true, violating $((\neg a \wedge \neg b) \vee c)$. Conversely, if we started with the latter formula and treated c as an auxiliary variable, we would turn a false formula into a true one. Other examples can be found in [16].

Note that the second condition assumes a strict ordering on Q, and not the relaxed variant where variables on the same quantifier level can be interchanged. This way, the conditions ensure that there are no cyclic dependencies among the selected definitions (i.e., this prevents cases where x is defined in terms of y while y is defined in terms of x). The last condition ensures that there is always a setting of x that satisfies the definition, and that there is only one such setting: i.e., the value of x is completely determined by the settings of V.

Let us say that the $S_x(A)$ is the subformula to which the set of clauses A forces x to correspond. It never needs to be explicitly computed, but for theoretical results it can be defined as a disjunction of all assignments to the variables in A which make x true. Let $\gamma|_{x \leftarrow B}$ mean the result of taking the formula γ and substituting every occurrence of x by the formula B. We can note that if A is a definition of x in $Q.\gamma$, then $Q.\gamma \equiv Q.\gamma|_{x \leftarrow S_x(A)}$.

Algorithm 2. Partial-solve

Data: $Q.\phi$: QBF in PCNF

1 $C = \phi$;
2 $(X, D) = \text{extractDefs}(Q.\phi)$
3 $(Q', X_u) = \text{insertUniversal}(X, Q)$
4 $U = \neg(\text{replace}(X, X_u, D))$;
5 **return** $\text{QDPLL}^p(Q', C, U, D, X, X_u)$

Algorithm 2 outlines our approach. The function **extractDefs** is the gate search algorithm. Any gate extraction algorithm can be used here, as long as it returns X, the set of variables found to be auxiliary, and D, which is a CNF obtained by conjoining the disjoint definitions for variables in X.

The function **insertUniversal** takes a set of variables X and a quantifier prefix Q. For each variable $x \in X$ it inserts a fresh universal variable at the earliest universal level following x, and returns the new prefix and the list of newly inserted variables.

The function **replace** takes the set of definition clauses D, and replaces every occurrence of every variable in X by its corresponding variable from X_u. Then, it returns a CNF formula which is the conjunction of the clauses of the result.

Then, **QDPLL**^{p} is ran on the CNF formula $Q'.C$ with cubes U. Here QDPLL^p is a modification of QDPLL which explicitly takes the initial clause and cube databases, and also takes the information about the definitions. The only difference between $QDPLL^p$ and $QDPLL$, except for the way the databases are initialized, is in the way new cubes are obtained. Instead of selecting a subset of variables to satisfy all clauses in C, QDPLL^p only satisfies the clauses in $C - D$ obtaining a cube c. Then $c' = replace(X, X_u, \{c\})$ is used as the starting cube.

Note that whenever a new cube is generated under a complete assignment π, the newly added cube c' is guaranteed to be made true by π. To show that, it suffices to show that at the point of cube generation, π sets $x = x'$ for any $x \in X$ that corresponds to $x' \in X_u$. Suppose not. Let x be the quantifier-earliest variable such that $x \neq x'$ under π. Let A be the clausal definition of x, and let A_u be A with all variables from X replaced by their counterparts from X_u. Since all earlier variables are equivalent, $S_x(A) \equiv S_x(A_u)$. Since no cubes are true, and the negation of A_u is among the cubes, then all clauses of A_u are satisfied by π. So, x' is set to $S_x(A_u)$. Also, since all clauses are true, so is A, and $x \equiv S_x(A)$. So, $x \equiv x'$. This ensures that QDPLL^p can use c' as a starting cube for learning.

Theorem 2. *Partial-solve is guaranteed to return the value of its input $Q.\phi$.*

Proof. We shall use the same notation as in Alg. 2.

If $QDPLL^p(Q', C, U, D, X, X_u)$ discovers an empty clause, then $Q'.C$ has no models. Then neither did $Q.C = Q.\phi$, so it is $\bot$.

Let ϕ^- be the formula obtained from $\phi - D$ by replacing every $x \in X$ by $S_x(A)$, where A is the definition of x. Note that ϕ^- might no longer be in CNF, and that no variables from X appear in ϕ^-. We note that $\phi \equiv (\phi - D) \wedge (x_1 \equiv S_{x_1}(A_1)) \wedge (x_2 \equiv S_{x_2}(A_2)) \wedge \ldots$ for all variables $x_i \in X$ and corresponding definitions A_i. So, $Q.\phi^- \equiv Q.\phi$ by simple equivalence reasoning.

So $Q.\phi \equiv \bot$, then so is $Q.\phi^-$, and it has a countermodel. Because ϕ^- does not contain any variables from X, then it must have a countermodel where for every node n splitting on a variable from X, the subtrees under n are identical. Let M_1 be such a countermodel for $Q.\phi^-$.

Consider the cubes $U = \neg(replace(X, X_u, D))$, and the prefix $(Q', X_u) =$ insertUniversal (X, Q). By adding dummy variables, we have $Q'.\phi^- \equiv Q.\phi$. Let M_2 be the countermodel of $Q'.\phi^-$ obtained from M_1 by picking those values of the auxiliary universal variables as to satisfy their definitions in $replace(X, X_u, D)$.

Every leaf node in M_2 would falsify all the cubes in U, so this countermodel is a countermodel for $Q'.U$. Also, note that $\phi \to \phi^-$, so this is also a countermodel for $Q'.C$.

Now, consider a new cube c obtained by $QDPLL^p$ at cube gathering phase. Suppose c is true some leaf of M_2 corresponding to an assignment π. By construction of M_2, all universal auxiliary variables are set according to their definitions in π. We can then generate an assignment π' from π by setting all the auxiliary existentials to the values of their definition. By choice of M_1, the leaf corresponding to π' must also be in M_2. Since c contains no auxiliary existentials, $c|_{\pi'} \equiv \top$. Also note that in π' the values of variables in X_u match their corresponding values from X. So, let c' be c with variables from X_u replaced by the corresponding ones from X. Then $c'|_{\pi'} \equiv \top$. However, by construction of c, c' contains one literal from every clause in $C - D$, and, since all variables in X are set according to their definitions, $D|_{\pi'} \equiv \top$. Then $\phi|_{\pi'} \equiv \top$, and $\phi|_{\pi'} \to \phi^-|_{\pi'}$. So, this leaf could not have been in a countermodel of $Q'.\phi^-$. So, any new cube c generated by $QDPLL^p$ will be $\bot$ at any leaf of M_2.

So, no cube generated by $QDPLL^p(Q', C, U, D, X, X_u)$ could be $\top$ at any leaf of M_2. This means that if $QDPLL^p(Q', C, U, D, X, X_u)$ discovers an empty cube, then M_2 could not exist, and neither could M_1. Then, $Q.\phi$ has no countermodels and is $\top$.

4   Handling the Plaisted-Greenbaum Transformation

The approach detailed in Section 3 only works when the encoding used completely specifies the value of auxiliary variables. However, the Plaisted-Greenbaum encoding only uses one side of the implications, making it impossible to handle the gate as described above. In this section we will describe how it is possible to reconstruct structure from problems encoded this way.

Let x be an existential literal from a QBF $Q.\phi$. Let $C_1 \subseteq \phi$ contain all clauses with x, and $C_2 \subseteq \phi$ contain all clauses with $\neg x$. If at least one of these sets contains exclusively variables that scope x, then the formula can be modified without changing its value so that x is functionally dependent on the variables before it.

Suppose C_1 only contains variables that scope x. Then, $C_1 = \{c_1, c_2, c_3, ...\}$ can be rewritten as $\gamma \to x$, where γ is a DNF that can be composed by the disjunction $\bigvee_{c \in C_1} \neg(c - \{x\})$.

Example 3. For example, if $C = (a \lor x) \land (b \lor \neg a \lor x) \land (d \lor x)$, then $\gamma \equiv (\neg a) \lor (\neg b \land a) \lor \neg d$.

Theorem 3. *If γ is obtained as stated, and $\phi_2 = (\phi - C_1) \land (\gamma \equiv x)$, then $Q.\phi \equiv Q.\phi_2$.*

Proof. First we note that $\phi \land (x \to \gamma) \equiv \phi_2$, so $\phi_2 \to \phi$ and thus $Q.\phi_2 \to Q.\phi$.

Now, assume $Q.\phi$. Let i be the position of x in Q. Consider the leaf nodes of the model of $Q.\phi$. If $\phi_2 \equiv \bot$ at any of them, then let n be its ancestor at level $i + 1$. Since $\phi \equiv \top$ and $\phi_2 \equiv \bot$, then $x \equiv \top$ and $\gamma \equiv \bot$. We also note that $\phi - C_1$ is a CNF which contains only negative occurrences of x. Also, if $\gamma \equiv \bot$, then $C_1 \equiv \top$. So, flipping x will not change the value of ϕ if $\gamma \equiv \bot$. So, if n is in a model for $Q.\phi$, then there is an alternative model which includes the sibling of n. By applying the same reasoning for each occurrence of x, it is possible to construct a valid Q-model of ϕ_2.

Using this property, we can repeatedly replace any such implication by an equivalence (obviously, the ordering constraints on Q prevent cycles). After that, we can apply Algorithm 2 to solve the problem utilizing these equivalences.

Example 4. Suppose $Q.\phi = \forall u \exists ab.(b \lor a) \land (b \lor u) \land \psi$, where ψ only contains negative occurrences of b. Then $Q.\phi \equiv Q.(b \lor a) \land (b \lor u) \land (\neg b \lor \neg a \lor \neg u) \land \psi$. So, we could solve $Q \forall c.(b \lor a) \land (b \lor u) \land (\neg b \lor \neg a \lor \neg u) \land \psi$ while initializing the cube database to $(\neg c \land \neg a) \lor (\neg c \land \neg u) \lor (c \land a \land u)$, where c is a fresh universal corresponding to b. The new cubes would then only be gathered from ψ.

We can further improve the algorithm by observing that the same properties that allowed removing some of the definition clauses still hold. Thus, the full equivalence is not needed in either the clause or the cube database.

Note that now there are clauses encoding $x_i \equiv \gamma$, and in the rest of the clauses x_i only appears negatively. Also, there are cubes encoding $x'_i \equiv \gamma$, and in the rest of the cubes (including those that will be generated during solving) x'_i only appears negatively. In both of those cases, one side of the implication can be removed, resulting in $\gamma \to x_i$ in the clauses and $\neg\gamma \to \neg x'_i$ in the cubes.

The intuition behind this idea is that by the time x needs to be assigned, if it has not been forced to be true, then all clauses with its positive polarity are true. If under this assignment a solution is found where x is true (thus violating the other half of the definition clauses), then by flipping x we get a solution where those clauses are not violated. So, removing these clauses does not change the value of the QBF. The similar reasoning works for cubes.

In terms of correctness of QDPLL^p, one can note that removing cubes cannot affect it. Removing clauses is also sound as long as the value of the QBF is not affected (which is true, as shown above).

Example 5. For the example above, it is sufficient to solve $Q\forall c.(b \vee a) \wedge (b \vee u) \wedge \psi$ while initializing the cube database to $(c \wedge a \wedge u)$.

So, the simplified technique comes down to the following:

- Find a literal x_i such that in the set of clauses C_1 which contain x_i, it is tailing. Reformulate C_1 as $\gamma \rightarrow x_i$.
- Create a corresponding auxiliary variable x_i', and add $\neg\gamma \rightarrow \neg x_i'$ to the cube database
- Remove C_1. from consideration when creating a new cube.

The above can be repeated for all suitable variables, as long as no cycles are created. Note that negating γ might in general be an expensive operation. This can be avoided by introducing auxiliary variables. In fact, the Tseitin or the Plaisted-Greenbaum transformation can be used to encode them. After the transformation is completed, QDPLL^p can be used to solve the problem.

Related Concepts. If no new variables are introduced in encoding $\neg\gamma \rightarrow \neg x$, then all the clauses that encode it can be shown to be **blocked**, with the blocking literal $\neg x$ [1]. It is sound to introduce blocked clauses, which provides alternative intuition for Theorem 3. However, this syntactic reasoning might not generalize to more complicated versions of γ, for example, those obtained by reasoning on propagation.

The conditions on the variable x to which this method can be applied are actually the same as the conditions for being **depth monotonic** [5]. Intuitively, they mean that the value of x can be determined without considering the clauses containing $\neg x$.

Our approach could potentially be strengthened to include dependency reasoning. By determining functional dependencies of a variable, we could rule out other dependencies to strengthen universal reduction. This might be done by identifying where our conditions agree with dependency reasoning, for example, the triangle dependency scheme [18], and might be an interesting avenue for future work.

5   Experiments

We have implemented a no-frills barebones DPLL QBF solver with clause and cube learning, VSIDS variable heuristic and phase saving. This solver was used as a baseline (B).

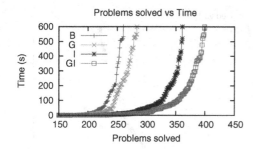

Fig. 1. Different versions of the algorithm

Then we equipped it with the ability to perform partial dual propagation. We have implemented a simple pattern matching gate extraction mechanism which recognizes the simple encodings of AND, OR and equivalence gates. The quantifier is initially treated as a partial order, where variables on the same quantifier level are unordered. As new gates are discovered and added, additional constraints are added to prevent cycles. Initial VSIDS value is set to 1 for the input variables, and to 0 for the auxiliary variables, no further distinction is made after the initialization. This was the second variant of the solver (G).

The third variant, I, was as G was, except instead of gate recognition it used the implication reasoning from Section 4. This is the third variant of our solver, aimed primarily at reconstructing the Plaisted-Greenbaum transformation.

The last variant (GI) was the same as the previous one, except both gate reconstruction and implication reasoning were applied. First, the solver extracted all the gates, after which it applied implication reasoning.

All experiments were ran on a 2.8GHz machine with 12GB of RAM under a time limit of 600 CPU seconds per instance. The benchmark set was taken from the CNF track of QBFEval'2010 [6].

Figure 1 compares the different versions of the algorithm. The gate recognition algorithm is clearly helpful, but the dataset did not seem to have a lot of problems that were amenable to that approach. Implications reasoning greatly improved the solving time on many more problems. Combined together, these approaches offer the best performance.

We then compared our result with the state-of-the-art QBF solvers: depQBF [14] the winner of QBFEval'10[1], and Qube (version 7.2) [8] – the results of running it under default settings are reported as *Qube7.2-d*. We also compared with GhostQ (version of 2010), which is a non-CNF solver provided with a script for gate recognition [13]. We omitted that result from the plot because GhostQ only solved 136 problems.

The graph clearly shows how the addition of our approach turned a simple baseline solver B, which is clearly inferior to other solvers, into GI which confidently outperforms the stand-alone solvers.

Preprocessing. The state of the art QBF solving is not limited to stand-alone solvers. Before being given to the solver, the formula can be reduced and simplified. In fact, Qube comes with a built-in preprocessor sQueezeBF [7]; depQBF works best when combined with Bloqqer [1]. We report the results of these two methods under the names *Qube7.2-all* and *Bloqqer + depqbf*, respectively. We can see that preprocessing has a strong impact on performance.

---

[1] These are results for the version of 2010; the latest available performed slightly worse.

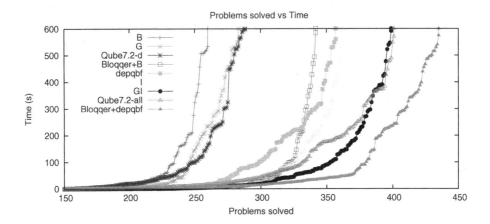

Fig. 2. Comparison with state of the art QBF solvers

We have also combined Bloqqer with B to get *Bloqqer+B*. We note the dramatic impact of this, however, it was not enough to compare with GI. Unfortunately, applying preprocessing out of the box negates our approach, since preprocessing breaks down the structure we aim to reconstruct. The solvers B, I, G and GI with Bloqqer solved 342, 340, 343 and 341 problems, respectively.

Although the naive application of preprocessing does not work with our method, it is interesting to see if the approaches can be combined. For example, the unrecovered portion of the CNF could be preprocessed (while taking into account the recovered clauses); the CNF could be preprocessed after the recovery but before solving; or, further yet, a duality-preserving version of the preprocessor might be developed. All these require modifications to the existing preprocessors and are an interesting avenue for future work.

To indicate the potential usefulness of our approach when combined preprocessing, we show a more detailed comparison between GI and Qube7.2-all and Bloqqer+depqbf. Figure 3 displays GI solving time on the y axis and the competitors on the x axis (any problems below the bisecting line are solved faster by GI). Color indicates how well much structure of the formula was recovered. The number is natural log of the fraction of the formula recovered. A value of 0 indicates a formula where nothing was recovered, while −10 is the one which was recovered (almost) completely. The problems which timed out for a solver were placed above the timeout value of 600, beyond the red line.

The figures indicate that while GI lags on a number of instances (most of which show a poorer reconstruction rate), for some problems it also offers substantial speedups, and is able to solve a number of problems unsolved by other approaches.

Of the four approaches: GhostQ, Qube7.2-all, Bloqqer+depqbf and GI, the number of uniquely solved instances was 0, 24, 38 and 13, respectively.

Finally, Figure 4 gives an indication of the reconstructive power of the three algorithms. We look at how many clauses remain in the problem and were not

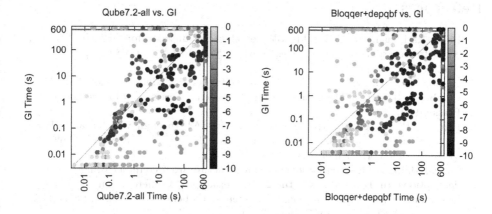

Fig. 3. Comparison of *GI* with fully preprocessed methods

identified as part of any definition. On the y axis is the fraction of the original problem remaining after the extraction (the smaller it is, the better). On the x axis is the number of problems reconstructed with this percentage or better.

We note that the gate extraction seems to be able to reconstruct a small number of problems well, and is close to useless on many others. The implications, on the other hand, are almost always present in problems, even if they were not originally encoded using the Plaisted-Greenbaum transformation. It is able to yield at least some reductions in the majority of cases. Combining the two methods, as expected, leads to even better results than any one of the methods separately. The new method is now able to fully reconstruct problems encoded using vanilla Tseitin or Plaisted-Greenbaum approaches, and can significantly reduce many other problems.

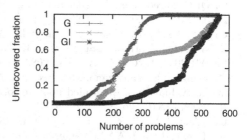

Fig. 4. Analysis of reconstructive power

6   Conclusion

We have presented an approach for utilizing partial dual propagation in QBF solving. Rather than requiring a complete dual representation of the formula, we can utilize partial definitions recovered from the CNF problem. The framework is relatively general, and could be be used with other reconstruction techniques. We have also developed a method to recover structure from problems encoded using the Plaisted-Greenbaum transformation.

Empirical evidence demonstrates the benefits of this technique, which could also be used in conjunction with many orthogonal approaches.

References

1. Biere, A., Lonsing, F., Seidl, M.: Blocked clause elimination for QBF. In: Bjørner, N., Sofronie-Stokkermans, V. (eds.) CADE 2011. LNCS, vol. 6803, pp. 101–115. Springer, Heidelberg (2011)
2. Buning, H.A.K., Lettmann, T.: Propositional logic: deduction and algorithms. Cambridge Tracts in Theoretical Computer Science Series. Cambridge University Press (1999)
3. Eén, N., Biere, A.: Effective preprocessing in SAT through variable and clause elimination. In: Bacchus, F., Walsh, T. (eds.) SAT 2005. LNCS, vol. 3569, pp. 61–75. Springer, Heidelberg (2005)
4. Fu, Z., Malik, S.: Extracting logic circuit structure from conjunctive normal form descriptions. In: Proceedings of the 20th International Conference on VLSI Design held jointly with 6th International Conference: Embedded Systems, VLSID 2007, pp. 37–42. IEEE Computer Society, Washington, DC (2007)
5. Van Gelder, A.: Contributions to the theory of practical quantified boolean formula solving. In: Milano, M. (ed.) CP 2012. LNCS, vol. 7514, pp. 647–663. Springer, Heidelberg (2012)
6. Giunchiglia, E., Narizzano, M., Tacchella, A.: Quantified Boolean Formulas satisfiability library (QBFLIB) (2001), http://www.qbflib.org
7. Giunchiglia, E., Marin, P., Narizzano, M.: sQueezeBF: An effective preprocessor for QBFs based on equivalence reasoning. In: Strichman, O., Szeider, S. (eds.) SAT 2010. LNCS, vol. 6175, pp. 85–98. Springer, Heidelberg (2010)
8. Giunchiglia, E., Narizzano, M., Tacchella, A.: QUBE: A system for deciding Quantified Boolean Formulas satisfiability. In: Goré, R.P., Leitsch, A., Nipkow, T. (eds.) IJCAR 2001. LNCS (LNAI), vol. 2083, p. 364. Springer, Heidelberg (2001)
9. Giunchiglia, E., Narizzano, M., Tacchella, A.: Clause/term resolution and learning in the evaluation of quantified boolean formulas. Journal of Artificial Intelligence Research (JAIR) 26, 371–416 (2006)
10. Goultiaeva, A., Bacchus, F.: Exploiting QBF duality on a circuit representation. In: Twenty-Fourth AAAI Conference on Artificial Intelligence (2010)
11. Goultiaeva, A., Seidl, M., Biere, A.: Bridging the gap between dual propagation and CNF-based QBF solving. In: Proceedings of the Conference on Design, Automation and Test in Europe (2013)
12. Grégoire, É., Ostrowski, R., Mazure, B., Saïs, L.: Automatic extraction of functional dependencies. In: Hoos, H.H., Mitchell, D.G. (eds.) SAT 2004. LNCS, vol. 3542, pp. 122–132. Springer, Heidelberg (2005)
13. Klieber, W., Sapra, S., Gao, S., Clarke, E.: A non-prenex, non-clausal QBF solver with game-state learning. In: Strichman, O., Szeider, S. (eds.) SAT 2010. LNCS, vol. 6175, pp. 128–142. Springer, Heidelberg (2010)
14. Lonsing, F., Biere, A.: DepQBF: A dependency-aware QBF solver. Journal on Satisfiability, Boolean Modeling and Computation 7(2-3), 71–76 (2010)
15. Ostrowski, R., Grégoire, É., Mazure, B., Saïs, L.: Recovering and exploiting structural knowledge from CNF formulas. In: Van Hentenryck, P. (ed.) CP 2002. LNCS, vol. 2470, pp. 185–199. Springer, Heidelberg (2002)
16. Pigorsch, F., Scholl, C.: Exploiting structure in an AIG based QBF solver. In: Proceedings of the Conference on Design, Automation and Test in Europe, DATE 2009, pp. 1596–1601. European Design and Automation Association, Leuven (2009)
17. Plaisted, D.A., Greenbaum, S.: A structure-preserving clause form translation. Journal of Symbolic Computation 2(3), 293–304 (1986)

18. Samer, M., Szeider, S.: Backdoor sets of quantified boolean formulas. Journal of Automated Reasoning (JAR) 42(1), 77–97 (2009)
19. Tseitin, G.: On the complexity of proofs in poropositional logics. In: Siekmann, J., Wrightson, G. (eds.) Automation of Reasoning: Classical Papers in Computational Logic 1967–1970, vol. 2. Springer (1983)
20. Zhang, L.: Solving QBF with combined conjunctive and disjunctive normal form. In: Twenty-First Conference on Artificial Intelligence (2006)
21. Zhang, L., Malik, S.: Towards a symmetric treatment of satisfaction and conflicts in quantified boolean formula evaluation. In: Van Hentenryck, P. (ed.) CP 2002. LNCS, vol. 2470, pp. 200–215. Springer, Heidelberg (2002)

Efficient Clause Learning for Quantified Boolean Formulas via QBF Pseudo Unit Propagation[*]

Florian Lonsing[1], Uwe Egly[1], and Allen Van Gelder[2]

[1] Technische Universität Wien
http://www.kr.tuwien.ac.at/staff/{lonsing,egly}
[2] University of California, Santa Cruz
http://www.cse.ucsc.edu/~avg

Abstract. Recent solvers for quantified boolean formulas (QBF) use a clause learning method based on a procedure proposed by Giunchiglia et al. (JAIR 2006), which avoids creating tautological clauses. Recently, an exponential worst case for this procedure has been shown by Van Gelder (CP 2012). That paper introduced QBF Pseudo Unit Propagation (QPUP) for non-tautological clause learning in a limited setting and showed that its worst case is theoretically polynomial, although it might be impractical in a high-performance QBF solver, due to excessive time and space consumption. No implementation was reported.

We describe an enhanced version of QPUP learning that is practical to incorporate into high-performance QBF solvers, being compatible with pure-literal rules and dependency schemes. It can be used for proving in a concise format that a QBF formula is either unsatisfiable or satisfiable (working on both proofs in tandem). A lazy version of QPUP permits non-tautological clauses to be learned without actually carrying out resolutions, but this version is unable to produce proofs.

Experimental results show that QPUP is somewhat faster than the previous non-tautological clause learning procedure on benchmarks from QBFEVAL-12-SR.

1 Introduction

Solvers for Quantified Boolean Formulas (QBFs) are rapidly increasing in strength, partly due to increased understanding of how to incorporate conflict-driven clause learning (CDCL), which found great practical success in propositional satisfiability. Several current solvers are patterned after the Q-resolution method described by Giunchiglia *et al.* [3]. A thorough survey of the field through 2005 may be found in this paper.

We present a formal framework for solvers to search for proofs that an instance is true or false in tandem, and to use information found in one proof search to assist in the other proof search. The framework is closely related to the work of Zhang and Malik [19], and Giunchiglia *et al.* [3], but uses the idea of a *Complete Guard Formula* and a closely related *Basic Guard Formula* to prove some key properties straightforwardly. This approach differs from that used for CirQit2 [4] and GhostQ [7] in that those solvers use a representation in which the entire negated formula is known, whereas the guard formula is only partially known.

---

[*] The first two authors are supported by the Austrian Science Fund under grant S11409-N23.

M. Järvisalo and A. Van Gelder (Eds.): SAT 2013, LNCS 7962, pp. 100–115, 2013.
© Springer-Verlag Berlin Heidelberg 2013

This paper then presents a prototype implementation of QBF Pseudo Unit Propagation (QPUP) non-tautological clause learning. The main innovation is to identify a cut in the conflict graph such that the learned clause associated with the cut will be *non-tautological* and *asserting*: that is, after backtracking, the learned clause will enable a new literal to be implied immediately. After the cut is identified, resolution proceeds from clauses on the boundary of the cut toward the conflicting clause, in an order such that tautologies cannot occur. Previous techniques performed resolutions beginning at the conflicting clause and working back toward the boundary of the cut, without ever explicitly defining the cut. When the latter order is used, tautologies can occur and require possibly expensive special treatment (exponential time in the worst case).

All clauses derived during the QPUP procedure satisfy an invariant that permits an asserting learned clause to be constructed lazily, without actually performing the resolutions. This shortcut saves substantial clause-learning time, but does not permit a detailed Q-resolution proof to be output as a by-product, and occasionally constructs a slightly weaker learned clause.

An implementation of QPUP learning was incorporated in the open-source solver DepQBF [8,10,9]. Experimental results (Section 6) illustrate the potential of QPUP learning compared to traditional clause learning in terms of more solved formulas, fewer backtracks and reduced run times.[1]

2   Preliminaries

In this section, we collect specific notation for later use. In general, *quantified boolean formulas* (QBFs) generalize propositional formulas by adding universal and existential quantification of boolean variables. See [6] for a thorough introduction. For this paper QBFs are in prenex conjunctive normal form (PCNF): $\Psi = \vec{Q}.\mathcal{F}$ consists of prenex $\vec{Q}$ and clause-matrix $\mathcal{F}$ (the *original* clauses). The prenex $\vec{Q}$ is partitioned into maximal contiguous subsequences of the same quantifier type, called *quantifier blocks*. Each quantifier block has a different *qdepth* with the outermost block having *qdepth* $= 1$. If p and q are variables of opposite quantifier types, we say that q is **inner** to p if $qdepth(p) < qdepth(q)$.

Clauses may be written as literals enclosed in square brackets (e.g., $[p, q, \overline{r}]$), and $[]$ denotes the empty clause. Where the context permits, letters e and others near the beginning of the alphabet denote existential literals, while letters u and others near the end of the alphabet denote universal literals. Letters like p, q, r denote literals of unspecified quantifier type. We use $\perp$ to denote the constant false in the role of a literal. The variable underlying literal p is denoted by $|p|$.

A *closed* QBF (all variable occurrences are quantified) evaluates to either 0 (false) or 1 (true), as defined by induction on its principal operator: **(1)** $(\exists x\, \phi(x)) = 1$ iff $\phi(0) = 1$ or $\phi(1) = 1$. **(2)** $(\forall x\, \phi(x)) = 0$ iff $\phi(0) = 0$ or $\phi(1) = 0$. **(3)** Other operators have the same semantics as in propositional logic. This definition emphasizes the connection of QBF to two-person games with complete information, in which player E (Existential) tries to set existential variables to make the QBF evaluate to 1, and

---

[1] Please visit http://www.kr.tuwien.ac.at/staff/lonsing/sat13submission.tar.gz for a longer version of this paper, binaries, and some logs.

player A (Universal) tries to set universal variables to make the QBF evaluate to 0. Players set their variable when it is outermost, or for non-prenex, when it is the root of a subformula (see [7] for more details). Only one player has a winning strategy.

The proof system known as *Q-resolution* consists of two operations, *resolution* and *universal reduction* (some papers combine them into one operation). Q-resolution is of central importance for QBFs because it is a sound and complete proof system [5].

Definition 2.1. *Resolution* is defined as usual. The *resolvent* of C_1 and C_2 on existential e is denoted as $\mathsf{res}_q(C_1, C_2) = [(C_1 - \overline{e}), (C_2 - e)]$ or as $C_1 *_e C_2$. (We drop set-forming braces around singleton sets where obvious.) The resolvent must be non-tautologous for Q-resolution. *Universal reduction* is special to QBF. The notation is $\mathsf{unrd}_q(C_3) = (C_3 - u)$, where C_3 is non-tautologous and u is tailing for C_3. Here, a universal literal u is said to be *tailing* for C_3 if no existential literal in C_3 is inner to u. A postfix operator notation for the same expression is $C_3 \, \Delta_u$. Performing all possible universal reductions on C_3 is denoted by $\mathsf{unrd}_*(C_3)$ or $C_3 \, \Delta_*$, and the resulting clause is said to be *fully reduced*. The operators are left-associative, like $+$ and $-$, so that compound expressions without parentheses can be read left to right.

A *Q-derivation* of a clause is a proof using the Q-resolution operations; a *Q-refutation* is a Q-derivation of the empty clause. ☐

Definition 2.2. An *assignment* (sometimes called a *partial assignment*) is a partial function from variables to truth values, and is usually represented as the set of literals that it maps to true. A *total assignment* assigns a truth value to every variable. Assignments are denoted by σ, τ, etc. Applications of σ to logical expressions are denoted by $q\lceil_\sigma$, $C\lceil_\sigma$, $\mathcal{F}\lceil_\sigma$, etc., and consist of replacing assigned variables in the expression by their truth values in σ, then simplifying with truth-value identities (but not propagating unit clauses). If σ assigns variables that are quantified in Ψ, those quantifiers are deleted in $\Psi\lceil_\sigma$, and their variables receive the assignment specified by σ. ☐

3   Guard Formulas

This section describes *guard formulas*, and states their main properties that are important for QBF solving. We begin with some terminology.

Definition 3.1. Let the initial PCNF formula be $\Psi = \overrightarrow{Q} . \mathcal{F}$, where $\overrightarrow{Q}$ is the quantifier prefix and $\mathcal{F}$ is the matrix of clauses. A *consistent minimal hitting set* (**cmhs**) for a set of clauses $\mathcal{F}$ is a partial assignment σ (regarded as a consistent set of literals) such that every clause $C \in \mathcal{F}$ is satisfied by σ and no proper subset of σ has this property. Note that $\mathcal{F}$ has no hitting set if it is propositionally unsatisfiable.

Let $\widetilde{Q}$ be the same as $\overrightarrow{Q}$ except that the quantifier type of each variable is inverted. The *Complete Guard Formula* for Ψ is the PCNF $\Gamma_* = \widetilde{Q} . \mathcal{G}_*(\mathcal{F})$, where $\mathcal{G}_*(\mathcal{F})$ is the set of clauses defined as follows:

$$C \in \mathcal{G}_*(\mathcal{F}) \text{ if and only if } \neg(C) \text{ is a cmhs for } \mathcal{F} \qquad (1)$$

Here, $\neg(C)$ is the partial assignment consisting of the negations of all literals in C. (The negation of a clause is often called a *cube*.) ☐

Ψ	$\forall u$	$\exists d$	$\exists a$	$\forall v$	$\exists b$	$\exists c$	$\exists e$
C_1	u	$\bar{d}$	a				
C_2	$\bar{u}$		a	$\bar{v}$		$\bar{c}$	$\bar{e}$
C_3	$\bar{u}$						e
C_4		$\bar{d}$	a	v	$\bar{b}$		
C_5		d	$\bar{a}$	$\bar{v}$	b		
C_6		d				c	
C_7			$\bar{a}$	v	$\bar{b}$	$\bar{c}$	
C_8				v	b		
C_9				$\bar{v}$	$\bar{b}$		
C_{10}				v		c	

Γ_*	$\exists u$	$\forall d$	$\forall a$	$\exists v$	$\forall b$	$\forall c$	$\forall e$
D_1	u	d	a	v	$\bar{b}$	$\bar{c}$	
D_2	u	d	a	$\bar{v}$	b	$\bar{c}$	
D_3	$\bar{u}$	$\bar{d}$		$\bar{v}$	b	c	$\bar{e}$
D_4		$\bar{d}$	$\bar{a}$	$\bar{v}$	b		$\bar{e}$
D_5		d	a	v	$\bar{b}$	$\bar{c}$	$\bar{e}$

Γ_B	$\exists u$	$\forall d$	$\forall a$	$\exists v$	$\forall b$	$\forall c$	$\forall e$
D_1	u	d	a	v			
D_2	u	d	a	$\bar{v}$			
D_3	$\bar{u}$	$\bar{d}$		$\bar{v}$			
D_4		$\bar{d}$	$\bar{a}$	$\bar{v}$			
D_5		d	a	v			

Fig. 1. QCNFs Ψ, Γ_*, and Γ_B, discussed in Example 3.6 and later examples

Since Γ_* is a PCNF we may exploit the soundness and completeness of Q-resolution for PCNF refutations. In general discussions we call Ψ the *original formula* and call Γ_* and related formulas *guard formulas*.

Definition 3.2. If two propositional formulas $\mathcal{F}_1$ and $\mathcal{F}_2$ evaluate to the same truth value for every total assignment, then $\mathcal{F}_1 \equiv \mathcal{F}_2$, read as $\mathcal{F}_1$ and $\mathcal{F}_2$ are **propositionally equivalent**. If every total assignment that satisfies $\mathcal{F}_1$ also satisfies $\mathcal{F}_2$, then $\mathcal{F}_1 \models \mathcal{F}_2$, read as $\mathcal{F}_1$ **logically implies** $\mathcal{F}_2$. $\qquad\square$

The idea of cubes is familiar in QBF, but Γ_* has this important property, which has not been enunciated before, as far as we know (proofs are omitted to save space):

Lemma 3.3. With the notation of Definition 3.1: **(A)** $\mathcal{G}_*(\mathcal{F}) \equiv \neg\mathcal{F}$; **(B)** Γ_* has the opposite truth value from Ψ. $\qquad\blacksquare$

Definition 3.4. Let Ψ and Γ_* and $\mathcal{F}$ and $\mathcal{G}_*$ be as defined in Definition 3.1. The **Basic Guard Formula** for Ψ is the PCNF $\Gamma_B = \widetilde{Q}.\,\mathcal{G}_B(\mathcal{F})$, where $\mathcal{G}_B(\mathcal{F})$ is the set of clauses arising by performing all possible universal reductions on clauses in $\mathcal{G}_*(\mathcal{F})$. (Recall that the quantifier type of each variable in $\widetilde{Q}$ is the opposite of its quantifier type in $\overrightarrow{Q}$.) $\qquad\square$

Lemma 3.5. With the notation of Definition 3.4: **(A)** Γ_B has the opposite truth value from Ψ; **(B)** Γ_B has a Q-refutation if and only if the truth value of Ψ is **true**; **(C)** For any $C \in \mathcal{G}_B(\mathcal{F})$, $\neg(C)$ can be extended to a cmhs for $\mathcal{F}$ by adding only literals e_i such that e_i is inner to some $u \in \neg(C)$, where e_i is existential and u is universal in $\overrightarrow{Q}$. $\qquad\blacksquare$

Example 3.6. Let Ψ be the QCNF shown in chart form on the left of Figure 1. The corresponding Γ_* and Γ_B are shown on the right. Γ_B does not admit a Q-refutation, so we conclude by Lemma 3.5 that Ψ is **false**. $\qquad\square$

It is not practical for a solver to construct $\mathcal{G}_*(\mathcal{F})$ or $\mathcal{G}_B(\mathcal{F})$ explicitly since the sizes of these clause sets might be anywhere from empty to exponentially larger than $\mathcal{F}$. Instead, solvers (ideally) discover clauses in $\mathcal{G}_*(\mathcal{F})$ as the proof search goes along, reduce them to clauses in $\mathcal{G}_B(\mathcal{F})$, and record them. (Some implementations may discover non-minimal consistent hitting sets and not extract a cmhs.) Any partial assignment that satisfies all clauses in $\mathcal{F}$ is a consistent hitting set. If at any point the solver discovers a Q-refutation of whatever guard clauses have been discovered, Ψ is proven to be **true**.

4   QBF Conflict-Driven Clause-Learning Solvers (Review)

The **QBF conflict-driven clause-learning** (QCDCL) strategy for PCNF solving is inspired by the great success of conflict-driven clause-learning (CDCL) in propositional SAT solving [11,12,2]. Although CDCL is often described in the literature as a variant of DPLL with learning added, it has been argued that the idea of CDCL is actually quite different [11,16].

Due to universal quantifiers, the QCDCL strategy becomes considerably more technical. To date, the most in-depth treatment in the literature is found in Giunchiglia *et al.* [3], although they call it Q-DLL-LN. This section reviews the main ideas so that QPUP clause learning may be placed in context.

4.1   QCDCL Rounds

A QCDCL solver proceeds by *rounds*. Initially a PCNF $\Psi = \vec{Q}.\mathcal{F}$ is known and $\Gamma_* = \vec{Q}.\mathcal{G}_*(\mathcal{F})$ (see Section 3) is unknown. As clauses related to $\mathcal{G}_*(\mathcal{F})$ are discovered, they are added to an (initially empty) set of *guard clauses* $\mathcal{G}$. Each round proceeds by *decision levels* until a terminating event occurs. Assignments accumulate throughout a round in a sequence often called the **trail**, τ. Each assignment is applied to Ψ (and is implicitly applied to Γ_*), giving $\Psi\lceil_\tau$ and $\Gamma_*\lceil_\tau$, and is appended to τ before the next assignment is made. Assignments have categories, such as "assumption," or one of the safe assignments detailed in Section 4.2, to facilitate clause learning after a conflict. The *alevel* (assignment level) of a literal is the decision level at which it was assigned.

At decision-level 0, beginning with an empty τ, *safe* assignments are applied to $\Psi\lceil_\tau$ and $\mathcal{G}\lceil_\tau$, where $\mathcal{G}$ is whatever part of Γ_B has been discovered and recorded. *Safe* assignments are those that cannot change the truth value of $\Psi\lceil_\tau$ and $\Gamma_*\lceil_\tau$. Safe variable assignments continue until no more are found, or until a terminating event occurs.

At positive decision levels, the first variable assignment is an *assumption*, which is *unsafe* (may change the truth value of $\Psi\lceil_\tau$ or $\Gamma_*\lceil_\tau$). For soundness, the assumption literal must not be inner to any unassigned literal.

Subsequent variable assignments on the same decision level are *safe* for $\Psi\lceil_\tau$ and $\Gamma_*\lceil_\tau$. Safe variable assignments continue until no more are found, or until a terminating event occurs. For all decision levels, if no terminating event has occurred, the decision level is increased by 1, a new assumption is made, and the higher decision level continues with additional safe variable assignments, as just described.

4.2   Safe QCDCL Assignments

Perhaps the major complication in going from CDCL to QCDCL is the number of safe assignments to be managed. The first complication is that there are two quite different PCNF formulas being updated and a safe assignment needs to be safe for both formulas. In many cases, it is obviously safe for one formula, but not so clear for the other.

The safe assignments typically implemented in QCDCL-based solvers are:

Unit-clause implication: If $C \in \mathcal{F}$ and $C\lceil_\tau$, followed by universal reductions, contains the single existential literal e, append e to τ.

Guard unit implication: If $C \in \mathcal{G}$ and $C\lceil_\tau$, followed by universal reductions, contains the single existential literal u, based on $\widetilde{Q}$, append u to τ.
Existential pure-literal rule: If the existential literal e appears in some clause in $\mathcal{F}\lceil_\tau$, and $\overline{e}$ does not appear in $\mathcal{F}\lceil_\tau$, then append e to τ.
Universal pure-literal rule: If the universal literal u appears in some clause in $\mathcal{F}\lceil_\tau$, and $\overline{u}$ does not appear in $\mathcal{F}\lceil_\tau$, then append $\overline{u}$ to τ.

4.3   QCDCL Terminating Events

There are three kinds of terminating events for a round:

(1) All clauses in $\mathcal{F}$ are satisfied; i.e., $\mathcal{F}\lceil_\tau$ is empty. Then the assignments in τ comprise a consistent hitting set for $\mathcal{F}$ and a clause of $\mathcal{G}_*(\mathcal{F})$ may be discovered by finding a cmhs and negating it. (In practice an implementation might settle for an over-approximation of the cmhs.) This clause is simplified by universal reductions, keeping in mind that the quantifier types are now dictated by $\widetilde{Q}$, yielding a new guard clause G. If a cmhs was used, $G \in \Gamma_B$, otherwise it is subsumed by some clause in Γ_B. G is added to $\mathcal{G}$. Applying τ falsifies G, since it was a negated hitting set before the universal reductions. Next, assignments are retracted from τ, a complete decision level at a time, from highest to lower, until the remaining assignments no longer falsify G.

(2) A *conflict* occurs in some clause $C \in \mathcal{F}$, meaning that τ, coupled with universal reductions in C, have *falsified* C, i.e., $C\lceil_\tau$, followed by universal reductions, is an empty clause. In this case a new clause D is derived from Ψ by Q-resolution and added to $\mathcal{F}$. $D\lceil_\tau$, followed by universal reductions, also is an empty clause. Next, assignments are retracted, a complete decision level at a time, from highest to lower, until the remaining assignments, followed by universal reductions, no longer falsify D.

(3) A conflict occurs in some clause $C \in \mathcal{G}$, meaning that $C\lceil_\tau$, followed by universal reductions, is an empty clause. Note that the only clauses in $\mathcal{G}$ are those added by earlier instances of case 1 and this case. In this case a new guard clause E is derived from $\widetilde{Q}.\mathcal{G}$ by Q-resolution[2] and added to $\mathcal{G}$. $E\lceil_\tau$, followed by universal reductions, also is an empty clause. Next, assignments are retracted, a complete decision level at a time, from highest to lower, until the remaining assignments, followed by universal reductions, no longer falsify E.

If D in case 2 is an empty clause, the truth value of Ψ can be proven to be **false**. If E in case 3 is an empty clause, the truth value of Ψ can be proven to be **true**. It is straightforward in principle to extract either Q-refutation from a trace of the solver's actions [1,13]. If G in case 1 is an empty clause, the truth value of Ψ can be shown to be true simply by applying the hitting set underlying G to Ψ, because the hitting set contains only existential variables (based on $\overrightarrow{Q}$). In these cases, the instance is solved.

If no empty clause has been derived, another round is started. The unretracted part of τ from the previous round is the initial value of τ for the next round. Whatever decision levels were not retracted remain in place. Safe assignments continue on the highest unretracted decision level.

---

[2] Previous descriptions of this general strategy in the literature speak of "cubes," and "term resolution," as separate concepts, but the formulation as a guard formula needs only Q-resolution.

Definition 4.1. An *asserting clause* in the context of a trail σ is a clause C such that $C\lceil_\sigma$ satisfies the conditions for unit-clause implication in the original formula or for guard unit implication in the guard formula (Section 4.2). I.e., $C\lceil_\sigma$ has only one existential literal e (possibly $\perp$), called the *asserting literal*, and e is not inner to any universal literals in $C\lceil_\sigma$. This terminology is primarily used when σ is the trail after backtracking from τ; i.e., σ is a proper prefix of τ.　　　□

An important optimization is that the learned clause, D in case 2 or E in case 3, is asserting after backtracking, so at least one safe assignment is available before a new assumption is needed. When the learned clause is asserting, it is usual to backtrack as many decision levels as possible, while maintaining the asserting property.

By the nature of a round, an asserting learned clause cannot be subsumed by an already known clause, or else it would have become a unit-clause implication or guard unit implication before the completion of the decision level to which the round backtracked after learning this clause. We know this did not happen because at least one higher decision level was started before the asserting clause was learned. Since every round learns a clause, it follows that the number of rounds is finite (as long as all learned clauses are remembered).

5　Learning with QBF Pseudo Unit Propagation (QPUP)

We introduce a practical version of QPUP which can be used to derive a learned clause from a conflict graph. A simple version of QPUP uses the entire conflict graph and was introduced as a theoretical, rather than practical construct [17]. Its point was to show that a non-tautological asserting clause (Definition 4.1) could be learned in time polynomial in the size of the conflict graph (whereas the published methods of learning a non-tautological clause might take exponential time).

This section describes how to selectively apply QPUP after a conflict has occurred, keeping operations confined to recent decision levels as far as practical, in the spirit of the propositional CDCL strategy. This more sophisticated version of QPUP learning is deferred to Section 5.2, until additional nomenclature has been introduced. The procedure is essentially the same for conflicts in the original formula and the guard formula.

5.1　QCDCL Conflict Graphs

During a QCDCL search suppose a falsified clause is encountered after a sequence of assumptions, unit-clause implications, and otherwise-assigned literals, as described in Section 4. Each literal in the trail τ is either an *assumption* or an *implied literal* or an *otherwise-assigned literal* (any other safe assignment).

Definition 5.1. A clause is *effectively unit* in the context of τ if the restriction based on τ, followed by universal reductions (see Definition 2.1), makes the clause a unit clause. A clause is said to be *effectively empty* (or *falsified*) in the context of a partial assignment τ if the restriction based on τ, followed by any applicable universal reductions, makes the clause an empty clause. The *antecedent* clause of an implied literal p (denoted $ante(p)$) is the (unique) clause that became effectively unit earliest to imply p. If a clause became effectively empty, we say that $\perp$ is the "implied literal".

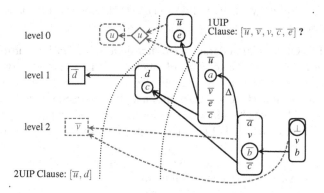

Fig. 2. QCDCL conflict graph; see Example 5.2 and later examples. Circles enclose implied literals. Boxes at the left enclose assumptions. Diamonds enclose otherwise-assigned literals. Rounded boxes enclose an antecedent clause. The antecedent clause of a universal implied literal is in the opposite formula and cannot be used for resolution. The rounded box has dashed lines to denote this. Dashed arrows show connections to universal literals. The dotted arcs show various cuts. A solid arrow crossing a cut goes to an existential literal whose negation is definitely in the QCDCL learned clause associated with that cut (cf. propositional CDCL). A dashed arrow always crosses the cut and goes to a universal literal, but its negation may not be in the learned clause.

The *conflict graph* associated with a falsified clause is the rooted directed acyclic graph (DAG) in which $\perp$ is the root vertex and its antecedent is the falsified clause. The remaining vertices are the assumptions, implied literals, and otherwise-assigned literals reachable from $\perp$, based on the directed edges.

The *directed edges* of the conflict graph are (p, q), where p and q are vertices in the conflict graph, p is existential, and $\overline{q} \in ante(p)$. See Figure 2.

Otherwise-assigned literals in the conflict graph for the original formula can arise through the *universal pure-literal rule* and through *guard unit implication*. Like assumptions, otherwise-assigned literals have no antecedent. Existential pure literals are not vertices in the conflict graph of an original formula.

The only differences in a conflict graph for the guard formula are that the vertices with antecedents are guard unit implications (which are existential literals in the context of the guard formula) and the otherwise-assigned literals are *existential pure literals* in the original formula or *unit-clause implications* in the original formula. Both of these literal types are *universal* in the context of the guard formula. Universal pure literals are not in the guard conflict graph.   □

Example 5.2. For concreteness we suppose that the QCDCL proof search assumes negative literals. Figure 2 shows a conflict graph for the original formula of Figure 1 that occurs in round 4. In the first three rounds the guard clauses D_1 and D_2 were discovered and the guard clause $[u]$ was derived, causing all decision levels to be retracted.

At the beginning of round 4, u is a guard unit implication at level 0, so it is an otherwise-assigned literal for purposes of the original formula. Now e is implied via C_3. Round 4 continues as follows: Level 1: assume $\overline{d}$; imply c via C_6; imply a via C_2.

Level 2: assume $\overline{v}$; imply $\overline{b}$ via C_7; imply $\perp$ via C_8. Discussion is continued in several subsequent examples. □

Definition 5.3. In a QCNF conflict graph a *conflict-generating cut* is a partition of the vertices in the *reason side* and the *conflict side* such that: **(A)** $\perp$ is on the conflict side; **(B)** every vertex on the conflict side is reachable from $\perp$ by a directed path using only vertices on the conflict side; **(C)** every assumption and otherwise-assigned literal is on the reason side. We abbreviate "conflict-generating cut" to "cut" in later discussions.

A *unique implication point* **(UIP)** p is an existential vertex such that all edges from existential literals that are reachable from $\perp$ and are assigned later than p go to: **(D)** p or **(E)** an existential literal assigned later than p or **(F)** an existential literal assigned at a decision level less than p or **(G)** a universal literal. □

In propositional CDCL, the clause associated with a cut consists of the negations of literals on the reason side that are reached by a single edge from some literal on the conflict side. In QCNF this clause might not be derivable in Q-resolution because universal literals can result in tautologous resolvents. Moreover, some universal literals might be removable by universal reduction.

The *UIP cut* for a UIP p in traditional CDCL places all existential literals that are assigned later than p and are in the conflict graph on the conflict side, and places all other literals in the conflict graph on the reason side. This paper refines the definition of the *UIP cut* in the context of QCDCL to permit certain existential literals that are assigned earlier than p to appear on the conflict side. This might be necessary to be able to associate a *non-tautological* clause with the cut.

Example 5.4. Figure 2 shows a typical situation where the traditional UIP cut would derive a tautological clause. The dotted arcs show some conflict-generating cuts. Vertex c is a UIP. The traditional associated 1UIP cut is shown by the rightmost dotted arc. The clause associated with this cut, obtained by resolving the clauses on the conflict side, is tautological. Neither v nor $\overline{v}$ can be eliminated by universal reduction using only clauses on the conflict side because they are always blocked by b or $\overline{c}$ or $\overline{e}$.

Long distance resolution was proposed by Zhang and Malik [19] to accommodate such tautological clauses. Giunchiglia and co-authors pioneered the derivation of learned clauses using Q-resolution and avoiding tautological clauses [3]. □

5.2   QPUP Clauses

The idea of QPUP is that resolutions are performed that mimic the derivation of the implied literal by unit-clause resolution, treating certain earlier-assigned existential literals as unit clauses. However, the *negations* of those earlier-assigned literals are included in the QPUP clause. In this sense, QPUP extends propositional Pseudo Unit Propagation (PUP) [18].

Definition 5.5. Let a conflict graph and a conflict-generating cut be given, as described in Section 5.1. Recall *alevel* (assignment level) from Section 4.1. Each assumption increases the *alevel* by one and subsequently assigned literals up to the next assumption take this value for their *alevel*. Let *dlevel* be the decision level at which the conflict occurred.

A function $qpup(e,\ a\ell)$ is any function that returns a clause for each existential literal e on the conflict side of the cut, including $\perp$, with certain properties:

1. If all edges leaving e go to the reason side, then $qpup(e,\ a\ell) = ante(e)$.
2. If $\overline{r} \in qpup(e,\ a\ell)$, then r is on the reason side and is reachable by a single edge from some vertex on the conflict side.
3. Let A be the set of existential literals in $qpup(e,\ a\ell)$ with $alevel \le a\ell$. If $\overline{r} \in qpup(e,\ a\ell)$ and r is universal and r would be unassigned after retracting all assignments with $alevel > a\ell$, then r is tailing in $(qpup(e,\ a\ell) - A - e)$.
4. A clause that subsumes $qpup(e,\ a\ell)$ can be derived by Q-resolution from $ante(e)$ and the clauses $qpup(r_i,\ a\ell)$ such that $\overline{r_i} \in ante(e)$ and r_i is on the conflict side. (Note that a clause subsumes itself.) $\quad\square$

The properties are well defined because the conflict graph is acyclic. Note that properties 1 and 4 ensure that $qpup(e,\ a\ell)$ is not tautological.

Not every cut permits $qpup$ to be defined. Suppose $qpup$ can be defined for a given cut and a backtrack level $a\ell < dlevel$. The important invariant for $D = qpup(e,\ a\ell)$ is

$$\mathsf{unrd}_*(qpup(e,\ a\ell) - \{e\}) \subseteq qpup(\perp,\ a\ell) \tag{2}$$

I.e., if e is removed from D and all possible universal reductions are performed in $D - e$, then the remaining literals are in $qpup(\perp,\ a\ell)$. Note that the invariant holds also when $e = \perp$. Given a conflict graph, the goal is to identify a suitable conflict-generating cut such that the clause $qpup(\perp,\ a\ell)$ is non-tautological and asserting and hence can be used as a learned clause.

Example 5.6. Referring again to Figure 2, it is easy to see by inspection that $qpup$ cannot be defined for the rightmost cut for any $a\ell < dlevel$, because a is on the conflict side and $\overline{a}$ cannot be resolved out without creating a tautology.

For the leftmost cut, $qpup$ is easily defined for $a\ell = 0$. For c and e their $qpup$ is their antecedent. Then $qpup(a, 0) = ante(a) *_e qpup(e, 0) *_c qpup(c, 0) \Delta_{\overline{v}} = [\overline{u}, d, a]$. This can be resolved against the clause with $\overline{a}$, etc., without creating a tautology. Finally, $qpup(\perp, 0) = [\overline{u}, d]$, which is asserting for level 0.

Note that $alevel(e) < alevel(\overline{d})$, but including e on the conflict side is necessary to enable tautologies to be avoided. Finally, it is worth noting that traditional QCDCL based on [3] derives the weaker clause $[\overline{u}, d, \overline{v}, \overline{e}]$ by starting at the falsified clause $[v, b]$ and resolving on b, c, a, and again on c. $\quad\square$

Our principal contribution is the description (in the next section) and experimental evaluation of a practical procedure to identify a cut and a value of $a\ell$ such that $qpup$ is efficient to compute and has the further property that $qpup(\perp,\ a\ell)$ is non-tautological and asserting after backtracking.

5.3 Cuts and Backtrack Levels for QPUP

This section provides an abstract description of our procedure to identify a suitable cut and backtrack level for QPUP after a conflict. For simplicity a conflict in the original formula is assumed.

The key idea is that an **agenda** is constructed and processed. The agenda is a sequence of literals in trail-assignment order that are relevant to deriving the needed *qpup* clauses. As an exception, unassigned universal literals appear with the same clause that caused them to be inserted. Processing the agenda corresponds to finding a suitable cut in iterative fashion. The procedure is illustrated by Examples 5.7 and 5.8 to make it easier to follow the general description.

Starting from an empty cut, existential literals are put on the conflict side until a UIP cut is found. If that cut cannot be associated with a non-tautological clause, processing continues and further literals are put on the conflict side, thus modifying the cut. During processing, no clauses are constructed explicitly. The state of the agenda is inspected to check if the clause associated with the current cut is non-tautological and asserting.

Literals are annotated with some status information. We use these graphical mnemonics: p/k denotes that $alevel(p) = k$ (∞ denotes it is unassigned); $\textcircled{e}$ means that e must be on the conflict side and its *qpup* must eventually be computed to produce a Q-derivation of the learned clause; $\boxed{q}$ denotes that q is a suitable UIP, i.e., *qpup* will be computed with $a\ell = (alevel(q) - 1)$ and $\overline{q}$ will be the asserting literal in $qpup(\bot, a\ell)$, the learned clause.

Literals are processed from right to left, i.e., in reverse trail order. Universal literals are skipped over and each existential literal is processed just once. Processing may cause other literals to be inserted into the agenda, but such inserts are always to the left of the literal being processed, i.e., in the part of the agenda yet to be processed.

In general terms, the processing has two phases: In the first phase a suitable UIP literal is identified and this establishes the value of $a\ell$ for which *qpup* is needed. In the second phase the complete cut is identified.

The agenda is initialized with the literals in $ante(\bot)$, with $\bot$ rightmost, unassigned universals immediately to its left, and remaining literals further to the left, in trail order. For phase 1, proceeding right to left, examine the next existential literal, say $\overline{e}/k$.

(1) If some literal to the left of $\overline{e}$ has $alevel = k$, e must go on the conflict side. **(2)** If e is inner to some pair of complementary universal literals in the agenda, e must go on the conflict side. **(3)** If e is inner to some universal literal u/m in the agenda where $m \geq k$, e must go on the conflict side.

If none of the above tests determines that e must go on the conflict side, then change the notation to $\overline{\mathbf{e}}/k$, denoting that $\overline{e}$ will be the asserting literal, set $a\ell = k - 1$, and proceed to phase 2.

Otherwise, replace $\overline{e}/k$ with $ante(e)$ in the agenda, as follows: $\textcircled{e}$ replaces $\overline{e}/k$; unassigned universal literals that are not already in the agenda are inserted immediately to the left of $\textcircled{e}$; remaining literals of $ante(e)$ that are not already in the agenda are inserted in correct trail order (somewhere to the left of $\textcircled{e}$), annotated with their *alevels*. Continue phase 1 to the left of $\textcircled{e}$.

If phase 1 reaches the left end of the agenda without identifying an asserting literal, then $qpup(\bot, 0)$ will reduce to the empty clause.

Example 5.7. Referring again to Figure 2, Table 1 shows the evolution of the agenda through phase 1. After step 1, $\overline{a}$ cannot be the asserting literal because $\overline{c}$ is earlier on the same *alevel*. After step 2, $\overline{c}$ cannot be the asserting literal because it is inner to v

Table 1. Agenda processing discussed in Examples 5.7 and 5.8

Step					Agenda				
0							$v/2$	$b/2$	⊥⃝
1					$\bar{c}/1$	$\bar{a}/1$	$v/2$	$\bar{b}$⃝	⊥⃝
2	$\bar{u}/0$	$\bar{e}/0$	$\bar{c}/1$		$\bar{v}/2$	a⃝	$v/2$	$\bar{b}$⃝	⊥⃝
3	$\bar{u}/0$	$\bar{e}/0$	$d/1$	c⃝	$\bar{v}/2$	a⃝	$v/2$	$\bar{b}$⃝	⊥⃝
4	$\bar{u}/0$	$\bar{e}/0$	d⃞$/1$	c⃝	$\bar{v}/2$	a⃝	$v/2$	$\bar{b}$⃝	⊥⃝
					end phase 1				
5	$\bar{u}/0$	e⃝	d⃞$/1$	c⃝	$\bar{v}/2$	a⃝	$v/2$	$\bar{b}$⃝	⊥⃝

and $\bar{v}$. After step 3, d can be the asserting literal because it is *not* inner to v and $\bar{v}$. Step 4 terminates phase 1. □

Phase 2 begins immediately to the left of the asserting literal in the agenda. Proceeding right to left as in phase 1, examine the next existential literal, say $\bar{f}/j$. We know $j \leq a\ell$. If f is inner to some pair of complementary universal literals in the agenda, f must go on the conflict side. Otherwise, $\bar{f}/j$ remains unchanged in the agenda and processing moves left. If f must go on the conflict side, replace $\bar{f}/j$ with $ante(f)$ in the agenda, following the same procedure as above for replacing $\bar{e}/k$ with $ante(e)$. In particular f⃝ now appears in the agenda and phase 2 continues to the left of f⃝. Phase 2 terminates after the leftmost literal in the agenda has been processed. At this point, the UIP cut has been found and is associated with the non-tautological asserting clause $qpup(\bot, a\ell)$, which is to be learned.

Example 5.8. Continuing from Example 5.7, the end of Table 1 shows the evolution of the agenda through phase 2. After step 4, $\bar{e}$ cannot be on the reason side because it is inner to v and $\bar{v}$, so it is replaced by $ante(e)$. After step 5, no existential literals inner to v and $\bar{v}$ remain on the reason side and processing terminates. Literal d is the UIP. Literals e, c, a and $\bar{b}$ are on the conflict side. □

After termination of agenda processing, clauses $qpup(e, a\ell)$ are computed for every existential literal e on the conflict side. The computations are done *in* trail order, i.e., from left to right in the agenda. The order is important because the computation of some $qpup(e_2, a\ell)$ might use $qpup(e_1, a\ell)$ if e_1 is left of e_2 in the agenda, i.e., earlier on trail. Finally, the clause $qpup(\bot, a\ell)$ is computed. Example 5.6 illustrates the computation of the QPUP clauses referring to the agenda from Example 5.8.

Lemma 5.9. The agenda-based procedure presented above **(A)** has worst-case run time which is polynomial in the size of the conflict graph and **(B)** allows to derive a non-tautological asserting learned clause $C = qpup(\bot, a\ell)$. ■

5.4 Lazy QPUP

The agenda-based approach from the previous section allows for a lazy form of QPUP learning where *no* resolutions are carried out. In effect, it inspects the final agenda to

Table 2. Running times in seconds on *qdpllexp* family. "segv" denotes "segmentation violation".

family index	18	19	20	21	22	23
QuBE 1.3	10	22	47	105	segv	segv
DepQBF 0.1	8	16	32	69	140	298
CirQit 3.15	1	1	3	5	11	21
DepQBF QPUP	.00	.00	.00	.00	.00	.00

Note: DepQBF QPUP does not register any CPU time, even up to level 99. The run logs show that the resolution count increases by 8 for each level in the family.

determine which literals will be in $qpup(\bot, a\ell)$, i.e., the learned clause. No literals on the conflict side (those enclosed in circles) will appear; all existential literals on the reason side *must* appear. For universal literals, those that can be reduced out at the end and complementary pairs definitely will not appear. To be safe, other universal literals are kept, but they cannot prevent the learned clause from being asserting after backtracking to $a\ell$.

6   Experimental Results

We implemented a prototype version of QPUP learning and lazy QPUP as described in Sections 5.2 and 5.4 in the open-source search-based QBF solver DepQBF[3] for comparisons with traditional QCDCL. QPUP learning is applied to the original formula as well as to the guard formula. It is compatible with all the sophisticated techniques already in DepQBF, including pure literal detection, proof generation for certificate extraction [1,13], and the standard dependency scheme.

It has been reported that QCDCL clause learning, published as *Q-DLL-LN* [3] and used in other solvers (often under the label *QDPLL*), can spend time that is exponential in the size of the conflict graph to learn a single clause [17]. A family of small instances was given that elicits exponential behavior. Since the original motivation for QPUP was to avoid this behavior, we checked it on this family. The first three lines of Table 2, reproduced from [17], provide empirical confirmation of the theoretical analysis that the running time for traditional QDPLL learning doubles for each increase of one level in the family. The note shows that QPUP does not experience exponential growth.

To compare DepQBF with QPUP learning, lazy QPUP and traditional QCDCL, we considered the 276 preprocessed instances used for *QBFEVAL-12 Second Round* (QBFEVAL-12-SR). We did not apply any further preprocessing to these instances.[4] For additional tests we used preprocessed instances from QBFEVAL-10 which were not solved by preprocessing. We ran experiments on 64-bit Linux AMD Opteron 6176 SE with a time limit of 900 seconds and a memory limit of 7 GB.[5]

Table 3 shows a comparison of DepQBF with traditional QCDCL, QPUP learning and lazy QPUP. Figure 3 shows these results in a cactus plot. Lazy QPUP solves the largest number of instances, both overall and individually with respect to satisfiable and

---

[3] Please visit http://lonsing.github.com/depqbf/ for released versions. The version reported in this section is available from the first author.

[4] Visit http://fmv.jku.at/seidl/qbfeval2012r2/ and select "eval12bloqqer,"

[5] Please visit http://www.kr.tuwien.ac.at/staff/lonsing/sat13submission.tar.gz for binaries, logs and a longer report.

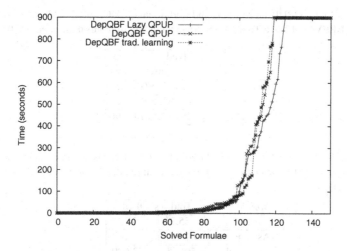

Fig. 3. Cactus plot of run times related to Table 3 for QBFEVAL-12-SR

unsatisfiable ones. The PAR10 time (i.e., average time with timeouts multiplied by 10) of lazy QPUP is moderately smaller than the time of the other two configurations.

Table 4 shows detailed statistics on instances solved by both QPUP learning and traditional QCDCL. Due to the large overlap in solved instances these statistics are comparable. (Lazy QPUP counts closely match QPUP.) QPUP is higher on most counts, but is 12% lower on backtracks. Additional tests on the 373 preprocessed QBFEVAL-10 instances that were not solved during preprocessing showed a similar pattern, and are omitted to save space.

When all three methods are forced to use the same variable re-weighting policy, it becomes clear that QPUP produces shorter learned clauses than traditional QCDCL. It is noteworthy that the *logical computation* (meaning the sequence of assumptions, safe assignments, learned clauses and backtracks) was exactly the same on 228 of the 273 instances solved by both methods on preprocessed QBFEVAL-10 instances. This confirms that QPUP almost always produces the same learned constraint as traditional QCDCL. For the 45 instances that did differ in their logical computations Table 5 shows some statistics. In one striking instance,[6] both methods learned the same first four clauses, but then the fifth learned clause by traditional QCDCL was a superset of that learned by QPUP, and contained 128 additional literals (534 vs. 406).

Lazy QPUP performed the same logical computation as QPUP on 97% of the 273 instances reported in Table 5 and used 111 CPU Seconds on average. In addition it solved one additional instance on which QPUP timed out, which we attribute to a faster procedure, not a different learning strategy.

The overall picture in Table 3, the reduced numbers of backtracks in Tables 4 and 5, and the reduced learned-clause lengths in Table 5 show the potential of QPUP learning. Higher resolution counts for QPUP are expected when QCDCL does not encounter conflicting universal literals, based on the analysis of PUP [18], but other benefits of QPUP appear to compensate.

---

[6] TOILET16.1.iv.32-shuffled

Table 3. DepQBF with QPUP, lazy QPUP, and traditional QCDCL on the 276 preprocessed instances from QBFEVAL-12-SR. Times are average including timeouts multiplied by 10 (PAR10).

	Solved (sat,unsat)	Time (avg.)
Trad. QCDCL	119 (62, 57)	5,148
QPUP	119 (63, 56)	5,151
Lazy QPUP	125 (65, 60)	4,963

Table 4. Comparison of QPUP learning and traditional QCDCL based on the runs from Table 3. Averages are based on 113 instances solved by both configurations. "Resolutions" and "length" are per learned clause.

Both Solved	Trad. vs. QPUP	
Time	55.21	51.93
Assignments	$9.1 \cdot 10^6$	$11.1 \cdot 10^6$
Backtracks	59,000	52,000
Resolutions	23.50	34.05
Length	53.58	82.50

Table 5. Comparison of traditional QCDCL and QPUP learning on the 45 instances with differing logical computation among the 273 preprocessed QBFEVAL-10 instances solved by both configurations. Statistics are average values and standard deviation (σ) of the difference.

	Trad. QCDCL	QPUP	Difference	σ of Diff.
No. Learned Clauses	228,666	154,379	74,287	210,332
Learned Clause Length	191.8	131.4	60.4	141.7
No. Clause Resolutions	$3.96 \cdot 10^6$	$4.80 \cdot 10^6$	$-0.84 \cdot 10^6$	$4.93 \cdot 10^6$
No. Learned Cubes	79,555	89,723	-10,168	54,825
Learned Cube Length	409.2	408.3	0.9	7.2
No. Cube Resolutions	51,674	107,167	-55,493	237,015
Backtracks	268,528	209,942	58,586	216,325
CPU Seconds	208	186	22	111

7   Conclusion

This paper presented *QPUP learning*, a novel approach to conflict-driven clause-learning (QCDCL) in QBF solvers. Given a conflict graph, the idea is to resolve on variables in the *same* order as they were assigned, rather than in reverse order. In contrast to traditional QCDCL, QPUP learning is a polynomial-time procedure.

The implementation of QPUP learning in DepQBF is compatible with sophisticated techniques like pure literals, dependency schemes [14,15,8], proof generation and hence also certificate extraction. Experimental results show the potential of QPUP learning but, at the same time, identified several procedural optimizations as future work.

Further research directions are comparison of certificates obtained by resolution proofs based on traditional QCDCL and QPUP learning, a detailed analysis of the learned clauses, and formally proving the correctness of QPUP learning in the context of guard formulas and dependency schemes, building upon the framework of *guard formulas*.

References

1. Balabanov, V., Jiang, J.R.: Unified QBF certification and its applications. Formal Methods in System Design 41, 45–65 (2012)
2. Eén, N., Sörensson, N.: An extensible SAT-solver. In: Giunchiglia, E., Tacchella, A. (eds.) SAT 2003. LNCS, vol. 2919, pp. 502–518. Springer, Heidelberg (2004)
3. Giunchiglia, E., Narizzano, M., Tacchella, A.: Clause/term resolution and learning in the evaluation of quantified boolean formulas. JAIR 26, 371–416 (2006)
4. Goultiaeva, A., Bacchus, F.: Exploiting QBF duality on a circuit representation. In: AAAI (2010)
5. Kleine Büning, H., Karpinski, M., Flögel, A.: Resolution for quantified boolean formulas. Information and Computation 117, 12–18 (1995)
6. Kleine Büning, H., Lettmann, T.: Propositional Logic: Deduction and Algorithms. Cambridge University Press (1999)
7. Klieber, W., Sapra, S., Gao, S., Clarke, E.: A non-prenex, non-clausal QBF solver with game-state learning. In: Strichman, O., Szeider, S. (eds.) SAT 2010. LNCS, vol. 6175, pp. 128–142. Springer, Heidelberg (2010)
8. Lonsing, F., Biere, A.: A compact representation for syntactic dependencies in QBFs. In: Kullmann, O. (ed.) SAT 2009. LNCS, vol. 5584, pp. 398–411. Springer, Heidelberg (2009)
9. Lonsing, F., Biere, A.: DepQBF: A dependency-aware QBF solver: System description. JSAT 7, 71–76 (2010)
10. Lonsing, F., Biere, A.: Integrating dependency schemes in search-based QBF solvers. In: Strichman, O., Szeider, S. (eds.) SAT 2010. LNCS, vol. 6175, pp. 158–171. Springer, Heidelberg (2010)
11. Marques-Silva, J.P., Sakallah, K.A.: GRASP–a search algorithm for propositional satisfiability. IEEE Transactions on Computers 48, 506–521 (1999)
12. Moskewicz, M., Madigan, C., Zhao, Y., Zhang, L., Malik, S.: Chaff: Engineering an efficient SAT solver. In: 39th Design Automation Conference (June 2001)
13. Niemetz, A., Preiner, M., Lonsing, F., Seidl, M., Biere, A.: Resolution-based certificate extraction for QBF (tool presentation). In: Cimatti, A., Sebastiani, R. (eds.) SAT 2012. LNCS, vol. 7317, pp. 430–435. Springer, Heidelberg (2012)
14. Samer, M.: Variable dependencies of quantified CSPs. In: Cervesato, I., Veith, H., Voronkov, A. (eds.) LPAR 2008. LNCS (LNAI), vol. 5330, pp. 512–527. Springer, Heidelberg (2008)
15. Samer, M., Szeider, S.: Backdoor sets of quantified boolean formulas. JAR 42, 77–97 (2009)
16. Van Gelder, A.: Generalized conflict-clause strengthening for satisfiability solvers. In: Sakallah, K.A., Simon, L. (eds.) SAT 2011. LNCS, vol. 6695, pp. 329–342. Springer, Heidelberg (2011)
17. Van Gelder, A.: Contributions to the theory of practical quantified boolean formula solving. In: Milano, M. (ed.) CP 2012. LNCS, vol. 7514, pp. 647–663. Springer, Heidelberg (2012)
18. Van Gelder, A.: Producing and verifying extremely large propositional refutations: Have your cake and eat it too. AMAI 65(4), 329–372 (2012)
19. Zhang, L., Malik, S.: Conflict driven learning in a quantified boolean satisfiability solver. In: Proc. ICCAD, pp. 442–449 (2002)

Concurrent Clause Strengthening

Siert Wieringa and Keijo Heljanko[*]

Aalto University, School of Science
Department of Information and Computer Science
P.O. Box 15400, FI-00076 Aalto, Finland
{siert.wieringa,keijo.heljanko}@aalto.fi

Abstract. This work presents a novel strategy for improving SAT solver performance by using concurrency. Rather than aiming to parallelize search, we use concurrency to aid a conventional CDCL search procedure. More concretely, our work extends a conventional CDCL SAT solver with a second computation thread, which is solely used to strengthen the clauses learned by the solver. This provides a simple and natural way to exploit the availability of multi-core hardware.

We have employed our technique to extend two well established solvers, MiniSAT and Glucose. Despite its conceptual simplicity the technique yields a significant improvement of those solvers' performances, in particular for unsatisfiable benchmarks. For such benchmarks an extensive empirical evaluation revealed a remarkably consistent reduction of the wall clock time required to determine unsatisfiability, as well as an ability to solve more benchmarks in the same CPU time.

The proposed technique can be applied in combination with existing parallel SAT solving techniques, including both portfolio and search space splitting approaches. The approach presented here can thus be seen as orthogonal to those existing techniques.

1 Introduction

Propositional satisfiability (typically abbreviated SAT) is the problem of finding a satisfying truth assignment for a given propositional logic formula, or determining that no such assignment exists. This classifies the formula as respectively *satisfiable* or *unsatisfiable*. SAT is an important theoretical problem as it was the first problem ever to be proven NP-complete [8].

Despite the theoretical hardness of SAT, current state-of-the-art decision procedures for SAT, so called *SAT solvers,* have become surprisingly efficient. The introduction of *Conflict Driven Clause Learning* (CDCL) [23] was a crucial step in the process of making these algorithms into industrial strength problem solvers. However, modern SAT solvers are not just efficient implementations of the CDCL search procedure. Instead, they implement several forms of extra reasoning. For example, formula simplification before CDCL search, so called

---

[*] This work was partially funded by the Helsinki Institute for Information Technology (HIIT) and the Academy of Finland, project 139402.

M. Järvisalo and A. Van Gelder (Eds.): SAT 2013, LNCS 7962, pp. 116–132, 2013.

preprocessing, is commonly used. An important work for the widespread adaptation of this technique was the introduction of an efficient preprocessor called SatELite [10]. Some modern solvers, such as Lingeling [6], use *inprocessing*, which is the sequential interleaving of search and simplification procedures. For a recent and extensive overview of pre- and inprocessing techniques, as well as a concise set of rules formalizing such techniques please refer to [21].

Another technique that is crucial for performance, but not part of the core CDCL search procedure is *conflict clause strengthening* (e.g. [11,14,29]). This is usually performed during conflict analysis (see Sec. 2). The length of a conflict clause can be efficiently reduced to a *minimal* clause implied by the set of clauses used in its derivation [14,29]. Hence, the name *conflict clause minimization* is also commonly used [11]. However, reducing a conflict clause to a *minimal* conflict clause implied by all clauses in the formula is NP-hard, as this clause is of length zero iff the formula is unsatisfiable. In [14] the related *Generalized Conflict-Clause Strengthening* problem was defined and proven NP-hard.

Sequentially interleaving a more expensive conflict clause strengthening procedure with the core CDCL search procedure may provide performance improvements, but it is hard to develop good heuristics for deciding when to switch between searching and conflict clause strengthening. As the majority of computer hardware is nowadays equipped with multi-core CPUs conflict clause strengthening can instead be performed *in parallel* with the core search procedure. We have developed a novel solving architecture that uses two computation threads, one for CDCL search and one solely for strengthening conflict clauses. The conflict clause strengthening algorithm used is similar to existing algorithms for problem clause strengthening [25,17]. In extensive experimental results we show the performance of two implementations of our architecture, based on two different well established SAT solvers, MiniSAT [12] and Glucose [2].

For all experiments we will present results regarding *wall clock time* and *CPU time*. Wall clock time is defined as the amount of time that passes from the start to the finish of the solving process, and this measure is independent of the amount of resources that are used during that time. CPU time on the other hand is the sum of the time spend by each of the cores used, i.e. if a single program uses all the computation power of two CPU cores concurrently then the CPU time grows twice as fast as the wall clock time.

The presented two threaded solver maintains all the features of a normal SAT solver, including for example its interface for incremental SAT solving [13]. Hence, our new solver could in principle replace the conventional solver inside the parallel incremental solver that we presented in recent work [30]. Although the performance of the technique presented here in combination with incremental SAT is interesting this is left for further work.

One may consider a parallelization of an algorithm as a strategy for assigning any number of simultaneously available computation resources to performing a single task. By that definition this work does not present a parallelization of a SAT solver, as only the use of exactly two concurrent computation threads is considered. However, existing techniques for parallelizing SAT algorithms can be used in combination with our two threaded solver, in order to obtain a generic parallelization. Even running multiple copies of our two threaded solver is a practical proposition for such environments, given its good performance regarding CPU time. Hence, we will provide a short overview of relevant work on parallelizing SAT solvers.

Two major approaches for parallelizing SAT algorithms can be distinguished [19]. The first is the classic divide-and-conquer approach, which aims to partition the formula to divide the total workload evenly over multiple SAT solver instances [7,28,32]. The second approach is the so called *portfolio* approach [16,22]. Rather than partitioning the formula, portfolio systems run multiple solvers in parallel each of which attempt to solve the same formula. The system finishes whenever the fastest solver is done. Both approaches can be extended with some form of conflict clause sharing between the solver threads.

Although other techniques have recently been developed (e.g. [18,19]) portfolio solvers have received the majority of the research attention in recent years. Some insight into the good performance of these approaches is provided in [20]. ManySAT [16] is a well known adaptation of the portfolio strategy. It employs conflict clause sharing and is thus a so called *cooperative portfolio*. It is build around running multiple copies of the sequential solver MiniSAT [12] in parallel. Although each of the solver threads may be given different settings the threads are largely *homogeneous*.

Other portfolio solvers are *non-cooperative*, but use truly *heterogeneous* solver threads. Examples are ppfolio[1], SATzilla [31], and 3S [22]. These all use a collection of different SAT solvers from several developers. Whereas SATzilla and 3S try to be clever about which solvers to use for solving a particular formula ppfolio is completely naïve. In fact, ppfolio is a very simple program that just executes multiple solvers in parallel. It was meant to illustrate a lower bound on what is achievable using portfolios[1], but it turned out to be one of the strongest solvers at the SAT Competition[2] in 2011.

In several recent cooperating portfolios using homogeneous solver threads, such as Plingeling [6], cooperation is limited to sharing unit clauses only. The relatively weak performance of portfolios sharing more than just unit clauses was the motivation for work presenting a new set of clause sharing heuristics [1]. The solver implementing those heuristics, called PeneLoPe, won a silver medal at the SAT Challenge in 2012 [4]. The winner, solving one instance more, was a noncooperative portfolio called pfolioUZK [4]. PeneLoPe is based on ManySAT and its performance is remarkable, considering its use of homogeneous solver threads.

2   Definitions

A *literal* l is either a Boolean variable x or its negation $\neg x$. Double negations cancel out, hence $\neg\neg l = l$. An *assignment* ρ is a set of literals such that if $l \in \rho$ then $\neg l \notin \rho$. If $l \in \rho$ we say that literal l is assigned the value **true**. If $\neg l \in \rho$

---

[1] www.cril.univ-artois.fr/~roussel/ppfolio

[2] www.satcompetition.org

it is said that l is assigned the value **false**, or equivalently that l is *falsified*. If for some literal l neither l nor $\neg l$ is in the assignment ρ then l is *unassigned*. For an assignment ρ we denote by $\neg\rho$ the set $\{\neg l \mid l \in \rho\}$. A *clause* c is a set of literals $c = \{l_0, l_1, \cdots, l_n\}$ representing the disjunction $\bigvee c = l_0 \vee l_1 \cdots \vee l_n$. Hence, clause c is *satisfied* by assignment ρ iff $c \cap \rho \neq \emptyset$. A clause consisting of exactly one literal is called a *unit clause*.

As typical in work on SAT solvers, we only consider formulas in *conjunctive normal form* (CNF). Such formulas are formed as conjunctions of disjunctions, and hence can be represented as sets of clauses. The formula $\mathcal{F}$ *under the assignment* ρ is denoted $\mathcal{F}^\rho$ as in [5]. It is defined as the formula $\mathcal{F}$ after removing all clauses satisfied by ρ, followed by shrinking the remaining clauses by removing literals that are falsified by ρ. Formally:

$$\mathcal{F}^\rho = \{ c \setminus \neg\rho \mid c \in \mathcal{F} \text{ and } c \cap \rho = \emptyset \}$$

Let $iup(\mathcal{F}, \rho)$ be the assignment ρ that is the result of executing the following *iterative unit propagation* loop:

$$\textbf{while } \{l\} \in \mathcal{F}^\rho \textbf{ do } \rho = \rho \cup \{l\}$$

Moreover we define $\mathcal{F}|_\rho = \mathcal{F}^{iup(\mathcal{F}, \rho)}$, which is the result of simplifying formula $\mathcal{F}$ under assignment ρ by iterative unit propagation. If $\emptyset \in \mathcal{F}|_\rho$ we say that a *conflict* has been reached. If on the other hand $\mathcal{F}|_\rho = \emptyset$ then assignment ρ satisfies $\mathcal{F}$. The DPLL algorithm [9] is the classical algorithm for determining the satisfiability of CNF formulas. It starts from the formula $\mathcal{F}$ and an empty assignment ρ, and alternates between iterative unit propagation and *branching decisions*. During a branching decision, or simply *decision*, the algorithm picks a *decision variable* x_d that is unassigned by ρ and assigns it to either **true** or **false**. Whenever iterative unit propagation leads to a conflict the algorithm backtracks to the last decision to which it had not backtracked before, and negates the assignment made at that decision. This backtracking search continuous until either all variables of $\mathcal{F}$ are assigned, or all branches of the search tree have been unsuccessfully explored. In the former case ρ satisfies $\mathcal{F}$, in the latter case $\mathcal{F}$ is unsatisfiable.

Most modern SAT solvers are so called *Conflict Driven Clause Learning* (CDCL) solvers [23,24]. Just like the basic DPLL procedure the search for a satisfying assignment proceeds by alternating between iterative unit propagation and branching decisions. The crucial difference is what happens when a conflict is reached. In this case, a CDCL solver will analyze the sequence of decisions and implications that lead to the conflict. During this *conflict analysis* the solver derives a *conflict clause*, which is a clause implied by the input formula that gives a representation of the "cause" of the conflict. By including the conflict clause in the set of clauses on which iterative unit propagation is performed hitting another conflict with the same cause can be avoided.

An important property of the most popular clause learning scheme for CDCL solvers, called *first unique implication point* (1-UIP) [24], is that each conflict clause contains exactly one literal that was falsified by the last decision or the subsequent unit propagation. This literal is called the *asserting literal*.

After conflict analysis the CDCL solver must backtrack. Unlike the DPLL procedure CDCL solvers use non-chronological backtracking, which is driven by the conflict clauses. By definition all literals in a conflict clause are assigned the value **false** by assignment ρ when it is derived. After learning conflict clause c, the solver backtracks until the earliest decision at which all literals of c except the asserting literal l_a are assigned **false**. The literal l_a is then assigned the value **true**, as this is required to satisfy c. Subsequent unit propagation may yield a new conflict which is handled in the same way.

Any conflict clause c, derived by a CDCL solver from the formula $\mathcal{F}$ with the aid of previously derived conflict clauses P, can be derived using a so called *trivial resolution derivation* [5]. This implies that if the value **false** is assigned to all literals of a conflict clause, then the result of simplifying formula $\mathcal{F} \cup P$ under that assignment by iterative unit propagation is guaranteed to reach a conflict, i.e. the following property holds:

$$\emptyset \in \mathcal{F}'|_{\rho} \text{ for } \mathcal{F}' = \mathcal{F} \cup P \text{ and } \rho = \{\neg l \mid l \in c\} \tag{1}$$

Another important property of conflict clauses derived using the 1-UIP scheme is their *1-empowerment* property [27]. Informally this means that if all literals of a conflict clause c except its asserting literal l_a are assigned to **false**, then iterative unit propagation on the set $\mathcal{F} \cup P$ does not yield the necessary truth assignment **true** to l_a, i.e. the following property holds:

$$l_a \notin iup(\mathcal{F}', \rho) \text{ for } \mathcal{F}' = \mathcal{F} \cup P \text{ and } \rho = \{\neg l \mid l \in c \text{ and } l \neq l_a\} \tag{2}$$

It is said that c is *1-empowering* with respect to $\mathcal{F} \cup P$ via its asserting literal l_a. This property implies that adding the conflict clause c to the learnt clause set P strictly extends the propagation abilities of the solver. In other words, the *deductive power* of the CNF formula $\mathcal{F} \cup P$ is strictly increased [17].

3   The Solver-Reducer Architecture

We propose an architecture using two concurrently executing threads, which are called the SOLVER and the REDUCER. The SOLVER acts like any conventional SAT solver, except for its interaction with the REDUCER. The interaction between the SOLVER and the REDUCER is limited to passing clauses through two shared-memory data structures called the *work set* and the *result queue*. The work set is used to pass clauses from the SOLVER to the REDUCER, the result queue is used for passing clauses in the opposite direction, as illustrated in Fig. 1.

Whenever the SOLVER learns a clause it writes a copy of that clause to the work set. The REDUCER reads clauses from the work set and tries to strengthen them. When the REDUCER successfully reduces the length of a clause, it places the new shorter clause in the result queue. The SOLVER checks frequently whether there are any clauses in the result queue. If this is the case the SOLVER enters the clauses from the result queue in its learnt clause set. It is possible that all of the literals in such a clause are assigned the value **false** in the current assignment

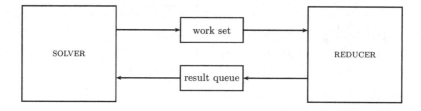

Fig. 1. The solver-reducer architecture

ρ of the SOLVER. In this case the SOLVER must backtrack before entering the clause in the set. Our implementation of *on-the-fly* addition of "foreign" clauses in the SOLVER tries to keep backtracking to a minimum. If the clause is asserting then this is handled in the same way as for normal conflict clauses. We could have chosen to introduce the clauses only when the SOLVER's assignment ρ contains no decision variables (as in e.g. [30]), but a more dynamic approach was considered more appropriate here. There is no mechanism to ensure that the clause c is removed from the SOLVER's learnt clause set when a clause $c' \subset c$ is obtained from the result queue.

Our proposed architecture is conceptually simple, and we will show that it can provide substantial performance improvements. An unfortunate side-effect of our approach is that the behavior of the solver becomes non-deterministic, as the execution order is determined by the operating system's thread scheduling policy. This means that runs of our two threaded solver are not reproducible, and performance may vary drastically in between two different runs on the same formula. The same problem occurs also when using more conventional paral-lelizations of SAT solvers. In [15] it was shown that a deterministic version of ManySAT, using *synchronization barriers* and a dynamic heuristic for deciding when to perform synchronization, on average performed almost as well as the original version. We believe that a similar technique could be applied successfully in an implementation of the solver-reducer architecture.

3.1 The Reducer

The REDUCER continuously checks the work set for new input clauses, and then runs its reduction algorithm. The algorithm is based on unit propagation and conflict clause learning. Basically, the algorithm assigns the literals of the input clause c_{in} to **false** one by one until iterative unit propagation leads to a conflict, or all variables are assigned. The pseudocode for this algorithm is given in Fig. 2, where P_R represents the set of learnt clauses maintained by the REDUCER.

Consider the case where the algorithm returns c_{out} from Line 7. This case occurs iff there is no trivial resolution derivation of c_{in} from $\mathcal{F} \cup P_R$. This is possible, because the set P_R does not necessarily contain all the clauses P_S that were contained in the SOLVER when c_{in} was derived. The returned clause $c_{out} \subseteq c_{in}$ is obtained by removing with respect to c_{in} any literals $l \in c_{in}$ for

1. $c_{out} = \emptyset; \rho = \emptyset$
2. $\mathcal{F}' = \mathcal{F} \cup P_R$
3. for $l \in (c_{in} \setminus c_{out})$ s.t. $\neg l \notin iup(\mathcal{F}', \rho)$
4.   if $\emptyset \in \mathcal{F}'|_\rho$ then $(P_R, c_{new}) = $ ANALYZE$(\mathcal{F}, P_R, \rho)$; return c_{new}
5. $c_{out} = c_{out} \cup \{l\}; \rho = \rho \cup \{\neg l\}$
6. $P_R = P_R \cup \{c_{out}\}$
7. return c_{out}

Fig. 2. Pseudocode for the REDUCER's algorithm

which the value **false** of the corresponding literal l was implied rather than assigned. This implements *self-subsumption resolution* [10]. It is sound because the forced falsification of l means that for some $c \subset c_{in}$ it holds that $\mathcal{F} \models c \cup \{\neg l\}$, and by resolution on the clauses c_{in} and $c \cup \{\neg l\}$ it follows that $\mathcal{F} \models c_{in} \setminus \{l\}$. Because c_{out} is 1-empowering with respect to $\mathcal{F} \cup P_R$ it is added to P_R.

Now consider the case were the algorithm returns after calling the function ANALYZE on Line 4 of the pseudocode. The function ANALYZE analysis the conflicting assignment ρ. Until ρ is non-conflicting or $\rho = \emptyset$ it performs backtracking by removing literals from ρ, conflict clause learning by adding clauses to P_R, and iterative unit propagation. Because for each clause added to P_R at least one literal is removed from ρ the number of new conflict clauses is bounded by $|\rho| \leq |c_{in}|$. These conflict clauses are crucial for the performance of the REDUCER, but they are never shared with the SOLVER. The function ANALYZE returns a clause[3] c_{new} such that $c_{new} \subseteq c_{out}$. Consider the assignment ρ after the backtracking performed by ANALYZE. If $\rho = \emptyset$ then $c_{new} = \emptyset$. Else, for some $l \in c_{out}$ and $\rho' \subseteq \rho$ it holds that $l \in iup(\mathcal{F}', \rho')$. In this case $c_{new} = \{l' \mid \neg l' \in \rho'\} \cup \{l\}$.

Our REDUCER's algorithm is very similar to the *vivification algorithm* of [25]. The vivification algorithm aims to find redundant literals in the problem clauses $c \in \mathcal{F}$ by assigning the value **false** for the literals in c, and performing unit propagation on the formula $\mathcal{F} \setminus \{c\}$. In case a conflict arises the algorithm learns only exactly one new conflict clause. The order in which the literals are assigned is heuristically controlled in the vivification algorithm. In our REDUCER implementation the literals are assigned in the order in which they appear in the clause. Due to the organization of the conflict clause analysis procedure of the SOLVER this means that the asserting literal of the conflict clause is always assigned first. Note that the clause $c \cup \{l_a\}$, where l_a is the asserting literal, is equivalent to the implication $(\neg \bigvee c) \rightarrow l_a$. Although the clause can be rewritten as an implication with any one of its literals as the consequent, this implication of l_a is the one that guaranteed the 1-empowering property of the clause in the SOLVER. By starting from l_a, the REDUCER aims to reduce the size of the antecedent $\neg c$ of this deduction power increasing implication.

---

[3] In the implementation c_{new} is obtained using MiniSAT's ANALYZEFINAL function, where ρ is regarded as the set of *assumptions* [13].

The REDUCER can not do anything that a conventional solver could not also do in the same number of steps. This means that the solver-reducer architecture does not implement a stronger proof system (see, e.g. [5]) than a conventional CDCL solver. In fact, a modern SAT solver using the VSIDS heuristic [24], phase saving [26], and frequent restarts (e.g. [3]) has a tendency to "run towards conflicts" just like the REDUCER does. Consider a SAT solver using phase saving and extreme parameter settings: It restarts after every conflict, and uses a VSIDS activity decay so large that the set of variables involved in the most recent conflict always have larger activities than any other variables. In this case, after every conflict and the subsequent restart, the VSIDS heuristic will pick as decision variables those variables that occur in the most recent conflict clause. Combined with phase saving this will lead to the same sequence of assignments as the REDUCER would make to reduce that conflict clause.

3.2   The Work Set

A set of clauses stored in a shared-memory data structure called the work set is forming the inputs of the REDUCER. It is possible to implement the work set as a simple unbounded FIFO queue. This may be sufficient if the REDUCER has only very few clauses in its learnt clause set, as in this case it can often perform unit propagation fast enough to keep up with the conflict clause generation of the SOLVER. However, the clause learning in the REDUCER is crucial to the strength of the reduction procedure. As the size of REDUCER's learnt clause set increases it is able to provide stronger reduced clauses, but at a lower speed.

If the REDUCER can not keep up with the SOLVER then a work set implemented as a FIFO queue will just keep growing. As the REDUCER lags behind it will only strengthen "old" clauses, that are less likely to be of use to the SOLVER. An unbounded LIFO queue would make the REDUCER focus on reducing recent clauses first, but strong clauses may shift to the back of such a queue quickly if the REDUCER is momentarily busy. Giving preference to strengthening clauses that are likely to be "important" to the SOLVER is natural. The "quality" of a conflict clause can be crudely approximated by its length, or alternatively by its *Literal Blocks Distance* (LBD) [2]. Hence, an alternative work set implementation could keep an unbounded set of clauses sorted by their length. However, as the average conflict clause length changes during the search, a clause that was relatively long ("bad") at the time it was learned may seem relatively short ("good") after some time has passed. Thus, this unbounded sorted set also leads to reducing outdated clauses. The same argument holds when the LBD is used for sorting the set, as the LBD measure of a clause is bounded by its length.

We achieved the best performance using work set with a limited capacity. If the SOLVER enters a clause into a full work set then this clause will replace the oldest clause in the set. If the REDUCER requests a clause from the work set it is given the "best" clause from the set. In this way, the REDUCER's inputs are kept both "fresh" and "good".

3.3   Implementation

We have implemented our solver-reducer architecture on top of two well established existing SAT solvers, MiniSAT 2.2.0[4] [12] and Glucose 2.1[5] [2]. MiniSAT is often used as a basis for the development of new solving techniques, as witnessed by the existence of a "MiniSAT hack track" at the SAT competitions. Glucose won the SAT Challenge in 2012 [4]. Older Glucose versions won at the applications tracks of the SAT competitions in 2009 (for unsatisfiable benchmarks) and in 2011 (for mixed benchmarks). Because Glucose is based on MiniSAT the solver-reducer architecture was easy to port to Glucose once it had been developed inside MiniSAT. We will refer to the two open-source[6] solver-reducer implementations we created as respectively MiniRed and GlucoRed.

Both the SOLVER and the REDUCER of MiniRed are build as extensions to the MiniSAT solver. All the default settings and heuristics of MiniSAT were maintained in the SOLVER and the REDUCER. Similarly, the SOLVER and the REDUCER of GlucoRed maintain the default settings of the Glucose solver. An example of a heuristic that concerns both the SOLVER and the REDUCER is the heuristic for deciding when to reduce the size of their learnt clause sets. GlucoRed uses the LBD measure of a clause as a sorting metric for the work set, i.e. when the REDUCER requests a clause from the work set it is given the clause with the smallest LBD. Because MiniSAT does not compute LBD values MiniRed uses clause length as a sorting metric instead. The result queue is implemented as an unbounded FIFO queue.

The two threads interact solely by passing clauses, or more precisely pointers to clauses, through the work set and the result queue. Exclusive access to those datastructures is achieved by the use of a single *lock*. To reduce the number of times the lock must be acquired the SOLVER and REDUCER always combine read and write operations. In the REDUCER this is straightforward: If the length of a clause is reduced, then the new shorter clause is written to the result queue once the lock has been obtained to read a new input clause from the work set. The SOLVER combines reading and writing by checking the result queue for new reduced clauses whenever it has acquired the lock to write a new clause to the work set, i.e. whenever it derives a conflict clause. The SOLVER always postpones the addition of reduced clauses from the result queue to its learnt clause database until just before its next branching decision.

4   Experimental Evaluation

All experiments in this work were performed in a computing cluster, using machines that each have two six core Intel Xeon X5650 processors[7]. A memory limit of 7GB was enforced.

---

[4] http://www.minisat.se

[5] http://www.lri.fr/~simon

[6] http://users.ics.aalto.fi/swiering/solver_reducer

[7] These resources were provided by the Aalto Science-IT project.

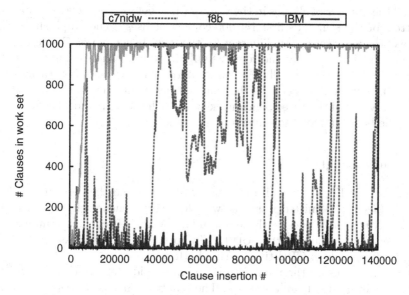

Fig. 3. Erratic use of the work set during three different runs

The MiniSAT [12] distribution provides a version of the solver with an integrated preprocessor, which is similar to SatELite [10]. During the SAT Challenge [4] the developers of Glucose [2] did not use such an internal preprocessor. Instead, they provided a script that first ran SatELite, and then ran Glucose on the resulting formula. For a fair comparison of the strength of the solvers we chose to run all solvers without enabling their integrated preprocessors, and provide them with both original and simplified benchmarks.

The first set of benchmarks we used is named *Competition*, and contains in total 547 benchmarks. The set combines 300 benchmarks from the application track of the SAT competition held in 2011, and the 247 application track benchmarks from the SAT Challenge 2012 that were marked as unused in previous competitions [4]. The set *Simplified* contains 501 benchmarks that are the result of running SatELite on the set *Competition*. The difference in size between the *Competition* set and the *Simplified* set is caused by leaving out benchmarks that were proven unsatisfiable by SatELite, and benchmarks that could not be simplified in 15 minutes.

4.1 Capacity of the Work Set

All experiments used a work set with a capacity of 1000 clauses. The average performance of our implementation is not particularly sensitive to this setting. It is however hard to make any general statements about the typical use of the work set. We illustrate this using a small experiment for which we solved three unsatisfiable benchmarks from the *Simplified* set using MiniRed. Each of these benchmarks takes just over thirty seconds to solve using conventional MiniSAT.

In Fig. 3 the number of clauses in the work set just before a new clause is inserted is plotted for the SOLVER's first 140 000 conflict clauses[8]. The graph shows that the use of the work set is very different for the three benchmarks. For f8b the set fills up almost immediately and remains full afterwards, whereas for c7nidw the size keeps varying dramatically. For benchmark IBM[9] the REDUCER easily keeps up with the supply of clauses, as the work set never fills. Interestingly, IBM was also the only one of the three benchmarks for which the added REDUCER did not seem to be beneficial for the solver's performance.

4.2   Clause Length

The numbers in Fig. 4 were obtained using MiniRed and the benchmarks from the *Simplified* set. MiniRed was run twice for every benchmark, once with the default settings, and once with the standard MiniSAT conflict minimization procedure [11] disabled in the SOLVER. In total 367 benchmarks were solved within 1800 seconds of CPU time during both runs.

Let us first focus on the numbers printed in a bold font, which represent the results for MiniRed's default settings. The numbers on the arrows indicate the average length of all the clauses that passed it during the 367 runs. Note that the absolute values of these numbers are meaningless, as they are averaged over a large set of independent and very different runs. The relation between these numbers nevertheless provides some insight in the operation of our architecture.

The arrow that points up out of the work set represents the clauses that are deleted from the work set because of its limited capacity, as described in Sec. 3.2. During this experiment on average 34.6% of the clauses placed in the work set were deleted. The average length of those clauses is large (91.3) compared to the average length of the clauses that are entering the work set (56.8). This was expected, as the work set delivers the shortest clauses first to the REDUCER. The average length of the clauses passing through the REDUCER dropped from 38.1 to 27.6 literals. On average 30.2% of the clauses remain the exact same length after passing through the REDUCER. These clauses are not placed in the result queue, as represented by the arrow that points down at the bottom of Fig. 4. Unsurprisingly the average length of those clauses is rather short (15.3).

The results for the experiment in which MiniRed was run with the SOLVER's conflict clause minimization disabled are printed in an italic font in the figure. The total number of conflict clauses generated by the SOLVER over the 367 runs grew by 17%, and those clauses were on average 2.5 times longer. However, the clauses that are actually delivered from the work set to the REDUCER are not much longer than those in the first experiment, and after reduction they are even slightly shorter. Surprisingly, the average overall performance of MiniRed was almost identical in both experiments. Note that disabling the conflict clause minimization in conventional MiniSAT results in a substantial degradation of

---

[8] For the benchmarks c7nidw and IBM the total number of conflict clauses generated by the SOLVER was slightly over 300 000 clauses, for f8b the total was around 150 000.

[9] IBM abbreviates IBM_FV_2004_rule_batch_26_SAT_dat.k95.

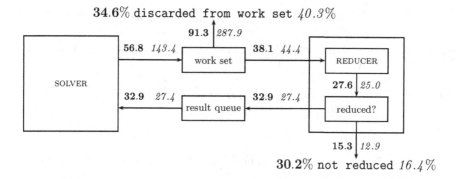

Fig. 4. Average clause lengths over 367 benchmarks

the performance [11], hence MiniRed's consistent good performance is made possible by the REDUCER. It apparently did not harm the SOLVER that the length of the longest clauses was never reduced at all, not even using the conflict clause minimization routine. In the remainder of this work we will only use the default settings, in which the SOLVER's conflict clause minimization is enabled.

4.3   Performance

Table 1 contains the number of benchmarks solved by the four different solvers within 900 seconds of wall clock time. The numbers in the table that are printed in a smaller font inside brackets represent the number of benchmarks solved within 1800 seconds of CPU time. The column VBS (*Virtual Best Solver*) provides the total number of benchmarks solved by at least one of the four solvers. The columns labelled Δ underline the difference between the number of benchmarks solved with- and without REDUCER.

Note the impressive performance improvement the solver-reducer architecture provides for unsatisfiable benchmarks. MiniRed improves the number of unsatisfiable benchmarks solved for the *Simplified* set by 58 benchmarks, and even regarding CPU time still provides a 31 benchmark improvement over MiniSAT. The gaps are smaller but still significant for the Glucose based implementation. The results for the unsatisfiable benchmarks from the *Competition* set are presented as cactus plots in Fig. 5. Comparison is made based on wall clock time in Fig. 5a and based on CPU time in Fig. 5b. The same is done for the unsatisfiable benchmarks from the *Simplified* set in Fig. 6a and Fig. 6b. The logarithmic-scale scatter plots in Fig. 7 and Fig. 8 provide another presentation of the wall clock time performance of MiniSAT versus MiniRed, and Glucose versus GlucoRed. The remarkable consistency of the improvement for unsatisfiable benchmarks can be clearly seen.

It is not surprising that the REDUCER does not contribute much to the average performance for *satisfiable* benchmarks, and that thus for such benchmarks the addition of the REDUCER results in worse performance regarding the CPU

Table 1. Number of benchmarks solved

Set		VBS	MiniSAT	MiniRed	Δ	Glucose	GlucoRed	Δ
Competition	UNSAT	239 (251)	151 (171)	208 (208)	57 (37)	207 (234)	235 (235)	28 (1)
	SAT	177 (179)	166 (174)	168 (168)	2 (-6)	158 (167)	155 (157)	−3 (-10)
Simplified	UNSAT	246 (249)	164 (191)	222 (222)	58 (31)	220 (232)	237 (237)	17 (5)
	SAT	166 (168)	150 (157)	159 (159)	9 (2)	155 (157)	147 (149)	−8 (-8)

Table 2. Number of benchmarks in the *Simplified* set solved by PeneLoPe

	GlucoRed	PeneLoPe-2	PeneLoPe-4	PeneLoPe-8
UNSAT	237 (237)	227 (227)	231 (221)	247 (217)
SAT	147 (149)	142 (142)	160 (154)	164 (149)

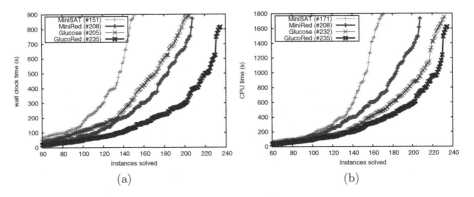

Fig. 5. Results for unsatisfiable benchmarks from the *Competition* set

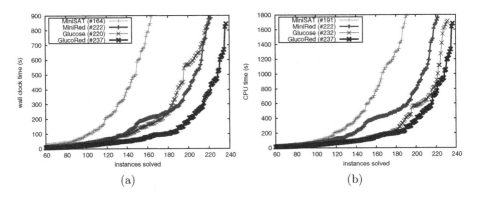

Fig. 6. Results for unsatisfiable benchmarks from the *Simplified* set

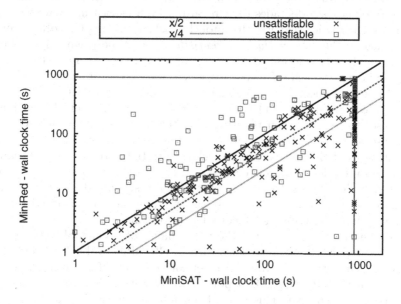

Fig. 7. MiniSat versus MiniRed on benchmarks from the *Simplified* set

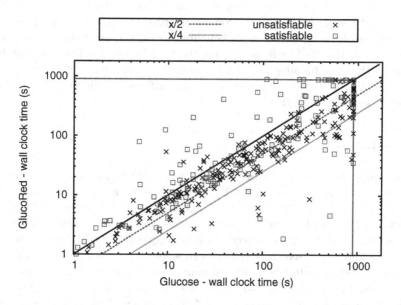

Fig. 8. Glucose versus GlucoRed on benchmarks from the *Simplified* set

time. Between Glucose and GlucoRed, the number of satisfiable benchmarks from the *Simplified* set degrades even regarding wall clock time. Glucose uses many tunable heuristics that we left untouched when creating GlucoRed. Some of these, such as the *restart blocking heuristic* [3], may be negatively affected by the on-the-fly introduction of REDUCER result clauses. An important measure for the heuristics inside Glucose is the average LBD of its conflict clauses. GlucoRed does not incorporate the clauses provided by the REDUCER in this average. Moreover, Glucose and GlucoRed use the LBD measure for deciding which clauses to remove when reducing their learnt clause sets. MiniSAT and MiniRed use an activity based heuristic for this purpose. We expect that for implementations of the solver-reducer architecture the latter is better because it has a natural tendency to delete subsumed clauses. This is important as the REDUCER provides clauses to the SOLVER that are subsumed by clauses that are (or were) already in its learnt clause set. Adaptation of the heuristics from PeneLoPe [1] may also improve GlucoRed's performance.

It would be interesting to study the performance of a PeneLoPe style portfolio of solver-reducer implementations such as GlucoRed. Table 2 presents the number of benchmarks in the *Simplified* set solved by PeneLoPe using 2, 4 and 8 cores. Recall that PeneLoPe is a portfolio solver using homogeneous solver threads and clause sharing. This type of portfolio is expected to perform best on formulas that are satisfiable, as compared to unsatisfiable formulas the run time deviations between multiple runs of a similar solver are larger [19]. PeneLoPe witnesses this by solving more satisfiable benchmarks using four threads than it does using two threads, given the same amount of CPU time. Clearly, GlucoRed and PeneLoPe have orthogonal strengths. Given the same amount of CPU time GlucoRed can prove more benchmarks unsatisfiable than PeneLoPe, regardless of whether 2, 4 or 8 threads are used for PeneLoPe. For unsatisfiable benchmarks the two threaded solver GlucoRed is so much more efficient that in 900 seconds of wall clock time it solves six unsatisfiable benchmarks more than PeneLoPe does using four threads.

5   Conclusions

This work presents the solver-reducer architecture, which employs strengthening of conflict clauses in parallel with CDCL search in a modern SAT solver. An extensive empirical evaluation showed the good performance of this conceptually simple idea, which can be combined with conventional parallelization strategies.

The use of concurrency to aid conventional sequential CDCL search, rather than to parallelize that search, has not been suggested before. This simple but novel idea can be exploited in many different ways. For example, a logical next step would be to consider concurrent formula simplification. This would be a natural way of employing concurrency in recent solvers that use *inprocessing*, i.e. the sequential interleaving of solving and simplifying procedures.

References

1. Audemard, G., Hoessen, B., Jabbour, S., Lagniez, J.-M., Piette, C.: Revisiting clause exchange in parallel SAT solving. In: Cimatti, A., Sebastiani, R. (eds.) SAT 2012. LNCS, vol. 7317, pp. 200–213. Springer, Heidelberg (2012)
2. Audemard, G., Simon, L.: Predicting learnt clauses quality in modern SAT solvers. In: Boutilier, C. (ed.) IJCAI, pp. 399–404 (2009)
3. Audemard, G., Simon, L.: Refining restarts strategies for SAT and UNSAT. In: Milano, M. (ed.) CP 2012. LNCS, vol. 7514, pp. 118–126. Springer, Heidelberg (2012)
4. Balint, A., Belov, A., Diepold, D., Gerber, S., Järvisalo, M., Sinz, C. (eds.): Proceedings of SAT Challenge 2012: Solver and Benchmark Descriptions. Department of Computer Science Series of Publications B, vol. B-2012-2. University of Helsinki, Helsinki (2012)
5. Beame, P., Kautz, H.A., Sabharwal, A.: Towards understanding and harnessing the potential of clause learning. J. Artif. Intell. Res. (JAIR) 22, 319–351 (2004)
6. Biere, A.: Lingeling, Plingeling, PicoSAT and PrecoSAT at SAT Race 2010. FMV Technical Report 10/1, Johannes Kepler University, Linz, Austria (2010)
7. Böhm, M., Speckenmeyer, E.: A fast parallel SAT-solver - efficient workload balancing. Ann. Math. Artif. Intell. 17(3-4), 381–400 (1996)
8. Cook, S.A.: The complexity of theorem-proving procedures. In: STOC, pp. 151–158. ACM (1971)
9. Davis, M., Logemann, G., Loveland, D.W.: A machine program for theorem-proving. Commun. ACM 5(7), 394–397 (1962)
10. Eén, N., Biere, A.: Effective preprocessing in SAT through variable and clause elimination. In: Bacchus, F., Walsh, T. (eds.) SAT 2005. LNCS, vol. 3569, pp. 61–75. Springer, Heidelberg (2005)
11. Eén, N., Sörensson, N.: MiniSat v1.13 - a SAT solver with conflict-clause minimization. Poster for SAT 2005 (2005), http://www.minisat.se
12. Eén, N., Sörensson, N.: An extensible SAT-solver. In: Giunchiglia, E., Tacchella, A. (eds.) SAT 2003. LNCS, vol. 2919, pp. 502–518. Springer, Heidelberg (2004)
13. Eén, N., Sörensson, N.: Temporal induction by incremental SAT solving. Electr. Notes Theor. Comput. Sci. 89(4), 543–560 (2003)
14. Van Gelder, A.: Generalized conflict-clause strengthening for satisfiability solvers. In: Sakallah, K.A., Simon, L. (eds.) SAT 2011. LNCS, vol. 6695, pp. 329–342. Springer, Heidelberg (2011)
15. Hamadi, Y., Jabbour, S., Piette, C., Sais, L.: Deterministic parallel DPLL. JSAT 7(4), 127–132 (2011)
16. Hamadi, Y., Jabbour, S., Sais, L.: ManySAT: A parallel SAT solver. JSAT 6(4), 245–262 (2009)
17. Han, H., Somenzi, F.: Alembic: An efficient algorithm for CNF preprocessing. In: DAC, pp. 582–587. IEEE (2007)
18. Heule, M.J.H., Kullmann, O., Wieringa, S., Biere, A.: Cube and conquer: Guiding CDCL SAT solvers by lookaheads. In: Eder, K., Lourenço, J., Shehory, O. (eds.) HVC 2011. LNCS, vol. 7261, pp. 50–65. Springer, Heidelberg (2012)
19. Hyvärinen, A.E.J., Junttila, T.A., Niemelä, I.: Partitioning search spaces of a randomized search. Fundam. Inform. 107(2-3), 289–311 (2011)
20. Hyvärinen, A.E.J., Manthey, N.: Designing scalable parallel SAT solvers. In: Cimatti, A., Sebastiani, R. (eds.) SAT 2012. LNCS, vol. 7317, pp. 214–227. Springer, Heidelberg (2012)

21. Järvisalo, M., Heule, M.J.H., Biere, A.: Inprocessing rules. In: Gramlich, B., Miller, D., Sattler, U. (eds.) IJCAR 2012. LNCS, vol. 7364, pp. 355–370. Springer, Heidelberg (2012)

22. Malitsky, Y., Sabharwal, A., Samulowitz, H., Sellmann, M.: Parallel SAT solver selection and scheduling. In: Milano, M. (ed.) CP 2012. LNCS, vol. 7514, pp. 512–526. Springer, Heidelberg (2012)

23. Marques-Silva, J.P., Sakallah, K.A.: GRASP - a new search algorithm for satisfiability. In: ICCAD, pp. 220–227 (1996)

24. Moskewicz, M.W., Madigan, C.F., Zhao, Y., Zhang, L., Malik, S.: Chaff: Engineering an efficient SAT solver. In: DAC, pp. 530–535. ACM (2001)

25. Piette, C., Hamadi, Y., Sais, L.: Vivifying propositional clausal formulae. In: Ghallab, M., Spyropoulos, C.D., Fakotakis, N., Avouris, N.M. (eds.) ECAI. Frontiers in Artificial Intelligence and Applications, vol. 178, pp. 525–529. IOS Press (2008)

26. Pipatsrisawat, K., Darwiche, A.: A lightweight component caching scheme for satisfiability solvers. In: Marques-Silva, J., Sakallah, K.A. (eds.) SAT 2007. LNCS, vol. 4501, pp. 294–299. Springer, Heidelberg (2007)

27. Pipatsrisawat, K., Darwiche, A.: A new clause learning scheme for efficient unsatisfiability proofs. In: Fox, D., Gomes, C.P. (eds.) AAAI, pp. 1481–1484. AAAI Press (2008)

28. Schubert, T., Lewis, M.D.T., Becker, B.: PaMiraXT: Parallel SAT solving with threads and message passing. JSAT 6(4), 203–222 (2009)

29. Sörensson, N., Biere, A.: Minimizing learned clauses. In: Kullmann, O. (ed.) SAT 2009. LNCS, vol. 5584, pp. 237–243. Springer, Heidelberg (2009)

30. Wieringa, S., Heljanko, K.: Asynchronous multi-core incremental SAT solving. In: Piterman, N., Smolka, S.A. (eds.) TACAS 2013. LNCS, vol. 7795, pp. 139–153. Springer, Heidelberg (2013)

31. Xu, L., Hutter, F., Hoos, H.H., Leyton-Brown, K.: SATzilla: Portfolio-based algorithm selection for SAT. J. Artif. Intell. Res. (JAIR) 32, 565–606 (2008)

32. Zhang, H., Bonacina, M.P., Hsiang, J.: PSATO: a distributed propositional prover and its application to quasigroup problems. J. Symb. Comput. 21(4), 543–560 (1996)

Parallel MUS Extraction[*]

Anton Belov[1], Norbert Manthey[2], and Joao Marques-Silva[1,3]

[1] Complex and Adaptive Systems Laboratory University College Dublin
[2] Institute of Artificial Intelligence, Technische Universität Dresden, Germany
[3] IST/INESC-ID, Technical University of Lisbon, Portugal

Abstract. Parallelization is a natural direction towards the improvements in the scalability of algorithms for the computation of Minimally Unsatisfiable Subformulas (MUSes), and group-MUSes, of CNF formulas. In this paper we propose and analyze a number of approaches to parallel MUS computation. Just as it is the case with the parallel CDCL-based SAT solving, the communication, i.e. the exchange of learned clauses between the solvers running in parallel, emerges as an important component of parallel MUS extraction algorithms. However, in the context of MUS computation the communication might be unsound. We argue that the assumption-based approach to the incremental CDCL-based SAT solving is the key enabling technology for effective sound communication in the context of parallel MUS extraction, and show that fully unrestricted communication is possible in this setting. Furthermore, we propose a number of techniques to improve the quality of communication, as well as the quality of job distribution in the parallel MUS extractor. We evaluate the proposed techniques empirically on industrially-relevant instances of both plain and group MUS problems, and demonstrate significant (up to an order of magnitude) improvements due to the parallelization.

1 Introduction

A minimally unsatisfiable subformula (MUS) of an unsatisfiable CNF formula is any minimal, with respect to set inclusion, subset of its clauses that is unsatisfiable. MUSes, and the related *group-MUSes* [15,21], find a wide range of practical applications [21,3], and so the development of efficient MUS extraction algorithms is currently an active area of research (see [16] for a survey, [26,4,24] for recent developments). State-of-the-art MUS extraction algorithms use SAT solvers as $\mathcal{NP}$ oracles, and typically perform a large number of SAT solver calls — each call with a different subformula of the original input formula. The fact that many of these calls are independent suggests that MUS computation problem might be a good candidate for parallelization.

A number of successful approaches to the parallelization of SAT solving have been developed (see [13,19] for the recent overviews). In one of the widely used variants of parallel SAT solvers, namely *portfolio solvers*, several incarnations of a sequential solver, possibly with different configurations, are executed on the same input formula in

---

[*] The first and third authors are financially supported by SFI PI grant BEACON (09/IN.1/I2618), and by FCT grants ATTEST (CMU-PT/ELE/0009/2009) and POLARIS (PTDC/EIA-CCO/123051/2010).

M. Järvisalo and A. Van Gelder (Eds.): SAT 2013, LNCS 7962, pp. 133–149, 2013.
© Springer-Verlag Berlin Heidelberg 2013

parallel. An essential component of portfolio solvers built on top of CDCL-based SAT solvers is the mechanism for the exchange of learned clauses — the *communication* — between the sequential sub-solvers. As such, it is plausible, and, as we show, indeed the case, that communication is an important aspect of any parallel MUS extraction solution geared towards industrially-relevant problems. However, while the communication is sound in portfolio-based parallel SAT solvers (since the sub-solvers work on the same input formula), this is not necessarily the case in a parallel MUS extractor, since now the formulas might differ.

In this paper we analyze a number of approaches to the parallel MUS computation. We notice that some of the simpler of these approaches result in performance degradation with respect to a sequential solution, however we do confirm the importance of communication for the parallel MUS extraction. More importantly, we argue that the *assumption-based approach* to the incremental CDCL-based SAT solving, introduced in [9], is the key enabling technology for effective sound communication in scalable parallel MUS extraction algorithms, and suggest that this might also be the case in more general settings where CDCL-based SAT solvers work on different related formulas in parallel. We carefully analyze the communication aspect in the context of parallel MUS extraction, and show that in this setting fully *unrestricted* communication is possible. Furthermore, we propose a number of effective clause filtering techniques, and an improved job distribution scheme based on the analysis of unsatisfiable cores. We evaluate the proposed algorithms and techniques empirically, and demonstrate significant speed-ups and scalability (e.g. median 2.94x, with up to 132x, speed up on 4 cores) on a set of industrially-relevant MUS and group-MUS extraction benchmarks.

2   Preliminaries and Background

We assume the familiarity with propositional logic, its clausal fragment, and commonly used terminology of the area of SAT. We focus on formulas in CNF (*formulas*, from hence on), which we treat as (finite) (multi-)sets of clauses. We assume that clauses do not contain duplicate variables. Given a formula F we denote the set of variables that occur in F by $Var(F)$, and the set of variables that occur in a clause $C \in F$ by $Var(C)$. An *assignment* τ for F is a map $\tau : Var(F) \rightarrow \{0, 1\}$. Assignments are extended to formulas according to the semantics of classical propositional logic. If $\tau(F) = 1$, then τ is a *model* of F. If a formula F has (resp. does not have) a model, then F is *satisfiable* (resp. *unsatisfiable*). By $F|_\tau$ we denote the *reduct* of the formula F wrt. the assignment τ – that is the formula obtained from F by removing the satisfied clauses and falsified literals from the remaining clauses. The *resolution rule* states that, given two clauses $C_1 = (x \vee A)$ and $C_2 = (\neg x \vee B)$, the clause $C = (A \vee B)$, called the *resolvent* of C_1 and C_2, can be inferred by *resolving* on the variable x. We write $C = C_1 \otimes_x C_2$. Note that $\{C_1, C_2\} \models C$.

A CNF formula F is *minimally unsatisfiable* if (i) F is unsatisfiable, and (ii) for any clause $C \in F$, the formula $F \setminus \{C\}$ is satisfiable. The set of minimally unsatisfiable CNF formulas is denoted by MU. A CNF formula F' is a *minimally unsatisfiable subformula (MUS)* of a formula F if $F' \subseteq F$ and $F' \in$ MU. The set of MUSes of a CNF formula F is denoted by MUS(F). A clause $C \in F$ is *necessary* for F (cf. [14])

if F is unsatisfiable and $F \setminus \{C\}$ is satisfiable. Necessary clauses are often referred to as *transition* clauses. The set of all necessary clauses of F is precisely $\bigcap \text{MUS}(F)$. Thus $F \in \text{MU}$ if and only if every clause of F is necessary. The problem of deciding whether a given CNF formula is in MU is DP-complete [23]. Motivated by several applications, minimal unsatisfiability and related concepts have been extended to CNF formulas where clauses are partitioned into disjoint sets called *groups* [15,21].

The basic deletion-based MUS extraction algorithm operates in the following manner. Starting from an unsatisfiable formula F, the algorithm picks a clause $C \in F$, and tests the formula $F \setminus \{C\}$ for satisfiability. If the formula is unsatisfiable, C is removed from F, i.e. we let $F = F \setminus \{C\}$. Otherwise, C is necessary for F, and so for every unsatisfiable subformula of F, and hence C is included in the computed MUS. Once all clauses of the input formula F are tested for necessity in this manner, the remaining clauses constitute an MUS of F. While the basic deletion algorithm is neither theoretically nor empirically effective (see, for example, [16,4]), the addition of *clause-set refinement* and *model rotation*[17] makes it the top performing algorithm for industrially relevant instances. Clause-set refinement takes advantage of the capability of modern SAT solvers to produce an unsatisfiable core: since a core includes at least one MUS, all clauses outside the core can be removed from the formula after a single UNSAT outcome. Model rotation, on the other hand, allows to detect multiple necessary clauses in the case of SAT outcome: when $F \setminus \{C\} \in \text{SAT}$, the model τ returned by the SAT solver serves as a witness of the necessity of C in F, and model rotation attempts to (cheaply) modify τ to obtain a witnesses for other clauses of F, possibly declaring multiple clauses of F necessary after a single SAT outcome.

Although the modern sequential CDCL-based SAT solvers are rooted in the DPLL algorithm [8], the addition of *clause learning* [18] and *back-jumping*, drastically changes the behaviour of the algorithm. We refer the reader to a tutorial introduction of CDCL in [25] that illuminates these changes. Additional enhancements to CDCL present in the modern SAT solvers include *restarts* [10], advanced data-structures and decision heuristics [20], and sophisticated heuristics to control the quality of learned clauses, based, for example, on the idea of *literal block distance* [2]. Many details of the modern CDCL-based SAT solving can be found in [7]. Although the specifics of the clause learning mechanism in CDCL-based SAT solvers are not crucial for the understanding of this paper, one aspect that is important is that new learned clauses are derived from the clauses of the input formula and the previously learned clauses used in a conflict via a sequence of resolutions steps. This ensures that the learned clauses are logically entailed by the clauses of the input formula.

A detailed overview of the modern parallel SAT solving systems can be found in [13,19]. Here, we focus on a widely used variant of parallel SAT solvers, namely *portfolio solvers*, as for example MANYSAT [12], PLINGELING [6] or PENELOPE [1]. This type of solvers execute several incarnations of a sequential solver, possibly with different configurations, on the same input formula in parallel. An important component of this type of solvers is the exchange of learned clauses between the sequential sub-solvers — the *communication*. The quality of the exchanged clauses becomes particularly important, and several heuristics have been proposed. In [12] only the clauses of a fixed size have been shared, while in [11] this *sharing filter* has been improved further to a *quality based heuristic*: sharing limits are relaxed if not sufficiently many clauses are exchanged, and when too many clauses are shared, the sharing limits are tightened again.

Incremental SAT Solving. In many practical applications SAT solvers are invoked on a sequence of related formulas. The *incremental* SAT solving paradigm is motivated by the fact that the clauses learned from subformulas common to the successive formulas can be reused. The most widely used approach to the incremental SAT is the so-called *assumption-based* approach introduced in [9]. In this approach, a SAT solver provides an interface to add clauses, and an interface to determine the satisfiability of the set of currently added clauses, F, together with a user-provided set of *assumption* literals $A = \{a_1, \ldots, a_k\}$ – that is, to test the satisfiability of the formula $F \cup \bigcup_{a_i \in A}\{(a_i)\}$. An important feature of the approach of [9] is that the assumptions are *not* added to the formula prior to solving, but, instead, are used as the top-level decisions. As a result, any clause learned while solving the formula F under the assumptions A can be used to solve a formula $F' \supseteq F$ under a possibly different set of assumptions A'.

Assumption-based incremental SAT solving is often used to emulate arbitrary modifications to the input formula. Given a formula $F = \{C_1, \ldots, C_n\}$, the set $A = \{a_1, \ldots, a_n\}$ of fresh *assumption variables* is constructed (i.e. $Var(F) \cap A = \emptyset$), and the formula $F_A = \{(a_i \vee C_i) \mid C_i \in F\}$ is loaded into an incremental SAT solver. Then, for example, in order to establish the satisfiability of $F' \subseteq F$, the formula F_A is solved under assumptions $\{\neg a_i \mid C_i \in F'\} \cup \{a_j \mid C_j \notin F'\}$. In effect, the assumptions *temporarily* remove clauses outside of F'. If the outcome is sat, the model, restricted to $Var(F)$, is a model of F'. If the outcome is unsat, SAT solvers return a set of literals $A_{core} \subseteq A$ such that F_A is unsatisfiable under the assumptions $\{\neg a_i \mid a_i \in A_{core}\} \cup \{a_j \mid a_j \notin A_{core}\}$, and so the formula $\{C_i \mid a_i \in A_{core}\}$ is an *unsatisfiable core* of F'. Using the incrementality, any clause $C_i \in F$ can be removed *permanently* by adding the unit (a_i) to the SAT solver. Conversely, the addition of the unit $(\neg a_i)$ permanently asserts, or *finalizes*, C_i. Importantly, in this setting the negated assumptions are *not* resolved out of the learned clauses, whereas the negated unit clauses *are*. For example, if $\neg a_1$ and $\neg a_2$ are assumptions, and $(\neg a_3)$ is a unit clause, and if D is a learned clause whose derivation used clauses containing a_1, a_2, a_3, then the literals a_1 and a_2 are in D, whereas a_3 is resolved out of D. If a unit (a_2) is later added to the formula, then D is satisfied and is not used for further reasoning.

An alternative to the approach of [9], discussed extensively in [22], is to add assumptions as *temporary* unit clauses to the SAT solver's formula. To be usable for the subsequent incremental invocations, the clauses learned from any of these units must be extended with the negation of the assumption literals used to derive them. Although for the applications discussed in [22] this post-processing step pays off, as we argue shortly the approach of [9] appears to be the key to the efficient parallel MUS extraction.

3   Parallel MUS Extraction Algorithm

In this paper we describe a *low-level* parallelization of a particular MUS extraction algorithm, that is, while the high-level flow of the algorithm is unchanged, we off-load various satisfiability tests required by the algorithm to multiple threads. Clearly, such low-level parallelization can be further integrated into a *high-level* parallel MUS extractor, that would run different (parallelized on the low-level) MUS extraction algorithms in parallel. Such high-level parallelization is a subject of future research.

The parallel MUS extraction algorithm proposed in this paper is based on the hybrid MUS extraction algorithm [17] (a variant of the deletion-based algorithm, augmented with the clause-set refinement and model rotation). The algorithm maintains a single master thread, and one or more worker threads. The master keeps a current snapshot of the working formula, and distributes the work items to the workers. A work item is simply a clause (or a group) that needs to be tested for necessity with respect to the current working formula. Thus, each worker owns a SAT solver, and given a work item $\langle F, C_i \rangle$, the responsibility of the worker is to invoke the SAT solver on the formula $F \setminus \{C_i\}$ and provide the result to the master. Once all available workers are started, the master waits for some or all of the workers to finish processing their work item. The results of finished workers are then aggregated, the master's working formula is updated, and any currently running workers whose work item's status has already been determined (the *redundant* workers) are aborted. Finally, the master proceeds to assign the next work items to the available workers, until no more work items are left.

There is a number of degrees of freedom within this framework: (i) *synchronous* vs. *asynchronous* execution — in the synchronous mode the master waits for all workers to finish their current task before advancing to the next iteration, while in the asynchronous execution the master processes the results as they come in from the workers; (ii) *work distribution* — whether the workers test the necessity of the same or a different clause; (iii) *communication between workers* — whether the workers are allowed to exchange the learned clauses or not. In the rest of the paper we denote various configurations by three letter acronyms: S (resp. A) for synchronous (resp. asynchronous) execution, followed by S (resp. D) for same (resp. different) clause distribution, followed by N (resp. C) for absence (resp. presence) of communication between the workers.

Perhaps the simplest reasonable configuration is the asynchronous execution on the same clause, i.e. all workers are given the same task $\langle F, C_i \rangle$ (the AS_ configurations). This configuration is akin to portfolio-like solutions for parallel SAT solving in that it takes advantage of the fact that the run times of different incarnations of the same SAT solver working on the same formula may vary significantly. Since all workers test the same clause, once some workers are finished (there may be more than one), all others become redundant. Clearly, this configuration should benefit from communication — a comparison of the results for ASN and ASC configurations (Table 1 in Sec. 5) confirms that this is the case. Since in ASC all workers work on exactly the same formula $F \setminus \{C_i\}$, they can freely exchange learned clauses. The main drawback of ASC is that, despite the communication, the workers largely duplicate each other efforts. As a result, and since the parallel execution incurs a non-trivial overhead on the system (mostly due to memory accesses), this configuration performs worse than a sequential solution.

The next configurations we consider are those with synchronous execution on different clauses (the SD_ configurations). Here the workers are given the tasks $\langle F, C_1 \rangle$, $\langle F, C_2 \rangle$, ... with the goal to distribute the work of checking the necessity of the clauses between the workers. Note that since the master algorithm employs both the clause set refinement and the model rotation, some of the workers will end up executing redundant tasks. We will address this drawback shortly, however meanwhile let us discuss the communication aspect of SDC. The main observation is that the communication between workers is *not* sound. To be specific, consider two workers W_1, W_2 working on the tasks $\langle F, C_1 \rangle$, and $\langle F, C_2 \rangle$, respectively. That is, W_1 is running its SAT solver on the formula $F \setminus \{C_1\}$, while W_2 is working on the formula $F \setminus \{C_2\}$. The problem now is that some clauses derived from C_1 by W_2 might not be valid logical consequences of the formula $F \setminus \{C_1\}$ solved by W_1. One could envision a number of ways to circumvent this problem. For example, we could prohibit W_1 from sending any clause derived from C_2 to W_2 (and vise versa), however in this case the workers need to be aware of what other workers are doing. Another option would be to force W_1 to refuse any clause derived from C_1 — this solution would require to augment every exchanged clause with some information regarding its origin, and to analyze every received learned clause. Finally, one could resort to an ($\mathcal{NP}$-complete) implication test for each received clause. Clearly, neither of these solutions are satisfactory. Yet, a natural solution does exist: use *assumption-based incremental SAT solvers*. Before we come back to this important point, we note another drawback of SD_ configurations: due to the fact that the run times of the SAT calls executed by workers may vary significantly, the algorithm waits for the completion of the longest running call while other workers are idle. Worse, it might be that the longest call is redundant given the results of some of the workers that finished their SAT calls faster. Thus, SD_ configurations might be hampered by a low CPU utilization, and a large percentage of "wasted" SAT calls.

It should be no surprise then, that a scalable parallel MUS extraction algorithm requires both the asynchronous execution, a work distribution strategy that reduces the duplication of workers' efforts, and sound communication — this the configuration ADC. The problem of sound communication in the asynchronous context not only remains, but becomes exacerbated, as we now might have a situation where W_1 is processing a work item $\langle F, C_1 \rangle$, while W_2 is working on $\langle F', C_2 \rangle$ with $F' \subset F$, and so the clauses derived from $F \setminus F'$ by W_1 might not be valid for W_2.

We now argue that assumption-based incremental SAT solving (i.e. the approach introduced in [9]), often seen as simply an implementation technique, is in fact a key enabling technology for the scalable parallel MUS extraction algorithms. Recall from Sec. 2 that in the incremental SAT setting the test of the satisfiability of some subformula F_1 of an input formula F can be performed without removing any of the clauses of F. Instead, the clauses outside of F_1 can be temporarily disabled using assumptions. As a result, clauses that are learned while analyzing a different subformula F_2 of F are valid (though might also be temporarily disabled) for the analysis of F_1. Consider now the organization of the parallel MUS extraction algorithm described above, whereby the workers are provided with *incremental* SAT solvers. During the initialization, all workers are given the same augmented formula $F_A = \{(a_i \vee C_i) \mid C_i \in F\}$ constructed by adding assumption literals to the clauses (or groups) of the input formula F. The workers invoke their incremental SAT solvers on F_A under a set of assumptions that represent the subformula assigned to them by the master. As the construction of an MUS of F progresses, the master determines that some clauses of F need to be either permanently removed or finalized — to achieve this, the master adds the corresponding unit clauses to the SAT solvers of the workers. Now, consider again the configuration

SDC — notice that since the execution is synchronous, all workers have *exactly the same* input formula $F_A \cup U$, where U is the set of unit clauses added to the formula in order to delete and to finalize some clauses. Since the only difference between the worker's execution is the set of assumptions under which they test the satisfiability of $F_A \cup U$, the workers are free to exchange the learned clauses, without any limitations and any additional reasoning. In the asynchronous setting, ADC, the situation is not quite straightforward, since then, while W_1 is working on the formula $F_A \cup U$, the worker W_2, which "ran ahead" of W_1, might be working on the formula $F_A \cup U \cup U'$, where U' is a set of unit clauses added by the master since the beginning of execution of W_1. Since the input formula of W_1 is a subformula of the input formula of W_2, the clauses learned by W_1 will be valid for W_2. However, it is not clear that W_2 can send its learned clauses "back" to W_1, since W_2 has additional clauses in its formula. We will prove that fully *unrestricted* communication is sound in the asynchronous setting as well.

Notice that in the assumption-based incremental SAT setting the assumption literals provide an automatic way of *tagging* the learned clauses with the information of their origins. At the same time, SAT solving under assumptions automatically ensures that any clause previously learned from a currently disabled clause is disabled. The alternative approach to the incremental SAT, whereby the assumptions are added as temporary unit clauses (cf. Sec. 2), would, in our setting, require the reconstruction step described in [22] for every learned clause sent to another solver, which is not likely to scale. Our observations suggest that the approach of [9] to the assumption-based incremental SAT solving might be the key to the effective communication in more general scenarios where the CDCL-based SAT solvers work on different related formulas in parallel.

Formal Description of the Algorithm. By $F = \{C_1, \ldots, C_n\}$ we denote the input CNF formula, whose MUS M is to be computed. Let $A_F = \{a_1, \ldots, a_n\}$ be a set of fresh *assumption* variables (or, *assumptions*). Each assumption variable a_i will be implicitly associated with the clause C_i. As in Sec. 2, by F_A we will denote the formula $F_A = \{(a_1 \vee C_1), \ldots, (a_n \vee C_n)\}$. Given any $A \subseteq A_F$, let $cls(A) = \{C_i \mid a_i \in A\}$. Throughout the execution, the master maintains two sets of assumptions: the set of *necessary* assumptions $A_{nec} \subseteq A_F$ that corresponds to clauses that are declared to be necessary (i.e. part of the computed MUS), and the set of *unnecessary* assumptions $A_{unnec} \subseteq A_F$ that corresponds to the clauses that will not be included in the computed MUS. The sets A_{nec} and A_{unnec} are disjoint. For convenience, we let $A_{unk} = A_F \setminus (A_{nec} \cup A_{unnec})$ to denote the set of assumptions that correspond to clauses whose status is unknown. The state of the master is described by a pair $\langle A_{nec}, A_{unnec} \rangle$ of the sets of the necessary and the unnecessary assumptions. Given such a state pair (or, simply, a *state*) $S = \langle A_{nec}, A_{unnec} \rangle$, by $F(S)$ we denote the formula

$$F(S) = F_A \cup \{(\neg a_i) \mid a_i \in A_{nec}\} \cup \{(a_j) \mid a_j \in A_{unnec}\}. \tag{3.1}$$

The pseudocode of the algorithm is presented in Alg. 1. Each of the workers W_w, $w = 1, \ldots, nw$, runs on its own thread, and has its own incremental SAT solver which is initialized with the formula F_A. As the master progresses, it adds negative units to finalize the necessary clauses, and positive units to remove the unnecessary clauses from the workers' SAT solvers. Since in the asynchronous configurations the formulas inside the SAT solvers may diverge, it will be convenient to view each worker W_w as

---

Algorithm 1. Parallel MUS extraction algorithm (with incremental SAT)

Input : F — unsatisfiable CNF Formula; nw — number of worker threads
Output: $M \in \mathsf{MUS}(F)$

```
1  initializeWorkers(F, {W₁,...,Wₙw})              // each Wᵢ is a worker
2  ⟨A_nec, A_unnec⟩ ← ⟨∅, ∅⟩                        // initial state

3  while A_unk ≠ ∅ or there are running workers do  // A_unk ≜ A_F \ (A_nec ∪ A_unnec)
4  |  if A_unk ≠ ∅ then                             // there are untested clauses
5  |  |  foreach idle W_w do
6  |  |  |  aᵢ ← pickAssumption(A_unk)
7  |  |  |  W_w.updateState(⟨A_nec, A_unnec⟩)
8  |  |  |  W_w.startTask(aᵢ)
9  |  sleepUntilFinished()
10 |  Res = { W_w.getResult() | W_w is finished }
11 |  ⟨Δ_{A_nec}, Δ_{A_unnec}⟩ ← mergeResults(Res)
12 |  ⟨A_nec, A_unnec⟩ ← ⟨A_nec ∪ Δ_{A_nec}, A_unnec ∪ Δ_{A_unnec}⟩
13 |  abortRedundantWorkers()
14 return M ≜ cls(A_nec)                            //  M ∈ MUS(F)
```

---

having its own version of a state-pair $S^w = \langle A^w_{nec}, A^w_{unnec} \rangle$, with $F(S^w)$ (as per 3.1) being exactly the set of input clauses in $\mathtt{W}_w$'s SAT solver. Details of the functions used in Alg. 1 are discussed below:

$\mathtt{pickAssumption}(A_{unk})$: for _S_ configurations, the function picks $a_i \in A_{unk}$, and returns the same a_i for each invocation in the **foreach** loop on line 5; for _D_ configurations, for each invocation the function returns a different $a_i \in A_{unk}$, if possible, and the last picked a_i if not.

$\mathtt{W}_w.\mathtt{updateState}(\langle A_{nec}, A_{unnec} \rangle)$: sets S^w to be identical to S; on the implementation level this causes the addition of unit clauses to $\mathtt{W}_w$'s SAT solver.

$\mathtt{W}_w.\mathtt{startTask}(a_i)$: starts $\mathtt{W}_w$'s SAT solver. If $S^w = \langle A^w_{nec}, A^w_{unnec} \rangle$ is the state of the worker $\mathtt{W}_w$ at the moment of invocation, the set of clauses in $\mathtt{W}_w$'s SAT solver is $F(S^w)$, and the SAT solver is invoked under assumptions $\{a_i\} \cup \{\neg a_j \mid a_{j \neq i} \in A_{unk}\}$. Thus, the worker tests whether the clause C_i, associated with the assumption a_i, is necessary for the formula $cls(A^w_{nec} \cup A^w_{unk})$. We will say that a_i is $\mathtt{W}_w$'s *task literal*, and that, until the SAT solver run has finished, $\mathtt{W}_w$ is processing its *task*.

$\mathtt{sleepUntilFinished}()$: for synchronous configurations (S__) this function waits until all workers finished their tasks; for asynchronous configurations (A__) this function waits until at least one worker has finished its task (but there might still be more than one finished worker when this function returns).

$\mathtt{W}_w.\mathtt{getResult}()$: retrieves the outcome of the SAT test performed by a finished worker $\mathtt{W}_w$. Since a worker's state might be out of sync with the master's, for notational convenience we assume that the returned result is a tuple $R^w = \langle S^w, a_i, st, \tau, A^w_{core} \rangle$, where S^w is the state of the worker, a_i is the worker's task literal, st is the outcome of the SAT call (sat or unsat), τ is the model returned by the SAT solver in case

$st = \mathtt{sat}$, and $A^w_{core} \subseteq A^w_{unk}$ is a set of assumption literals in the final conflict clause in case $st = \mathtt{unsat}$.

$\mathtt{mergeResults}(Res)$: is responsible for the analysis of the set Res of the results of finished workers. The function returns a tuple $\langle \Delta_{A_{nec}}, \Delta_{A_{unnec}} \rangle$ of assumptions that correspond to the newly discovered necessary and unnecessary clauses, initialized with $\langle \emptyset, \emptyset \rangle$. For each $R^w = \langle S^w, a_i, st, \tau, A^w_{core} \rangle \in Res$ with $st = \mathtt{sat}$, the function appends a_i to $\Delta_{A_{nec}}$, and executes the model rotation algorithm on the formula $cls(A_{nec} \cup A_{unk})$ with the assignment τ to detect additional necessary clauses. For each such clause, the corresponding assumption variable is added to $\Delta_{A_{nec}}$. For each result tuple with $st = \mathtt{unsat}$ the function first checks whether A^w_{core} is a subset of A_{unk} — in the asynchronous mode this might not be the case, since the state of the worker S^w might be out-of-date with respect to the state of the master. All $\mathtt{unsat}$ results for which $A^w_{core} \not\subseteq A_{unk}$ are discarded, and from the remaining $\mathtt{unsat}$ results one set A^w_{core} is selected[1]. The function then sets $\Delta_{A_{unnec}}$ to be $A_{unk} \setminus A^w_{core}$.

$\mathtt{abortRedundantWorkers}()$: aborts all workers whose task literal is not in A_{unk}. The learned clauses accumulated by a worker's SAT solver remain in the solver.

Proof of Correctness and the Soundness of Unrestricted Communication. The correctness of the presented algorithm hinges on the following loop invariant.

Invariant 3.1. *For $v = 1, \ldots$, let $S_v = \langle A^v_{nec}, A^v_{unnec} \rangle$ denote the state of the master prior to the v-th test of the main loop guard (line 3 of Alg. 1). Then, the formula $cls(A^v_{nec} \cup A^v_{unk})$ is unsatisfiable, and every clause in $cls(A^v_{nec})$ is necessary for it.*

To prove the invariant we need to establish a correctness property of the results returned by the finished workers. The property holds trivially in the configurations without communication, however a subtlety arises when the *unrestricted* communication is enabled. This is best demonstrated by the following example.

Example 1. Let the (already augmented) formula F_A be $\{(a_1 \vee \neg b \vee \neg c \vee \neg d), (a_2 \vee \neg b \vee c \vee d), (a_3 \vee b \vee c \vee \neg d), (a_4 \vee b \vee c \vee d), (a_5 \vee \neg c \vee \neg d), (a_6 \vee c \vee \neg d), (a_7 \vee \neg c \vee d)\}$. Assume that there are three workers, and $\mathtt{W}_1$ already determined that C_1 is unnecessary:

$\mathtt{W}_w$	Task	Assumptions	A^w_{nec}	A^w_{unnec}
$\mathtt{W}_1$	a_4	$\{\neg a_2, \neg a_3, a_4, \neg a_5, \neg a_6, \neg a_7\}$	$\{\}$	$\{a_1\}$
$\mathtt{W}_2$	a_2	$\{\neg a_1, a_2, \neg a_3, \neg a_4, \neg a_5, \neg a_6, \neg a_7\}$	$\{\}$	$\{\}$
$\mathtt{W}_3$	a_3	$\{\neg a_1, \neg a_2, a_3, \neg a_4, \neg a_5, \neg a_6, \neg a_7\}$	$\{\}$	$\{\}$

Assume $\mathtt{W}_1$ finishes its task first: it returns $\mathtt{sat}$, and a model τ that witnesses the clause C_4. In $\mathtt{mergeResults}()$ the master applies model rotation, and determines that C_2 is also necessary. The master now has $A_{nec} = \{a_2, a_4\}$ and $A_{unnec} = \{a_1\}$. Since $\mathtt{W}_2$'s task is a_2, it becomes redundant and is aborted. Assume $\mathtt{W}_1$ is given a_5, and $\mathtt{W}_2$ some other task (not shown). Note that the master adds the units $\{(\neg a_2), (\neg a_4)\}$ to $\mathtt{W}_1$'s solver, and the units $\{(a_1), (\neg a_2), (\neg a_4)\}$ to $\mathtt{W}_2$'s, prior to the call.

$\mathtt{W}_1$	a_5	$\{\neg a_3, a_5, \neg a_6, \neg a_7\}$	$\{a_2, a_4\}$	$\{a_1\}$

Then, during conflict analysis the learned clause $(c \vee d)$ can be generated by $\mathtt{W}_1$ from $(a_2 \vee \neg b \vee c \vee d), (a_4 \vee b \vee c \vee d)$, and the two units $(\neg a_2), (\neg a_4)$. When this clause

---

[1] In our implementation we select a set A^w_{core} of the smallest size.

is received by W_3, it learns $(a_5 \vee a_6 \vee a_7)$ from $F(S^3)$ by resolving $(c \vee d)$ with $(a_5 \vee \neg c \vee \neg d)$, $(a_6 \vee c \vee \neg d)$ and $(a_7 \vee \neg c \vee d)$, resulting in the set $A^3_{core} = \{a_5, a_6, a_7\}$. But, the formula $cls(A^3_{core})$ is satisfiable ! As we see shortly, the master must conjoin A^3_{core} with A_{nec} to get the "real" unsatisfiable core. ∎

Lemma 1. *Let* $S^w = \langle A^w_{nec}, A^w_{unnec} \rangle$ *be the state of a worker* W_w *at the time of the invocation of the function* $W_w . \mathtt{startTask}(a_i)$. *Let* A *be any subset of* $A^w_{unk} \setminus \{a_i\}$, *and* D *be any set of clauses implied by the formula* $F(S^w) \cup \{(\neg a_j) \mid a_j \in A\}$. *Furthermore, let* $\langle st, \tau, A^w_{core} \rangle$ *be the outcome of a SAT solver call on the formula* $F(S^w) \cup D$ *under the set of assumptions* $P = \{a_i\} \cup \{\neg a_j \mid a_{j \neq i} \in A^w_{unk}\}$. *Then,*

(i) *if* $st = \mathtt{sat}$, *then the formula* $F(S^w)$ *is satisfiable* P, *and* τ *is a model of* $F(S^w)$ *(that respects* P).

(ii) *if* $st = \mathtt{unsat}$, *then the formula* $F(S^w)$ *is unsatisfiable under the assumptions* $P' = \{\neg a_j \mid a_j \in A \cup A^w_{core}\}$.

The set D in Lemma 1 is intended to represent the set of "extra" clauses that a worker W_w has received from other workers during the execution of its task, and the set $A \subseteq A^w_{unk}$ of assumptions to correspond to the clauses that were discovered to be necessary by the master since W_w started its task. Notice the addition of the set A to the set of assumptions P' in part (ii) of the lemma (cf. Example 1). We now argue that all clauses received by any worker W_w satisfy the condition of Lemma 1.

Lemma 2. *Let* $S^w = \langle A^w_{nec}, A^w_{unnec} \rangle$ *be the state of a worker* W_w *that has successfully completed its task* a_i *(i.e. it has not been aborted by the master), and let* $S = \langle A_{nec}, A_{unnec} \rangle$ *be the state of the master by the time it calls* $W_w . \mathtt{getResult}()$ *(line 10, Alg. 1). Then, for every clause* C *in* W_w'*s SAT solver,*

$$F(S^w) \cup \{(\neg a_j) \mid a_j \in A_{nec} \setminus A^w_{nec}\} \models C. \qquad (3.2)$$

Proof (sketch). The complete proof involves an inductive argument on the global sequence of generated learned clauses. The base case is the non-trivial part of the proof and is established using the following observations. Let W_1 and W_2 be two workers with the states S^1 and S^2 respectively, such that $S^1_{unk} \supset S^w_{unk} \supset S^2_{unk}$, i.e. W_1 is "behind" W_w and W_2 is "ahead" of W_w. Since $F(S^1) \subset F(S^w)$, any clause C learned by W_1 from its input formula satisfies (3.2). Let C be a clause learned by W_2 from its input formula $F(S^2)$, i.e. C is implied by the formula

$$F(S^2) = F(S^w) \cup \{(\neg a_j) \mid a_j \in A^2_{nec} \setminus A^w_{nec}\} \cup \{(a_k) \mid a_k \in A^2_{unnec} \setminus A^w_{unnec}\}.$$

The assumptions a_k occur only positively in $F(S^2)$, so, due to the units (a_k), clauses with a_k will not be used as conflict. Hence, a_k do not occur in C, and so C is implied by

$$F(S^w) \cup \{(\neg a_j) \mid a_j \in A^2_{nec} \setminus A^w_{nec}\}.$$

We conclude that (3.2) holds by taking into account that $A^2_{nec} \subseteq A_{nec}$. □

Lemma 3 below establishes the correctness of the results returned by the workers by putting together Lemmas 1 and 2, and taking into account the fact that for any worker

W_w with state S^w that has *successfully* completed a task a_i, we have that $a_i \in A_{unk}$, as otherwise the master would have aborted W_w during abortRedundantWorkers() call at the end of the *previous iteration*. In particular, $a_i \notin (A_{nec} \setminus A_{nec}^w)$, and so the set $A_{nec} \setminus A_{nec}^w$ from (3.2) satisfies the condition imposed on the set A in Lemma 1. Notice that in Example 1, if W_2 was *not* aborted when a_2 was found necessary, on the reception of the clause $(c \vee d)$ from W_1 it would return unsat, instead of sat.

Lemma 3. *Let $S^w = \langle A_{nec}^w, A_{unnec}^w \rangle$ be the state of a worker W_w that has successfully completed its task a_i, and let $S = \langle A_{nec}, A_{unnec} \rangle$ be the state of the master by the time it obtains W_w's result tuple $R = \langle S^w, a_i, st, \tau, A_{core}^w \rangle$. Then, if $st = $ sat, then τ is a model of the formula $cls(A_{nec} \cup A_{unk}) \setminus \{C_i\}$; if $st = $ unsat, then the formula $cls(A_{nec} \cup A_{core}^w)$ is unsatisfiable.*

Using Lemma 3 and the definition of mergeResults() we establish Invariant 3.1.

Theorem 1. *Algorithm 1 terminates on any unsatisfiable input formula F, and the set $M = cls(A_{nec})$ returned by the algorithm is an MUS of F.*

SAT Solver Modifications. To exchange the learned clauses, a globally accessible clause pool, implemented as a ring buffer, is created. Each incremental SAT solver incarnation adds its learned clauses to the pool and receives the clauses submitted by other solvers. A solver incarnation sends a learned clause immediately after its generation if the clause passes the heuristic filters (discussed below). Clauses are received from the pool prior to a decision on decision level 0.

Improving Communication. Although, as shown above, the unrestricted communication is sound, clause sharing has to be restricted due to following reasons: (i) the received clauses might be redundant; (ii) additional clauses slow down the reasoning of a SAT solver incarnation; (iii) the usefulness of the new clauses cannot be determined in advance. Thus, we restrict the communication to the clauses that appear to be *promising* by adding two sharing filters to the system: learned clauses are shared if (i) their size or (ii) the *literal block distance* (LBD) [2] are less than a certain threshold. The default configuration uses a size limit of 10 literals and an LBD limit of 5. Following the ideas in [11] we also added a configuration DYN, in which these sharing thresholds are controlled dynamically. As previously discussed, the exchanged learned clauses may include a large number of assumption literals, which affect both the size and the LBD value of the clauses. For example, a clause with a single non-assumption literal and a large number of assumption literals might be filtered out due to its size. Clearly, such clause will be extremely useful to other solvers, as it might trigger unit propagation once the assumptions are assigned. Thus, from the filtering point of view, the assumption literals are superfluous, and so we added the configuration PRASS in which these literals are ignored ("protected") in the analysis by the sharing filters. Since the activity of learned clauses is initialized, we also allow this for received clauses (BUMP).

Core-Based Scheduling. Work duplication remains an important problem in our algorithm. To reduce the duplication we implemented a scheduling scheme based on the analysis of unsatisfiable cores returned by the workers. The scheme relies on the intuitive observation that clauses that appear in the intersection of unsatisfiable cores during the execution of the algorithm (including those discarded by `mergeResults()`) are likely to be necessary. In *core-based scheduling* CBS we prioritize clauses based on their core membership count.

4   Related Work

To our knowledge the only published work on the parallelization of MUS extraction algorithms is the recently published papers by Wieringa [26] and Wieringa and Heljanko [27]. In both papers an MUS extractor is built on top of a parallel incremental SAT solver. As the focus of [26] is the analysis of model rotation, the parallelization aspects, and importantly, the communication aspects, are not discussed in sufficient detail. In [27] the authors present a parallel incremental SAT solver `Tarmo`, and use the MUS extraction problem as a case study to demonstrate its effectiveness. To this extent, the authors implemented the MUS extraction algorithm described in [26] on top of `Tarmo` — we will refer to this combination as `TarmoMUS`, as this is the name of the MUS extractor distributed by the authors. The essential difference between the algorithm proposed in this paper, and `TarmoMUS` is that in our algorithm the communication is unrestricted (modulo the filtering techniques discussed above), whereas in `TarmoMUS` the communication is restricted to be "forward" only. This restriction both incurs an additional implementational overhead, and reduces the quality and usefulness of exchanged clauses. In Sec. 5 we demonstrate that our algorithm scales significantly better that of [27] — we attribute this difference to the unrestricted communication. Additional important technique in our algorithm that is absent from `TarmoMUS` is the assumption "protection" during clause filtering.

5   Experimental Evaluation

The algorithm described in this paper was implemented in C++ with pthreads, and the resulting tool, `pMUSer2`, was evaluated on a subset of benchmarks used in the MUS track of SAT Competition 2011. The subset consists of 175 MUS and 201 group-MUS instances on which the sequential MUS extractor `MUSer2` [5] takes more than 10 seconds of CPU time. The experiments were performed on an HPC cluster, where each node is a dual quad-core Intel Xeon E5450 3 GHz with 32 GB of memory. All tools were run with a timeout of 1800 seconds and a memory limit of 16 GB per input instance. All communicating configurations use the PRASS option by default.

Table 1 summarizes the results of various configurations of the parallel, as well as the sequential, algorithms. Clearly, adding communication to the asynchronous configurations of the increases the performance in terms of solved instances and overall solving time. The high average run time of the ADC configuration can be explained with the overhead introduced of less useful shared clauses. The table also shows clearly that using incremental SAT as the basis for parallel MUS is essential – without this technique only the configurations AS_ would be possible, but these two configurations show the worst performance. For the AS_ and SD_ configurations, also a slight decrease in the calls to the SAT solver can be recognized. Note, that this number of calls includes all solved instances, so that the increase from ADN to ADC is still plausible: the additional

Table 1. The table compares different algorithm configurations and shows the following statistic; the number of *solved instances* from the benchmark set; the *average run time* for the full benchmark set (including timeout instances) and, for parallel solvers, the *average CPU utilization*; the *penalized runtime*, i.e. the runtime of solving the full benchmark set with a penalty of factor 10 added for all instances that could not be solved; the total number of *sent clauses* shows how many learned clauses are provided for other solver incarnations; the total number of *SAT solver calls*; the percentage of SAT calls whose result has been ignored (*wasted calls*); the percentage of calls that have been *aborted*. The last column indicates whether the results are for plain MUS or for group MUS instances.

Configuration	Solved Instances	average run time	average cpu utilization	penalized run time	sent clauses	SAT solver calls	wasted solver calls	aborted solver calls	groups
MUSer2	144	186.46	100	590 K	–	413 K	–	–	–
ASN(4)	135	157.01	80.30	747 K	–	1458 K	21.80	54.13	–
ASC(4)	137	146.37	80.65	709 K	433 K	1453 K	21.88	54.04	–
SDN(4)	143	154.93	47.27	603 K	–	517 K	15.91	–	–
SDC(4)	141	138.31	46.70	636 K	433 K	512 K	16.19	–	–
ADN(4)	146	126.45	91.90	544 K	–	488 K	5.76	11.70	–
ADC(4)	150	154.09	90.79	476 K	1186 K	602 K	5.84	12.26	–
ADC-PRASS(4)	148	104.95	90.15	504 K	548 K	585 K	5.15	12.12	–
ADC+DYN+BUMP(4)	153	133.98	91.55	419 K	7071 K	661 K	7.01	11.83	–
ADC(8)	151	128.05	87.76	454 K	867 K	672 K	7.78	13.30	–
ADC+CBS(8)	151	117.22	86.71	452 K	1283 K	552 K	4.10	3.53	–
ADC+DYN+BUMP+CBS(8)	155	136.35	88.27	383 K	5564 K	635 K	5.56	3.13	–
MUSer2	194	123.56	100	150 K	–	325 K	–	–	✓
ADC(4)	198	106.92	90.19	75 K	487 K	530 K	14.35	33.39	✓
ADC+DYN+BUMP+CBS(4)	197	110.11	93.79	94 K	1688 K	350 K	11.68	7.29	✓
ADC(8)	198	100.76	86.99	74 K	333 K	692 K	20.93	33.26	✓
ADC+DYN+BUMP+CBS(8)	198	88.52	88.39	71 K	1178 K	439 K	9.92	8.66	✓

SAT solver calls depend on the additionally solved instances. Since the communication speeds up single SAT calls, more solver calls are wasted, because a competing solver has finished its task faster and thus aborts redundant solvers. Note, that the CPU utilization of the SD_ (synchronous) configurations is almost half of that of asynchronous configurations. Therefore, using asynchronous SAT solver calls is important to the scalability of the parallel algorithm and further motivates the analysis of communication for this setting. Comparing the results of the basic configurations with the sequential solver MUSer2 we note that only the configurations AD_ improve the performance from 144 to 150 solved instances with an improved average run time.

In Sec. 3 we discussed several improvements of the algorithm, which are also evaluated in Table 1. Disabling the PRASS option reduces the performance, and also reduces the number of shared clauses significantly – underlining that ignoring assumption literals is a must for successful clause sharing. Optimizing clause sharing with BUMP and DYN improves ADC further to 153 solved instances and an improved average run time. Note that this configuration also shares significantly more clauses than ADC, but

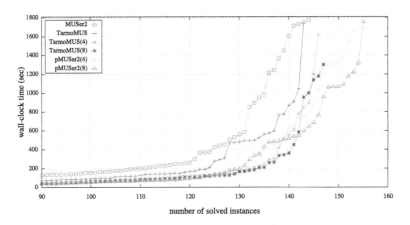

Fig. 1. Comparison of extractors on plain MUS instances

wastes more SAT calls. The core-based scheduling heuristic CBS does not improve the performance on 4 cores. The best four core configuration ADC+DYN+BUMP is referred as pMUSer2(4). In the eight core setting, the addition of CBS improves the number of SAT calls and reduces the percentage of wasted and aborted calls, without improving the overall performance significantly compared to ADC. However, adding sharing optimizations DYN and BUMP to ADC+CBS improves the overall performance: four more instances can be solved, the average run time decreases and the number of wasted and aborted SAT calls is also smaller than for any other eight core configuration. The best eight core configuration we found is ADC+DYN+BUMP+CBS, which we refer to as pMUSer(8). The behaviour of the parallel MUS extraction algorithm in the context of group-MUS extraction is quite similar. Again, the configuration ADC gives the best results and can solve already 198 out of 201 instances. For four cores, adding sharing or scheduling optimizations does not increase the performance, but again CBS helps to reduce the number of SAT calls as well as wasted and aborted SAT calls and DYN increases the number of shared clauses. When adding more cores, also 198 instances can be solved also by ADC+DYN+BUMP+CBS – suggesting that if the timeout were increased slightly, the four core variant could have solved these instances as well.

Figure 1 depicts the comparative behaviour of MUSer2 and pMUSer with 4 and 8 cores on the plain MUS instances. In addition, we evaluated the parallel MUS extractor TarmoMUS [27], discussed in Sec. 4. While in the sequential mode TarmoMUS is notably faster than MUSer2[2], the plot demonstrates that our algorithm scales significantly better with the number of cores, than TarmoMUS. For example, already a 4-core configuration of pMUSer2 outperforms the 4-core configuration of TarmoMUS. The statistics to compare the scalability of the algorithms are presented in Table 2. For both average and median speedup pMUSer2 gives much better results on plain instances. From sequential to four cores, MUSer2 scales linear in average. Obtained speedups range from 0.49 up to 132.59, showing that the parallelization can result in super-linear speedups

---

[2] Our analysis suggests that this is due to different versions of the SAT solvers used by the tools.

Table 2. Relative speedup with the addition of parallel resources: *minimum, average, maximum* and *median*. As a basis for the calculation the *commonly* solved instances have been used.

Solver 1	Solver 2	Min.	Avg.	Max.	Median	Common	Groups
TarmoMUS	TarmoMUS(4)	0.55	1.44	4.40	1.17	141	–
TarmoMUS	TarmoMUS(8)	0.41	1.74	6.92	1.29	141	–
MUSer2	pMUSer2(4)	0.49	4.09	132.59	2.94	143	–
MUSer2	pMUSer2(8)	0.28	4.01	97.66	3.38	142	–
MUSer2	pMUSer2(4)	0.46	1.41	4.07	1.33	194	✓
MUSer2	pMUSer2(8)	0.44	1.88	9.20	1.49	194	✓

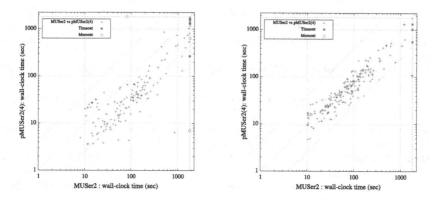

Fig. 2. Wall-clock time, sequential vs. 4 cores: left — plain MUS; right – group MUS

(consider also the scatter plots in Figure 2). For `TarmoMUS` the average, maximum and median speed-ups are lower, and when more resources are added, the performance increases only slightly. Neither `pMUSer2` nor `TarmoMUS` scale well to eight cores.

6   Conclusion

We argued that assumption-based incremental SAT solving is essential to ensuring the scalability of the proposed parallel MUS extraction algorithm. We proved the soundness of unrestricted communication in our algorithm, and proposed a number of optimizations focused on improving the quality of communication and job distribution. While the algorithm scales extremely well from a single-core to the 4-core setting, we did not observe similar improvements going from the 4-core setting to the 8-core. In our view, the main obstacle to the scalability to a high number of cores is the fact that as the number of cores grows, the workers are more likely to duplicate each others efforts. While the situation is somewhat improved by the core-based scheduling, the solution is not yet satisfactory, and requires further research. Additional avenue for improvement might lie in the high-level parallelization, whereby different (parallelized on the low-level) MUS extraction algorithms are executed in parallel.

Acknowledgements. We thank Allen Van Gelder and the anonymous referees, for their comments and suggestions that helped us to improve the presentation of our ideas.

References

1. Audemard, G., Hoessen, B., Jabbour, S., Lagniez, J.-M., Piette, C.: Revisiting clause exchange in parallel SAT solving. In: Cimatti, A., Sebastiani, R. (eds.) SAT 2012. LNCS, vol. 7317, pp. 200–213. Springer, Heidelberg (2012)
2. Audemard, G., Simon, L.: Predicting learnt clauses quality in modern sat solvers. In: Proceedings of the 21st International Jont Conference on Artifical Intelligence, IJCAI 2009, pp. 399–404. Morgan Kaufmann Publishers Inc., San Francisco (2009)
3. Belov, A., Chen, H., Mishchenko, A., Marques-Silva, J.: Core minimization in SAT-based abstraction. In: DATE (2013)
4. Belov, A., Lynce, I., Marques-Silva, J.: Towards efficient MUS extraction. AI Communications 25(2), 97–116 (2012)
5. Belov, A., Marques-Silva, J.: MUSer2: An efficient MUS extractor. Journal of Satisfiability 8, 123–128 (2012)
6. Biere, A.: Lingeling, Plingeling, PicoSAT and PrecoSAT at SAT Race 2010. FMV Report Series Technical Report 10/1, Johannes Kepler University, Linz, Austria (2010)
7. Biere, A., Heule, M., van Maaren, H., Walsh, T. (eds.): Handbook of Satisfiability. Frontiers in Artificial Intelligence and Applications, vol. 185. IOS Press (2009)
8. Davis, M., Logemann, G., Loveland, D.: A machine program for theorem-proving. Communications of the ACM 5, 394–397 (1962)
9. Eén, N., Sörensson, N.: Temporal induction by incremental SAT solving. Electr. Notes Theor. Comput. Sci. 89(4), 543–560 (2003)
10. Gomes, C.P., Selman, B., Crato, N., Kautz, H.: Heavy-tailed phenomena in satisfiability and constraint satisfaction problems. Journal of Automated Reasoning 24(1-2), 67–100 (2000)
11. Hamadi, Y., Jabbour, S., Sais, L.: Control-based clause sharing in parallel sat solving. In: Proceedings of the 21st International Jont Conference on Artifical Intelligence, pp. 499–504. Morgan Kaufmann Publishers Inc., San Francisco (2009)
12. Hamadi, Y., Jabbour, S., Sais, L.: ManySAT: a parallel SAT solver. JSAT 6(4), 245–262 (2009)
13. Hölldobler, S., Manthey, N., Nguyen, V.H., Stecklina, J., Steinke, P.: A short overview on modern parallel SAT-solvers. In: Wasito, I., et al. (eds.) Proceedings of the International Conference on Advanced Computer Science and Information Systems, pp. 201–206 (2011) ISBN 978-979-1421-11-9
14. Kullmann, O., Lynce, I., Marques-Silva, J.: Categorisation of clauses in conjunctive normal forms: Minimally unsatisfiable sub-clause-sets and the lean kernel. In: Biere, A., Gomes, C.P. (eds.) SAT 2006. LNCS, vol. 4121, pp. 22–35. Springer, Heidelberg (2006)
15. Liffiton, M.H., Sakallah, K.A.: Algorithms for computing minimal unsatisfiable subsets of constraints. J. Autom. Reasoning 40(1), 1–33 (2008)
16. Marques-Silva, J.: Minimal unsatisfiability: Models, algorithms and applications. In: IS-MVL, pp. 9–14 (2010)
17. Marques-Silva, J., Lynce, I.: On improving MUS extraction algorithms. In: Sakallah, K.A., Simon, L. (eds.) SAT 2011. LNCS, vol. 6695, pp. 159–173. Springer, Heidelberg (2011)
18. Marques Silva, J.P., Sakallah, K.A.: Grasp: A search algorithm for propositional satisfiability. IEEE Trans. Computers 48(5), 506–521 (1999)
19. Martins, R., Manquinho, V., Lynce, I.: An overview of parallel SAT solving. Constraints 17, 304–347 (2012)

20. Moskewicz, M.W., Madigan, C.F., Zhao, Y., Zhang, L., Malik, S.: Chaff: Engineering an efficient SAT solver. In: Proc. 38th Annual ACM/IEEE Design Automation Conf. (DAC), pp. 530–535. ACM (2001)
21. Nadel, A.: Boosting minimal unsatisfiable core extraction. In: FMCAD, pp. 121–128 (October 2010)
22. Nadel, A., Ryvchin, V.: Efficient SAT solving under assumptions. In: Cimatti, A., Sebastiani, R. (eds.) SAT 2012. LNCS, vol. 7317, pp. 242–255. Springer, Heidelberg (2012)
23. Papadimitriou, C.H., Wolfe, D.: The complexity of facets resolved. J. Comput. Syst. Sci. 37(1), 2–13 (1988)
24. Ryvchin, V., Strichman, O.: Faster extraction of high-level minimal unsatisfiable cores. In: Sakallah, K.A., Simon, L. (eds.) SAT 2011. LNCS, vol. 6695, pp. 174–187. Springer, Heidelberg (2011)
25. Van Gelder, A.: Generalized conflict-clause strengthening for satisfiability solvers. In: Sakallah, K.A., Simon, L. (eds.) SAT 2011. LNCS, vol. 6695, pp. 329–342. Springer, Heidelberg (2011)
26. Wieringa, S.: Understanding, improving and parallelizing MUS finding using model rotation. In: Milano, M. (ed.) CP 2012. LNCS, vol. 7514, pp. 672–687. Springer, Heidelberg (2012)
27. Wieringa, S., Heljanko, K.: Asynchronous multi-core incremental SAT solving. In: Piterman, N., Smolka, S.A. (eds.) TACAS 2013. LNCS, vol. 7795, pp. 139–153. Springer, Heidelberg (2013)

A Modular Approach to MaxSAT Modulo Theories[*]

Alessandro Cimatti[1], Alberto Griggio[1],
Bastiaan Joost Schaafsma[1,2], and Roberto Sebastiani[2]

[1] FBK-IRST, Trento, Italy
[2] DISI, University of Trento, Italy

Abstract. In this paper we present a novel "modular" approach for (weighted partial) MaxSAT Modulo Theories. The main idea is to combine a lazy SMT solver with a purely-propositional (weighted partial) MaxSAT solver, by making them exchange information iteratively: the former produces an increasing set of theory lemmas which are used by the latter to progressively refine an approximation of the final subset of the soft clauses, which is eventually returned as output.

The approach has several practical features. First, it is independent from the theories addressed. Second, it is simple to implement and to update, since both SMT and MaxSAT solvers can be used as blackboxes. Third, it can be interfaced with external MaxSAT and SMT solvers in a plug-and-play manner, so that to benefit for free of tools which are or will be made available.

We have implemented our approach on top of the MATHSAT5 SMT solver and of a selection of external MaxSAT solvers, and we have evaluated it by means of an extensive empirical test on SMT-LIB benchmarks. The results confirm the validity and potential of this approach.

1 Introduction

MaxSAT [19] is the problem of determining the maximum number of clauses, of a given Boolean formula, that can be satisfied by some assignment. Its *weighted* and *partial* variants allow to associate fixed weights to clauses, and to search only for solutions that satisfy a given subset of the clauses. (In this paper, unless otherwise specified, by "MaxSAT" we always consider the general case of weighted partial MaxSAT; thus, we often omit the adjectives "weighted" and "partial".) In recent years, the solvers for MaxSAT have demonstrated substantial improvements [20,6,17,18,21,3], and have now important practical applications (e.g. Formal Verification, Automatic Test Pattern Generation, Field Programmable Gate Array routing).

The MaxSAT problems can be generalized from the Boolean case to the case of Satisfiability Modulo Theories (SMT) [8], where first order formulas are interpreted with respect to some (combinations of) background theories. Theories of interest are,

[*] We are very grateful to Carlos Ansótegui, Bruno Dutertre and Leonardo de Moura for providing to us precious information about their respective solvers. Alberto Griggio is supported by Provincia Autonoma di Trento and the European Community's FP7/2007-2013 under grant agreement Marie Curie FP7 - PCOFUND-GA-2008-226070 "progetto Trentino", project ADAPTATION; Bas Schaafsma and Roberto Sebastiani are supported in part by Semiconductor Research Corporation under GRC Research Project 2012-TJ-2266 WOLF.

M. Järvisalo and A. Van Gelder (Eds.): SAT 2013, LNCS 7962, pp. 150–165, 2013.

e.g., those of bit vectors ($\mathcal{BV}$), of arrays ($\mathcal{AR}$), of linear arithmetic ($\mathcal{LA}$) on the rationals ($\mathcal{LA}(\mathbb{Q})$) or on the integers ($\mathcal{LA}(\mathbb{Z})$).

Because of the increase in expressiveness of SMT, the MaxSAT Modulo Theory problem (MaxSMT hereafter) has many important applications (e.g., formal verification of timed & hybrid systems and of parametric systems, planning with resources, radio frequency assignment problems.) However, MaxSMT—and, more generally, the optimization problems in SMT— have received relatively little attention in the literature. To some extent, this can be explained with the technical difficulties associated with the combination of two non-trivial components, namely an SMT engine (that requires the integration of constraints into SAT) and a MaxSAT optimization procedure.

In this paper, we propose a novel and comprehensive approach to (weighted partial) MaxSMT. The approach is highly modular, in that it combines, as black boxes, two components: (i) a lazy SMT solver, and (ii) a purely-propositional (weighted partial) MaxSAT solver. During the search, these two components exchange information iteratively: the SMT solver produces an increasing set of theory lemmas, which are used by the MaxSAT solver to progressively refine an approximation of the final subset of the soft clauses, which is eventually returned as output.

Basically, the SMT solver is used to dynamically lift the suitable amount of theory information to the Boolean level, where the MaxSAT solver performs the optimization process. We call the approach *Lemma-Lifting (LL)*, similarly to the LL approach for the extraction of unsatisfiable cores in SMT [13].

The approach has several interesting features. First, it is independent from the theories addressed: the lemmas returned by the SMT solver during the search are abstracted into Boolean formulas before being passed to the MaxSAT solver. Second, the LL algorithm is general and simple to implement: it imposes no restriction on the MaxSAT solver, while the only requirement on the SMT solver is that it is able to return the lemmas constructed during search. Third, the LL algorithm can be realized by interfacing external MaxSAT and SMT solvers in a plug-and-play manner. In this way, we can use all the available approaches and tools, and benefit of future advances in lazy SMT and MaxSAT technology.

We have proved the formal properties of the LL MaxSMT algorithm. We implemented LL on top of the MATHSAT5 SMT solver [12], and of a selection of external MaxSAT tools. We have evaluated and compared the performances of the various LL configurations, and of every MaxSMT or MaxSMT-like solver we are aware of, by means of an extensive empirical test on MaxSMT-modified SMT-LIB benchmarks. The results confirm the validity and potential of this approach.

Content. The paper is organized as follows. After having provided some background knowledge on SMT, MaxSAT and MaxSMT in §2, we present and discuss our new approach and algorithm in §3. We proceed with a discussion of related work in §4. In §5 we present and comment empirical tests. We conclude and suggest some future developments in §6.

2 Background

Terminology and Notation. We consider some decidable first-order theory $\mathcal{T}$ (or a combination $\bigcup_i \mathcal{T}_i$ of theories). We call $\mathcal{T}$-*atom* (resp. -literal, -clause, -formula) a ground atomic formula (resp. literal, clause, formula) in $\mathcal{T}$. (Notice that a Boolean atom can be seen as a subcase of $\mathcal{T}$-atom, etc.) We distinguish the space of $\mathcal{T}$-formulas ($\mathcal{T}$) from that of plain Boolean formulas ($\mathcal{B}$) by denoting them with the "$\mathcal{T}$" and the "$\mathcal{B}$" superscripts respectively; we use no superscript when we make no such distinction. Given a $\mathcal{T}$-formula (-clause, -literal, -assignment etc.) $\varphi^{\mathcal{T}}$, we call *Boolean abstraction* of $\varphi^{\mathcal{T}}$ the formula $\varphi^{\mathcal{B}} \stackrel{\text{def}}{=} \mathcal{T}2\mathcal{B}(\varphi^{\mathcal{T}})$ obtained by rewriting each non-Boolean $\mathcal{T}$-atom in $\varphi^{\mathcal{T}}$ into a fresh Boolean atom; vice versa, $\varphi^{\mathcal{T}} \stackrel{\text{def}}{=} \mathcal{B}2\mathcal{T}(\varphi^{\mathcal{B}}) \stackrel{\text{def}}{=} \mathcal{T}2\mathcal{B}^{-1}(\varphi^{\mathcal{B}})$ is the *refinement* of $\varphi^{\mathcal{B}}$. (To this extent, if not otherwise specified, when some symbol $\langle sym \rangle$ is used with both the "$\mathcal{T}$" and the "$\mathcal{B}$" superscripts, then $\langle sym \rangle^{\mathcal{B}}$ denotes the Boolean abstraction of $\langle sym \rangle^{\mathcal{T}}$, and vice versa.) We say that a truth assignment $\mu^{\mathcal{T}}$ *propositionally satisfies* $\varphi^{\mathcal{T}}$, written $\mu^{\mathcal{T}} \models_p \varphi^{\mathcal{T}}$, iff $\mu^{\mathcal{B}} \models \varphi^{\mathcal{B}}$.

In both the $\mathcal{T}$- and $\mathcal{B}$- spaces, we assume all formulas are in CNF, and we represent them as sets of clauses; we represent truth assignments as sets of literals. The symbols $\varphi_{...}, \psi_{...}, \phi_{...}$ denote formulas, and $\mu_{...}, \eta_{...}$ denote truth assignments, regardless their subscripts or superscripts. A *weighted clause* is a clause C which is augmented with a value $w \in \mathbb{N} \cup \{+\infty\}$, which is called the *weight* of C, denoted by $\mathsf{Weight}(C)$; a weighted clause is called *hard*, iff its weight is $+\infty$, *soft*, otherwise. Sets of hard and soft clauses are denoted with the subscript $._h$ and $._s$ respectively. $\mathsf{Weight}(\psi_s)$ denotes the sum of the weights of the clauses in ψ_s.

2.1 Satisfiability Modulo Theories

We call a *theory solver for* $\mathcal{T}$, $\mathcal{T}$-Solver, a tool able to decide the $\mathcal{T}$-satisfiability of a conjunction/set $\mu^{\mathcal{T}}$ of $\mathcal{T}$-literals. If $\mu^{\mathcal{T}}$ is $\mathcal{T}$-unsatisfiable, then $\mathcal{T}$-Solver returns UNSAT and the subset η of $\mathcal{T}$-literals in $\mu^{\mathcal{T}}$ which was found $\mathcal{T}$-unsatisfiable; (η is hereafter called a $\mathcal{T}$-*conflict set*, and $\neg\eta$ a $\mathcal{T}$-*conflict clause*.) if $\mu^{\mathcal{T}}$ is $\mathcal{T}$-satisfiable, then $\mathcal{T}$-Solver returns SAT; it may also be able to return some unassigned $\mathcal{T}$-literal $l \notin \mu^{\mathcal{T}}$ s.t. $\{l_1, ..., l_n\} \models_{\mathcal{T}} l$, where $\{l_1, ..., l_n\} \subseteq \mu^{\mathcal{T}}$. We call this process $\mathcal{T}$-*deduction* and $(\bigvee_{i=1}^n \neg l_i \vee l)$ a $\mathcal{T}$-*deduction clause*. Notice that $\mathcal{T}$-conflict and $\mathcal{T}$-deduction clauses are valid in $\mathcal{T}$. We call them $\mathcal{T}$-*lemmas*.

In a lazy $\mathrm{SMT}(\mathcal{T})$ solver, the Boolean abstraction $\varphi^{\mathcal{B}}$ of the input formula φ is given as input to a CDCL SAT solver, and whenever a satisfying assignment $\mu^{\mathcal{B}}$ is found s.t. $\mu^{\mathcal{B}} \models \varphi^{\mathcal{B}}$, the corresponding set of $\mathcal{T}$-literals $\mu^{\mathcal{T}}$ is fed to the $\mathcal{T}$-Solver; if $\mu^{\mathcal{T}}$ is found $\mathcal{T}$-consistent, then φ is $\mathcal{T}$-consistent; otherwise, $\mathcal{T}$-Solver returns the $\mathcal{T}$-conflict set η causing the inconsistency, so that the clause $\neg\eta^{\mathcal{B}}$ (the Boolean abstraction of $\neg\eta$) is used to drive the backjumping and learning mechanism of the SAT solver.

Important optimizations are *early pruning* and $\mathcal{T}$-*propagation*: the $\mathcal{T}$-Solver is invoked also on an intermediate assignment $\mu^{\mathcal{T}}$: if it is $\mathcal{T}$-unsatisfiable, then the procedure can backtrack; if not, and if the $\mathcal{T}$-Solver is able to perform a $\mathcal{T}$-deduction $\{l_1, ..., l_n\} \models_{\mathcal{T}} l$, then l can be unit-propagated, and the $\mathcal{T}$-deduction clause $(\bigvee_{i=1}^n \neg l_i \vee l)$ can be used in backjumping and learning. Another technique is *static learning*, where $\mathcal{T}$-lemmas expressing "obvious" constraints on $\mathcal{T}$-atoms occurring in the input formula (e.g. mutual-exclusion, transitivity constraints) are learned a priori.

The above schema is a coarse abstraction of the procedures underlying all the state-of-the-art lazy SMT tools. The interested reader is pointed to, e.g., [23,8] for details and further references.

2.2 MaxSAT

A *(weighted partial)*[1] *MaxSAT formula* is a set of weighted clauses in the form $\varphi^\mathcal{B} \stackrel{\text{def}}{=} \varphi_h^\mathcal{B} \cup \varphi_s^\mathcal{B}$, s.t. $\varphi_h^\mathcal{B}$ and $\varphi_s^\mathcal{B}$ are sets of hard and soft clauses respectively. A *MaxSAT problem* consists in finding a maximum-weight clause set $\psi_s^\mathcal{B}$ s.t. $\psi_s^\mathcal{B} \subseteq \varphi_s^\mathcal{B}$ and $\varphi_h^\mathcal{B} \cup \psi_s^\mathcal{B}$ is satisfiable. (Notice that such $\psi_s^\mathcal{B}$ is not unique in general.) MaxSAT$(\varphi_h^\mathcal{B}, \varphi_s^\mathcal{B})$ denotes a function computing one such $\psi_s^\mathcal{B}$, and MaxWeight$(\varphi_h^\mathcal{B}, \varphi_s^\mathcal{B})$ denotes Weight$(\psi_s^\mathcal{B})$.

Notice that Weight$(C^\mathcal{B})$ can be considered as the "cost" of non-satisfying the soft clause $C^\mathcal{B}$, and MaxSAT can be seen as the problem of minimizing such cost over all the soft clauses. To this extent, a *MaxSAT Solver* is a function s.t. MaxSAT$(\varphi_h^\mathcal{B}, \varphi_s^\mathcal{B})$ returns a maximum-weight clause set $\psi_s^\mathcal{B}$ s.t. $\psi_s^\mathcal{B} \subseteq \varphi_s^\mathcal{B}$ and $\varphi_h^\mathcal{B} \cup \psi_s^\mathcal{B}$ is satisfiable.

The MaxSAT problem can be generalized to the case in which $\varphi_h^\mathcal{B}$ and $\varphi_s^\mathcal{B}$ are sets of arbitrary formulas rather than sets of single clauses.[2] Let $\lambda_s \stackrel{\text{def}}{=} \{S_i\}_i$ be a set of fresh selection variables, one for each constraint $\phi_i^\mathcal{B}$ in $\varphi_s^\mathcal{B}$, let $\varphi_s'^\mathcal{B} \stackrel{\text{def}}{=} \{\neg S_i \vee \phi_i^\mathcal{B} \mid \phi_i^\mathcal{B} \in \varphi_s^\mathcal{B}\}$, and let $\psi_h^\mathcal{B}$ be the set of clauses resulting from conversion of $\varphi_h^\mathcal{B} \cup \varphi_s'^\mathcal{B}$ into CNF. Thus the generalized MaxSAT problem $(\varphi_h^\mathcal{B}, \varphi_s^\mathcal{B})$ can be reduced to a standard MaxSAT problem on the sets of clauses $(\psi_h^\mathcal{B}, \lambda_s)$, in which all soft clauses are unit clauses.

Current state-of-the-art MaxSAT solvers can be roughly divided into 3 categories. Solvers based on branch & bound, such as [20,17], employ specialized inference rules while performing a standard branch and bound search for MaxWeight$(\varphi_h^\mathcal{B}, \varphi_s^\mathcal{B})$. Iterative solvers, like e.g. [6], work by adding to each soft clause $C_j^\mathcal{B} \in \varphi_s^\mathcal{B}$ a fresh literal R_j (called a *relaxation literal*), and by imposing bounds on the number of relaxation literals that can be assigned to true, using cardinality constraints. The space of such bounds is typically explored using binary search. Finally, core-guided solvers, such as e.g. [18,21], improve upon iterative solvers by exploiting unsatisfiable cores to decide if/when to add a relaxation literal to a soft clause, and to minimize the number of cardinality constraints needed.

2.3 MaxSAT Modulo Theories and SMT with Cost Optimization

The MaxSAT problem generalizes straightforwardly to SMT level. Given a background theory $\mathcal{T}$ as before, a *(weighted partial) MaxSAT Modulo Theories (MaxSMT) formula* is a set of weighted $\mathcal{T}$-clauses in the form $\varphi^\mathcal{T} \stackrel{\text{def}}{=} \varphi_h^\mathcal{T} \cup \varphi_s^\mathcal{T}$. A *MaxSAT Modulo Theories*

---

[1] A MaxSAT formula is not "weighted" iff Weight$(C_j^\mathcal{B}) = 1$ for every $C_j^\mathcal{B} \in \varphi_s^\mathcal{B}$, and it is not "partial" iff $\varphi_h^\mathcal{B}$ is empty. Hereafter, unless otherwise specified, we consider the general case ignoring this distinction, hence dropping the adjectives "weighted" and "partial".

[2] This includes also the so-called *Block MaxSAT* problem, where each (weighted) soft constraint is itself a conjunctions of clauses, representing a "block" of clauses subject to the same weight, s.t. it suffices to violate one such clause to pay the cost of the constraint.

(MaxSMT) problem consists in finding a maximum-weight clause set $\psi_s^{\mathcal{T}}$ s.t. $\psi_s^{\mathcal{T}} \subseteq \varphi_s^{\mathcal{T}}$ and $\varphi_h^{\mathcal{T}} \cup \psi_s^{\mathcal{T}}$ is $\mathcal{T}$-satisfiable. As with the Boolean case, MaxSMT$(\varphi_h^{\mathcal{T}}, \varphi_s^{\mathcal{T}})$ denotes a function computing one such $\psi_s^{\mathcal{T}}$, and MaxWeight$(\varphi_h^{\mathcal{T}}, \varphi_s^{\mathcal{T}})$ denotes Weight$(\psi_s^{\mathcal{T}})$. (The same considerations and conventions on "weighted", "partial", and "generalized" MaxSAT in §2.2 hereafter apply for MaxSMT.)

Importantly, a MaxSMT problem can be encoded into an SMT problem with cost minimization $\langle \varphi^{\mathcal{T}'}, \text{cost} \rangle$, either with Pseudo-Boolean (PB) cost functions [22,10]:

$$\varphi^{\mathcal{T}'} = \varphi_h^{\mathcal{T}} \cup \bigcup_{C_j^{\mathcal{T}} \in \varphi_s^{\mathcal{T}}} \{(A_j \vee C_j^{\mathcal{T}})\}; \quad \text{cost} \overset{\text{def}}{=} \sum_{C_j^{\mathcal{T}} \in \varphi_s^{\mathcal{T}}} w_j \cdot A_j \tag{1}$$

where $w_j \overset{\text{def}}{=} \text{Weight}(C_j^{\mathcal{T}})$ and the A_j's are fresh Boolean atoms, or with $\mathcal{LA}$ cost functions [22,24]:

$$\varphi^{\mathcal{T}'} = \varphi_h^{\mathcal{T}} \cup \bigcup_{C_j^{\mathcal{T}} \in \varphi_s^{\mathcal{T}}} (\{(A_j \vee C_j^{\mathcal{T}}), (\neg A_j \vee x_j = w_j), (A_j \vee x_j = 0)\});$$

$$\text{cost} \overset{\text{def}}{=} \sum_{C_j^{\mathcal{T}} \in \varphi_s^{\mathcal{T}}} x_j. \tag{2}$$

where the x_j's are $\mathcal{LA}$ variables.

3 A Novel Modular MaxSMT Algorithm

In what follows, we consider a MaxSMT problem $\varphi^{\mathcal{T}} \overset{\text{def}}{=} \varphi_h^{\mathcal{T}} \cup \varphi_s^{\mathcal{T}}$, and w_{max} denotes MaxWeight$(\varphi_h^{\mathcal{T}}, \varphi_s^{\mathcal{T}})$. The symbols $\Theta^{\mathcal{T}}$ and $\Theta_i^{\mathcal{T}}$ denote sets of $\mathcal{T}$-lemmas on $\mathcal{T}$-atoms occurring in $\varphi_h^{\mathcal{T}} \cup \varphi_s^{\mathcal{T}}$, whilst $\Theta_*^{\mathcal{T}}$ denotes the set of *all* such $\mathcal{T}$-lemmas.

Observe that $\Theta_*^{\mathcal{T}}$ is a finite set, since $\Theta^{\mathcal{T}}$, $\Theta_i^{\mathcal{T}}$ and $\Theta_*^{\mathcal{T}}$ are defined to be sets of $\mathcal{T}$-lemmas containing only atoms in the input formula. In general, modern SMT solvers might introduce new atoms during search, which can thus appear in some $\mathcal{T}$-lemmas. This scenario is not considered here to keep the presentation simple. However, it can be covered under the additional assumption that $\mathcal{T}$-lemmas are generated from a finite set of atoms, which is typically the case for modern SMT solvers (see e.g. [9,7]).

3.1   The Basic Algorithm

Algorithm 1 reports a "modular" procedure for MaxSMT. Intuitively, an SMT and a MaxSAT solver are used as guided enumerators of, respectively:[3]

- a finite sequence of $\mathcal{T}$-lemma sets $\Theta_0^{\mathcal{T}}, \Theta_1^{\mathcal{T}}, ..., \Theta_n^{\mathcal{T}}$ s.t. $\Theta_0^{\mathcal{T}} = \emptyset$,

$$\Theta_0^{\mathcal{T}} \subseteq \Theta_1^{\mathcal{T}} \subset \Theta_2^{\mathcal{T}} \subset ... \subset \Theta_n^{\mathcal{T}}, \tag{3}$$

$$\Theta_n^{\mathcal{T}} \subseteq \Theta_*^{\mathcal{T}}, \tag{4}$$

which progressively rule out all the $\mathcal{T}$-unsatisfiable truth assignments which propositionally satisfy $\varphi_h^{\mathcal{T}}$ and some subset $\psi_{s,i}^{\mathcal{T}}$ of $\varphi_s^{\mathcal{T}}$ s.t. Weight$(\psi_{s,i}^{\mathcal{T}}) > w$;

---

[3] When referring to Algorithm 1, the index "$._i$" in $\Theta_i^{\mathcal{T}}$, $\psi_{s,i}^{\mathcal{T}}$ etc. refers to the values of $\Theta^{\mathcal{T}}$, $\psi_s^{\mathcal{T}}$ etc. at the end of the i-th cycle in the while loop.

Algorithm 1. A Lemma-Lifting procedure for MaxSMT($\varphi_h^\mathcal{T}, \varphi_s^\mathcal{T}$)

Input:
  $\varphi_h^\mathcal{T}$: a set of hard $\mathcal{T}$-clauses;
  $\varphi_s^\mathcal{T}$: a set of (weighted) soft $\mathcal{T}$-clauses;
Output:
  a maximum-weight set of soft $\mathcal{T}$-clauses $\psi_s^\mathcal{T}$ s.t. $\psi_s^\mathcal{T} \subseteq \varphi_s^\mathcal{T}$ and $\varphi_h^\mathcal{T} \cup \psi_s^\mathcal{T}$ is $\mathcal{T}$-satisfiable

  1: $\langle \varphi_h^\mathcal{B}, \varphi_s^\mathcal{B} \rangle \leftarrow \mathcal{T}2\mathcal{B}\,(\langle \varphi_h^\mathcal{T}, \varphi_s^\mathcal{T} \rangle)$;
  2: $\Theta^\mathcal{T} \leftarrow \emptyset$;
  3: $\psi_s^\mathcal{T} \leftarrow \varphi_s^\mathcal{T}$;
  4: **while** (SMT.Solve $(\varphi_h^\mathcal{T} \cup \psi_s^\mathcal{T} \cup \Theta^\mathcal{T})$ = UNSAT) **do**
  5:    $\Theta^\mathcal{T} \leftarrow \Theta^\mathcal{T} \cup$ SMT.GetTLemmas ();
  6:    $\Theta^\mathcal{B} \leftarrow \mathcal{T}2\mathcal{B}\,(\Theta^\mathcal{T})$;
  7:    $\psi_s^\mathcal{B} \leftarrow$ MaxSAT($\varphi_h^\mathcal{B} \cup \Theta^\mathcal{B}, \varphi_s^\mathcal{B}$);
  8:    $\psi_s^\mathcal{T} \leftarrow \mathcal{B}2\mathcal{T}\,(\psi_s^\mathcal{B})$;
  9: **end while**
 10: **return** $\psi_s^\mathcal{T}$;
 11:
 12: SMT.Solve $(\varphi^\mathcal{T})$ checks whether $\varphi^\mathcal{T}$ is $\mathcal{T}$-satisfiable
 13: SMT.GetTLemmas () returns the $\mathcal{T}$-lemmas computed by the latest call to SMT.Solve

---

– (the Boolean abstraction of) a finite sequence of soft-clause sets $\psi_{s,0}^\mathcal{T}, \ldots, \psi_{s,i}^\mathcal{T}, \ldots \psi_{s,n}^\mathcal{T}$ where $\psi_{s,0}^\mathcal{T} = \varphi_s^\mathcal{T}$, $\psi_{s,i}^\mathcal{T} \subseteq \varphi_s^\mathcal{T}$ for every i, $\psi_{s,n}^\mathcal{T} = $ MaxSMT($\varphi_h^\mathcal{T}, \varphi_s^\mathcal{T}$), and

$$\text{Weight}(\psi_{s,n}^\mathcal{T}) \leq \ldots \leq \text{Weight}(\psi_{s,i+1}^\mathcal{T}) \leq \text{Weight}(\psi_{s,i}^\mathcal{T}) \leq \ldots . \tag{5}$$

$$\text{MaxWeight}(\varphi_h^\mathcal{T} \cup \varphi_s^\mathcal{T}) = \text{Weight}(\psi_{s,n}^\mathcal{T}). \tag{6}$$

Notice that neither $\psi_{s,i+1}^\mathcal{T} \subseteq \psi_{s,i}^\mathcal{T}$ nor $\text{Weight}(\psi_{s,i+1}^\mathcal{T}) < \text{Weight}(\psi_{s,i}^\mathcal{T})$ hold in general.

Each $\Theta_{i+1}^\mathcal{T}$ results from adding to $\Theta_i^\mathcal{T}$ the $\mathcal{T}$-lemmas computed by an SMT solver to prove the $\mathcal{T}$-unsatisfiability of $\varphi_h^\mathcal{T} \cup \psi_{s,i}^\mathcal{T} \cup \Theta_i^\mathcal{T}$. Each $\psi_{s,i}^\mathcal{T}$ is obtained by invoking a MaxSAT solver on the Boolean abstraction of $\varphi_h^\mathcal{T} \cup \Theta_i^\mathcal{T}$ and $\varphi_s^\mathcal{T}$ as hard and soft component respectively.

The termination, correctness, and completeness of Algorithm 1 is formally proved in [11]. Intuitively, at every loop $i > 0$ s.t. $\varphi_h^\mathcal{T} \cup \psi_{s,i}^\mathcal{T} \cup \Theta_i^\mathcal{T}$ is found $\mathcal{T}$-unsatisfiable by SMT.Solve, since its Boolean abstraction $\varphi_h^\mathcal{B} \cup \psi_{s,i}^\mathcal{B} \cup \Theta_i^\mathcal{B}$ is satisfiable by construction of $\psi_{s,i}^\mathcal{B}$, then SMT.GetTLemmas returns at least one new $\mathcal{T}$-lemma; thus (3) holds, (4) holds by definition of $\Theta_*^\mathcal{T}$, hence (5) holds by construction of $\psi_{s,.}^\mathcal{B}$. By (3), (4), and since $\Theta_*^\mathcal{T}$ is finite and it contains all the possible theory information related to $\varphi_h^\mathcal{T} \cup \varphi_s^\mathcal{T}$, we have that, for some loop index n, $\Theta_n^\mathcal{T} \subseteq \Theta_*^\mathcal{T}$ and $\Theta_n^\mathcal{T}$ contains all $\mathcal{T}$-lemmas which rule out all $\mathcal{T}$-inconsistent truth assignments propositionally satisfying $\varphi_h^\mathcal{B} \cup \psi_{s,n}^\mathcal{B}$. Then $\varphi_h^\mathcal{T} \cup \psi_{s,n}^\mathcal{T} \cup \Theta_n^\mathcal{T}$ is $\mathcal{T}$-satisfiable, because $\varphi_h^\mathcal{B} \cup \psi_{s,n}^\mathcal{B} \cup \Theta_n^\mathcal{B}$ is satisfiable by construction of $\psi_{s,n}^\mathcal{B}$, so that the procedure terminates. From this, it is easy to show that (6) holds.

Notice that, in general, in the call SMT.Solve $(\varphi_h^{\mathcal{T}} \cup \psi_s^{\mathcal{T}} \cup \Theta^{\mathcal{T}})$ the "$\cup\, \Theta^{\mathcal{T}}$" element is not necessary from the logic viewpoint, but it prevents the SMT solver to re-generate from scratch previously-computed $\mathcal{T}$-lemmas in $\Theta^{\mathcal{T}}$.

Example 1. Let $\varphi_h^{\mathcal{T}}, \varphi_s^{\mathcal{T}}$ be as follows (values $[v]$ denote clause weights):

$$\varphi_h^{\mathcal{T}} \stackrel{\text{def}}{=} \emptyset \qquad\qquad \varphi_h^{\mathcal{B}} \stackrel{\text{def}}{=} \emptyset$$

$$\varphi_s^{\mathcal{T}} \stackrel{\text{def}}{=} \begin{Bmatrix} C_0 : ((x \le 0))\ [4] \\ C_1 : ((x \le 1))\ [3] \\ C_2 : ((x \ge 2))\ [2] \\ C_3 : ((x \ge 3))\ [6] \end{Bmatrix} \quad \varphi_s^{\mathcal{B}} \stackrel{\text{def}}{=} \begin{Bmatrix} (A_0)\ [4] \\ (A_1)\ [3] \\ (A_2)\ [2] \\ (A_3)\ [6] \end{Bmatrix} \ where: \begin{matrix} A_0 \stackrel{\text{def}}{=} (x \le 0), \\ A_1 \stackrel{\text{def}}{=} (x \le 1), \\ A_2 \stackrel{\text{def}}{=} (x \ge 2), \\ A_3 \stackrel{\text{def}}{=} (x \ge 3). \end{matrix}$$

Notice that the set of all possible $\mathcal{T}$-lemmas on the $\mathcal{T}$-atoms of $\varphi_h^{\mathcal{T}} \cup \varphi_s^{\mathcal{T}}$ is:

$$\Theta_*^{\mathcal{T}} = \begin{Bmatrix} \theta_1 : (\neg(x \le 0) \vee (x \le 1)) \\ \theta_2 : (\neg(x \ge 3) \vee (x \ge 2)) \\ \theta_3 : (\neg(x \le 0) \vee \neg(x \ge 2)) \\ \theta_4 : (\neg(x \le 0) \vee \neg(x \ge 3)) \\ \theta_5 : (\neg(x \le 1) \vee \neg(x \ge 2)) \\ \theta_6 : (\neg(x \le 1) \vee \neg(x \ge 3)) \end{Bmatrix} \quad \Theta_*^{\mathcal{B}} = \begin{Bmatrix} (\neg A_0 \vee A_1) \\ (\neg A_3 \vee A_2) \\ (\neg A_0 \vee \neg A_2) \\ (\neg A_0 \vee \neg A_3) \\ (\neg A_1 \vee \neg A_2) \\ (\neg A_1 \vee \neg A_3) \end{Bmatrix}$$

Then, one possible execution of the algorithm is:

i	$\Theta_i^{\mathcal{T}}$	$\psi_{s,i}^{\mathcal{T}}$	$\text{Weight}(\psi_{s,i}^{\mathcal{T}})$	$SMT(\varphi_h^{\mathcal{T}} \cup \psi_{s,i}^{\mathcal{T}} \cup \Theta_i^{\mathcal{T}})$
0	{}	$\{C_0, C_1, C_2, C_3\}$	15	UNSAT
1	$\{\theta_4\}$	$\{C_1, C_2, C_3\}$	11	UNSAT
2	$\{\theta_4, \theta_6\}$	$\{C_0, C_1, C_2\}$	9	UNSAT
3	$\{\theta_4, \theta_6, \theta_3\}$	$\{C_2, C_3\}$	8	SAT

from which $\psi_s^{\mathcal{T}} = \{C_2, C_3\}$ and $\text{Weight}(\psi_s^{\mathcal{T}}) = 8$. A faster execution (which may be obtained, e.g., by enforcing the generation of extra $\mathcal{T}$-lemmas in the SMT solver) is:

i	$\Theta_i^{\mathcal{T}}$	$\psi_{s,i}^{\mathcal{T}}$	$\text{Weight}(\psi_{s,i}^{\mathcal{T}})$	$SMT(\varphi_h^{\mathcal{T}} \cup \psi_{s,i}^{\mathcal{T}} \cup \Theta_i^{\mathcal{T}})$
0	{}	$\{C_0, C_1, C_2, C_3\}$	15	UNSAT
1	$\{\theta_1, \theta_2, \theta_5\}$	$\{C_2, C_3\}$	8	SAT

◇

3.2   Optimizations

Algorithm 1 is very simple in principle, and it can be implemented using an SMT solver and a MaxSAT solver as black boxes in a plug-and-play manner.[4] Moreover, this allows for benefiting for free of any advanced tool available from the shelf, or for choosing the most suitable tools for a given problem.

Under the hypothesis of using the two solvers as black boxes, we consider some implementation issues which may further improve its efficiency.

---

[4] Provided that the SMT solver, like MATHSAT5, offers a way of retrieving the set of $\mathcal{T}$-lemmas which it used to prove the $\mathcal{T}$-inconsistency of the input formula, or, like most lazy SMT solvers, it can provide an SMT resolution proof, from which the latter set can be extracted.

Incrementality of MaxSAT. Since MaxSAT is invoked sequentially on incremental sets of hard clauses and on the same set of soft ones, it is natural to conjecture that having an incremental implementation of MaxSAT, which "remembers" the status of the search from call to call, should improve the efficiency of the overall procedure.

Reuse of SMT calls. SMT.Solve is not invoked incrementally in the classic "push-and-pop" sense because —apart from the fact that $\psi^{\mathcal{T}}_{s,i} \subseteq \varphi^{\mathcal{T}}_s$ for every i— there is no set-theoretic relation between the $\psi^{\mathcal{T}}_{s,i}$'s. However, it is possible to use SMT solving *under assumptions*: each soft clause $C^{\mathcal{T}}_j$ in $\varphi^{\mathcal{T}}_s$ is augmented with a fresh selection Boolean variable S_j (i.e., $\varphi^{\mathcal{T}}_s$ is rewritten into $\varphi'^{\mathcal{T}}_s \stackrel{\text{def}}{=} \{(\neg S_j \lor C^{\mathcal{T}}_j) \mid C^{\mathcal{T}}_j \in \varphi^{\mathcal{T}}_s\}$) and the proper set of selection variables $\lambda_s \stackrel{\text{def}}{=} \{S_j \mid C^{\mathcal{T}}_j \in \psi^{\mathcal{T}}_s\}$ is assumed at each call. This allows for "remembering" and reusing learned clauses from call to call. (Notice that, as long as the $\mathcal{T}$-lemmas are remembered from call to call, it is possible to drop the "$\cup \, \Theta^{\mathcal{T}}$" in the call SMT.Solve $(\varphi^{\mathcal{T}}_h \cup \psi^{\mathcal{T}}_s \cup \Theta^{\mathcal{T}})$.)

In a "white-box" integration scenario, in which it is possible to modify either or both the solvers involved, the following considerations may be of interest.

Generation of extra $\mathcal{T}$-lemmas. As illustrated in the second execution of Example 1, generating and storing extra $\mathcal{T}$-lemmas inside the SMT-solving phase —not only these explicitly involved in the conflict analysis— enlarges the $\mathcal{T}$-lemma pool and may possibly reduce the number of cycles. This can be obtained by means of SMT techniques like static learning and by storing *all* the $\mathcal{T}$-deduction clauses inferred by $\mathcal{T}$-propagation[5] (see [23,8]). Notice that, to avoid introducing overhead for the underlying SAT solver, it suffices to *store* such $\mathcal{T}$-lemmas, without *learning* them.

4 Related Work

Maximum satisfiability in SMT was first studied in [22], in the context of a general framework for optimization in SMT using "progressively stronger theories". An implementation for MaxSMT of this framework is described, but it is not publicly available.

An explicit reference to MaxSMT is found in [4], which describes the evaluation of an implementation of the WPM procedure [6] based on the YICES [2] SMT solver. This implementation is not publicly available. Another reference is in [5], where weighted Constraint Satisfaction Problems are translated into weighted MaxSMT instances.

The YICES solver provides also native support for MaxSMT. The approach used is based on incrementally invoking the solver in a mixed linear/binary-search fashion onto an SMT encoding of the MaxSMT problem, similar to that described in §2.1. The algorithm is not described in any publication, but we could obtain such information from personal communications with the authors.

The source distribution of the Z3 [15] solver provides an example implementation of an SMT version of the core-guided MaxSAT algorithm of [16], using the Z3 API. The algorithm is based on enumerating and counting unsatisfiable subformulas.

---

[5] In many SMT solvers implementing $\mathcal{T}$-propagation, $\mathcal{T}$-deduction clauses are generated on demand, only if they are needed by the underlying CDCL SAT solver for conflict analysis.

Also related are the works on optimization in SMT [10,24], that can be used to encode the various MaxSMT problems. The work in [10] introduces the notion of "Theory of Costs" $\mathcal{C}$ to handle Pseudo-Boolean (PB) cost functions and constraints by an ad-hoc and independent "$\mathcal{C}$-solver" in the standard lazy SMT schema. MaxSMT can be handled by encoding it straightforwardly into a PB optimization problem (see §2.1). The implementation is available. The work in [24] introduced a wider notion of optimization in SMT, OMT($\mathcal{LA}(\mathbb{Q}) \cup \mathcal{T}$), with cost functions on variables on the *reals*, which allows for encoding also MaxSMT and SMT with PB cost functions (see §2.1). Some OMT($\mathcal{LA}(\mathbb{Q}) \cup \mathcal{T}$) procedures combining lazy SMT and standard LP minimization techniques are presented. The implementation, done on top of the MATHSAT5 [12] SMT solver, is available.

Davies and Bacchus [14] proposed a MaxSAT algorithm (hereafter "DB") which, similarly to LL, works by iteratively ruling out subsets of the soft clauses of the input problem. In particular, DB builds iteratively a set $\mathcal{K}$ of unsat cores for $\varphi_h^{\mathcal{B}} \cup \varphi_s^{\mathcal{B}}$, i.e., at each loop iteration: (i) computes a new subset of soft clauses hs to drop as *minimum-cost hitting set* of $\mathcal{K}$; (ii) computes a new unsat core κ of $\varphi_h^{\mathcal{B}} \cup \varphi_s^{\mathcal{B}} \setminus hs$; this is repeated as long as $\varphi_h^{\mathcal{B}} \cup \varphi_s^{\mathcal{B}} \setminus hs$ is unsatisfiable.

Although [14] does not mention SMT, in principle this algorithm could be leveraged to SMT level (hereafter "DB-SMT"), by substituting SAT-level solving and unsat-core extraction with SMT-level ones. (Notice, however, that unlike with the SAT domain, efficiently finding minimal or nearly-minimal unsat cores in SMT is still an open research problem, see [13].) If so, LL and DB-SMT would be based on similar principles:[6]

- both algorithms would be based on constraint generation, producing constraints at every loop iteration which rule out subsets of the soft clauses;
- both would decouple solving and minimizing into two different subroutines.

The technical differences, however, would be manifold:

- Unlike with DB-SMT, LL is not a generalization of DB to SMT: unlike with DB, if it is fed a pair of purely-Boolean formulas, then it terminates in one iteration.
- DB-SMT would be driven by the combinatorics of the unsat cores to rule out, whilst LL is driven by the theory-information to be provided.
- The $\mathcal{T}$-lemma sets $\Theta_i^{\mathcal{T}}$ in LL are not the SMT counterpart of the unsat cores κ_i in DB-SMT: the former contain only *novel* clauses, the latter do not; there is not one-to-one correspondence between the generated sets of $\mathcal{T}$-lemmas and unsat cores. [7]
- MaxSAT is not the SMT counterpart of minimum-cost hitting set extraction: the latter starts from more fine-grained information, in the form of sets of unsat cores.
- It is easy to see that it would take at least N cycles to DB-SMT to rule out N clauses from $\psi_s^{\mathcal{B}}$. With LL the number of soft clauses discharged at each loop depends only on the quantity and quality of the $\mathcal{T}$-lemmas generated: in many cases (see Fig. 1) one iteration is enough to generate all the necessary $\mathcal{T}$-lemmas. [7]

---

[6] We are grateful to an anonymous reviewer who pointed out an analogy between DB and LL.

[7] For example, consider the second execution in Example 1: with DB-SMT there would be no unsat core κ_1 "equivalent" to $\Theta_1^{\mathcal{T}} \overset{\text{def}}{=} \{\theta_1, \theta_2, \theta_5\}$, allowing to directly pass from step 0 to step 1, since one needs 2 cores (and hence 2 loops) to generate a minimum-cost hs of size 2.

– the LL schema requires no SMT unsat-core extraction, nor minimum-cost hitting-set computation. (The MaxSAT subroutine is not committed to any MaxSAT schema.)

Finally, and importantly, DB/DB-SMT and LL radically differ in the *context* they were conceived (MaxSAT vs. MaxSMT), in their *usability* (the two schemas would pose very different constraints to a MaxSMT implementer) and *goals* (DB was conceived to address some efficiency issues in MaxSAT solvers [14], whilst LL is proposed as a modular approach to build MaxSMT solvers).

5   Experimental Evaluation

We have implemented our LL MaxSMT approach on top of our MATHSAT5 SMT solver [12] and of a selection of external MaxSAT tools. We have evaluated and compared the performances of the various LL instances and of every MaxSMT or MaxSMT-like solver available we are aware of, by means of an extensive empirical test on MaxSMT-modified SMT-LIB benchmarks.[8] In this section we present such evaluation.

5.1   Test Description

Benchmarks. As benchmark problems, we took the unsatisfiable SMT instances in the $\mathcal{LA}(\mathbb{Q})$ and $\mathcal{LA}(\mathbb{Z})$ categories of SMT-COMP [1], and we converted them into two groups of MaxSMT problems —partial MaxSMT and weighted partial MaxSMT respectively— by a random partition into hard and (weighted) soft clauses. In order to handle both CNF and non-CNF formulas, for each instance, we created the set of soft constraints by randomly selecting 20% assigning them a weight of 1 for the partial MaxSMT experiments, and a weight uniformly selected in the range $1 \ldots 100$ for the weighted partial MaxSMT ones, and then applied the process described in §2.2.

Competing MaxSMT Solvers. Before starting our evaluation, we have asked to the main scientists of the MaxSAT community about the existence of *incremental* MaxSAT procedures, obtaining a negative answer. Thus, we decided to produce ourselves an implementation of the WPM MaxSAT algorithm [6] on top of MINISAT, and we also tried to enhance it with some degree of incrementality. Recently, Carlos Ansótegui has kindly sent us the code of a modified version of the WPM, which he had also adapted to get some incrementality, which invokes the YICES-1.0.36 as an external solver. Moreover, to guarantee a more interesting comparison, we have implemented on top of MATHSAT5 the same MaxSMT extension of the core-guided algorithm of [16] implemented in Z3. Finally, since the original implementation of SMT with PB cost functions and constraints of [10] was implemented on top of MATHSAT4, and in order to have a more significant comparison, we have recently ported it into MATHSAT5.

Thus, in this evaluation the first competitors were four instances of our Lemma-Lifting (LL) implementation, each using a different external MaxSAT solver:

---

[8] We also asked the authors of the related MaxSMT papers of §4 for other benchmarks, but none provided any meaningful benchmark.

LL$_{WPM}$, which uses the publicly-available WPM [6] implementation used in the 2012 MaxSAT-evaluation (which uses PICOSAT);

LL$_{YICES-WPM}$, which uses the above-mentioned non-public implementation of WPM provided to us by Carlos Ansótegui;

LL$_{OWPM}$, which uses our own incremental WPM implementation;

LL$_{NI-OWPM}$, as before, non-incremental version.

Other competitors were the following MaxSMT solvers:

YICES, the MaxSMT extension of YICES (see §4);

Z3, the MaxSMT extension of Z3 (see §4);

MATHSAT5-MAX, our own implementation on top of MATHSAT5 of the core-guided algorithm of [16], as with Z3.

Notice that the last two solvers handle only unweighted partial MaxSMT problems. The final competitors were the following solvers based on SMT with cost optimization (see §4), using the encodings described in §2.3:

MATHSAT4+$\mathcal{C}$(L), the tool from [10], using linear-search mode;

MATHSAT5+$\mathcal{C}$(L), the porting of the above procedure into MATHSAT5, using linear-search mode;

MATHSAT5+$\mathcal{C}$(B), as before, using binary-search mode;

OPTIMATHSAT, the OMT($\mathcal{LA}(\mathbb{Q}) \cup \mathcal{T}$) tool of [24] described in §4, using its adaptive binary/linear search heuristics.

Notice that, if $\mathcal{T}$ is the $\mathcal{LA}(\mathbb{Z})$ theory, it is not possible to encode MaxSMT into OMT($\mathcal{LA}(\mathbb{Q}) \cup \mathcal{T}$) because the current implementation of OPTIMATHSAT cannot handle $\mathcal{LA}(\mathbb{Q}) \cup \mathcal{LA}(\mathbb{Z})$.[9]

The Experiments. The experiments we performed can be divided into two groups. In the first group, we tested and compared the performances of all the eleven MaxSMT solvers on the benchmarks described above. This was performed on Intel(R) Xeon(R) CPU E5650 2.67GHz platform, with a 4GB memory limit and a 20 minute time limit for each run. In the second group, we made some more accurate analysis of the behaviour of the LL implementations. This was performed on a Intel(R) Xeon(R) CPU E5520 2.27GHz platform, using the same memory and time limits.

Check of the Results. We have checked the correctness of the results of all our own tools (i.e., those based on MATHSAT4 or MATHSAT5) by checking the models returned and by independently proving the unsatisfiability of the formula $\varphi^{\mathcal{T}'} \cup \{\text{cost} < k\}$, where $\varphi^{\mathcal{T}'}$ and cost are defined as in (2) and k is the value of the cost returned by the tool. All results agreed with one another and were found correct by the above test.

---

[9] This is due to the fact that the OMT($\mathcal{LA}(\mathbb{Q}) \cup \mathcal{T}$) framework requires that $\mathcal{LA}(\mathbb{Q})$ and $\mathcal{T}$ are signature-disjoint theories [24], which is not the case if $\mathcal{T}$ is $\mathcal{LA}(\mathbb{Z})$.

Table 1. Results of the eleven MaxSMT solvers on partial MaxSMT instances

Solver	$\mathcal{LA}(\mathbb{Z})$		$\mathcal{LA}(\mathbb{Q})$		Total	
	#Solved	time (sec)	#Solved	time (sec)	#Solved	time (sec)
MATHSAT5-MAX	**95 / 106**	**6575.60**	88 / 93	2274.69	**183 / 199**	**8850.29**
LLOWPM	92 / 106	5942.20	88 / 93	1785.48	180 / 199	7727.68
YICES	92 / 106	14478.43	87 / 93	5537.47	179 / 199	20015.9
LLNI-OWPM	89 / 106	4439.98	88 / 93	1780.97	177 / 199	6220.95
LLYICES−WPM	89 / 106	4937.91	87 / 93	1855.45	176 / 199	6793.36
LLWPM	88 / 106	7154.19	88 / 93	2071.27	176 / 199	9225.46
MATHSAT5+$\mathcal{C}$(L)	84 / 106	7112.43	87 / 93	2175.34	171 / 199	9287.77
MATHSAT4+$\mathcal{C}$(L)	83 / 106	5220.14	85 / 93	1944.48	168 / 199	7164.62
Z3	89 / 106	4066.92	76 / 93	2427.59	165 / 199	6494.51
MATHSAT5+$\mathcal{C}$(B)	78 / 106	5030.85	87 / 93	2545.69	165 / 199	7576.54
OPTIMATHSAT	—	—	**89 / 93**	**1360.05**	—	—

Table 2. Results of the eleven MaxSMT solvers on partial weighted MaxSMT instances

Solver	$\mathcal{LA}(\mathbb{Z})$		$\mathcal{LA}(\mathbb{Q})$		Total	
	#Solved	time (sec)	#Solved	time (sec)	#Solved	time (sec)
LLWPM	**90 / 106**	**5194.73**	87 / 93	3033.66	**177 / 199**	**8228.39**
LLNI-OWPM	86 / 106	1672.41	88 / 93	2062.35	174 / 199	3734.76
MATHSAT5+$\mathcal{C}$(L)	89 / 106	5501.38	84 / 93	2359.61	173 / 199	7860.99
LLOWPM	85 / 106	1304.13	87 / 93	1836.53	172 / 199	3140.66
MATHSAT4+$\mathcal{C}$(L)	87 / 106	3105.01	85 / 93	2541.83	172 / 199	5646.84
LLYICES−WPM	82 / 106	1423.53	87 / 93	2350.02	169 / 199	3773.55
YICES	83 / 106	12305.88	80 / 93	9804.16	163 / 199	22110.04
MATHSAT5+$\mathcal{C}$(B)	79 / 106	9482.61	83 / 93	2627.35	162 / 199	12109.96
OPTIMATHSAT	—	—	**88 / 93**	**1947.06**	88 / 93	1947.06
Z3	—	—	—	—	—	—
MATHSAT5-MAX	—	—	—	—	—	—

5.2 Results

The results of the evaluation of the eleven MaxSMT solvers are presented in Tables 1 and 2, reporting for each solver the number of instances solved within the timeout and the total runtime taken to solve them. (Rows are sorted according to total ($\mathcal{LA}(\mathbb{Q})$ + $\mathcal{LA}(\mathbb{Z})$) performance, best performances for each category are in **bold**.) Note that, for the reasons highlighted above, Z3 and MATHSAT5-MAX were not run on weighted instances, and OPTIMATHSAT was not ran on the $\mathcal{LA}(\mathbb{Z})$ instances (this is marked with a "—").

Looking at the data in Tables 1 and 2 some considerations are in order.

(i) MATHSAT5-MAX is the overall best performer on the unweighted group, whilst LLWPM is the overall best performer on the weighted one. If we restrict to $\mathcal{LA}(\mathbb{Q})$ theory, OPTIMATHSAT is the winner in both unweighted and weighted groups.

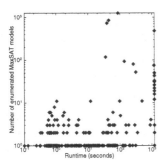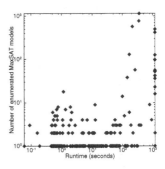

Fig. 1. Number of while-cycles wrt. runtime for LL$_{OWPM}$ (left) and LL$_{NI\text{-}OWPM}$ (right).

However, there is no hands-down winner, and the performance gaps among the eleven solvers are not dramatic.

(ii) Overall, the LL tools behave quite well, all being in the highest part of the ranking. Among them, there is not an absolute winner: LL$_{OWPM}$ is the best performer on the unweighted group, whilst LL$_{WPM}$ is the best performer on the weighted one.

(iii) There is no definite winner between LL$_{OWPM}$ and LL$_{NI\text{-}OWPM}$: the former is better on unweighted test, the latter on weighted ones, and the performance gaps are very limited. Thus, incrementality does not seem to pay as much as one could expect. Similarly, there is no definite winner between LL$_{YICES\text{-}WPM}$ and LL$_{WPM}$. (Notice, however, that these two call two different backend solvers, YICES and PICOSAT.)

Overall, the results are too limited and heterogeneous to infer the superiority of one approach wrt. another. However, we can safely conclude that, despite its simplicity, the LL approach is competitive wrt state-of-the-art ones.

In Figure 1 and Table 3 we analyze the behaviour of the LL solvers on all tests (weighted/unweighted, $\mathcal{LA}(\mathbb{Q})/\mathcal{LA}(\mathbb{Z})$).

Figure 1 shows the number of while-cycles performed by the LL algorithm with LL$_{OWPM}$ (left) and LL$_{NI\text{-}OWPM}$ (right), plotted against runtime. We notice that a very high percentage of instances is solved at the first loop, and the vast majority of instances is solved in less than 10 loops. This induces us to conjecture that SMT.Solve in many cases is able to produce very soon all the $\mathcal{T}$-lemmas which are necessary to MaxSAT to rule out the wrong truth assignments.

Table 3 analyzes the percentage of CPU time spent inside MaxSAT calls for the four LL tools.[10] We notice that for most solvers and most instances, the solver spends less than 20% and this fact is particularly evident in the easiest problems. Thus, the overall CPU time is mostly dominated by the time spent inside SMT.Solve. In particular, in the samples in which the solution is found in one loop, most time is taken by SMT.Solve to enumerate the necessary $\mathcal{T}$-lemmas in one shot.

This also explains in part the low effect of incrementality in our experiments, since the cost of MaxSAT calls does not seem to represent the actual bottleneck of the process.

---

[10] For instance (1st block, 3rd column): out of 116 instance problem for which LL$_{OWPM}$ took less than one second to execute, with 81 instances MaxSAT calls required less than 20% 40%

Table 3. An overview of runtime spent on MaxSAT calls compared to total runtime for LL solvers

		Runtime LL$_{\text{OWPM}}$ (in seconds)				
		$[0, 1[$	$[1, 10[$	$[10, 100[$	$[100, 1000[$	≥ 1000
	$0 \leq p < 20$	81	41	15	5	13
% Time	$20 \leq p < 40$	19	12	9	17	27
MaxSAT	$40 \leq p < 60$	10	6	5	19	12
	$60 \leq p < 80$	1	0	6	15	3
	$80 \leq p \leq 100$	5	15	0	9	0
	Total	116	74	35	65	55
		Runtime LL$_{\text{OWPM}}$ (in seconds)				
		$[0, 1[$	$[1, 10[$	$[10, 100[$	$[100, 1000[$	≥ 1000
	$0 \leq p < 20$	115	53	43	49	59
% Time	$20 \leq p < 40$	5	6	1	1	9
MaxSAT	$40 \leq p < 60$	1	1	2	0	1
	$60 \leq p < 80$	0	1	0	0	0
	$80 \leq p \leq 100$	0	1	4	0	0
	Total	121	62	50	50	69
		Runtime LL$_{\text{NI-OWPM}}$ (in seconds)				
		$[0, 1[$	$[1, 10[$	$[10, 100[$	$[100, 1000[$	≥ 1000
	$0 \leq p < 20$	95	49	32	48	53
% Time	$20 \leq p < 40$	11	9	8	2	4
MaxSAT	$40 \leq p < 60$	4	0	5	0	0
	$60 \leq p < 80$	8	1	0	0	1
	$80 \leq p \leq 100$	4	2	11	0	4
	Total	122	61	56	50	62
		Runtime LL$_{\text{WPM}}$ (in seconds)				
		$[0, 1[$	$[1, 10[$	$[10, 100[$	$[100, 1000[$	≥ 1000
	$0 \leq p < 20$	115	58	16	43	57
% Time	$20 \leq p < 40$	1	1	11	8	1
MaxSAT	$40 \leq p < 60$	4	1	9	2	0
	$60 \leq p < 80$	8	0	2	1	1
	$80 \leq p \leq 100$	1	0	6	3	4
	Total	129	60	44	57	63

Notice that, if we compare the data on LL$_{\text{OWPM}}$ and LL$_{\text{NI-OWPM}}$ in Table 3, we notice that indeed in the incremental version the percentage of time spent inside MaxSAT is smaller than in the non-incremental one. However, since the total cost is mostly dominated by SMT.Solve, the benefits of this fact are not significant. (Also, we must recall that, since we are not expert MaxSAT developers, our implementation of incrementality is quite naive).

6   Conclusions and Future Work

In this paper we have presented a novel "modular" Lemma-Lifting approach for MaxSMT, which combines a lazy SMT solver with a purely-propositional MaxSAT solver. Despite its simplicity, LL proves competitive with previous approaches.

Depending on one's expertise on and access to SMT and MaxSAT technology, we see different ways the LL approach can be implemented into a MaxSMT tool.

– Whoever cannot or does not want to put the hands on either solver's code, can take both an SMT and a MaxSAT solver off-the-shelf and implement our algorithm on top of their API (or even interface with them via file exchange). In this case, implementation is straightforward.
– MaxSAT-solver developers can leverage to SMT level the expressiveness of their own tool by interfacing with one SMT solver, without implementing any SMT functionality in-house. They can also customize their own MaxSAT tool to improve the synergy of the two tools (in particular, by making it as incremental as possible).
– SMT-solver developers can extend their own tool with MaxSMT functionality by interfacing with one or more MaxSAT solvers off-the-shelf, with no need of implementing MaxSAT functionalities in-house. They can also customize their own solver (e.g., by maximizing the generation of theory lemmas).
– A person with access to, and enough expertise on, both SMT- and MaxSAT- solver development can adopt our approach to produce a highly efficient MaxSMT tool, with the possibility of customizing both tools. Notice that, in this case, our approach can also be combined with other SMT optimization techniques (e.g., those described in [22,10,24,4]).

We believe that this paper opens novel research avenues in MaxSMT. In particular, we see many directions along which the LL approach can be improved and extended.

Customizing SMT and MaxSAT solvers. The LL approach would strongly benefit from more effective $\mathcal{T}$-lemma generators and incremental MaxSAT solvers.
Interleaving of SMT- and MaxSAT-solving steps. Algorithm 1 interleaves *complete* calls to SMT.Solve and MaxSAT. This can be generalized to more fine-grained interleaving schemas, in which *steps* of such executions can be interleaved. For instance, it is possible to interrupt SMT.Solve as soon as some amount of $\mathcal{T}$-lemmas has been generated, and invoke MaxSAT afterwords. Vice versa, it is possible to interrupt MaxSAT as soon as a non-optimal $\psi_s^{\mathcal{B}}$ is generated, and feed it back to SMT.Solve. (As an extreme case, one could feed to SMT.Solve only the assignment $\mu^{\mathcal{T}}$ produced by MaxSAT: if so, SMT.Solve would be simply used as a $\mathcal{T}$-Solver.)
Combination with other approaches. The lemma-Lifting approach can be combined with other approaches in various ways. For instance, one could enhance the use of SMT by exploiting SMT with cost constraints [10] and extraction of $\mathcal{T}$-unsatisfiable cores [13] to further prune the search.

Overall, novel strategies and heuristics can be investigated to extend and improve Algorithm 1 along the above directions.

References

1. SMT-COMP, http://www.smtcomp.org/2010/
2. Yices, http://yices.csl.sri.com/
3. Max-SAT 2013, Eighth Max-SAT Evaluation (2013), http://maxsat.ia.udl.cat

4. Ansótegui, C., Bofill, M., Palahí, M., Suy, J., Villaret, M.: Satisfiability Modulo Theories: An Efficient Approach for the Resource-Constrained Project Scheduling Problem. In: SARA (2011)
5. Ansótegui, C., Bofill, M., Palahí, M., Suy, J., Villaret, M.: Solving weighted CSPs with meta-constraints by reformulation into Satisfiability Modulo Theories. Constraints 18(2) (2013)
6. Ansótegui, C., Bonet, M.L., Levy, J.: SAT-based MaxSAT algorithms. Artif. Intell. 196 (2013)
7. Barrett, C., Nieuwenhuis, R., Oliveras, A., Tinelli, C.: Splitting on Demand in SAT Modulo Theories. In: Hermann, M., Voronkov, A. (eds.) LPAR 2006. LNCS (LNAI), vol. 4246, pp. 512–526. Springer, Heidelberg (2006)
8. Barrett, C., Sebastiani, R., Seshia, S.A., Tinelli, C.: Satisfiability Modulo Theories. In: Biere, et al. (eds.) Handbook of Satisfiability, ch. 26, IOS Press (2009)
9. Bozzano, M., Bruttomesso, R., Cimatti, A., Junttila, T.A., Ranise, S., van Rossum, P., Sebastiani, R.: Efficient Theory Combination via Boolean Search. Information and Computation 204(10) (2006)
10. Cimatti, A., Franzén, A., Griggio, A., Sebastiani, R., Stenico, C.: Satisfiability modulo the theory of costs: Foundations and applications. In: Esparza, J., Majumdar, R. (eds.) TACAS 2010. LNCS, vol. 6015, pp. 99–113. Springer, Heidelberg (2010)
11. Cimatti, A., Griggio, A., Schaafsma, B., Sebastiani, R.: A Modular Approach to MaxSAT Modulo Theories (2013), Extended version
   http://disi.unitn.it/~rseba/sat13/extended.pdf
12. Cimatti, A., Griggio, A., Schaafsma, B.J., Sebastiani, R.: The MathSAT 5 SMT Solver. In: Piterman, N., Smolka, S.A. (eds.) TACAS 2013. LNCS, vol. 7795, pp. 93–107. Springer, Heidelberg (2013)
13. Cimatti, A., Griggio, A., Sebastiani, R.: Computing Small Unsatisfiable Cores in SAT Modulo Theories. JAIR 40 (2011)
14. Davies, J., Bacchus, F.: Solving MAXSAT by solving a sequence of simpler SAT instances. In: Lee, J. (ed.) CP 2011. LNCS, vol. 6876, pp. 225–239. Springer, Heidelberg (2011)
15. de Moura, L., Bjørner, N.: Z3: An Efficient SMT Solver. In: Ramakrishnan, C.R., Rehof, J. (eds.) TACAS 2008. LNCS, vol. 4963, pp. 337–340. Springer, Heidelberg (2008)
16. Fu, Z., Malik, S.: On solving the partial Max-SAT problem. In: Biere, A., Gomes, C.P. (eds.) SAT 2006. LNCS, vol. 4121, pp. 252–265. Springer, Heidelberg (2006)
17. Heras, F., Larrosa, J., Oliveras, A.: Minimaxsat: An Efficient Weighted Max-SAT solver. JAIR 31 (2008)
18. Heras, F., Morgado, A., Marques-Silva, J.: Core-guided binary search algorithms for maximum satisfiability. In: AAAI (2011)
19. Li, C.M., Manyà, F.: MaxSAT, Hard and Soft Constraints. In: Biere, et al. (eds.) Handbook of Satisfiability, ch. 19. IOS Press (2009)
20. Li, C.M., Manyà, F., Planes, J.: New inference rules for Max-SAT. JAIR 30 (2007)
21. Morgado, A., Heras, F., Marques-Silva, J.: Improvements to core-guided binary search for MaxSAT. In: Cimatti, A., Sebastiani, R. (eds.) SAT 2012. LNCS, vol. 7317, pp. 284–297. Springer, Heidelberg (2012)
22. Nieuwenhuis, R., Oliveras, A.: On SAT Modulo Theories and Optimization Problems. In: Biere, A., Gomes, C.P. (eds.) SAT 2006. LNCS, vol. 4121, pp. 156–169. Springer, Heidelberg (2006)
23. Sebastiani, R.: Lazy Satisfiability Modulo Theories. JSAT 3(3-4) (2007)
24. Sebastiani, R., Tomasi, S.: Optimization in SMT with LA(Q) Cost Functions. In: Gramlich, B., Miller, D., Sattler, U. (eds.) IJCAR 2012. LNCS (LNAI), vol. 7364, pp. 484–498. Springer, Heidelberg (2012)

Exploiting the Power of MIP Solvers in MAXSAT

Jessica Davies and Fahiem Bacchus

Department of Computer Science, University of Toronto,
Toronto, Ontario, Canada, M5S 3H5
{jdavies,fbacchus}@cs.toronto.edu

Abstract. MAXSAT is an optimization version of satisfiability. Since many practical problems involve optimization, there are a wide range of potential applications for effective MAXSAT solvers. In this paper we present an extensive empirical evaluation of a number of MAXSAT solvers. In addition to traditional MAXSAT solvers, we also evaluate the use of a state-of-the-art Mixed Integer Program (MIP) solver, CPLEX, for solving MAXSAT. MIP solvers are the most popular technology for solving optimization problems and are also theoretically more powerful than SAT solvers. In fact, we show that CPLEX is quite effective on a range of MAXSAT instances. Based on these observations we extend a previously developed hybrid approach for solving MAXSAT, that utilizes both a SAT solver and a MIP solver. Our extensions aim to take better advantage of the power of the MIP solver. The resulting improved hybrid solver is shown to be quite effective.

1 Introduction

MAXSAT is an optimization version of satisfiability (SAT). Both problems deal with propositional formulas expressed in CNF. The goal of SAT is to find a setting of the propositional variables that satisfies all clauses. MAXSAT, on the other hand, tries to find a setting of the variables that maximizes the number of satisfied clauses.

MAXSAT is complete for the class FP^{NP} (the set of function problems computable in polynomial time using an NP oracle). FP^{NP} includes many practical optimization problems, and by completeness all of them can be compactly encoded into MAXSAT. Hence, MAXSAT solvers that are effective on a wide range of inputs would be able to solve a variety of practical problems through the simple device of encoding into MAXSAT. This is already the case with SAT, where many real-world problems in NP can be effectively solved by encoding into SAT and applying current SAT solvers. Work on developing widely applicable MAXSAT solvers is still ongoing, and this paper aims to make a contribution to this effort.

Many important industrial applications involve solving optimization problems, and many powerful solution techniques for such problems have been developed. Problems with Boolean or integer variables (like MAXSAT) are most often solved using sophisticated Mixed Integer Program (MIP) solvers. MIP solvers solve problems expressed as a set of linear inequalities and a linear objective function,

M. Järvisalo and A. Van Gelder (Eds.): SAT 2013, LNCS 7962, pp. 166–181, 2013.

a representation that is more expressive than CNF. One common technique they employ is to utilize linear programming algorithms to solve the linear relaxation derived by allowing the integer variables to take on non-integral values. Cutting plane computations are then used to drive the linear relaxation towards integral solutions. The technique of cutting planes is theoretically more powerful than resolution [5], and thus these solvers potentially have access to more powerful inference methods than standard SAT solvers. In contrast, current MAXSAT solvers have almost exclusively used resolution-based SAT technology.

In this paper we perform an extensive empirical evaluation of a number of previous solvers. Our evaluation uses many more problem instances than any previously reported study, in part because we are interested in widely applicable MAXSAT solvers. We also evaluate the performance of a state-of-the-art MIP solver, IBM's CPLEX system, on these instances. Our evaluation, reported on in Sec. 3, provides a number of interesting insights. For example, we show that CPLEX is a very effective solver for MAXSAT. We also show that in the current state-of-the-art, the notion of a single best algorithmic approach for solving MAXSAT is suspect, as is the notion of a single best solver. Our experiments do however indicate that the solvers tested can be divided into two subsets with one subset arguably dominating the other in terms of performance. However, within the high performance subset no single solver dominates.

This variance in performance among the different solvers across the problem instances indicates that each of these solvers embeds ideas that are effective on some problems. Hence, one possible direction for future research is to investigate ways of combining these ideas to uncover new algorithmic insights.[1] In previous work we had developed such a hybrid approach, MAXHS [7], that utilized a MIP solver, CPLEX, along with a SAT solver, MINISAT [8]. Each solver was given a subset of the MAXSAT problem, and information was communicated between the solvers so as to solve the combined problem. The strong performance of CPLEX in our experiments lead us to investigate ways of taking better advantage of the MIP solver. In particular, in the second part of the paper, reported on in Sec. 4, we develop a number of techniques for increasing the amount and effectiveness of information supplied to CPLEX, thus allowing it to make stronger inferences. Our new techniques yield a considerable performance improvement to the MAXHS solver (Sec. 6), and as shown in Sec. 3 the resulting improved MAXHS+ solver is clearly placed in the set of top performing MAXSAT solvers. We conclude the paper with some ideas for further work.

2   Background

In this paper we address **weighted partial MAXSAT** problems (WPMS). This is the most general type of MAXSAT problem and it includes as special cases all of the other types of MAXSAT problems studied in the literature. (All of the

---

[1] Our empirical evaluation shows that many instances remain unsolvable by any solver. Hence, although portfolio approaches could yield useful performance improvements, significant advances will also require new algorithmic ideas.

solvers we experiment with can solve all of these special cases as well as general WPMS problems). WPMS problems are CNF formulas in which some clauses are classified as being hard while others are classified as being soft. Any solution must satisfy all of the hard clauses, but can falsify the soft clauses. However, each soft clause has a weight and a truth assignment will incur a penalty or cost equal to the clause weight if it falsifies that clause.

More formally, a MAXSAT problem $\mathcal{F}$ is specified by a CNF formula in which each clause has an associated weight.[2] Let $wt(c)$ denote the weight of clause c. We require that $wt(c) > 0$ for every clause.[3] If $wt(c) = \infty$ we say that c is a **hard** clause, otherwise $wt(c) < \infty$ and c is a **soft** clause. We use $hard(\mathcal{F})$ to indicate the hard clauses of $\mathcal{F}$ and $soft(\mathcal{F})$ to denote the soft clauses. Note that $\mathcal{F} = hard(\mathcal{F}) \cup soft(\mathcal{F})$.

We define the function $cost$ as follows: (a) if H is a set of clauses then $cost(H)$ is the sum of the weights of the clauses in H $(cost(H) = \sum_{c \in H} wt(c))$; and (b) if π is a truth assignment to the variables of $\mathcal{F}$ then $cost(\pi)$ is the sum of the weights of the clauses falsified by π $(\sum_{\{c \mid \pi \not\models c\}} wt(c))$.

A **solution** to the MAXSAT problem $\mathcal{F}$ is a truth assignment π to the variables of $\mathcal{F}$ with minimum cost that satisfies all of the clauses in $hard(\mathcal{F})$. We let $mincost(\mathcal{F})$ denote the cost of a solution to $\mathcal{F}$. If $hard(\mathcal{F})$ is UNSAT then $\mathcal{F}$ has no solution. Testing for this case is simply a SAT problem, hence from here on we will **assume that** $hard(\mathcal{F})$ **is satisfiable.**

A **core** κ for a MAXSAT formula $\mathcal{F}$ is a subset of $soft(\mathcal{F})$ such that $\kappa \cup hard(\mathcal{F})$ is unsatisfiable. That is, all truth assignments falsify at least one clause of $\kappa \cup hard(\mathcal{F})$. Since every solution satisfies $hard(\mathcal{F})$, every solution must falsify at least one clause in κ.

A common technique in MAXSAT solving is to add a unique **blocking variable** to each soft clause. Assigning *true* to a clause's blocking variable (b-variable) immediately satisfies the clause. This allows the solver to "turn off" or relax various soft clauses as it tries to solve the MAXSAT problem.

Definition 1. *If $\mathcal{F}$ is a MAXSAT problem, then its b-**variable relaxation** is a SAT problem $\mathcal{F}^b = \{(c_i \vee b_i) : c_i \in soft(\mathcal{F})\} \cup hard(\mathcal{F})$ where all clause weights are removed. The b-variable b_i appears in the relaxed clause $(c_i \vee b_i)$ and **no where else** in $\mathcal{F}^b$.*

Each truth assignment π to the variables of $\mathcal{F}^b$ has a cost $bcost(\pi)$: if $\pi \not\models \mathcal{F}^b$ then $bcost(\pi) = \infty$, otherwise $bcost(\pi) = \sum_{b_i : \pi \models b_i} wt(c_i)$. The minimum $bcost$ satisfying assignments for $\mathcal{F}^b$ correspond to solutions of $\mathcal{F}$.

Proposition 1. *$mincost(\mathcal{F}) = \min_\pi bcost(\pi)$, where the minimum is taken over all truth assignments π to the variables of $\mathcal{F}^b$. Furthermore, if π achieves a minimum value of $bcost(\pi)$, then π restricted to the variables of $\mathcal{F}$ is a solution for $\mathcal{F}$.*

---

[2] Only integer clause weights are used in our experiments since most MAXSAT solvers require this restriction.

[3] Clauses with weight zero can be removed from $\mathcal{F}$ without impacting the solution. Clauses with negative weight yield a different problem from MAXSAT.

The observation behind the proposition is that for π to achieve a minimum value of $bcost(\pi)$ it must set b_i to *false* whenever it satisfies the soft clause c_i.

MIP Encoding: It is simple to encode a MAXSAT instance $\mathcal{F}$ as a MIP[4]. First, the clauses of the relaxed formula $\mathcal{F}^b$ are encoded as linear inequalities, using the standard method where a clause c is converted to the linear inequality $\sum_{j:p_j \in c} p_j + \sum_{i:\neg p_i \in c}(1 - p_i) \geq 1$. For example, the clause $(x \vee y \vee \neg z \vee b_1)$ becomes the linear inequality $x + y + (1 - z) + b_1 \geq 1$. Second, the objective function is to minimize $\sum_i wt(c_i) \times b_i$. The MIP thus tries to set the propositional variables so as to satisfy all clauses of $\mathcal{F}^b$ with minimum $bcost$.

Assumption Reasoning: The SAT solver MINISAT provides an **assumption** interface to test whether a given set of literals can be extended to a satisfying assignment. MINISAT can take as input a set of assumptions $\mathcal{A}$, specified as a set of literals, along with a CNF formula $\mathcal{F}$ and then determine if $\mathcal{F} \wedge \mathcal{A}$ is satisfiable. It will return a satisfying truth assignment for $\mathcal{F} \wedge \mathcal{A}$ if one exists (this truth assignment necessarily extends $\mathcal{A}$). Otherwise it will report UNSAT and return a learnt clause c which is a disjunction of negated literals of $\mathcal{A}$. This clause has the property that $\neg c$ specifies a subset of $\mathcal{A}$ such that $\mathcal{F} \wedge \neg c$ is unsatisfiable. This means $\mathcal{F} \models c$.

2.1 Existing MAXSAT Solvers

There have been two main approaches to building MAXSAT solvers. The first approach is to perform Branch and Bound search where a lower bound is computed by exploiting the logical structure of the CNF input, e.g., [9,14]. The second approach is to solve the MAXSAT problem as a sequence of SAT problems.

In previous work these SAT problems typically encode the decision problem: "$mincost(\mathcal{F}) = k$?". This encoding is based on adding blocking variables to the soft clauses, and then translating linear inequality constraints over the blocking variables to CNF.[5] Starting with $k = 0$, if the answer from the SAT solver is "no" (i.e., the formula is unsatisfiable), the next lowest possible value for k, k^+, is computed from information extracted from the core returned by the SAT solver. Then the decision problem $mincost(\mathcal{F}) = k^+$ is encoded as the next SAT problem to be solved. The previously computed cores are also exploited in this decision problem by requiring that at least one clause from every previously extracted core is falsified. Many variations on this concept have been recently proposed [2,10,1]. The main disadvantage of these approaches is that as the k gets larger, the SAT instances that need to be solved become larger and harder.

MAXHS: The MAXHS solver attempts to reduce the burden placed on the SAT solver by also employing a MIP solver (CPLEX) [7]. MAXHS's algorithm also involves solving a sequence of SAT problems, however, these SAT problems are always subsets of the original MAXSAT formula $\mathcal{F}$ and are thus usually easy for

---

[4] The origins of this encoding are not clear. However, it is well known.

[5] Some solvers, notably WBO [15], reason with these linear constraints directly instead of converting them to CNF.

the SAT solver to refute. MAXHS uses the assumptions mechanism of MINISAT to test subsets of $\mathcal{F}$ and derive cores. MINISAT is given the formula $\mathcal{F}^b$ and some setting of the b-variables as the assumptions. If MINISAT returns UNSAT, a clause $c = (b_{i_1} \vee \cdots \vee b_{i_k})$ such that $\mathcal{F}^b \models c$ will also be returned. Note that c will only contain positive b-variables since the b-variables only appear positively in $\mathcal{F}^b$ and thus no clause involving negative b-variables is entailed by $\mathcal{F}^b$. It is easy to see that the clause c corresponds to a core of $\mathcal{F}$.

Proposition 2. *If* $\mathcal{F}^b \models (b_{i_1} \vee \cdots \vee b_{i_k})$ *for some set of* b-*variables* $\{b_{i_j}\}|_{j=1}^k$, *then* $\kappa = \{c_{i_1}, ..., c_{i_k}\}$ *is a core of* $\mathcal{F}$. *We call* $(b_{i_1} \vee \cdots \vee b_{i_k})$ *a **core constraint**.*

Starting with an initial set of core constraints in the MIP model (Sec. 5.2), CPLEX is used to find a solution to them that minimizes the cost of the *true* b-variables. The CPLEX solution (a setting of the b-variables) is then given to MINISAT as the next set of assumptions. If MINISAT finds a satisfying solution π^b then π (its restriction to the variables of $\mathcal{F}$) is an optimal solution for $\mathcal{F}$. Otherwise MINISAT will return another core constraint that is added to the CPLEX model and the cycle is repeated.

The problem that CPLEX solves at each iteration can be interpreted as a hitting set problem: find a minimum cost collection of soft clauses sufficient to block all of the refutations (cores) that have been derived from $\mathcal{F}$ so far. In fact, the MAXHS approach is closely related to the implicit hitting set (IHS) problem as described in [12,6]. In IHS problems one is trying to compute a minimum cost hitting set without knowing ahead of time the collection of sets that need to be hit. Instead, one is provided with an oracle that when given the current candidate hitting set, either declares the candidate to be a correct hitting set or returns a new un-hit set from the implicit collection. In the MAXHS algorithm, the cores of $\mathcal{F}$ form the collection of sets to be hit, CPLEX computes candidate hitting sets, and the SAT assumption test acts as the oracle deciding if the current candidate hitting set is correct, returning a new un-hit core if it is not. However, the SAT assumption test may take exponential time, while the oracle in IHS is assumed to run in polynomial time.

The disadvantage of the MAXHS approach is that sometimes a large number of iterations have to be performed during which CPLEX returns different hitting sets. Each of these hitting sets must be ruled out by another core, which increases the size of the MIP model.

3   Empirical Evaluation of Current MAXSAT Solvers

We performed an empirical study of nine existing MAXSAT solvers: CPLEX (version 12.2), WPM1 (with the latest 2012 improvements [1]), WPM2 (version 2 [2]), BINCD [10], WBO [15], MINIMAXSAT [9], SAT4J [4], AKMAXSAT [13], and MAXHS-Orig [7]. All of these solvers are able to solve MAXSAT in its most general form, i.e., weighted partial MAXSAT, and thus have the widest range of applicability. Our study included recently developed solvers utilizing a sequence of SAT approach (BINCD, WPM1, WPM2 and MAXHS-Orig), some older solvers (SAT4J and

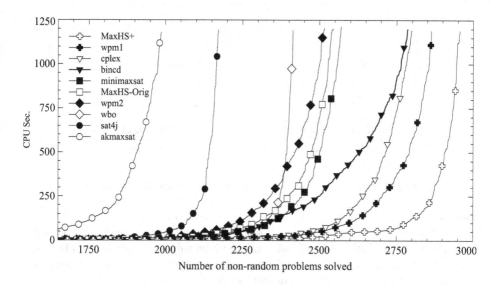

Fig. 1. Performance of solvers on all non-random problems

WBO), and two prominent Branch and Bound based solvers (AKMAXSAT and MINIMAXSAT). Also included was the MIP solver CPLEX using the encoding of MAXSAT specified in Sec. 2, and our original hybrid solver, MAXHS-Orig. We also compared against our newly developed solver MAXHS+. MAXHS+ extends the original MAXHS-Orig using the best overall combination of our newly developed techniques described in Sec. 4.

We obtained **all** problems from the previous seven MAXSAT evaluations [3]. We first discarded all instances in the Random category. After removing duplicate problems (as many as we could find) we ended up with 4502 problems divided up into 58 families. We then removed 17 of these families that in our judgement had little practical application. These included random problems, graph problems on random graphs, e.g., the maxcut, maxclique, "frb" and "kbtree" families, and pure math problems, e.g., the Ramsey and spin glass problems.

The remaining 3826 problems either fell into the "industrial" category or were problems that we felt had application to real problems. For example, MAXSAT has applications in automated planning [17], so we kept the crafted planning problems. Similarly, the "KnotPipatsrisawat" problems involve computing MPE (most probable explanation) which is heavily used in areas like computer vision. When in doubt we erred on the side of keeping the problems, as we are in general interested in applying MAXSAT solvers as widely as possible. It should be noted that our evaluation used many more non-random problems than any previously reported evaluation (including the prior MAXSAT evaluations).

Figure 1 shows a cactus plot of the solvers running on the 3826 non-random problem we kept. Our experiments were performed on 2.1 GHz AMD Opteron machines with 98GB RAM shared between 24 cores (about 4GB RAM per core).

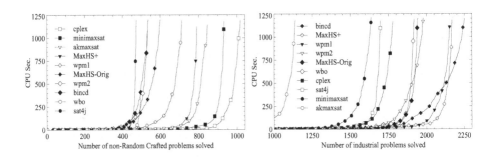

Fig. 2. Performance of solvers on Crafted and Industrial problems

Each problem was run under a 1200 sec. timeout and with a memory limit of 2.5GB. The data shows that our new solver MAXHS+ solved the most problems and that it significantly outperforms our previous solver MAXHS-Orig. The data also shows that BINCD, WPM1, and CPLEX have good performance in terms of the number of problems solved. The performance of CPLEX is particularly noteworthy. Although MIP solvers have been widely available for some time (before most MAXSAT solvers) very little has previously been reported about their performance on MAXSAT. Our data shows that CPLEX is a surprisingly good MAXSAT solver.

Figure 2 shows a break down between industrial and non-random crafted problems. It should be noted, however, that despite the labeling the non-random crafted problems also contain problems of practical (industrial) interest. The data shows that BINCD solves the most problems from the industrial class, with our new solver MAXHS+ and WPM1 having similar but not as good performance on these problems. On these problems, which tend to involve large CNF formulas, CPLEX does not perform as well. On the crafted problems the Branch and Bound solver MINIMAXSAT performed well, but interestingly CPLEX was best overall.

From this data we select MAXHS+, CPLEX, MINIMAXSAT, BINCD, and WPM1 as being in our class of top performers. These solvers dominate the others either on all problems, or on the industrial problems, or on the non-random crafted problems. Hence, we restrict our further attention to these solvers.

One bias in counting the number of problems solved is that the problem families are not equally sized. Hence, this metric will be skewed if a solver is good at solving a particular family and that family contains many problems. Table 1 shows for each of the top solvers and problem categories, the number of problems solved (from the cactus plots), the family score, and the number of families the solver was the best on. The family score for solver s is the sum over all families f of the percentage of problems in f that s solves (this attempts to normalize for the size of the family). There are 41 families in the category **All**, 22 in **Industrial**, and 19 in **Crafted**, so these are the maximum possible family scores. A solver s is best on a family f if it solves as many problems in f as any of the other top solvers. (There can be more than one solver best on a family.)

Table 1. The number of instances solved, Family Scores, and number of Families where each solver is best categorized by all, industrial, and crafted (non-random) problems

Solver	All			Industrial			Crafted		
	Solved	F-Score	F-Best	Solved	F-Score	F-Best	Solved	F-Score	F-Best
MAXHS+	**2956**	**26.75**	**20**	2165	13.98	9	791	12.77	**11**
WPM1	2863	25.92	13	2152	14.68	9	711	11.24	4
CPLEX	2798	25.69	17	1779	11.70	7	**1019**	**13.98**	10
BINCD	2785	25.38	12	**2251**	**14.97**	7	534	10.41	5
MINIMAXSAT	2570	22.80	13	1637	9.45	2	933	13.35	**11**

Table 1 shows, e.g., that although BINCD solves more problems in the industrial category and has the highest family score, it is best on fewer families than WPM1 and MAXHS+. It also shows that although WPM1 solves almost as many crafted problems as MAXHS+ it is best on fewer families.

Finally, Table 2 shows that each of these top solvers has quite a diverse coverage. The table shows for each pair of top solvers s and s' how many problems s solves that s' fails to solve. This number is shown in the cell at row s and column s'. In fact the table contains two number in each cell. The first is the number of industrial problems s solves but s' doesn't while the second number is the number of crafted problems s solves but s' doesn't. This metric is influenced by the timeout, as s might have solved a problem in 1200 seconds that s' would have solved in 1210 seconds if it hadn't timed out. To avoid this issue, to count a problem p as solved by s and not solved by s' we require that s solves p in less than 600 seconds while s' fails to solve p.

The data shows, e.g., that on this metric CPLEX dominates MINIMAXSAT, solving more problems that MINIMAXSAT fails to solve in both the industrial and crafted categories. Similarly MAXHS+ dominates WPM1 on this metric, and BINCD dominates all other solvers on this metric for industrial problems. The main message from this data, however, is that all of these solvers dominates each other solver on some problems (typically a non-trivial number of problems).

Table 2. The entry (nI,nC) located at row i and column j shows the number of industrial problems nI (crafted problems nC) solved by solver i within 600 sec. that j fails to solve

	MAXHS+	WPM1	CPLEX	BINCD	MINIMAXSAT
MAXHS+		**191/208**	420/19	61/**264**	**547**/37
WPM1	171/121		481/27	89/**235**	**636**/120
CPLEX	69/**234**	128/**321**		52/**494**	351/178
BINCD	**124**/19	**169**/63	**483**/19		**544**/22
MINIMAXSAT	18/**169**	115/**342**	192/97	11/**410**	

4   Exploiting CPLEX More Effectively

The MAXHS algorithm decomposes the MAXSAT problem into a series of SAT problems and hitting set problems. Neither the SAT solver nor CPLEX alone has enough information to solve the entire MAXSAT problem, since the SAT solver does not have any information about the clause weights, and the CPLEX model, which is only over b-variables, knows nothing about the original variables and clauses. The CPLEX model is also restricted to constraints of a specific form, i.e. core constraints which are clauses over positive b-variables. In the remainder of the paper we propose several techniques to overcome these limitations, in order to take better advantage of the MIP solver.

Many sound constraints exist over the soft clauses that do not take the form of core constraints, as illustrated by the following example.

Example 1. Let $\mathcal{F} = \{(x), (\neg x), (x \lor y), (\neg y), (\neg x \lor z), (\neg z \lor y)\}$ where each clause has weight 1. $\mathcal{F}^b$ is therefore the set of clauses $\{(b_1 \lor x), (b_2 \lor \neg x), (b_3 \lor x \lor y), (b_4 \lor \neg y), (b_5 \lor \neg x \lor z), (b_6 \lor \neg z \lor y)\}$. Suppose that the three cores $\kappa_1 = \{(x), (\neg x)\}$, $\kappa_2 = \{(\neg x), (x \lor y), (\neg y)\}$, and $\kappa_3 = \{(x \lor y), (\neg y), (\neg x \lor z), (\neg z \lor y)\}$ have been found. These cores correspond to the core constraints $\mathcal{K} = \{(b_1 \lor b_2), (b_2 \lor b_3 \lor b_4), (b_3 \lor b_4 \lor b_5 \lor b_6)\}$. We see that to satisfy these core constraints at least two b-variables in $\mathcal{F}^b$ must be set to *true*, and at least two soft clauses will be falsified by the MAXSAT solution. Given this lower bound, we can use a SAT solver to search over truth assignments that assign at most two b-variables *true*, looking for a cost-2 relaxation that satisfies $\mathcal{F}^b$. The search will benefit from the three core constraints, since they help to prune the search space. However, not all cost-2 relaxations that satisfy the core constraints need to be considered. For example, as soon as b_1 is assigned on any branch, $\neg b_2$ could be inferred because it is impossible to falsify both (x) and $(\neg x)$ at the same time. Therefore, b_1 and b_2 can not both belong to a minimum cost relaxation. Unfortunately, unit propagation in $\mathcal{F}^b \cup \mathcal{K}$ can not make this inference. Similarly, whenever $\neg b_2$ is assigned we obtain $\neg x$ and b_1 by unit propagation in $\mathcal{F}^b \cup \mathcal{K}$. However, we do not detect that $\neg b_5$ must hold as well since its soft clause is now satisfied. These two examples demonstrate that in addition to the core constraints $\mathcal{K}$, $\mathcal{F}$ also implies the constraints $(\neg b_1 \lor \neg b_2)$ and $(b_2 \lor \neg b_5)$. If these constraints could be discovered automatically, then the search over relaxations could be further constrained and potentially made more efficient.

In [7] a realizability condition was introduced. Realizability requires that there exists a truth assignment that falsifies all of the clauses in the hitting set and satisfies the hard clauses. This condition can be used to enforce some non-core constraints over the b-variables. However, it is insufficient to capture all constraints over the b-variables. For example, although the realizability condition would enforce the first non-core constraint in Example 1, $(\neg b_1 \lor \neg b_2)$, it would not capture the second, $(b_2 \lor \neg b_5)$. Therefore, we must look beyond the realizability condition for techniques to discover non-core constraints that the b-variables must satisfy.

4.1 b-Variable Equivalences

Relaxing a soft clause in $\mathcal{F}^b$ is not equivalent to falsifying it in $\mathcal{F}$. Example 1 indicates that although the b-variables of $\mathcal{F}^b$ are intended to represent the soft clauses of $\mathcal{F}$ this correspondence is not strictly enforced by $\mathcal{F}^b$. That is, $\mathcal{F}^b$ admits models that unnecessarily set b-variables to *true* even when the corresponding soft clause is satisfied. This is the reason that the inference $\neg b_2 \rightarrow \neg b_5$ was missed in Example 1.

Proposition 1 shows, however, that minimum cost models of $\mathcal{F}^b$ do obey a stricter correspondence of equivalence between the b-variable settings and the soft clauses satisfied. Since MAXSAT solving involves searching for minimum cost models, a natural and simple modification to $\mathcal{F}^b$ is to force the b-variables to be equivalent to the negation of their corresponding soft clauses.

Definition 2. *Let $\mathcal{F}$ be a MAXSAT formula. Then*

$$\mathcal{F}^b_{eq} = \mathcal{F}^b \cup \bigcup_{c_i \in soft(\mathcal{F})} \{(\neg b_i \vee \neg \ell) : \ell \in c_i\}$$

is the relaxation of $\mathcal{F}$ with b-variable equivalences.

We define a correspondence between the truth assignments for $\mathcal{F}$ and the truth assignments for $\mathcal{F}^b_{eq}$.

Definition 3. *If π is a truth assignment to the variables of $\mathcal{F}$ we let π^b denote its corresponding truth assignment to the variables of $\mathcal{F}^b_{eq}$, where*

$$\pi^b = \pi \cup \{\neg b_i : \pi \models c_i, c_i \in soft(\mathcal{F})\} \cup \{b_i : \pi \not\models c_i, c_i \in soft(\mathcal{F})\}.$$

If π^b is a truth assignment to the variables of $\mathcal{F}^b_{eq}$ we let π denote its corresponding truth assignment to the variables of $\mathcal{F}$ where π is simply π^b restricted to the variables of $\mathcal{F}$.

In this definition π^b is constructed so that it assigns each b-variable a truth value equivalent to the negation of the truth value π assigns to the corresponding soft clause. Thus π^b models the b-variable equivalences. Under this correspondence we obtain a 1-1 correspondence between the models of $\mathcal{F}^b_{eq}$ and the models of $hard(\mathcal{F})$.

Proposition 3. $\pi \models hard(\mathcal{F})$ *if and only if $\pi^b \models \mathcal{F}^b_{eq}$. Furthermore, if $\pi^b \models \mathcal{F}^b_{eq}$ then $cost(\pi) = bcost(\pi^b)$, and therefore π is a solution for the MAXSAT formula $\mathcal{F}$ if and only if π^b achieves minimum bcost over all satisfying truth assignments for $\mathcal{F}^b_{eq}$.*

This proposition shows that we can solve the MAXSAT problem $\mathcal{F}$ by searching for a *bcost* minimal satisfying assignment to $\mathcal{F}^b_{eq}$.

Algorithm 1. A MAXSAT algorithm that exploits non-core constraints

---

1 MAXSAT-**solver** $(\mathcal{F})$
2 $\mathcal{K} = \emptyset$
3 $obj = wt(c_i) * b_i + \ldots + wt(c_k) * b_k$
4 **while** *true* **do**
5       $\mathcal{A} = \text{Optimize}(\mathcal{K}, obj)$
6       $(\text{sat?}, \kappa) = \text{AssumptionSatSolver}(\mathcal{F}_{eq}^b, \mathcal{A})$
         // If SAT, κ contains the satisfying truth assignment.
         // If UNSAT, κ contains a clause over b-variables.
7   **if** *sat?* **then**
8   **break** // Exit While Loop, κ is a MAXSAT solution.
     // Add new constraint to the optimization problem,
9       $\mathcal{K} = \mathcal{K} \cup \{\kappa\}$
     // and to the SAT formula for better performance
10      $\mathcal{F}_{eq}^b = \mathcal{F}_{eq}^b \cup \{\kappa\}$
11 **return** $\big(\kappa,\ cost(\kappa)\big)$

4.2   Non-core Constraints in MAXHS

The extension to utilize non-core constraints in MAXHS is conceptually simple. We simply substitute the encoding $\mathcal{F}_{eq}^b$ for the weaker encoding $\mathcal{F}^b$. Now since in $\mathcal{F}_{eq}^b$ the b-variables are no longer pure, the SAT solver can return both core and non-core constraints. Each constraint is passed to CPLEX which operates as before. (A copy of the learnt constraint is also kept by the SAT solver). The resulting modified version of MAXHS is shown in Algorithm 1.

Initially, the set of b-variable constraints (clauses), $\mathcal{K}$, is empty (line 2). The objective function is defined on line 3 as the sum of the clause weights for b-variables that are assigned *true*. On line 5, an assignment to the b-variables, $\mathcal{A}$, is calculated that satisfies the current constraints $\mathcal{K}$ and minimizes the value of the objective function obj. This setting of the b-variables is passed as the set of assumptions to the SAT solver on line 6, along with the SAT instance $\mathcal{F}_{eq}^b$. If the SAT solver returns UNSAT, κ will be a clause over negated literals from $\mathcal{A}$. This constraint is added to $\mathcal{K}$ on line 9 and the process iterates until the SAT solver reports a solution.

Theorem 1. *Algorithm 1 returns a solution to the MAXSAT problem $\mathcal{F}$.*

Proof. First we observe if the κ returned by the SAT solver at line 6 is a clause then $\mathcal{F}_{eq}^b \models \kappa$ (as explained in Section 2). On the other hand, if κ is a satisfying assignment then $bcost(\kappa)$ is equal to the sum of the costs of the *true* b-variables in $\mathcal{A}$, the solution returned by the optimizer at line 5. This follows from the fact that κ extends $\mathcal{A}$ which has already set all of the b-variables. Let κ be the satisfying truth assignment causing the algorithm to terminate. All satisfying assignments of $\mathcal{F}_{eq}^b$ satisfy the constraints in $\mathcal{K}$ as each of these is entailed by $\mathcal{F}_{eq}^b$. Furthermore, $bcost(\kappa)$ is equal to the cost of an optimal solution to these constraints, thus κ achieves minimal *bcost* over all satisfying truth assignments for $\mathcal{F}^b_{eq}$, and by Proposition 3 κ restricted to the variables of $\mathcal{F}$ is a MAXSAT solution for $\mathcal{F}$.

Second, we observe that each iteration except the final one adds a constraint to $\mathcal{K}$ that eliminates at least one more setting of the *b*-variables. Since there are only a finite number of different settings, the algorithm must eventually terminate. ∎

The key difference with the original MAXHS algorithm is that the optimizer no longer deals with a pure hitting set problems as the constraints can now contain negative *b*-variables. This means that the paradigm of MAXHS changes from an implicit hitting set problem to something like a logic based Benders decomposition approach [11]. In particular, the optimization problem is being solved only over the *b*-variables while the SAT solver is being used to add additional constraints to the optimization model until its solution also satisfies the feasibility conditions. Although CPLEX is no longer solving a hitting set problem, we have found that it remains very effective in the presence of non-core constraints.

5 Other Improvements

In addition to the ability to learn non-core constraints, we propose two additional techniques that help to refine the constraints and exploit the strength of CPLEX.

5.1 Constraint Minimization

The first improvement is to more aggressively minimize the constraints before adding them to the CPLEX model. In general, shorter clausal constraints are stronger, so the quality of the constraints can be improved by using techniques to minimize their length. Therefore, we ensure that the constraints we add to CPLEX are minimal, in the sense that removing any literal from the clausal constraint leaves a clause that is no longer entailed by $\mathcal{F}^b_{eq}$. We use a simple destructive MUS algorithm, as described in [16], to achieve this. Empirically, we found that the minimization computation typically takes only a small percentage of the solver's runtime, so more sophisticated MUS algorithms are unlikely to yield a significant benefit in the current solver.

5.2 Disjoint Phase

Similar to the original MAXHS solver, we also use a disjoint core phase before Algorithm 1 begins to supply CPLEX with an initial set of core constraints. During this phase we run the SAT solver on $\mathcal{F}^b$ (rather than $\mathcal{F}^b_{eq}$) so that only core constraints are derived. Initially, the SAT solver is run under the assumption that all *b*-variables are *false*. This generates a core constraint (unless the MAXSAT problem has a solution of zero cost). After minimizing that constraint we run the SAT solver again under the assumption that all of the *b*-variables in the core constraints found so far are *true* and all other b-variables are *false*. This has the effect of removing all soft clauses that have participated in cores from the theory. Hence, the next core must be over a disjoint set of soft clauses. This process is repeated setting more and more of the b-variables to *true*, until the SAT solver can no longer find a contradiction. The collection of cores found are all disjoint and the corresponding linear constraints are initially added to CPLEX.

5.3   Seeding CPLEX with Constraints

Each call to CPLEX's solve routine incurs some overhead so it is desirable to reduce the number of calls to CPLEX. We propose to accomplish this by "seeding" the CPLEX model with a number of initially computed b-variable constraints. In this way each candidate solution (setting of the b-variables) returned by CPLEX is more informed about the constraints of the problem and thus more likely to be a true solution. We perform seeding after the disjoint core phase, but before the iterations of Algorithm 1 begin. We now describe several techniques to cheaply identify such additional b-variable constraints.

Eq-Seeding: In $\mathcal{F}_{eq}^b$, literals that appear in soft *unit* clauses of $\mathcal{F}$ are actually logically equivalent to their b-variables. To see this, recall that if $c_i = (x) \in soft(\mathcal{F})$ is a soft unit clause, then $\mathcal{F}_{eq}^b$ will contain clauses $(x \vee b_i)$ and $(\neg b_i \vee \neg x)$. These two clauses together imply that $b_i \equiv \neg x$. For a clause c of $\mathcal{F}^b$, if each variable in c has an equivalent b-variable (or is itself a b-variable), then we can derive a new constraint from c by replacing every original variable by its equivalent b-variable. This constraint is a clause over the b-variables that can be added to CPLEX.

Example 2. In Example 1, $b_1 \equiv \neg x$ due to the soft unit clause (x) and its relaxation by b_1. Similarly, $b_4 \equiv y$. Therefore, from the relaxed clause $(b_3 \vee x \vee y) \in \mathcal{F}^b$ we can obtain the b-variable constraint $(b_3 \vee \neg b_1 \vee b_4)$ by simply substituting the equivalent b-variable literals for the original literals.

Imp-Seeding: In $\mathcal{F}_{eq}^b$, each of the b-literals may imply other b-literals. We perform a trial unit propagation on each b-literal b_i in order to collect a set of implied b-literals $imp(b_i) = \{b_i^1, ..., b_i^k\}$. This represents a conjunction of k binary clauses $b_i \rightarrow b_i^j$ $(1 \leq j \leq k)$ over the b-variables. Although these k clauses could be individually added to CPLEX we can in fact encode their conjunction in a single linear constraint that can be given to CPLEX:

$$-k \times b_i + b_i^1 + \cdots + b_i^k \geq 0$$

Note that these are b-literals, so as is standard a negative literal b is encoded as $(1 - var(b))$ and a positive literal is encoded as $var(b)$ (Sec. 2). To understand this constraint note that if b_i is *true* (equal to 1) then all of the b_i^j variables must be 1 to make the sum non-negative.

Imp+Rev-Seeding: During the trial unit propagation of each b-literal b_i, we can also keep track of every original literal that is found to be implied by b_i in order to obtain sets of reverse implications: $rev(x) \subseteq \{b_i : \mathcal{F}_{eq}^b \wedge b_i \models x\}$. Then, for each clause $c_i \in \mathcal{F}^b$, we check if each of its original literals $x \in c_i$ has a non-empty $rev(\neg x)$. If so, a b-literal $b_{\neg x} \in rev(\neg x)$ is chosen for each x and its negation $\neg b_{\neg x}$ is substituted for x in c_i. The result is a new clause containing only b-variables that can be added to CPLEX. It is easy to see that this clause is sound by considering the following example.

Example 3. Suppose that $(x \vee y \vee b_1) \in \mathcal{F}^b$ where x, y are original literals and b_1 is a blocking variable. Suppose that $b_{\neg x} \in rev(\neg x)$ and $b_{\neg y} \in rev(\neg y)$. This means that clauses $(\neg b_{\neg x} \vee \neg x)$ and $(\neg b_{\neg y} \vee \neg y)$ are implied by $\mathcal{F}_{eq}^b$. Therefore $(\neg b_{\neg x} \vee \neg b_{\neg y} \vee b_1)$, which can be obtained in two resolution steps, is also implied by $\mathcal{F}_{eq}^b$ and can be added to CPLEX.

Since the b-literal implications $imp(b_i)$ are also available, we add the Imp-Seeding constraints as well if we are computing the Rev-Seeding constraints. Note that if $b \equiv x$, as in Eq-Seeding, we obtain at least as many seeded constraints as would be obtained by Eq-Seeding. If $rev(\neg x)$ contains more than one b-literal, we could choose any one of them to form the new clause. We simply use an equivalent b-literal if one exists, and otherwise we choose the first b-literal that was found to imply $\neg x$. In future work we could investigate different ways of choosing the member of $rev(\neg x)$, or methods for using them all.

6   Empirical Evaluation of Proposed MAXHS Improvements

In this section we examine the empirical behaviour of the improvements to the MAXHS algorithm proposed above. Our experiments with the original MAXHS algorithm showed that it spent most of its time in the MIP solver and relatively little in the SAT solver. In the improved versions of MAXHS, the MIP solver still dominates the CPU time. However, seeding and our other techniques provide better information to CPLEX, which means that fewer calls are required to converge on a solution.

We ran a number of different versions of MAXHS on the problems described in Sec. 3. Figure 3 shows a cactus plot of their performance running on all non-random problems. The data shows a number of things. First, adding core minimization (+min on the plot) yields a significant performance gain compared to the original MAXHS solver. When we add to this version the ability to generate non-core constraints via the $\mathcal{F}_{eq}^b$ relaxation (+noncore), there is another jump in the number of problems solved.

If we seed CPLEX with extra constraints (Eq, Imp and Imp+Rev Seeding) in addition to the previous two techniques, performance improves again. However, there is relatively little to choose in overall performance between the different types of CPLEX seeding we developed. When we looked at the time taken to solve various problems we did find that on some instances the more extensive seeding (Imp+Rev) yields a factor of 10 improvement in solving time. However, on some problems such seeding adds a very large number of additional constraints to

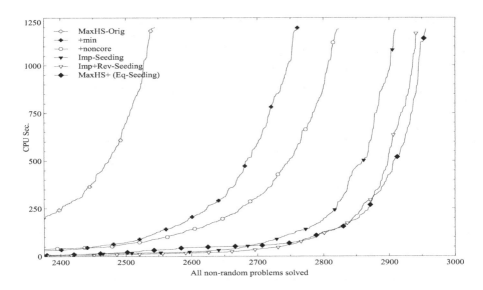

Fig. 3. Performance of MAXHS variants on all non-random problems

CPLEX, without much improvement in the solution candidates produced. These extra constraints sometimes produce an increase in CPLEX's runtime sufficient to cause a time-out. Future work will require examining particular families or problem instances to obtain understanding of the trade-offs involved with our different levels of seeding.

MAXHS+ The seeding method with a slight advantage in terms of overall number of problems solved is Eq-Seeding. We referred to the configuration that uses Eq-Seeding (as well as non-core constraints, minimization, etc.) as MAXHS+ throughout the paper.

7   Conclusion

We made two main contributions in this paper. First we have reported on the results of an extensive evaluation of current MAXSAT solvers. These results provide a number of insights into current solvers. Second, inspired by one of these insights, that the MIP solver CPLEX is more effective than expected, we developed a number of new techniques aimed at enhancing our previously developed hybrid MAXSAT solver MAXHS. These techniques were mainly aimed at improving the information given to CPLEX so as to better exploit it.

   In future work we aim to find out if some of the techniques used in other solvers, e.g., the binary search used in BINCD and the weight stratification method used in WPM1, which help them solve a number of distinct problem instances, can be exploited in our framework. We also plan to investigate the trade-offs we observed with the different types of seeding in more detail.

References

1. Ansótegui, C., Bonet, M.L., Gabàs, J., Levy, J.: Improving SAT-based weighted maxSAT solvers. In: Milano, M. (ed.) CP 2012. LNCS, vol. 7514, pp. 86–101. Springer, Heidelberg (2012)
2. Ansótegui, C., Bonet, M.L., Levy, J.: A new algorithm for weighted partial maxsat. In: Proceedings of the AAAI National Conference (AAAI), pp. 3–8 (2010)
3. Argelich, J., Li, C.M., Manyà, F., Planes, J.: The maxsat evaluations (2007-2011), http://www.maxsat.udl.cat
4. Berre, D.L., Parrain, A.: The sat4j library, release 2.2. JSAT 7(2-3), 59–64 (2010)
5. Buss, S.R.: Lectures on proof theory. Tech. rep. (1996), http://www.math.ucsd.edu/~sbuss/ResearchWeb/Barbados95Notes/reporte.pdf
6. Chandrasekaran, K., Karp, R., Moreno-Centeno, E., Vempala, S.: Algorithms for implicit hitting set problems. In: Proceedings of the Symposium on Discrete Algorithms (SODA), pp. 614–629 (2011)
7. Davies, J., Bacchus, F.: Solving MAXSAT by solving a sequence of simpler SAT instances. In: Lee, J. (ed.) CP 2011. LNCS, vol. 6876, pp. 225–239. Springer, Heidelberg (2011)
8. Eén, N., Sörensson, N.: An extensible SAT-solver. In: Giunchiglia, E., Tacchella, A. (eds.) SAT 2003. LNCS, vol. 2919, pp. 502–518. Springer, Heidelberg (2004)
9. Heras, F., Larrosa, J., Oliveras, A.: Minimaxsat: An efficient weighted max-sat solver. Journal of Artificial Intelligence Research (JAIR) 31, 1–32 (2008)
10. Heras, F., Morgado, A., Marques-Silva, J.: Core-guided binary search algorithms for maximum satisfiability. In: Proceedings of the AAAI National Conference (AAAI), pp. 36–41 (2011)
11. Hooker, J.N.: Planning and scheduling by logic-based benders decomposition. Operations Research 55(3), 588–602 (2007)
12. Karp, R.M.: Implicit hitting set problems and multi-genome alignment. In: Amir, A., Parida, L. (eds.) CPM 2010. LNCS, vol. 6129, p. 151. Springer, Heidelberg (2010)
13. Kügel, A.: Improved exact solver for the weighted Max-SAT problem. In: Workshop on the Pragmatics of SAT (2010)
14. Li, C.M., Manyà, F., Mohamedou, N.O., Planes, J.: Resolution-based lower bounds in maxsat. Constraints 15(4), 456–484 (2010)
15. Manquinho, V., Marques-Silva, J., Planes, J.: Algorithms for weighted boolean optimization. In: Kullmann, O. (ed.) SAT 2009. LNCS, vol. 5584, pp. 495–508. Springer, Heidelberg (2009)
16. Marques-Silva, J., Lynce, I.: On improving MUS extraction algorithms. In: Sakallah, K.A., Simon, L. (eds.) SAT 2011. LNCS, vol. 6695, pp. 159–173. Springer, Heidelberg (2011)
17. Zhang, L., Bacchus, F.: Maxsat heuristics for cost optimal planning. In: Proceedings of the AAAI National Conference (AAAI) (2012)

Community-Based Partitioning for MaxSAT Solving[*]

Ruben Martins, Vasco Manquinho, and Inês Lynce

IST/INESC-ID, Technical University of Lisbon, Portugal
{ruben,vmm,ines}@sat.inesc-id.pt

Abstract. Unsatisfiability-based algorithms for Maximum Satisfiability (Max-SAT) have been shown to be very effective in solving several classes of problem instances. These algorithms rely on successive calls to a SAT solver, where an unsatisfiable subformula is identified at each iteration. However, in some cases, the SAT solver returns unnecessarily large subformulas. In this paper a new technique is proposed to partition the MaxSAT formula in order to identify smaller unsatisfiable subformulas at each call of the SAT solver. Preliminary experimental results analyze the effect of partitioning the MaxSAT formula into communities. This technique is shown to significantly improve the unsatisfiability-based algorithm for different benchmark sets.

1 Introduction

Problem partitioning is a well-known technique used for general problem solving and it has already been proposed for Boolean optimization [1,2] formulations. The main goal of partitioning is to identify easier to solve subproblems such that it will help to solve the overall problem.

In recent years, several algorithms and solvers have been proposed for Maximum Satisfiability (MaxSAT). In particular, unsatisfiability-based MaxSAT solvers [3,4,5,6,7] have been shown to be very effective in tackling real-world problems [8]. These solvers are based on iteratively calling a SAT solver enhanced with the ability of providing a certificate of unsatisfiability. However, one drawback of these algorithms results from the SAT solver returning unnecessary large unsatisfiable subformulas as certificates. Instead of dealing initially with the whole formula, we start with a smaller formula that is extended at each iteration of the algorithm. The goal is to initially have smaller formulas that enable the SAT solver to provide smaller certificates of unsatisfiability.

In this paper we propose a new method for formula partitioning in an unsatisfiability-based algorithm for partial MaxSAT. A graph representation of the formula is used and graph communities are identified based on a modularity measure, thus allowing to build partitions of soft clauses in the MaxSAT formula. The paper is organized as follows. The next section introduces MaxSAT and briefly reviews the main approaches for MaxSAT solving. In section 3 an unsatisfiability-based algorithm with partitioning of soft clauses

---

[*] This work was partially supported by FCT under research projects iExplain (PTDC/EIA-CCO/102077/2008) and ASPEN (PTDC/EIA-CCO/110921/2009), and INESC-ID multiannual funding through the PIDDAC program funds (PEst-OE/EEI/LA0021/2011).

M. Järvisalo and A. Van Gelder (Eds.): SAT 2013, LNCS 7962, pp. 182–191, 2013.

is described. Section 4 proposes partition methods for partial MaxSAT, in particular a new approach based on the identification of graph communities. Experimental results are presented in section 5 and the paper concludes in section 6.

2   Preliminaries

The Maximum Satisfiability (MaxSAT) problem is an optimization version of the Propositional Satisfiability (SAT) problem which consists in finding an assignment to the variables of the CNF formula such that the number of unsatisfied (satisfied) clauses is minimized (maximized). In the remainder of the paper, it is assumed that MaxSAT is defined as a minimization problem.

MaxSAT has several variants such as partial MaxSAT, weighted MaxSAT and weighted partial MaxSAT. In a partial MaxSAT formula $\varphi = \varphi_h \cup \varphi_s$, some clauses are declared as hard (φ_h), while the rest are declared as soft (φ_s). The objective in partial MaxSAT is to find an assignment to formula variables such that all hard clauses φ_h are satisfied, while minimizing the number of unsatisfied soft clauses in φ_s. Finally, in the weighted versions of MaxSAT, soft clauses can have weights greater than or equal to 1 and the objective is to satisfy all hard clauses while minimizing the total weight of unsatisfied soft clauses.

2.1   MaxSAT Algorithms

In the last decade, several new techniques and algorithms for MaxSAT have been proposed [9], resulting in significant improvements in MaxSAT solvers. More recently, new algorithms have been devised that are more effective for solving industrial benchmark instances[1], namely linear search on the objective value of the MaxSAT instance and unsatisfiability-based algorithms.

In the linear search approach, a new relaxation variable is initially added to each soft clause and the resulting formula is solved by a SAT solver. Whenever a solution is found, a new constraint on the relaxation variables is added such that solutions with a higher or equal value are excluded. This new constraint is usually translated into a set of propositional clauses so that a SAT solver can handle the resulting formula [10]. Otherwise, a pseudo-Boolean solver must be used. The algorithm stops when the resulting formula becomes unsatisfied.

A different approach is trying to satisfy all hard and soft clauses using a SAT solver enhanced with the ability to produce certificates of unsatisfiability [3]. At each call of the SAT solver, an unsatisfiable subformula is identified and relaxed by adding a new relaxation variable to each soft clause in the unsatisfiable subformula. Additionally, a new constraint is added such that at most one of the new relaxation variables can be assigned value true. Again, in order to continue using a SAT solver, this new constraint must be encoded into a set of propositional clauses. Next, the SAT solver is called with the resulting formula. The algorithm stops when the formula becomes satisfiable. Several extensions of this approach have been proposed for solving MaxSAT and its variants [5,4,6,7].

---

[1] See results from MaxSAT Evaluations at http://maxsat.ia.udl.cat/

Algorithm 1. Unsatisfiability-based algorithm for partial MaxSAT enhanced with partitioning of soft clauses

---

Input: $\varphi = \varphi_h \cup \varphi_s$
Output: satisfiable assignment to φ or UNSAT

1 $(\mathsf{st}, \varphi_C) \leftarrow \mathrm{SAT}(\varphi_h)$ // check if the MaxSAT formula is UNSAT
2 **if** st $=$ UNSAT **then**
3 | **return** UNSAT
4 $\gamma \leftarrow \langle \gamma_1, \dots, \gamma_n \rangle \leftarrow \mathtt{partitionSoft}(\varphi_s)$
5 $\varphi_W \leftarrow \varphi_h$
6 **while** true **do**
7 | $\varphi_W \leftarrow \varphi_W \cup \mathtt{first}(\gamma)$
8 | $\gamma \leftarrow \gamma \setminus \mathtt{first}(\gamma)$
9 | $(\mathsf{st}, \varphi_C) \leftarrow \mathrm{SAT}(\varphi_W)$
10 | **while** st $=$ UNSAT **do**
11 | | $V_R \leftarrow \emptyset$
12 | | **foreach** $\omega \in (\varphi_C \cap \varphi_s)$ **do**
13 | | | $V_R \leftarrow V_R \cup \{r\}$ // r is a new variable
14 | | | $\omega_R \leftarrow \omega \cup \{r\}$ // relax soft clause
15 | | | $\varphi_W \leftarrow \varphi_W \setminus \{\omega\} \cup \{\omega_R\}$
16 | | $\varphi_W \leftarrow \varphi_W \cup \{\mathrm{CNF}(\sum_{r \in V_R} r = 1)\}$
17 | | $(\mathsf{st}, \varphi_C) \leftarrow \mathrm{SAT}(\varphi_W)$
18 | **if** $\gamma = \emptyset$ **then**
19 | | **return** satisfiable assignment to φ_W

---

3 Partition-Based MaxSAT Algorithm

In this section we review an unsatisfiability-based MaxSAT algorithm that takes advantage of partitioning. The original algorithm [2] was proposed for weighted variants of MaxSAT where partitions are built considering the different weights of soft clauses.

Algorithm 1 starts by checking if the MaxSAT instance φ is satisfiable by calling a SAT solver only with hard clauses φ_h. Next, the set of soft clauses φ_s is split into a list of partitions (line 4) such that each soft clause is assigned to one partition. Initially, the working formula only considers the hard clauses φ_h (line 5). At each iteration, a partition γ_i of soft clauses is added to the working formula (line 7) and removed from the partition list γ (line 8). A SAT solver is then applied to φ_W, returning a pair (st, φ_C) where st denotes the outcome of the solver: SAT or UNSAT. While the outcome is UNSAT, φ_C contains the unsatisfiable subformula identified by the SAT solver and the unsatisfiable subformula is relaxed as in the original algorithm [3]. Next, the SAT solver is applied to the modified working formula (line 17). After a given number of relaxations, the working formula becomes satisfiable[2] and a new partition of soft clauses is added to the working formula. If there are no more partitions in γ, then the solver found an optimal solution to the original MaxSAT formula (line 19).

---

[2] Notice that initially we already confirmed that the MaxSAT formula is not unsatisfiable due to the hard clauses. Since at each iteration at least one soft clause is relaxed, the working formula at some point becomes satisfiable.

In Algorithm 1, a partition method must be used to split the set of soft clauses. For weighted variants of MaxSAT, it was already shown that splitting the soft clauses according to its weight allows the solver to be much more effective [11,2]. However, for partial MaxSAT this partition method cannot be used since all soft clauses have weight 1. In the next section we present two graph-based methods for partitioning of soft clauses for partial MaxSAT.

4   Partial MaxSAT Partitioning

It has already been shown that partitioning can greatly boost the performance of unsatisfiability-based solvers for weighted MaxSAT [11,2]. However, using the weight of soft clauses for partitioning is not useful for partial MaxSAT. As a result, other methods based on the structure of the formula must be used. Next, we briefly review the hypergraph partitioning method first proposed for weighted MaxSAT. Afterwards, we propose a new partition method for MaxSAT based on a graph representation of the MaxSAT formula. The new partition method uses the graph representation to identify communities using a modularity measure. The methods described in this section represent different implementations of the `partitionSoft` procedure in line 4 of Algorithm 1.

4.1   Hypergraph Partitioning

Hypergraph partitioning has already been applied to SAT [12], as well as to weighted MaxSAT solving [2]. A hypergraph is a generalization of a graph where an edge, also called hyperedge, can connect any number of vertices. In our case, for each soft and hard clause there is a corresponding vertex in the hypergraph. Moreover, for each formula variable x_j there is an hyperedge connecting all vertices that represent soft or hard clauses containing variable x_j.

After building the hypergraph, the tool `hmetis` [13] is used as a black box to identify the partitions. In the experiments, `hmetis` is configured to identify 16 partitions in each problem instance [2]. Afterwards, for each partition only the soft clauses are considered. As a result, partitions containing only hard clauses are removed.

4.2   Community-Based Partitioning

The identification of communities in SAT instances has been previously proposed [14] and it was shown to be effective in characterizing industrial SAT instances. For that, SAT instances are first represented as undirected weighted graphs and partitions of vertices (communities) are identified using a modularity measure. In this paper we use both graph representations described in [14] for SAT instances, namely the Variable Incidence Graph (VIG) and the Clause-Variable Incidence Graph (CVIG) model. In addition, a different weighting function is proposed in this paper.

Graph Representations

We start by defining an incidence function I on the formula variables x_j in the soft clauses φ_s as follows:

$$I(x_j) = 1 + \sum_{x_j \in \omega \,\wedge\, \omega \in \varphi_s} \frac{1}{|\omega|} \tag{1}$$

Notice that $I(x_j) = 1$ if variable x_j does not occur in any soft clause.

In the Variable Incidence Graph (VIG) model, a graph G is built such that for each variable x_j in the problem instance there is a corresponding vertex in G. Moreover, if x_j and x_k belong to the same clause (hard or soft), then there is an edge between the vertices corresponding to these variables with the following weight:

$$w(x_j, x_k) = \sum_{x_j, x_k \in \omega \,\wedge\, \omega \in \varphi} \frac{I(x_j) \cdot I(x_k)}{\binom{|\omega|}{2}} \tag{2}$$

Observe that if we consider $I(x_j) = 1$ for all variables, then this weight function corresponds to the one proposed in [14], where all clauses are equally relevant. However, for MaxSAT one has to consider both soft and hard clauses. In our graph representation, more weight is given to clauses that establish edges between variables that occur in soft clauses. The motivation is to fortify the relationship between variables that occur in soft clauses.

In the Clause-Variable Incidence Graph (CVIG) model, for each variable x_j and for each clause $\omega_i \in \varphi$, there is a corresponding vertex in graph G. In this model, edges only connect vertices representing a variable and a clause where the variable occurs. Hence, if a variable x_j occurs in clause ω_i, then there is an edge between those vertices with weight:

$$w(x_j, \omega_i) = \frac{I(x_j)}{|\omega_i|} \tag{3}$$

Community Identification

After building a graph representation for the problem instance, we are interested in making explicit the hidden structure of the MaxSAT formula by identifying partitions in the graph. Clearly one can devise many different ways of partitioning. Therefore, it is necessary to evaluate the quality of a given set of partitions.

In recent years, the use of modularity measures became common for the identification of communities in graphs [15,16,17]. The main goal of the modularity measure is to evaluate the quality of the communities in a graph where vertices inside a community should be densely connected, while vertices assigned to different communities should be sparsely connected. Let $G = (V, w)$ denote a complete undirected weighted graph where V is the set of vertices and $w : V \times V \to \mathbb{R}$ is a weight function for each pair of vertices. If an edge does not occur in G, then it has weight 0. Let $C = \{C_1, C_2, \ldots, C_n\}$ denote a set of communities such that every vertex $u \in V$ is assigned to one and only one community in C. Hence, the modularity value Q of the set of communities C in graph G can be defined as follows [16]:

$$Q = \sum_{C_k \in C} \left[\frac{\sum_{i,j \in C_k} w(i,j)}{m} - \left(\frac{\sum_{i \in C_k} \sum_{j \in V} w(i,j)}{2m} \right)^2 \right] \qquad (4)$$

where $m = \sum_{i,j \in V} w(i,j)$ denotes the sum of the weights of all the edges in G.

One drawback of community identification using modularity measures is that finding a set of communities with an optimal modularity value is computationally hard [18]. As a result, several approximation algorithms have been proposed [19,20,21]. In this paper, the method proposed in [21] is used.

From Communities to Partitions

After identifying the communities in the graph, one must build the set of partitions to be used in Algorithm 1. When using the CVIG model, building partitions of soft clauses is straightforward since clauses are directly represented in the graph. For each community with vertices representing soft clauses, there is a corresponding partition containing the respective soft clauses in the community. After building the partitions, these are sorted by ascending size with respect to the number of soft clauses. Therefore, partitions with smaller size are considered first in Algorithm 1.

In the VIG model, only variables are represented in the graph. Therefore, given the set of communities C, we define that a soft clause ω belongs to the community C_k that maximizes $|C_k \cap \omega|$, i.e. C_k is the community with the most variables in ω. In case of a tie, ω is assigned to the community of the lowest index variable in ω. After assigning all soft clauses to communities, partitions to be used in Algorithm 1 are built as in the CVIG model.

5 Experimental Results

All experiments were run on the partial MaxSAT instances from the industrial category of the MaxSAT evaluation of 2012. The evaluation was performed on two AMD Opteron 6276 processors (2.3 GHz) running Fedora 18 with a timeout of 1,800 seconds and a memory limit of 16 GB.

The partitioning techniques described in the previous section were implemented on top of WBO [5]. Figure 1 compares the different partitioning solvers against the original WBO that does not use any partitioning techniques. The solver using hypergraph partitioning is denoted by *hyp*. *VIG* and *CVIG* correspond to the community-based partitioning using the VIG and CVIG graph representations described in section 4.2, respectively. To assess the quality of the new partitions, we have also implemented a random partitioning technique where each soft clause is placed randomly in one of the partitions. We denote this solver by *rdm*. Similarly to the hypergraph partitioning, 16 partitions are used. The clauses are distributed uniformly among the partitions.

The table on the left of Figure 1 shows the number of instances solved in each benchmark set by each solver. Randomly partitioning the soft clauses can have a detrimental effect on the performance of the solver for several classes of benchmarks. However, it

Bench	#I	WBO	hyp	VIG	CVIG	rdm
aes	7	0	0	0	0	0
fir	59	40	26	28	29	18
simp	17	10	9	10	10	9
su	38	11	6	11	10	6
msp	64	4	5	4	6	1
mtg	40	18	39	38	36	19
syn	74	32	31	35	32	30
circuit	4	1	2	2	2	2
haplotype	6	5	5	5	5	5
nencdr	84	19	67	68	68	66
nlogencdr	84	24	71	70	75	68
routing	15	15	6	15	6	3
protein	12	1	2	1	1	2
Total	504	180	269	287	280	229

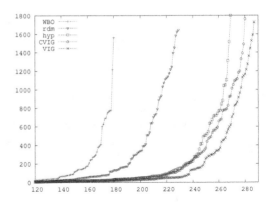

Fig. 1. Comparison between different partitioning solvers

can also significantly improve the performance of the solver on other classes of benchmarks, such as nencdr and nlogencdr. Nevertheless, other partition methods based on structural information of the formula are clearly better.

Community-based partitioning outperforms hypergraph partitioning. Note that hypergraph partitioning creates a fixed number of partitions. However, in community-based partitioning, the number of partitions is dynamic and depends on the structure of the formula. This may explain the effectiveness of community-based partitioning.

The cactus plot of Figure 1 results from the running times of the different solvers. The x-axis shows the number of solved instances, whereas the y-axis shows the running time in seconds. We can distinguish between three classes of solvers: (i) solvers that do not use partitioning (WBO), (ii) solvers that use random partitioning (rdm) and (iii) solvers that use the structure of the formula to create the partitions (hyp, CVIG and VIG). Even random partitioning improves the overall performance of the solver. However, when the structure of the formula is considered, the performance of the solver is further improved.

Even though partitioning approaches outperform WBO on most benchmarks, there are some benchmarks where partitioning may lead to a detrimental effect on the performance of the solver. Our motivation for partitioning is to identify easier to solve subproblems. As a side effect, this may lead to finding smaller unsatisfiable subformulas at each call of the SAT solver. On average, WBO finds unsatisfiable subformulas with 110 soft clauses, whereas unsatisfiable subformulas in VIG have 66 soft clauses. The other solvers using partitions also behave similarly and find on average smaller unsatisfiable subformulas than WBO. This behavior is particularly visible on the nencdr and nloencdr benchmarks [22]. For these benchmarks, WBO finds on average unsatisfiable subformulas with 167 soft clauses, whereas unsatisfiable subformulas in VIG have only 47 soft clauses.

However, if the partitions are not adequate, then we may split *related* soft clauses between different partitions. This may prevent the solver from finding small unsatisfiable subformulas. For example, this behavior is observed in the routing benchmarks [23]. On average, WBO finds unsatisfiable subformulas with 7 soft clauses, whereas CVIG finds unsatisfiable subformulas with 45 soft clauses. Due to an inadequate partitioning,

CVIG is only able to solve 6 out of 15 instances of `routing`. Note that VIG can solve all `routing` instances since the partitions used allowed to find on average unsatisfiable subformulas with 9 soft clauses. Example 1 shows the impact that inadequate partitioning may have on the performance of the solver.

Example 1. Consider a partial MaxSAT formula $\varphi = \varphi_h \cup \varphi_s$. Assume that φ_h contains the hard clause $\omega_1 = (x_1 \vee x_2 \vee x_3)$ and that φ_s contains the soft clauses ω_2, ω_3 and ω_4, where $\omega_2 = (\bar{x}_1), \omega_3 = (\bar{x}_2)$ and $\omega_4 = (\bar{x}_3)$. If these soft clauses are placed in the same partition then we can find a trivial unsatisfiable subformula with the hard clause ω_1. However, consider the worst case scenario where ω_2, ω_3 and ω_4 are placed in different partitions with other soft clauses. Let us assume that we first try to find unsatisfiable subformulas in the partition that contains ω_2. In this case, we may find unsatisfiable subformulas formed by ω_2 with other soft clauses. Note that each time a new unsatisfiable subformula φ_C is found, all soft clauses in φ_C are relaxed. Therefore, after several iterations, when the working formula finally contains ω_2, ω_3 and ω_4, we may have already relaxed these soft clauses several times. If this is the case, then we will no longer be able to find the small unsatisfiable subformula that could be identified if no partitioning was used.

It was observed that the inadequate partitioning presented in Example 1 occurs frequently in some classes of benchmarks, such as `fir` and `routing`. Moreover, if a random partitioning is used, then this problem is even more accentuated. This may explain why random partitioning deteriorates the performance of the solver for several benchmark sets. On the other hand, this shows that an adequate partitioning of the formula is essential for the effectiveness of the solver. Future work will focus on improving partitioning techniques to further reduce the probability of inadequate partitioning.

WBO is a state-of-the-art solver for weighted partial MaxSAT but is not as effective for partial MaxSAT. Even though partitioning significantly improves the unsatisfiability-based algorithm of WBO, it is still not enough to match the performance of state-of-the-art solvers for partial MaxSAT. However, the partitioning approaches presented in this paper are not limited to the unsatisfiability-based algorithm of WBO but can also be extended to other unsatisfiability-based algorithms [4,7]. As future work, we propose to implement the partitioning techniques described in this paper on top of other unsatisfiability-based algorithms for partial MaxSAT.

6   Conclusions

Partitioning the soft clauses has shown to significantly improve the unsatisfiability-based algorithm of WBO for most classes of benchmarks. Moreover, if the structure of the formula is taken into consideration when creating the partitions we can further improve the effectiveness of the solver. This supports the idea that using the structure of the formula to guide the search improves the performance of the solver and provides a strong stimulus for future research.

As future work, we propose to extend our modularity-based partitioning for weighted MaxSAT. Furthermore, the partitioning approaches proposed in this paper are not limited to the WBO algorithm and will be used in other unsatisfiability-based algorithms.

References

1. Coudert, O.: On Solving Covering Problems. In: Proceedings of the Design Automation Conference, pp. 197–202 (June 1996)
2. Martins, R., Manquinho, V., Lynce, I.: On Partitioning for Maximum Satisfiability. In: European Conference on Artificial Intelligence, pp. 913–914 (2012)
3. Fu, Z., Malik, S.: On Solving the Partial MAX-SAT Problem. In: Biere, A., Gomes, C.P. (eds.) SAT 2006. LNCS, vol. 4121, pp. 252–265. Springer, Heidelberg (2006)
4. Ansótegui, C., Bonet, M.L., Levy, J.: Solving (Weighted) Partial MaxSAT through Satisfiability Testing. In: Kullmann, O. (ed.) SAT 2009. LNCS, vol. 5584, pp. 427–440. Springer, Heidelberg (2009)
5. Manquinho, V., Marques-Silva, J., Planes, J.: Algorithms for Weighted Boolean Optimization. In: Kullmann, O. (ed.) SAT 2009. LNCS, vol. 5584, pp. 495–508. Springer, Heidelberg (2009)
6. Ansótegui, C., Bonet, M., Levy, J.: A New Algorithm for Weighted Partial MaxSAT. In: AAAI Conference on Artificial Intelligence, pp. 3–8 (2010)
7. Heras, F., Morgado, A., Marques-Silva, J.: Core-Guided Binary Search Algorithms for Maximum Satisfiability. In: AAAI Conference on Artificial Intelligence, pp. 36–41 (2011)
8. Janota, M., Lynce, I., Manquinho, V., Marques-Silva, J.: PackUp: Tools for Package Upgradability Solving. Journal on Satisfiability, Boolean Modeling and Computation 8(1/2), 89–94 (2012)
9. Li, C.M., Manyà, F.: MaxSAT, Hard and Soft Constraints. In: Handbook of Satisfiability, pp. 613–631. IOS Press (2009)
10. Koshimura, M., Zhang, T., Fujita, H., Hasegawa, R.: QMaxSAT: A Partial Max-SAT Solver. Journal on Satisfiability, Boolean Modeling and Computation 8, 95–100 (2012)
11. Ansótegui, C., Bonet, M.L., Gabàs, J., Levy, J.: Improving SAT-Based Weighted MaxSAT Solvers. In: Milano, M. (ed.) CP 2012. LNCS, vol. 7514, pp. 86–101. Springer, Heidelberg (2012)
12. Park, T.J., Gelder, A.V.: Partitioning Methods for Satisfiability Testing on Large Formulas. Information and Computation 162(1-2), 179–184 (2000)
13. Karypis, G., Aggarwal, R., Kumar, V., Shekhar, S.: Multilevel hypergraph partitioning: Application in VLSI domain. IEEE Transactions on VLSI Systems 7, 69–79 (1999)
14. Ansótegui, C., Giráldez-Cru, J., Levy, J.: The Community Structure of SAT Formulas. In: Cimatti, A., Sebastiani, R. (eds.) SAT 2012. LNCS, vol. 7317, pp. 410–423. Springer, Heidelberg (2012)
15. Girvan, M., Newman, M.E.J.: Community structure in social and biological networks. Proceedings of the National Academy of Sciences 99(12), 7821–7826 (2002)
16. Newman, M.E.J., Girvan, M.: Finding and evaluating community structure in networks. Physical Review E 69(026113) (2004)
17. Radicchi, F., Castellano, C., Cecconi, F., Loreto, V., Parisi, D.: Defining and identifying communities in networks. Proceedings of the National Academy of Sciences of the United States of America 101(9), 2658–2663 (2004)
18. Brandes, U., Delling, D., Gaertler, M., Goerke, R., Hoefer, M., Nikoloski, Z., Wagner, D.: Maximizing modularity is hard. arXiv: physics, 0608255 (2006)
19. Clauset, A., Newman, M.E.J., Moore, C.: Finding community structure in very large networks. Physical Review E 70(6), 066111 (2004)

20. Pons, P., Latapy, M.: Computing Communities in Large Networks Using Random Walks. Journal of Graph Algorithms and Applications 10(2), 191–218 (2006)
21. Blondel, V., Guillaume, J., Lambiotte, R., Lefebvre, E.: Fast unfolding of communities in large networks. Journal of Statistical Mechanics 2008(10), P10008 (2008)
22. Morgado, A., Marques-Silva, J.: Combinatorial Optimization Solutions for the Maximum Quartet Consistency Problem. Fundamenta Informaticae 102(3-4), 363–389 (2010)
23. Aloul, F.A., Ramani, A., Markov, I.L., Sakallah, K.A.: PBS: A backtrack search pseudo Boolean solver. In: Symposium on the Theory and Applications of Satisfiability Testing, pp. 346–353 (2002)

Experiments with Reduction Finding

Charles Jordan[1,*] and Łukasz Kaiser[2]

[1] ERATO Minato Project, JST & Hokkaido University
[2] LIAFA, CNRS & Université Paris Diderot
skip@ist.hokudai.ac.jp, kaiser@liafa.univ-paris-diderot.fr

Abstract. Reductions are perhaps the most useful tool in complexity theory and, naturally, it is in general undecidable to determine whether a reduction exists between two given decision problems. However, asking for a reduction on inputs of bounded size is essentially a Σ_2^p problem and can in principle be solved by ASP, QBF, or by iterated calls to SAT solvers. We describe our experiences developing and benchmarking automatic reduction finders. We created a dedicated reduction finder that does counter-example guided abstraction refinement by iteratively calling either a SAT solver or BDD package. We benchmark its performance with different SAT solvers and report the tradeoffs between the SAT and BDD approaches. Further, we compare this reduction finder with the direct approach using a number of QBF and ASP solvers. We describe the tradeoffs between the QBF and ASP approaches and show which solvers perform best on our Σ_2^p instances. It turns out that even state-of-the-art solvers leave a large room for improvement on problems of this kind. We thus provide our instances as a benchmark for future work on Σ_2^p solvers.

1 Introduction

Finding reductions between different decision problems is a central task in complexity theory. Polynomial-time reductions are perhaps the most traditional, and constructing such reductions generally involves creating certain gadgets or building some other form of structure on top of the instance that is to be reduced. The intuition that finding reductions resembles structured constructions can be captured formally: the reduction one finds is usually not only a polynomial-time function, but often a log-space one, or even a quantifier-free projection.

The class of quantifier-free projections, defined formally in the next section, is a very restricted subset of log-space functions. Still, they are sufficient to capture important complexity classes (see Chapter 11 of [13]). For example[1], P=NP iff SAT $\leq_{qfp}$CVP, and NL=NP iff SAT $\leq_{qfp}$REACH. The hope when focusing on weaker (but still sufficiently strong) reductions is that they will put new complexity-theoretic results within reach, and there are examples where

---

* Supported in part by KAKENHI No. 25106501.
[1] We write X $\leq_{qfp}$Y if there is a quantifier-free projection from X to Y. CVP is the P-complete Circuit Value Problem, REACH is the NL-complete problem of directed reachability, and SAT is the NP-complete propositional satisfiability problem.

M. Järvisalo and A. Van Gelder (Eds.): SAT 2013, LNCS 7962, pp. 192–207, 2013.
© Springer-Verlag Berlin Heidelberg 2013

this has actually been accomplished [1]. Quantifier-free projections also have a significant advantage when trying to derive them automatically. They are by definition formulas of a simple form, so one can enumerate them easily once their dimension is fixed. In fact, instead of enumerating, one can write them in symbolic form using propositional variables. This opens the way for using propositional solvers to find such reductions automatically.

The problem of determining whether a quantifier-free projection exists between two given decision problems is still undecidable in general. But when we fix the dimension of the reduction we are looking for and only ask for it to be correct on inputs of bounded size, the question becomes essentially a Σ_2^p problem – it is of the form $\exists \overline{X} \, \forall \overline{Y} \, \varphi$ where $\overline{X}$ and $\overline{Y}$ are sets of propositional variables and φ is a quantifier-free propositional formula. This problem can then in principle be solved by a QBF or ASP solver, or by iterated calls to a SAT solver.

This paper describes our experiments with this kind of automated reduction finding. We present both a dedicated reduction finder called DE[2] and a generator[3] that allows to construct instances for QBF and ASP solvers that are equivalent to the given reduction finding problem. It is therefore a source of instances for which the hardness depends on the chosen parameters. To make it easy to use these problems for benchmarking, we provide both the generator (all source code is available as open-source) and the collection of qdimacs, qpro, and lparse files for the set of parameters we used in our experiments.[3]

In the long term, automatic reduction finders may help obtain unexpected complexity-theoretic results or re-discover stunning reductions. For example, the coNL-to-NL reduction behind the Immerman-Szelepcsényi Theorem, awarded the Gödel Prize in 1995, is in fact a dimension-8 quantifier-free projection and can in principle be found by DE. But current solvers do not perform sufficiently on high-dimensional instances: even dimension 3 is beyond reach of DE or any other solver with the present approach. Still, none of these solvers has been tuned for Σ_2^p problems and DE is a young project. We believe that our benchmarks are a source of meaningful, challenging SAT and 2QBF instances and we will work to include them in the next SAT and QBF evaluations. If solvers can be tuned to perform well on these kinds of instances, and improve their performance on Σ_2^p problems in general, we may be able to obtain interesting complexity-theoretic results in this way – so we encourage the community to experiment.

Related Work. The idea of automatic reduction finding, together with the first automated ReductionFinder, was developed in [3]. ReductionFinder works on a database of decision problems specified in stratified Datalog and attempts to place the problems into classes based on the existence of reductions. It uses the ASP solver CMODELS to search for reductions, and it has not previously been compared to other reduction-finding attempts nor is it publicly available. We focus entirely on the problem of finding a reduction between two given problems,

---

[2] DE is available at http://www-alg.ist.hokudai.ac.jp/~skip/de

[3] The generator with instructions and the collection of generated files we used for testing are available from http://toss.sf.net/reductGen.html

and thanks to a private copy of ReductionFinder we also compare our results to this previous approach.

2   Background in Descriptive Complexity

Classically, one defines a decision problem as a set of words and the complexity of a problem as the amount of computational resources (time, space) required to check on a Turing machine whether a word belongs to the set. In descriptive complexity, we take a higher-level view of decision problems. Instances do not need to be encoded as words, but are directly relational structures, for example graphs. The role of a Turing machine is in turn played by a formula in some logic, and the complexity of a problem is the expressive power required by the formula. It turns out that different logics correspond to different complexity classes and that all major complexity classes have logical characterizations.

Descriptive complexity provides a particularly convenient framework for automatically finding reductions. The fact that instances do not need to be encoded as words allows us to express interesting reductions succinctly, and formulas, unlike Turing machines, have natural normal forms. In this section we introduce the background in descriptive complexity necessary for this paper; refer to [13] or Chapter 3 of [8] for a more detailed introduction and additional material.

A *relational signature* $\tau := (R_1^{a_1}, \ldots, R_r^{a_r}, c_1, \ldots, c_s)$ is a tuple of predicate symbols R_i with arities a_i and constant symbols c_j. A finite τ-*structure*

$$\mathfrak{A} := (U, R_1 \subseteq U^{a_1}, \ldots, R_r \subseteq U^{a_r}, c_1 \in U, \ldots, c_s \in U)$$

consists of a finite universe U, an a_i-ary relation for each predicate symbol of τ, and a definition – an element of U – for each constant symbol.

For example, the signature for directed graphs contains a single, binary predicate symbol E and so a directed graph consists of a finite set of vertices and a binary edge relation. For convenience, we generally identify an n-element universe U with the set $\{0, \ldots, n-1\}$.

Formulas of *first-order logic* over a signature τ have the form

$$\varphi := R_i(x_1, \ldots, x_{a_i}) \mid x_i = x_j \mid x_i = c_j \mid \neg\varphi \mid \varphi \vee \varphi \mid \varphi \wedge \varphi \mid \exists x_i\, \varphi \mid \forall x_i\, \varphi,$$

where $x_1, x_2, \ldots$ are first-order variables, and the semantics, given an assignment of the variables x_i to elements e_i of the structure, is defined in the natural way. For example, $\exists x_1 R(x_1, x_2)$ holds for an assignment $x_2 \to e_2$ in $\mathfrak{A}$ if, and only if, there exists an element e_1 of U such that (e_1, e_2) is in the relation R in $\mathfrak{A}$.

Formulas without free variables define *properties* in the natural way. For example, $\forall x, y\, \neg E(x, y)$ defines the property of having no edges, i.e. being an empty graph. That is, the property defined by a formula φ is the set of all structures on which φ holds. We use properties to specify decision problems.

Queries and Reductions. Reductions map σ-structures to τ-structures, defining the universe, relations, and constants by means of logical formulas. Reductions are a special kind of *query*, and so we begin by defining first-order queries.

A *first-order query* from σ-structures to τ-structures is an $r + s + 2$-tuple,

$$q := (k, \varphi_0, \varphi_1, \ldots, \varphi_r, \psi_1, \ldots, \psi_s).$$

The number $k \in \mathbb{N}$ is the *dimension* of the query. Each φ_i, ψ_j is a first-order formula over the signature σ. Let $\mathfrak{A}$ be a σ-structure with universe $U^{\mathfrak{A}}$. The formula φ_0 has free variables $x_1, \ldots, x_k$ and defines the universe U of $q(\mathfrak{A})$,

$$U := \{(u_1, \ldots, u_k) \mid u_i \in U^{\mathfrak{A}}, \mathfrak{A} \models \varphi_0(u_1, \ldots, u_k)\}.$$

That is, the new universe consists of k-tuples of elements of the old universe, where φ_0 determines which k-tuples are included.

Each remaining φ_i has free variables $x_1^1, \ldots, x_1^k, x_2^1, \ldots, x_{a_i}^k$ and defines

$$R_i := \{(u_1^1, \ldots, u_1^k), \ldots, (u_{a_i}^1, \ldots, u_{a_i}^k) \mid \mathfrak{A} \models \varphi_i(u_1^1, \ldots, u_{a_i}^k)\} \cap U^{a_i}.$$

That is, φ_i determines which of the a_i-tuples of U are included in R_i. Finally, each ψ_i has free variables $x_1, \ldots, x_k$ and defines c_i as the unique $(u_1, \ldots, u_k) \in U$ such that $\mathfrak{A} \models \psi_i(u_1, \ldots, u_k)$.

First-order queries therefore transform σ-structures into τ-structures. Given a property P of σ-structures and a property Q of τ-structures, a first-order *reduction* r from P to Q is a first-order query that satisfies an additional condition. Namely, reductions must satisfy $\mathfrak{A} \in P \iff r(\mathfrak{A}) \in Q$ for all σ-structures $\mathfrak{A}$.

There are various kinds of first-order reductions (see [13]). Quantifier-free projections are the weakest version usually considered. There, all formulas in the reduction must be quantifier-free, and the reduction must also be a *projection*, i.e., each bit of the output depends on at most one bit of the input, where the bit is selected in first-order.

First-order reductions are weaker than, for example, polynomial-time reductions, and quantifier-free projections are even more restricted. However, natural problems that are complete for natural complexity classes via polynomial-time reductions tend to remain complete via these weaker reductions (see, e.g., [13]). In this paper we consider parametrized classes of quantifier-free reductions where all formulas are in disjunctive normal form (DNF). Therefore, all formulas in the reductions we consider are quantifier-free, however, the reductions are not necessarily projections. We also consider only formulas ψ_i that directly define the constant c_i (i.e., a fixed tuple of constant symbols from σ and $U^{\mathfrak{A}}$) and require φ_0 to be always true. These restrictions are for simplicity, and have minimal impact from the complexity-theory perspective, given that constants can be omitted or replaced by monadic relations.

Extending First-Order Logic. So far, we have focused only on first-order logic. Although it is generally more than adequate for reductions, first-order logic has several drawbacks when used to express properties. First of all, it is not expressive enough to describe many relations that can easily be computed. This limitation stems from the *locality* of first-order formulas. This property implies that it is not possible to express the transitive closure of a relation in first-order logic, so, e.g., also the property that a graph is connected.

To remove this limitation of first-order logic, one extends it with various operators. For example, the *transitive closure operator* allows us to write formulas of the form $TC[x_1, x_2.\varphi(x_1, x_2)](y_1, y_2)$. This formula takes the transitive and reflexive closure of the (implicit) relation defined by $\varphi(x_1, x_2)$ and then evaluates it on (y_1, y_2). Adding the transitive closure operator removes some limitations of FO, but how can we know what other problems remain? Let us review some of the best-known correspondences between logics and complexity classes.

The oldest result [6] shows that the class NP is captured by existential second-order logic. More practically, polynomial-time computations are captured by the extension of FO by the least fixed-point operator (LFP) when a linear order relation is present [11,18]. The requirement of a linear order can be weakened when a counting mechanism is added to the logic, and LFP with counting captures P on many classes of structures, such as grids, planar graphs [9] and all classes that exclude a fixed minor [10]. Although LFP is presumably more expressive than the transitive closure logic (TC) we mentioned, TC captures all problems solvable in non-deterministic logarithmic space (NL) on ordered structures [12].

Example 1. Having introduced the transitive closure operator and our notion of reductions, let us give a simple example that our reduction-finding systems can find[4]. Consider the following formulas,

$$\text{Reach} := TC[x, y.E(x, y)](s, t) \quad \text{AllReach} := \forall x_1, x_2 \ (TC[y, z.E(y, z)](x_1, x_2)).$$

Here, Reach expresses the NL-complete problem of reachability (there exists a directed path from s to t) and AllReach expresses the NL-complete problem of all-pairs reachability (there is a directed path from x to y for all vertices x, y).

Using the notation for reductions introduced above, a correct reduction from Reach to AllReach is

$$(k := 1, \ \varphi_0 := true, \ \varphi_1 := x_1 = s \vee x_2 = t \vee E(x_2, x_1)).$$

This reduction reverses all edges in the original graph, adds directed edges from s to all vertices and also adds directed edges to t from all vertices. It is not difficult to see that the result is strongly connected if, and only if, the original graph has a directed path from s to t. Note that a similar reduction exists without reversing the edges – however the above is the actual output of our program.

3   Finding Reductions

The fundamental problem that we want to solve is the following. Given two logical formulas φ_P and φ_Q, is there a reduction from the property defined by φ_P to that defined by φ_Q? Unfortunately, this problem is undecidable. In fact, it is also undecidable to determine whether a given reduction is correct for two fixed properties, or even whether two given properties are logically equivalent.

---

[4] With parameters $k = 1, c = 3, n = 4$ in $< 3s$.

Our fundamental approach is to fix an "outline" of the reduction we hope to find. For example, we may assume that all formulas[5] in the reduction are quantifier-free, in DNF, and are a disjunction of exactly c conjunctions. Once the signatures are fixed, there is only a finite number of atoms that could occur in these conjunctions. This is because there are only finitely many variables (we cannot introduce new variables with quantifiers) and constants, and only finitely many ways to combine these symbols with the relation symbols and equality. For each atom and conjunction, we introduce a Boolean variable representing whether or not the atom occurs in that conjunction of the reduction. Intuitively, we can now express the existence of a reduction as a logical formula

$$\exists \mathbf{r} \, \forall \mathfrak{A} \, (\mathfrak{A} \models \varphi_P \leftrightarrow \mathbf{r}(\mathfrak{A}) \models \varphi_Q), \tag{1}$$

where $\mathbf{r}$ is the finite set of Boolean variables defining the reduction, and $\mathfrak{A}$ ranges over all structures having the same signature as φ_P.

Of course, there are infinitely many such structures, which explains why the problem is still undecidable. However, experience shows that it usually suffices to consider structures of fairly small size, at least for natural problems and natural classes of weak reductions. That is, although one can construct artificial properties where arbitrarily large examples are needed, it seems that if a simple reduction between natural problems is correct on all small instances, then it is usually correct on all instances.

Therefore we focus, as did [3], on finding reductions that are correct on all structures of size n. Here, n is a parameter and it is also possible to consider ranges of n. Once n is fixed, as well as the outline for $\mathbf{r}$, checking Formula (1) becomes decidable. In fact, it is natural to represent $\mathfrak{A}$ with Boolean variables, and so Formula (1) becomes a one-alternation quantified Boolean formula. Satisfiability of such formulas[6] is complete for Σ_2^p.

Example 2. Let us show how the QBF for Formula (1) is constructed in the following case. Let P be the class of non-empty graphs defined by $\varphi_P = \exists x, y \, E(x, y)$ and let Q be the class of non-complete graphs given by $\varphi_Q = \exists x, y \, \neg E(x, y)$. We ask whether there is a quantifier-free reduction from P to Q of dimension $k = 1$, with $c = 1$ conjunctions, and that is correct on all graphs of size $n = 2$.

First, let us fix the outline for our reduction with one conjunction. Note that $\sigma = \tau = \{E\}$, so in this case the reduction we are looking for has the following form: $(k := 1, \varphi_0 := true, \varphi_1(x_1, x_2))$ for some formula $\varphi_1(x_1, x_2)$ which is a conjunction of literals. What atoms are possible over the signature $\{E\}$ with variables x_1, x_2? There are exactly 5 atoms in the basic syntax: $E(x_1, x_1)$, $E(x_1, x_2)$, $E(x_2, x_1)$, $E(x_2, x_2)$, and $x_1 = x_2$. So, in this most basic case, the outline for φ_1 has the form $\varphi_1(x_1, x_2) =$

$X_1 E(x_1, x_1) \, \wedge \, X_2 E(x_1, x_2) \, \wedge \, X_3 E(x_2, x_1) \, \wedge \, X_4 E(x_2, x_2) \, \wedge \, X_5 x_1 = x_2 \, \wedge$
$Y_1 \neg E(x_1, x_1) \wedge Y_2 \neg E(x_1, x_2) \wedge Y_3 \neg E(x_2, x_1) \wedge Y_4 \neg E(x_2, x_2) \wedge Y_5 \neg x_1 = x_2.$

---

[5] As mentioned above, we always fix $\varphi_0 = true$ and define constants by fixed tuples.
[6] Note that Formula (1) has leading existential quantifiers; satisfiability of one-alternation formulas with leading universal quantifiers is complete for Π_2^p.

Above, X_i and Y_i are propositional variables that determine whether the literal after them will appear or not: $X_1 E(x_1, x_2)$ means "$E(x_1, x_2)$ if X_1 is set and *true* otherwise", as becomes clear below. An outline is thus a formula with these additional propositional variables used as guards.

In all our tests, we use an extended set of atoms, not only the basic ones presented above for readability. In the extended set, in addition to relations over variables and equality as above, we allow the following atoms: for a fixed enumeration of the elements of the structure, we say that x is the minimal one, the maximal one, or that $x = y + 1$. This allows to find more reductions with the same outline parameters.

Having constructed the outline, let $\mathfrak{A}$ be a 2-element structure and assume that the tuple (i, j), for $i, j \in \{0, 1\}$, is in the relation E in $\mathfrak{A}$ if, and only if, the propositional variable E_{ij} is set. Note that the part $\mathfrak{A} \models \varphi_P$ of Formula (1), in our case $\mathfrak{A} \models \exists x, y\, E(x, y)$, can now be written as a purely propositional formula: $\bigvee_{i,j \in \{0,1\}} E_{ij}$. To express that $\mathbf{r}(\mathfrak{A}) \models \varphi_Q$ we need to use the definition of E in $\mathbf{r}(\mathfrak{A})$ given by φ_1. In our case, for $\mathbf{r}(\mathfrak{A}) \models \exists x, y\, \neg E(x, y)$ we write $\bigvee_{i,j \in \{0,1\}} \neg\varphi_{ij}$. Here φ_{ij} is derived from the outline of φ_1 using the propositional variables E_{ij}. For the basic outline presented above, that means $\varphi_{ij} =$

$$(\neg X_1 \vee E_{ii}) \wedge (\neg X_2 \vee E_{ij}) \wedge (\neg X_3 \vee E_{ji}) \wedge (\neg X_4 \vee E_{jj}) \wedge (\neg X_5 \vee i = j) \wedge$$
$$(\neg Y_1 \vee \neg E_{ii}) \wedge (\neg Y_2 \vee \neg E_{ij}) \wedge (\neg Y_3 \vee \neg E_{ji}) \wedge (\neg Y_4 \vee \neg E_{jj}) \wedge (\neg Y_5 \vee i \neq j),$$

where $i = j$ and $i \neq j$ get substituted by *true* or *false* depending on i and j.

In this way, we obtain the following propositional formula, which we call the QBF corresponding to Formula (1) for $k = 1, c = 1, n = 2$.

$$\exists X_1 \ldots X_5\, Y_1 \ldots Y_5\, \forall E_{00} \ldots E_{11} \left(\bigvee_{i,j \in \{0,1\}} E_{ij} \leftrightarrow \bigvee_{i,j \in \{0,1\}} \neg\varphi_{ij} \right).$$

Observe that this formula is satisfied exactly if there is a reduction with the specified outline correct on all structures of the specified size. Moreover, if satisfied, the outer-most existentially quantified variables allow to extract a reduction.

3.1   Approaches to Solving Σ_2^p Problems

Several approaches have been used to solve problems in Σ_2^p. Actually constructing the QBF for Formula (1) as above is fairly tedious, but poses no serious difficulty. This results in a QBF instance with one quantifier alternation where a satisfying assignment of the existential variables gives a reduction, so QBF solvers can be immediately applied.

Although they are perhaps not yet as mature as SAT solvers, in recent years there has been a great deal of work on efficient QBF solvers. See, for example, the QBFEVAL series of evaluations [17]. QBFEVAL'10 also included a 2QBF track, and we believe our instances could be attractive candidates for inclusion in that track. Still, as will be clarified below, in our case it is often more efficient to present the formula as a 3QBF instance.

In addition to QBF solvers, there are other approaches that have been used to solve Σ_2^p problems. For example, ASP solvers that support disjunctive logic programs can solve such problems using the reduction to disjunctive logic programs given by [4,16]. Examples of such solvers are CMODELS and CLASPD, and some previous work [5] has indicated that ASP solvers may outperform QBF solvers on Σ_2^p problems.

Another option is to expand the universal quantifier block of the formula using a conjunction over all possible assignments, resulting in a SAT instance. This allows SAT solvers to be used directly, however it entails an exponential increase in instance size. In our experience, this is impractical for large instances.

Finally, some recent work [3,14,15] has noted that one can use repeated calls to SAT solvers to solve Σ_2^p instances, essentially an application of counter-example guided abstraction refinement (CEGAR) [2]. This approach is also a finitely-truncated implementation of limit-learning [7], using guesses for the leading existential quantifiers as hypotheses and assignments to the universal quantifiers as counter-examples. Our dedicated reduction finder DE uses this approach.

In our view, the connection to limit-learning gives a particularly nice perspective on our problem. For example, removing the finite-size restriction needed for decidability results is *exactly* an attempt to learn reductions from counter-examples in the limit. Of course, it is undecidable whether such a learner has converged to a correct hypothesis. Techniques from inductive inference may provide valuable guidance on efficient learning, i.e., minimizing total computation time, or number of counter-examples required.

4   Reduction Finding Using QBF and ASP Solvers

In this section, we compare the performance of various QBF and ASP solvers on our problem. Our approach to generating QBF instances for reduction finding was outlined above. Essentially, we view the problem from the perspective of Formula (1) as a QBF instance and apply the above-mentioned approaches.

Note that Formula (1) is of the form $\exists r \, \forall \mathfrak{A} \, \psi$. The reduction r contributes existential propositional variables, the structure $\mathfrak{A}$ is the source of universal variables, and ψ is a quantifier-free propositional formula. While it is quantifier-free, in general this formula is not in conjunctive normal form (CNF). Most QBF solvers require their input to be in CNF and CNF conversion usually involves introducing new existentially quantified variables, which must be innermost.

We investigate three approaches to dealing with the conversion to CNF. First, there are QBF solvers that only require the formula to be in negation normal form (NNF), not CNF. The most common input format for such solvers is called qpro and in this case we generate the formula directly.

The second approach is to convert the propositional part of Formula (1) to CNF adding new existential variables.[7] In this case, we produce output in

---

[7] We used the standard Plaisted-Greenbaum technique for this, which in our tests slightly outperformed the often used Tseitin method.

qdimacs format. The resulting formula is of the form $\exists \overline{X} \ \forall \overline{Y} \ \exists \overline{Z} \ \psi_{\mathrm{CNF}}$ – it has one more quantifier alternation than strictly necessary.

The third approach is to first negate Formula (1), which then has the form $\forall r \ \exists \mathfrak{A} \ \psi$. We then convert ψ to CNF as above, and generate a qdimacs file with only one quantifier alternation.

Finally, we also generate lparse files for ASP solvers, as described in Subsection 3.1. This results in the following four cases that we test and benchmark.

qpro Constructing QBF for Formula (1) and using QBF solvers that support non-CNF QBF (CIRQIT).

qdimacs Constructing QBF for Formula (1) and converting directly to CNF, then using QBF solvers (RAREQS, DEPQBF, QUBE7.2, SKIZZO, CIRQIT).

nqdimacs Negating Formula (1) before CNF conversion to avoid one quantifier alternation, and using QBF solvers (RAREQS, DEPQBF, QUBE7.2, SKIZZO, CIRQIT).

lparse Constructing Formula (1) and using ASP solvers that support disjunctive logic programs (LPARSE or GRINGO, and CMODELS, CLASPD or GNT2).

Some of the combinations listed above performed quite poorly even on very simple reduction finding instances. In particular, CMODELS and GNT2 almost always abort (but produce correct output if they do not abort), and using CIRQIT with nqdimacs is very slow (but produces correct output). We therefore omit these combinations from our experimental results.

4.1   Comparing the QBF Approaches

We first concentrate on the following question: which of the three approaches to using QBF solvers, qdimacs, nqdimacs, and qpro, performs best? It turns out that there is a clear answer: qdimacs is the best option.

Comparing qdimacs and nqdimacs. Given the explanation above, one might think that nqdimacs is a more promising formulation for our instances – it expresses the same problem but avoids one quantifier alternation. We were mildly surprised to see that all QBF solvers we tested consistently performed worse on nqdimacs than on qdimacs instances with one more quantifier alternation. In Table 1 below we present the number of timeouts for qdimacs and nqdimacs instances for the easiest set of parameters we tested: $k = 1$, $c = 1$, and $n = 3$. Clearly, both DEPQBF and QUBE performed much better on qdimacs than on nqdimacs, e.g. there were no timeouts (set to 120s in this section) on any qdimacs instance for these two solvers but several hundred (out of 2304[8]) for nqdimacs, and the completed instances also took longer to finish. For SKIZZO the situation is less clear but the general pattern is the same. The CEGAR-based solver RAREQS

---

[8] We used 48 decision problems from [3] to be able to compare to ReductionFinder. They range from very simple, like the empty graph, to the NL-complete reachability and coNL-complete unreachability problems.

Table 1. Number of timeouts for `qdimacs` and `nqdimacs` solvers

	qdimacs	nqdimacs
DEPQBF	0	300
QUBE	10	285
SKIZZO	522	706

had only 2 timeouts on this instance, both in the negated setting, but exhibits the same pattern for $c = 2$ (0 vs. 304 timeouts).

Comparing qdimacs and qpro. When comparing `qdimacs` and `qpro` we must note that most QBF solvers accept `qdimacs` input while we found only one, CIRQIT, that accepts `qpro` and is freely available. This is crucial: either DEPQBF or QUBE reading `qdimacs` outperform CIRQIT uniformly, on all instances, with 1s margin of error, and both these solvers have far fewer timeouts than CIRQIT. On the other hand, when comparing only CIRQIT on `qdimacs` input with CIRQIT on `qpro` input, neither has a clear advantage – there are numerous instances where one times out and the other does not, and also instances where the opposite occurs. In Table 2 we show the number of timeouts of different solvers on three parameter sets of increasing difficulty.

Table 2. Number of timeouts for `qdimacs` and `qpro` solvers

	$k = 1\ c = 1\ n = 3$	$k = 1\ c = 2\ n = 3$	$k = 1\ c = 3\ n = 3$
RAREQS (qdimacs)	0	0	16
DEPQBF (qdimacs)	0	142	547
QUBE (qdimacs)	10	536	949
CIRQIT (qdimacs)	58	673	1138
CIRQIT (qpro)	157	523	903

The behavior of CIRQIT on `qdimacs` and `qpro` instances shows that one can benefit from knowing the structure of the formula. Together with the fact that `qdimacs` outperforms `nqdimacs`, these seem to indicate that a more careful handling of the innermost formula could lead to more efficient solvers.

Comparing QBF Solvers. Having settled on `qdimacs`, we now compare the performance of the five QBF solvers on different parameter sets. In Table 3 we report the number of timeouts for each solver and each parameter set.

For non-CEGAR solvers, DEPQBF and QUBE outperform SKIZZO and CIRQIT. Between DEPQBF and QUBE the situation is less clear, some instances work much better with one of these solvers, others with the other. The comparison between SKIZZO and CIRQIT is difficult as well. As to the dominance of DEPQBF and QUBE over SKIZZO and CIRQIT, it holds for almost all queries. Still, there are a few outliers on which DEPQBF and QUBE time out, but SKIZZO answers almost immediately. This allows to hope that tailoring the strategies of QBF solvers towards Σ_2^p problems might still lead to significant performance gains.

Table 3. Number of timeouts for different QBF solvers, $k = 1$

	$c = 1\ n = 3$	$c = 2\ n = 3$	$c = 3\ n = 3$	$c = 1\ n = 4$	$c = 2\ n = 4$	$c = 3\ n = 4$
RAREQS	0	0	16	19	65	204
DEPQBF	0	142	547	16	297	711
QUBE	10	536	949	82	760	1082
CIRQIT	58	673	1138	511	1092	1357
SKIZZO	522	1058	1156	975	1327	1434

4.2   Comparing Different Approaches to Reduction Finding

Having discussed the QBF approaches and chosen the best QBF solvers, let us finally compare the QBF approach with the approach using ASP solvers, and also with our reduction finder DE and ReductionFinder from [3].

Different ASP Solvers. We consider three different ASP solvers (CMODELS, GNT2, and CLASPD) and two different grounding programs (LPARSE and GRINGO). Grounding is performed before the solver is started, and may take significant time. However, we time the total of grounding and solving and so it is possible for the grounding program to timeout before the solver begins.

Two of the solvers (CMODELS and GNT2) abort very frequently, even on simple instances and regardless of the choice of grounding program. Therefore we only show results for CLASPD with LPARSE and GRINGO. Interestingly, while the total number of timeouts was similar for the two grounders, there were numerous instances where one timed out and the other finished quickly. This may give some reason to hope that significantly better performance may be possible with this approach with more careful grounding.

Results. In Table 4 we compare the different reduction finding approaches that we tested. For the SAT-solver based DE runs, we have chosen DE-GMS[9] as it performs best on hard instances. For BDD-based DE runs, we show DE-CUDD, the only reduction finder we tested that used BDDs. For QBF solvers we have chosen the two best performers, RAREQS and DEPQBF. Note, that RAREQS is a CEGAR-based solver, more similar to DE-MS than DEPQBF. For ASP solvers, we show CLASPD both with LPARSE and GRINGO, as explained above.

We also include the results for ReductionFinder [3]. ReductionFinder only considers reductions of a slightly different form – this usually results in a simpler instance, so it is shown here only for comparison. All other approaches we present find reductions of exactly the same form – naturally, the answers (whether there is a reduction or not) agree on all parameter sets that we tested.

The CEGAR approach, whether in DE or in RAREQS, outperforms the others. Other QBF solvers (DEPQBF, QUBE) match the performance of the original ReductionFinder and in general perform better than the ASP approach.

---

[9] DE can use MiniSat2 (DE-MS), GlueMiniSat (DE-GMS), CryptoMiniSat (DE-CMS) or BDDs (DE-CUDD).

Table 4. Number of timeouts (% solved) for tested reduction finding approaches, $k = 1$

	$c = 1\ n = 3$	$c = 2\ n = 3$	$c = 3\ n = 3$	$c = 1\ n = 4$	$c = 2\ n = 4$	$c = 3\ n = 4$
DE-GMS	0 (100.0%)	0 (100.0%)	10 (99.6%)	0 (100.0%)	5 (99.8%)	103 (95.5%)
DE-CUDD	0 (100.0%)	116 (95.0%)	537 (76.7%)	0 (100.0%)	186 (91.9%)	722 (68.7%)
RAREQS	0 (100.0%)	0 (100.0%)	16 (99.3%)	19 (99.1%)	65 (97.1%)	204 (91.1%)
DEPQBF	0 (100.0%)	142 (93.8%)	547 (76.2%)	16 (99.3%)	297 (87.1%)	711 (66.1%)
GRINGO	40 (98.3%)	393 (82.9%)	590 (74.4%)	72 (96.9%)	593 (74.3%)	836 (63.7%)
LPARSE	51 (97.8%)	396 (82.8%)	605 (73.7%)	75 (96.7%)	635 (72.4%)	850 (63.1%)
RedFind	1 (99.9%)	152 (93.4%)	396 (82.8%)	2 (99.9%)	347 (84.9%)	547 (76.3%)

5   Reduction Finding Using CEGAR

We now compare CEGAR approaches to reduction finding. We begin by describing the reduction finding procedure used in DE and then compare the reduction-finding implementations in DE with RAREQS, focusing on difficult instances.

5.1   Finding Reductions in DE

DE finds reductions by first choosing a candidate reduction r_0, then searching for a counter-example (i.e., a structure $\mathfrak{A}$ such that $\mathfrak{A} \models \varphi_P \iff r_0(\mathfrak{A}) \not\models \varphi_Q$). If a counter-example is found, it searches for a reduction that is correct on all examples seen so far, i.e., if we have seen examples $\{\mathfrak{A}_0, \dots, \mathfrak{A}_i\}$, then we search for an assignment of the Boolean variables $\mathbf{r}$ such that

$$(\mathfrak{A}_0 \models \varphi_P \leftrightarrow r(\mathfrak{A}_0) \models \varphi_Q) \land \dots \land (\mathfrak{A}_i \models \varphi_P \leftrightarrow r(\mathfrak{A}_i) \models \varphi_Q) , \qquad (2)$$

and iterate until either no counter-example or no candidate reduction is found.

In Formula (2), whether $\mathfrak{A}_j \models \varphi_P$ is known, so some simplifications can be easily performed. In our experience (and that of [3]), finding candidate reductions is more difficult than finding counter-examples. We therefore focus mostly on finding candidate reductions. However, we have optional heuristics to help choose *good* counter-examples. The times reported for DE alternate greedy minimization and maximization of counter-examples[10], except for CryptoMiniSat.

We implement candidate-finding using incremental SAT solvers or BDDs (using CUDD). Formula (2) is a natural candidate for incremental SAT solving: there are comparatively few Boolean variables $\mathbf{r}$ which are re-used, and at each stage we simply add the restriction corresponding to the new counter-example.

Using BDDs for candidate-finding is similar. At each stage, we have a BDD representing the set of candidate reductions that are consistent with the previous counter-examples. Given a new counter-example, we build a BDD of the hypotheses consistent with it and take the intersection of the two sets. However, to acquire a new counter-example, we must have a particular hypothesis.

---

[10] Given a counter-example, we greedily remove or add tuples to relations while checking that it remains a counter-example.

We take the "simplest" candidate reduction as our hypothesis, i.e., we select from the set of candidates consistent with the examples we have seen a candidate with the *minimal* number of atomic formulas appearing. This has an advantage: there are often several correct reductions in a search space. In this case, the BDD implementation will give a simplest correct reduction. The difference in size (and clarity) between this reduction and others can be substantial. We therefore often prefer the output of the BDD version. However, it is usually slower and much more memory-intensive than the SAT versions (see Subsection 5.2 below).

There is a known bug[11] in MiniSat2 related to simplification. This affects us, and so we disable simplification in MiniSat2. The same bug appears to be present in Glucose, so we do not benchmark Glucose. CryptoMiniSat and GlueMiniSat appear to be unaffected, and we leave simplification enabled for them.

5.2   Performance Results

In this subsection, we focus on a particular difficult instance, searching for reductions from $\overline{\text{REACH}}$ (the coNL-complete problem of checking whether there is no directed path from s to t in a graph) to REACH (the NL-complete problem of checking whether there is such a path). Finding a (correct) reduction between these problems proves the Immerman-Szelepcsényi Theorem. It is known that a dimension-8 QFP exists, it is interesting to determine whether $k = 8$ is required.

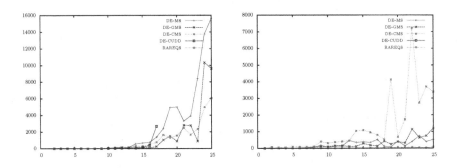

Fig. 1. CEGAR performance (in seconds) scaling n (left), c (right)

Our implementations do not approach $k = 8$. To begin, we fix simple parameters of $k = 1, c = 1, n = 3$ and examine performance when scaling k, c, n. The results are in Figure 1 and Table 5, with a (minimum) timeout of four hours. In our experience, RAREQS performs best when scaling c and the GlueMiniSat version performs best when scaling n. This is due to differences in how we handle common subformulas in DE and QBF generation – it is possible to unify these.

---

[11] https://github.com/niklasso/minisat/issues/3

Table 5. Times (in seconds) as k increases

	DE-MS	DE-GMS	DE-CMS	DE-CUDD	RAREQS
$k = 1, c = 1, n = 3$	0.05	0.06	0.08	0.07	0.03
$k = 2, c = 1, n = 2$	0.06	0.11	0.28	6.30	0.06
$k = 2, c = 1, n = 3$	3562.14	1696.26	1755.03	timeout	3267.10

When actually searching for reductions, we can range over counter-example sizes (starting with $n = 1$ or $n = 2$). The small counter-examples exclude many candidates, giving a large performance improvement[12]. For this example, no reduction in this space exists for $n = 5$, so our implementations would stop at that point. Increasing only n decreases the likelihood of a reduction existing, making the instance more strongly negative.

Scaling only c is similar; a "reduction" (correct on graphs of size 3 but not in general) exists at $c = 4$. In this example, the instance becomes more strongly positive as c increases. However, scaling these parameters shows the limit of our current approaches. Even with scaling, our implementations do not perform reasonably on hard instances (properties which use transitive closure) with $k > 2$.

Above we considered baseline settings of $k = 1, c = 1, n = 3$ and scaled a single parameter. In Figure 2, we scale n with a slightly-modified baseline of $k = 1, c = 2$ to show performance with more than one conjunction.

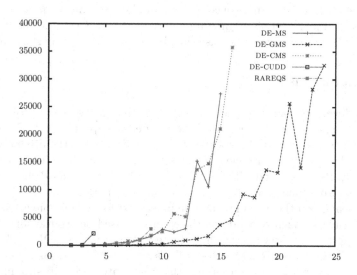

Fig. 2. CEGAR performance scaling n with $k = 1, c = 2$

---

[12] We do not include range benchmarks here. They give additional advantages to the CEGAR approach.

SAT vs BDD. The largest difference in performance between DE versions is between the BDD implementation and the SAT implementations. The BDD implementation maintains a structure explicitly representing all reductions consistent with the examples seen so far. This is memory-intensive and slower than the SAT versions, which only find a single explicit candidate at each stage.

However, the BDD version chooses a simplest candidate at each step. This usually results in the BDD version requiring fewer iterations of the counter-example/candidate loop, although each iteration is more expensive. For example, if we look for a reduction from the query $R(s)$ to the query $\exists x\, R(x)$ with parameters $k = 3, c = 4, n = 6$, DE with CryptoMiniSat uses 7 counter-examples and finds a correct, but unnecessarily complicated, reduction. DE with CUDD uses 3 examples and finds the simplest reduction in this search space.

Comparing SAT Solvers. The SAT-based DE implementations and RAREQS outperform the other approaches we consider. However, differences between the SAT solvers are visible in the results above. On the hardest examples we consider, DE with GlueMiniSat is best for large n. Our generator handles common subformulas better than DE, visible in the performance of RAREQS when scaling c.

On easy instances, DE using MiniSat often outperforms the others. This is likely because we disable simplification to avoid a bug in MiniSat and simplification is not needed for these instances. However, in practice we prefer the BDD-based implementation in such cases: any approach suffices for these instances and we prefer the more-understandable reductions found by this version.

When searching over ranges of counter-example sizes, GlueMiniSat usually performs best and we prefer GlueMiniSat on hard instances. CryptoMiniSat supports parallel SAT-solving – we do not use this, but it may improve performance.

6   Conclusions

We have developed and benchmarked several approaches to the problem of automatically discovering reductions between decision problems[13]. For each approach, it is possible to find instances where it outperforms the others. However, it is possible to state several clear conclusions. The dedicated CEGAR approach is generally the best, and GlueMiniSat performs best on hard instances. Our BDD-based approach gives the nicest output. The performance of QBF solvers depends heavily on the way in which the innermost formula is converted to CNF. For ASP solvers, much depends on the grounder used before the solver starts.

Our experiments show that each of the tested approaches – QBF solvers, ASP solvers, and the CEGAR method – still leaves a large room for improvements. We provide our testing instances and their generator, we will submit them to relevant competitions, and we encourage the community to use them for testing and optimization of all mentioned solvers. Moreover, we hope that it is possible to combine the strengths of each approach together with new improvements, in order to achieve better performance on hard, meaningful instances.

---

[13] Visit http://toss.sf.net/reduct.html to test the CEGAR approach online.

References

1. Allender, E., Balcázar, J.L., Immerman, N.: A first-order isomorphism theorem. SIAM J. Comput. 26(2), 539–556 (1997)
2. Clarke, E.M., Grumberg, O., Jha, S., Lu, Y., Veith, H.: Counterexample-guided abstraction refinement for symbolic model checking. J. ACM 50(5), 752–794 (2003)
3. Crouch, M., Immerman, N., Moss, J.E.B.: Finding reductions automatically. In: Blass, A., Dershowitz, N., Reisig, W. (eds.) Fields of Logic and Computation. LNCS, vol. 6300, pp. 181–200. Springer, Heidelberg (2010)
4. Eiter, T., Gottlob, G.: On the computational cost of disjunctive logic programming: Propositional case. Annals of Mathematics and Artificial Intelligence 15(3-4), 289–323 (1995)
5. Faber, W., Ricca, F.: Solving hard ASP programs efficiently. In: Baral, C., Greco, G., Leone, N., Terracina, G. (eds.) LPNMR 2005. LNCS (LNAI), vol. 3662, pp. 240–252. Springer, Heidelberg (2005)
6. Fagin, R.: Generalized first-order spectra and polynomial-time recognizable sets. In: Complexity of Computation, SIAM-AMS Proceedings, vol. 7, pp. 43–73. Amer. Mathematical Soc. (1974)
7. Gold, E.: Language identification in the limit. Inform. Control 10(5), 447–474 (1967)
8. Grädel, E., Kolaitis, P.G., Libkin, L., Marx, M., Spencer, J., Vardi, M.Y., Venema, Y., Weinstein, S.: Finite Model Theory and Its Applications. Texts in Theoretical Computer Science. Springer (2007)
9. Grohe, M.: Fixed-point logics on planar graphs. In: Proc. of LICS 1998, pp. 6–15. IEEE Computer Society (1998)
10. Grohe, M.: Fixed-point definability and polynomial time on graphs with excluded minors. J. ACM 59(5), 27:1–27:64 (2012)
11. Immerman, N.: Relational queries computable in polynomial time. Inform. Control 68, 86–104 (1986)
12. Immerman, N.: Languages that capture complexity classes. SIAM J. Comput. 16(4), 760–778 (1987)
13. Immerman, N.: Descriptive Complexity. Springer (1999)
14. Janota, M., Klieber, W., Marques-Silva, J., Clarke, E.: Solving QBF with counterexample guided refinement. In: Cimatti, A., Sebastiani, R. (eds.) SAT 2012. LNCS, vol. 7317, pp. 114–128. Springer, Heidelberg (2012)
15. Janota, M., Marques-Silva, J.: Abstraction-based algorithm for 2QBF. In: Sakallah, K.A., Simon, L. (eds.) SAT 2011. LNCS, vol. 6695, pp. 230–244. Springer, Heidelberg (2011)
16. Leone, N., Pfeifer, G., Faber, W., Eiter, T., Gottlob, G., Perri, S., Scarcello, F.: The DLV system for knowledge representation and reasoning. ACM Transactions on Computational Logic 7(3), 499–562 (2006)
17. Peschiera, C., Pulina, L., Tacchella, A., Bubeck, U., Kullmann, O., Lynce, I.: The seventh QBF solvers evaluation (QBFEVAL'10). In: Strichman, O., Szeider, S. (eds.) SAT 2010. LNCS, vol. 6175, pp. 237–250. Springer, Heidelberg (2010)
18. Vardi, M.Y.: The complexity of relational query languages. In: Proc. of STOC 1982, pp. 137–146. ACM (1982)

A Constraint Satisfaction Approach for Programmable Logic Detailed Placement

Andrew Mihal and Steve Teig

Tabula Inc., Santa Clara CA 95054, USA
{amihal,steve}@tabula.com

Abstract. This paper presents a Boolean SAT constraint satisfaction formulation of the detailed placement problem for programmable logic. The detailed placement problem is usually considered a poor candidate for a SAT-based solution due to complex timing constraints and the large size of the problem space. To overcome these challenges, we encode domain-specific knowledge into the problem formulation and add new features to the SAT solver. First, a Boolean encoding of timing constraints is presented that utilizes concepts from static timing analysis. Second, future cost clauses are added to the formulation to guide the SAT solver in a manner similar to A* search. Third, a dynamic clause generation approach is described that keeps the working problem size small by adding clauses on demand as the SAT solver explores the problem space. This includes dynamic cardinality clauses and dynamic addition of literals to cardinality clauses.

1 Introduction

SAT has had a long history of use in semiconductor design automation tools. It is used extensively in logic optimization, verification, and test generation [5, 10, 11, 14] and has been applied successfully on routing problems [16, 17, 21].

SAT-based placement has attracted less research interest. In the context of programmable logic, placers assign components from a netlist graph onto discrete sites on a programmable fabric. The result must satisfy a variety of constraints such as timing and routability. Devadas described placement via SAT-based bipartitioning in [4] but concluded that SAT solvers of the time were not yet powerful enough to solve more general 2-D placement problems.

Simulated annealing has been used effectively for placement, but this technique has disadvantages that can be addressed by a SAT-based approach. Given an initial placement, annealers make small perturbations (e.g. swapping the positions of two components) and measure the total delta cost reported by a set of cost functions. Downhill moves are accepted, and uphill moves are accepted with a probability determined by the magnitude of the cost increase and a decreasing temperature schedule. VPR is a well-known example of this approach [2].

The disadvantages of annealing are a result of the simplicity of the move set and the complexity of the cost functions. Any subproblem that requires a coordinated placement change involving several components to repair requires several independent annealing moves to be accepted sequentially. The cost functions must be carefully constructed to ensure that partial progress towards the final goal is seen as gradual improvement.

M. Järvisalo and A. Van Gelder (Eds.): SAT 2013, LNCS 7962, pp. 208–223, 2013.

This property can be difficult to arrange, especially if the repair requires moving unrelated, non-violating components out of the way in order to make room for the components that are actually violating constraints.

A constraint satisfaction approach can directly address this weakness. Instead of searching for small changes that gradually approach an acceptable solution, a placer based on constraint satisfaction could rearrange a large number of components simultaneously such that the final result satisfies all of the constraints. A SAT result directly gives an acceptable placement and an UNSAT result proves that no solution exists. This strategy has the potential to outperform simulated annealing if it can be made efficient.

There are numerous technical obstacles that make the placement problem appear to be poorly-suited to a SAT-based formulation. One is the large size of the problem space and the number of constraints. For example, a constraint that a particular netlist edge has a legal routing is dependent, at a minimum, on the placements of both the source and sink components of that edge. The number of constraints for this single edge thus grows as the product of the number of placement options for the source and sink components.

Timing constraints involving paths composed of multiple edges add even more complexity. The sum of placement-based routing and logic delays on a path between state elements must be less than the desired clock period for the design. These constraints involve numerous component placements, real numbers, addition, and inequalities which can be difficult to encode into Boolean formulas. Path-based constraints are also prone to deep searches where the solver discovers only after placing nearly the entire path that the last edge has no valid options remaining.

We have found solutions to these challenges by encoding domain-specific knowledge into the problem and by adding new SAT solver features to handle the large problem space. The general structure of our constraint satisfaction formulation is given in Section 3. We then present a set of contributions that build on this formulation to make a complete and efficient solution. Section 4 describes a technique for encoding timing constraints into Boolean clauses that uses concepts from static timing analysis. Section 5 introduces a domain-specific variable selection order. Future cost clauses are added using variables that occur early in this selection order to guide the solver away from conflicts that would otherwise be found only late in the search. Section 6 describes SAT solver extensions for dynamic clause generation that grow the problem formula on demand as the solver searches the problem space.

A complete detailed placer based on SAT is currently deployed at Tabula. This tool outperforms our previous placer based on simulated annealing both in run time and quality of results. Experimental results comparing these approaches are given in Section 7. We also measure the impact of each technique described in this paper in isolation. To get started, the following section briefly introduces the detailed placement problem for programmable logic.

2 Programmable Logic Detailed Placement

Programmable logic devices have long been compelling implementation platforms for digital electronics. They provide low up-front design costs and bypass the complexities of nanometer-scale transistor design. The general idea is that a prefabricated chip can

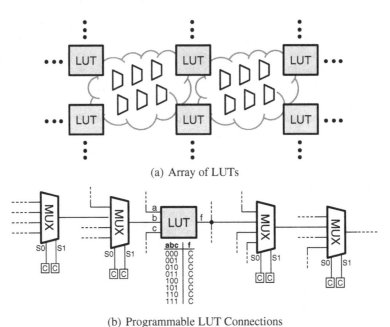

(a) Array of LUTs

(b) Programmable LUT Connections

Fig. 1. Generic Programmable Logic Device Architecture

be configured to implement any desired digital circuit by setting programmable bits on the chip.

Figure 1 shows an array of n-input lookup tables interconnected by a rich network of multiplexers. Each lookup table can implement any combinational function of n inputs by programming the 2^n bits of memory contained therein. Lookup tables are connected together by programming memory bits that drive the select signals of the multiplexers.

A modern device may contain more than one million such lookup tables, allowing very complex circuits to be realized. State elements such as flip-flops or latches are usually interspersed amongst the lookup tables to simplify the construction of sequential circuits. Coarser resources such as memories, arithmetic units, and even complete processor cores are sometimes included as well. Kuon et al. [12] provide a survey of modern architectures.

Engineers create a design for implementation on programmable logic by writing a register-transfer level circuit description in Verilog or VHDL. A suite of design automation tools provided by the programmable logic vendor compiles the design and produces the programmable bit values for the chip. The compilation process has four major phases: synthesis, global placement, detailed placement, and routing. Synthesis compiles the circuit description into an optimized netlist of lookup tables. Placement assigns the lookup tables in the netlist onto lookup tables on the physical chip. Routing configures the multiplexers so that the connections specified in the netlist are made on the chip. Each phase poses intriguing optimization challenges. Chen et al. [3] give a broad survey of the techniques commonly used.

This paper focuses on the detailed placement problem. Detailed placers assume that the netlist already has a placement that satisfies some global optimization criteria but still has problems of a more local nature that need to be repaired. For example, a global placer could use an analytical algorithm with a continuous coordinate system and an abstract geometric model of routing delay. The detailed placer refines this placement by snapping components to overlap-free discrete placement sites using a more accurate routing delay model based on actual paths through the interconnect network. We consider three problems in this paper that are not specific to any particular programmable logic architecture: placement overlaps, local routability, and timing.

2.1 Placement Overlaps

First, the placement must be free of overlaps. The global placer has already achieved a global density target, but there may remain scattered placement sites with more than one component. The detailed placer has to fix these problems by moving the overlapping components to nearby unoccupied sites.

2.2 Local Routability

Second, the placement must be locally routable. The standard assumption is that the router will be responsible for configuring multiplexers but will not consider moving netlist components to different sites. If the detailed placer can deduce that two netlist edges must use the same multiplexers with no possible detours, then the routing will fail.

The netlist and fabric shown in Figure 2 demonstrate the issue in its simplest form. Sites LUT1 and LUT2 are both directly driven by MUX3. If netlist components B and D are placed on sites LUT1 and LUT2, then the signal from component A is going to collide with the signal from component C at MUX3. The detailed placer must guarantee that the placement is free of this type of problem.

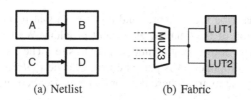

(a) Netlist   (b) Fabric

Fig. 2. Routing Collision Example

Local routability constraints are limited to multiplexers in the close vicinity of the source and sink. The detailed placer does not attempt to plan entire connections. After a few multiplexer hops, enough detour possibilities are visible to infer that the routing will be successful.

2.3   Timing Constraints

Third, the placement must meet timing constraints. The user specifies that the sequential logic must run at a particular clock frequency. This constraint requires that every netlist path between state elements has a total delay less than or equal to the target clock period τ.

Figure 3 shows a netlist with several such paths: $ABCD$, $ABECD$, $ABEF$, FEF, and $FECD$. Each path has logic delays (e.g. D_B) and routing delays (e.g. D_{AB}) that are dependent on the placement sites for the netlist components. The components on path $ABCD$, for example, must be placed in a way that satisfies the constraint $D_{AB} + D_B + D_{BC} + D_C + D_{CD} \leq \tau$.

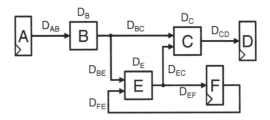

Fig. 3. Netlist Annotated With Delays

Paths that fail timing must be repaired by moving components to reduce routing delays or by using retiming to move logic across state elements. The detailed placer can assume that the global placer has already performed global delay optimization under an abstract model of routing delay. When the paths are reassessed using actual routing delays, some may violate timing. It is expected that only a minority of paths will require repair and that components will only have to move a relatively short distance away from their starting locations.

3   Constraint Satisfaction Formulation

In this section we describe the variables and constraints that make up the constraint satisfaction formulation of the detailed placement problem. We begin with variables that encode the placement of netlist components onto sites. The Boolean variable V_{AX} has the meaning that component A is placed on site X. These *placement variables* can be arranged into a sparse matrix where the components are the rows and the sites are the columns as follows:

$$
\begin{array}{c|cccc}
 & W & X & Y & Z & \cdots \\
\hline
A & V_{AW} & V_{AX} & V_{AY} & & \\
B & & V_{BX} & V_{BY} & & \\
C & & & V_{CY} & V_{CZ} & \\
\vdots & & & & & \ddots
\end{array}
\tag{1}
$$

The matrix is sparse because not every site is a candidate for every component. If the architecture contains a heterogeneous set of resources, the different types of netlist components (lookup tables, state elements, etc.) can be placed only on compatible sites. Matrix entries can also be omitted for geometric reasons. The placer may limit each component to consider only placement sites within a given radius of the initial location. Since the detailed placer is expected to make local changes and not global changes, this limitation is an acceptable way to bound the problem size.

For each row in the matrix, an *exactlyOne* constraint (a standard *atMostOne* cardinality constraint combined with a standard Boolean clause) is added to ensure that each component gets a placement. For each column in the matrix, an *atMostOne* constraint is added to prevent placement overlaps. These basic constraints are sufficient to produce legal placements.

Next, we define *pin permutation variables*. When a k-input LUT component is placed on an n-input LUT site, there are $_nP_k = n!/(n-k)!$ ways to assign the component's inputs to pins on the site. The configuration bits inside the LUT are correspondingly permuted to maintain the same logic function.

Pin permutations are useful when the architecture has asymmetrical connectivity to the pins. Given the placement sites for the source and sink components on a netlist edge, it may be preferable to route to one sink pin over another for either timing or routability reasons. The detailed placer therefore solves for pin placements in addition to component placements.

The pin permutation variable $P_{A\pi}$ has the meaning that component A is placed using permutation π. An *exactlyOne* constraint for each component creates a one-hot encoding. This encoding is reasonable because n and k are typically small numbers, e.g. 3.

To complete the problem formulation, routability and timing constraints for each netlist edge AB are created following this pattern:

$$(V_{AX} \wedge V_{BY} \wedge P_{B\pi}) \rightarrow C \tag{2}$$

Equation 2 states that a conjunction of placement decisions implies some consequence C. Sets of consequences that are pairwise incompatible are ruled out using clauses of the form $atMostOne(C_i, C_j, \ldots)$.

A local routability constraint is made as follows. The conjunction $(V_{AX} \wedge V_{BY} \wedge P_{B\pi})$ fully defines the placement of a netlist edge. It gives the source and sink component placements and the exact sink pin placement. We assume an algorithm that walks over the interconnect network starting from these pins and identifies local multiplexers that must be used to route the edge. We define a category of non-decision consequence variables M_{NA} to indicate that MUX N is occupied by the signal from component A. A clause of the form $(V_{AX} \wedge V_{BY} \wedge P_{B\pi}) \rightarrow M_{NA}$ is then added to the formulation for each multiplexer identified by the algorithm. These clauses state that particular placement choices for netlist edge AB imply that certain multiplexers are occupied by signal A. For each multiplexer N, an *atMostOne* clause over the consequence variables $M_{NA}, M_{NB}, \ldots$ prevents local routability conflicts.

Timing constraints follow the same pattern. The complete placement of netlist edge AB implies a routing delay D_{AB}. We assume there exists a timing model of the target architecture that can provide this delay value given the source and sink pin placement. This results in the clause:

$$(V_{AX} \wedge V_{BY} \wedge P_{B\pi}) \rightarrow (D_{AB} = d_{XY\pi}) \tag{3}$$

Logic delays depend only on the placement site and permutation choice for a single netlist component. This results in a similar clause:

$$(V_{BY} \wedge P_{B\pi}) \rightarrow (D_B = d_{Y\pi}) \tag{4}$$

The clauses that rule out mutually incompatible timing consequences are the inequalities $D_{AB} + D_B + \ldots \leq \tau$ for all paths between state elements in the netlist.

These clauses make a complete formulation of the detailed placement problem, but there are two obvious practical problems. First, the timing consequences are not Boolean variables and the timing constraints are not Boolean clauses. Second, the size of the formulation grows impractically large as the size of the netlist and architecture increases.

The number of timing constraint clauses grows with the number of paths in the netlist, which can be exponential in the number of netlist components. The number of local routability consequence variables is the product of the number of signals in the netlist and the number of multiplexers in the architecture. Programmable logic interconnect networks are over-provisioned, so the number of multiplexers can be an order of magnitude greater than the number of LUTs in the architecture.

The clauses that imply both types of consequence variables grow in number with the size of the placement space for each edge: the number of source component placements times the number of sink component placements times the number of sink component permutations. The following sections describe our solutions to these problems.

4   SAT Encoding of Timing Constraints

In this section, we improve the formulation of timing constraints to reduce the number of clauses and to produce a purely Boolean encoding. Previously, we described timing constraints of the form $D_{AB} + D_B + \ldots \leq \tau$. There is one such constraint for each path between state elements in the netlist, which can be exponential in the size of the netlist. Concepts from static timing analysis can be used to make a more efficient encoding.

Static timing analysis computes an *arrival time* and a *required time* at each component in the netlist graph [9]. The arrival time is the maximum path length to state elements backwards through the transitive fanin of a component. This value represents the time it takes from the beginning of the clock cycle for the data to propagate through the circuit and to become valid at the output of the component. The arrival time at the output of a state element is defined to be zero.

The required time is the clock period τ minus the maximum path length to state elements forwards through the transitive fanout of a component. This value is the latest time at which the component output can become valid and still make it to the downstream state elements before the end of the clock cycle. The required time at the input of a state element is defined to be τ.

The *slack* of a component is the required time minus the arrival time. A component with negative slack is on a path that fails timing.

The concept of slack leads to a better formulation of timing constraints that avoids enumerating all state-to-state paths in the netlist. Instead, we can simply constrain each component to have non-negative slack with a clause $\text{Arr}_A \leq \text{Req}_A$.

The arrival time Arr_A and required time Req_A are defined in terms of the immediate fanin and fanout components:

$$\text{Arr}_A = \max_{\text{fanin } Fi} (\text{Arr}_{Fi} + D_{FiA} + D_A) \tag{5}$$

$$\text{Req}_A = \min_{\text{fanout } Fo} (\text{Req}_{Fo} - D_{Fo} - D_{AFo}) \tag{6}$$

Using these formulas, equations 3 and 4 are rewritten as:

$$(V_{AX} \wedge V_{BY} \wedge P_{B\pi}) \rightarrow (\text{Arr}_B - \text{Arr}_A \geq d_{XY\pi} + d_{Y\pi})$$
$$\wedge (\text{Req}_A - \text{Req}_B \leq -d_{XY\pi} - d_{Y\pi}) \tag{7}$$

The result is a difference logic formulation. The only non-Boolean variables are the arrival and required times Arr_A and Req_A for each netlist component. The worst-case exponential number of constraints is avoided because these constraints scale with the number of netlist edges instead of the number of netlist paths.

To solve this formulation, a satisfiability-modulo-theory (SMT) solver that supports difference logic constraints could be used. Alternatively, the difference logic constraints can be encoded into an equivalent problem that is purely Boolean and then solved with a standard SAT solver. We will continue on the second path and briefly discuss the SMT option in the conclusion.

Two encoding options are the small-domain approach and the EIJ approach [19]. The EIJ approach requires enumeration of cycles amongst the difference logic constraints, which is undesirable because it is equivalent to enumerating paths between state elements in the netlist. The standard small-domain encoding approach requires instantiation of Boolean clauses that represent adder and comparator circuits, which can also be expensive.

Instead, we use a small-domain encoding variation due to Ohrimenko et al. [18]. The key idea is to define Boolean variables that represent upper and lower bounds on the difference logic variables instead of exact values of the variables. This approach is a natural fit for modeling arrival and required times and detecting negative slack.

For each netlist component A we create a number line subdivided into T discrete values representing times within the clock period τ. Each subdivision has an associated non-decision Boolean variable E_{At}. When true, this variable has the meaning that $\text{Req}_A \leq t\frac{\tau}{T}$. When false, this variable indicates that $\text{Arr}_A > t\frac{\tau}{T}$.

Adjacent variables on the number line are related by binary clauses that implement the transitivity property of inequalities:

$$E_{At_i} \rightarrow E_{At_{i+1}} \tag{8}$$

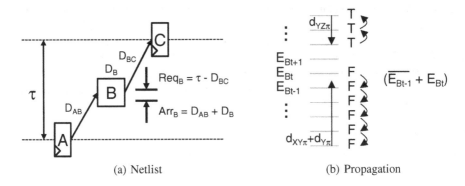

(a) Netlist   (b) Propagation

Fig. 4. Boolean Arrival/Required Variables

To see how this encoding works and how it enforces the timing constraint $\text{Arr}_A \leq \text{Req}_A$, consider the following example. Figure 4(a) shows a netlist with a single component between state elements. The placement state of edge AB determines the delays D_{AB} and D_B and results in the clause:

$$(V_{AX} \wedge V_{BY} \wedge P_{B\pi}) \rightarrow (\text{Arr}_B \geq d_{XY\pi} + d_{Y\pi}) \qquad (9)$$

The delay $d_{XY\pi} + d_{Y\pi}$ is rounded conservatively to a subdivision on the number line for component B. The right-hand side of equation 9 can then be replaced with the corresponding negative Boolean literal:

$$(V_{AX} \wedge V_{BY} \wedge P_{B\pi}) \rightarrow \overline{E_{Bt}} \quad \text{where } t = \lceil (d_{XY\pi} + d_{Y\pi})T/\tau \rceil \qquad (10)$$

Figure 4(b) shows what happens to the number line variables when this negative literal becomes asserted. Boolean constraint propagation will use the clauses described in equation 8 to cause all variables lower in the number line to become false as well. If the arrival time is greater than some quantized value t, then it must also be greater than all smaller values $t - 1, t - 2$, and so forth.

A similar sequence of events occurs for the required time at component B. The placement state of edge BC determines the delay D_{BC}:

$$\begin{aligned}(V_{BY} \wedge V_{CZ} \wedge P_{C\pi}) &\rightarrow (\text{Req}_B \leq \tau - d_{YZ\pi}) \\ &\rightarrow E_{Bt'} \quad \text{where } t' = \lfloor (\tau - d_{YZ\pi})T/\tau \rfloor\end{aligned} \qquad (11)$$

When the literal on the right-hand side of equation 11 becomes true, then equation 8 causes all variables higher in the number line to become true. If the required time is less than or equal to some quantized value t', then it must also be less than or equal to all higher values. We call equation 8 a *vertical chain* clause due to this cascading action.

Unassigned number line variables between the lowest true value and the highest false value represent the slack at the component. If the solver ever makes decisions that result in an arrival time becoming larger than a required time, then Boolean constraint propagation will try to assign a number line variable to be both true and false. Negative

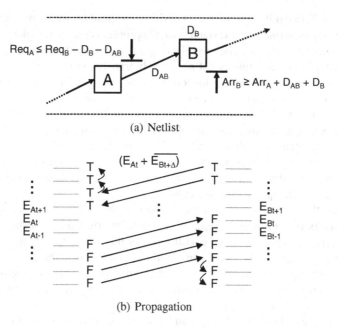

(a) Netlist

(b) Propagation

Fig. 5. Propagation Across a Netlist Edge

slack (i.e. failure to satisfy the timing constraints) is therefore detected as an ordinary Boolean SAT conflict.

Paths with more than one component are the most common case in practice and require an additional category of clauses. In Figure 5(a), neither the source nor the sink of edge AB is a state element. This case is handled by adding clauses that relate variables on the number line for component A to variables on the number line for component B:

$$(V_{AX} \wedge V_{BY} \wedge P_{B\pi}) \rightarrow \bigwedge_t (\overline{E_{At}} \rightarrow \overline{E_{B(t+\Delta)}}) \quad \text{where } \Delta = \lceil (d_{XY\pi} + d_{Y\pi})T/\tau \rceil \quad (12)$$

This collection of Boolean clauses has the action of propagating arrival times forward in the netlist and required times backward in the netlist with the addition of delay $D_{AB}+D_B$. These clauses are called *horizontal chain* clauses due to the cascading action between adjacent number lines. This is demonstrated graphically in Figure 5(b).

In summary, this approach mimics static timing analysis using Boolean constraint propagation. Path-based timing constraints are reduced to ordinary Boolean variables and clauses that scale linearly with the number of edges in the netlist.

5   Future Cost Clauses

The next technique addresses the issue of exploring the large problem space efficiently. Our experimentation has shown that customizing the variable selection order heuristic has a first-order effect on search efficiency.

In Section 3, two different categories of decision variables were presented: *placement* variables and *pin permutation* variables. In this problem domain it makes sense to select a site for a netlist component before selecting a permutation for that netlist component. The interconnect architecture is not necessarily translation invariant, so a permutation with good timing and routability characteristics at one site might not be appropriate at a different site. This selection order can also be seen as making coarse-grained decisions first and then making refinements later.

We utilize a custom two-stage variable selection heuristic to implement this ordering. All of the placement variables are selected first, followed by all of the pin permutation variables. Each stage is implemented with an independent Variable State Independent Decaying Sum (VSIDS) priority heap [15], so that the secondary selection criteria is still based on recent appearance in a conflict.

The VSIDS activity metrics for these two categories are seeded so that the solver initially favors placing netlist components on their original sites using their original permutations. We find that this heuristic guides the solver to finding a solution that makes fewer changes to the placement. Without this heuristic there is no other criteria by which one SAT solution is preferred over another, and the placer develops a tendency to move nonviolating components to different satisfactory locations in a way that causes "churn" but does not improve the placement.

There is an additional benefit that can be derived from this custom variable selection order. Recall that a complete netlist edge placement is described by the conjunction $(V_{AX} \wedge V_{BY} \wedge P_{B\pi})$. This conjunction contains two variables from the first stage of the variable selection order and one variable from the second stage. As an antecedent in a clause, it will not become true and imply its consequence variables until the search has entered the second stage. However, we know that exactly one permutation variable must eventually be selected. Writing out all of the clauses starting with $(V_{AX} \wedge V_{BY} \wedge P_{B\pi})$ for all n permutations, we have:

$$\text{exactlyOne}(P_{B1}, P_{B2}, \ldots P_{Bn}) \tag{13}$$

$$
\begin{aligned}
(V_{AX} \wedge V_{BY} \wedge P_{B1}) &\rightarrow C_1 \\
(V_{AX} \wedge V_{BY} \wedge P_{B2}) &\rightarrow C_2 \\
&\vdots \\
(V_{AX} \wedge V_{BY} \wedge P_{Bn}) &\rightarrow C_n
\end{aligned}
\tag{14}
$$

Applying hyper-resolution [1], we produce the clause:

$$(V_{AX} \wedge V_{BY}) \rightarrow \min_{1 \le i \le n} C_i \tag{15}$$

The "min" operator applied to a set of consequences means, intuitively, the least restrictive consequence of the set. In the case of timing constraints where the consequences are of the form $(\text{Arr}_B - \text{Arr}_A \ge d)$, this operator gives the minimum delay d. Although the exact pin permutation is not yet known, the edge delay must be at least the minimum delay of all the choices that will be available.

If the consequences are local routability variables M_{An}, the "min" operator would produce the conjunction of variables that are common to all C_i. In that scenario, there is one or more routing MUXes that must be used for the edge regardless of which sink component permutation is eventually chosen.

These clauses improve the search efficiency because they give the solver a lower bound on the consequences of decisions made in the first search stage. The clauses trigger conflicts that would otherwise only be found after the solver enters the second search stage. This behavior is similar to how an admissible future cost function provides guidance to A* search. To highlight this similarity we name these clauses *future cost clauses*.

6   Dynamic Clause Generation

The formulation described so far still requires an impractically large number of clauses. It is overly expensive both in memory and clause generation runtime. Each equation that starts with the antecedent $(V_{AX} \land V_{BY} \land P_{B\pi})$ expands to a number of clauses equal to the product of the number of source placement options, sink placement options, and sink permutations. For the horizontal chain clauses described in equation 12, this number is further multiplied by the number of subdivisions on the number line. This quantity is the number of constraints for a single netlist edge AB.

A solution to this scalability problem is to avoid adding all of these clauses to the formulation. Observe that the placement variable matrix (1) is mostly false due to the *exactlyOne* constraint on each row. Also, each clause $(V_{AX} \land V_{BY} \land P_{B\pi}) \rightarrow C$ starts with two negative literals $(\overline{V_{AX}} + \overline{V_{BY}} + \ldots)$. A majority of the clauses are therefore trivially satisfied during the search.

Furthermore, note that the detailed placer is responsible for repairing only local constraint violations and does not attempt to make global changes to the initial placement. It is expected that only a minority of netlist components will have to be moved to accomplish this task. The solver is likely to find a SAT or UNSAT solution after attempting only a small fraction of the possible placement options V_{AX} for each component.

Therefore, while all of the clauses are theoretically necessary for correctness, in practice most of them do not affect the search. We can take advantage of this fact to reduce the working size of the problem. Clauses of the form $(\overline{V_{AX}} + \overline{V_{BY}} + \ldots)$ can be left out until the search actually enters a subspace where V_{AX} and V_{BY} are both true.

We have constructed a SAT solver with support for dynamic clause generation based on MiniSAT version 1.12b [6]. The solver calls a method *decisionCallback* after assigning a placement variable $V_{AX} = T$ and after all unit propagations have completed without conflicts.

In the callback, the immediate fanin and fanout components B of component A are checked to see if they are placed according to the current solver state. If so, then some placement variable V_{BY} will be true on the solver trail. The placer can now generate clauses for fanin edge BA (or fanout edge AB) that are of the form $(\overline{V_{AX}} + \overline{V_{BY}} + \ldots)$.

The solver then processes the incoming clauses and makes its internal state consistent. This is straightforward when the new clauses still have one or more unassigned literals with respect to the current solver trail. If any one of the new clauses has all of its literals assigned false, the solver state needs to be repaired. The inconsistent clause should have become unit at some decision level in the past and the solver should not have entered the current subspace. We recover by backtracking.

6.1 Dynamic Cardinality Clauses

Our SAT solver also supports dynamic cardinality clauses and dynamic addition of literals to cardinality clauses. This feature is used to avoid creating the complete set of local routability variables M_{NA} with their associated *atMostOne* constraints. Because the solver is expected to search only a small fraction of the placement space, and clauses are only generated when necessary, only a small fraction of the possible M_{NA} variables will ever appear in any clause. The placer maintains a sparse matrix of these non-decision Boolean variables and fills it in on demand.

We add a new *dynamicAtMostOne* clause type to MiniSAT that supports dynamic addition of literals. MiniCARD's extension of the two-watched-literal scheme to $k-n+1$ watched literals for an *atMostN-of-K* clause is used as a foundation [13]. Since new literals appear in no clauses prior to the current *decisionCallback* call, dynamic addition can be accomplished with a minimum of disruption to the solver's internal state.

In the case where N existing literals are already true, the new literal must be false. Without backtracking, the new literal's assignment state is simply overwritten as if it had been assigned false at the proper decision level. Instead of inserting the new assignment into the middle of the solver's trail vector, MiniSAT's undo mechanism is re-purposed to perform this update lazily.

Assume that the new literal should have been assigned false at decision level d. An undo is placed on the first literal at decision level $d + 1$. The function of this undo is to push the new literal onto the trail when the solver cancels level $d + 1$. These steps result in the correct functional behavior as if the new literal had been assigned false at the proper time in the past.

6.2 Related Approaches

Eén and Sörensson describe a dynamic clause generation approach in [7] wherein the solver is restarted with additional clauses after a complete solution has been produced and examined. Our approach adds clauses while the solver is running and not only between invocations of an incremental solver.

The Lynx SAT solver includes a similar callback method for adding clauses after each propagation step [8]. That callback method examines the solver trail and adds clauses that conflict with the current assignment. In comparison, our approach allows clauses that do not conflict with the current assignment to be added as well.

Ohrimenko et al. [18] include lazy clause generation to reduce the number of Boolean clauses created for difference logic propagation. Our approach performs dynamic clause generation on a coarser level of abstraction. This is possible due to the natural subdivisions of the problem space and the low likelihood of actually searching the majority of these subspaces.

7 Experimental Results

Our SAT-based placer uses a search-and-repair strategy. Instead of turning the netlist into one large SAT instance that tries to repair all placement problems simultaneously, the placer runs a series of small instances that each try to repair a specific violation. Each subproblem covers a subset of the netlist with approximately 100 netlist components in the neighborhood of a violation. This subproblem size is sufficient to repair complex violations that require coordinated motion of dozens of components. Each SAT instance repairs not only the targeted violation but also other violations that are coincidentally included in the neighborhood. The timing constraints are gradually tightened towards the final target period τ over time as the placer makes progress.

To measure the improvement of the SAT-based approach over our previous simulated annealing placer, we ran experiments placing netlists from the OpenCores [20] database using both approaches. Table 1 compares the runtime and the achievable frequency of the two placers. Both measures are normalized to the results of the annealing-based placer. The SAT-based approach obtains better frequencies, and in general requires less runtime. In the cases where the SAT-based placer runs longer, the extra effort is rewarded with a better achievable frequency.

Table 2 measures the relative importance of the three main techniques described in this paper that enable our problems to be solved efficiently: A*-style future cost clauses, a domain-specific variable selection order, and dynamic clause generation. We built three configurations of our placer that each remove one of these features. We then compare the runtime and achievable frequency omitting one feature against the baseline configuration which has all three features turned on. The runtime and frequency numbers in Table 2 are normalized to the baseline SAT-based placer configuration. Our expectation is that a configuration with more than one feature disabled would be impractical to evaluate due to high runtime.

Table 2 shows that dynamic clause generation has the biggest impact on runtime. This is dominated by the time spent generating clauses. A comparison of the rightmost columns of tables 1 and 2 shows that only a small fraction of the total number of clauses are ever generated when dynamic clause generation is used. This improves the memory footprint as well as the runtime.

Table 1. Constraint Satisfaction vs. Annealing: Runtime and Frequency

Design	Comps	Annealing Placer		SAT-Based Placer			
		Runtime	Freq	Runtime	Freq	Avg Vars	Avg Clauses
camellia256	89341	1.0	1.0	0.054	1.14	751k	30k
sudoku	17784	1.0	1.0	0.266	1.49	758k	31k
dct	17199	1.0	1.0	2.526	1.52	1512k	116k
wishbone	12775	1.0	1.0	0.028	1.03	1081k	25k
fpu_double	12660	1.0	1.0	0.346	1.39	1598k	74k
aes	5818	1.0	1.0	2.179	1.20	908k	24k
r2000sc	5095	1.0	1.0	0.788	1.28	1404k	40k
sha256	3368	1.0	1.0	0.106	1.14	2100k	39k

Table 2. Relative Contribution of the A* Clauses, Variable Ordering, and Dynamic Clauses

Design	All Features Runtime	Freq	No A* Runtime	Freq	No Var Order Runtime	Freq	No Dynamic Runtime	Freq	Avg Clauses
camellia256	1.0	1.0	1.18	1.00	5.31	0.89	30.29	1.01	8099k
sudoku	1.0	1.0	1.08	1.00	5.25	0.97	7.40	1.00	743k
dct	1.0	1.0	1.65	1.06	8.37	0.94	4.49	0.87	8153k
wishbone	1.0	1.0	1.45	1.00	5.00	0.99	25.53	1.00	17449k
fpu_double	1.0	1.0	2.41	1.09	6.73	0.82	4.06	0.94	3730k
aes	1.0	1.0	1.17	1.00	7.45	0.99	2.71	0.93	5070k
r2000sc	1.0	1.0	1.84	1.00	8.71	0.99	7.95	0.93	22285k
sha256	1.0	1.0	1.19	1.00	10.96	1.00	11.68	1.00	16908k

8 Conclusions

Programmable logic detailed placement is a challenging problem domain for Boolean satisfiability that has not been attempted before. Using the techniques described in this paper, our formulation can be solved efficiently and used to construct a practical tool.

This approach outperforms our previous annealing-based placer in both run time and quality of results. All of the effort is focused on repairing actual violations instead of randomly searching for downhill moves. This placer is also able to repair problems that require coordinated changes to a large number of components. The annealing-based placer frequently terminates with inferior results because it is not able to repair these complex problems.

One drawback of our Boolean encoding of timing constraints is the quantization of time. Each delay sum $(D_{AB} + D_B)$ requires conservative rounding to discrete number line variables E_{At}. On a long state-to-state path through the netlist, these rounding errors accumulate and over-constrain the problem. This is especially problematic when the target clock period τ approaches the maximum frequency supported by the target architecture. The conservative rounding forces the placer to find a solution that exceeds the actual target, but the architecture does not have interconnect routes that are fast enough. Consequently, placements that would actually be acceptable are rejected.

To address this problem, our next project is to investigate the use of an SMT difference logic solver instead of a Boolean SAT solver. By natively supporting constraints such as those in equation 7, such a solver would no longer require discrete number line variables or horizontal and vertical chain clauses. This approach will increase the accuracy of the formulation, leading to better placement quality with higher achievable frequencies and faster run times. In order to be efficient, this solver would also have to support our custom variable selection order and dynamic clause generation techniques.

The recent advances in SAT solver capabilities have encouraged us all to apply SAT to new problem domains. By leveraging domain-specific knowledge to improve both the formulation and the solver, even a domain that was long considered unsuitable for SAT can now be solved efficiently.

References

1. Bacchus, F.: Enhancing Davis Putnam with extended binary clause reasoning. In: National Conference on Artificial Intelligence, pp. 613–619 (July 2002)
2. Betz, V., Rose, J.: VPR: A new packing, placement, and routing tool for FPGA research. In: Glesner, M., Luk, W. (eds.) FPL 1997. LNCS, vol. 1304, pp. 213–222. Springer, Heidelberg (1997)
3. Chen, D., Cong, J., Pan, P.: FPGA design automation: A survey. Foundations and Trends in Electronic Design Automation 1(3), 195–330 (2006)
4. Devadas, S.: Optimal layout via Boolean satisfiability. In: IEEE International Conference on Computer-Aided Design, pp. 294–297 (November 1989)
5. Drechsler, R., Eggersglüß, S., Fey, G., Tille, D.: Test Pattern Generation using Boolean Proof Engines. Springer (2009)
6. Eén, N., Sörensson, N.: An extensible SAT-solver. In: Giunchiglia, E., Tacchella, A. (eds.) SAT 2003. LNCS, vol. 2919, pp. 502–518. Springer, Heidelberg (2004)
7. Eén, N., Sörensson, N.: Temporal induction by incremental SAT solving. In: First Intl. Workshop on Bounded Model Checking, vol. 89, pp. 543–560 (2003)
8. Ganesh, V., O'Donnell, C.W., Soos, M., Devadas, S., Rinard, M.C., Solar-Lezama, A.: Lynx: A programmatic SAT solver for the RNA-folding problem. In: Cimatti, A., Sebastiani, R. (eds.) SAT 2012. LNCS, vol. 7317, pp. 143–156. Springer, Heidelberg (2012)
9. Kirkpatrick, T.I., Clark, N.R.: PERT as an aid to logic design. IBM Journal of Research and Development 10(2), 135–141 (1966)
10. Kuehlmann, A., Paruthi, V., Krohm, F., Ganai, M.: Robust Boolean reasoning for equivalence checking and functional property verification. IEEE Transactions on Computer-Aided Design 21(12), 1377–1394 (2002)
11. Kuehlmann, A.: Dynamic transition relation simplification for bounded property checking. In: International Conference on Computer-Aided Design, pp. 50–57 (2004)
12. Kuon, I., Tessier, R., Rose, J.: FPGA architecture: Survey and challenges. Foundations and Trends in Electronic Design Automation 2(2), 135–253 (2008)
13. Liffiton, M.H., Maglalang, J.C.: A cardinality solver: More expressive constraints for free. In: Cimatti, A., Sebastiani, R. (eds.) SAT 2012. LNCS, vol. 7317, pp. 485–486. Springer, Heidelberg (2012)
14. Mishchenko, A., Brayton, R., Jiang, J.R., Jang, S.: SAT-based logic optimization and resynthesis. In: Intl. Workshop on Logic and Synthesis, pp. 358–364 (May 2007)
15. Moskewicz, M., Madigan, C., Zhao, Y., Zhang, L., Malik, S.: Chaff: Engineering an efficient SAT solver. In: Design Automation Conference, pp. 530–535 (2001)
16. Nam, G., Aloul, F., Sakallah, K., Rutenbar, R.: A comparative study of two Boolean formulations of FPGA detailed routing constraints. IEEE Transactions on Computers 53(6) (June 2004)
17. Nam, G., Sakallah, K., Rutenbar, R.: Satisfiability-based layout revisited: Detailed routing of complex FPGAs via search-based Boolean SAT. In: Intl. Symposium on Field Programmable Gate Arrays, pp. 167–175 (1999)
18. Ohrimenko, O., Stuckey, P., Codish, M.: Propagation via lazy clause generation. Constraints 14(3), 357–391 (2009)
19. Seshia, S.: Adaptive Eager Boolean Encoding for Arithmetic Reasoning in Verification. PhD thesis, Carnegie Mellon University (2005)
20. Various. OpenCores open source hardware IP cores (April 2013), http://opencores.org
21. Glenn Wood, R., Rutenbar, R.: FPGA routing and routability estimation via Boolean satisfiability. IEEE Transactions on VLSI 6(2) (June 1998)

Minimizing Models
for Tseitin-Encoded SAT Instances

Markus Iser, Carsten Sinz, and Mana Taghdiri

Karlsruhe Institute of Technology (KIT), Germany
{markus.iser,carsten.sinz,mana.taghdiri}@kit.edu

Abstract. Many applications of SAT solving can profit from minimal models—a partial variable assignment that is still a witness for satisfiability. Examples include software verification, model checking, and counterexample-guided abstraction refinement. In this paper, we examine how a given model can be minimized for SAT instances that have been obtained by Tseitin encoding of a full propositional logic formula. Our approach uses a SAT solver to efficiently minimize a given model, focusing on only the input variables. Experiments show that some models can be reduced by over 50 percent.

1 Introduction

Many applications in logic and formal methods rely on SAT solvers as core decision procedures, and in most cases the application is not only interested in a yes/no answer, but also in a satisfying assignment (*model*) if one exists.

Models are used, for example, to represent counterexample traces in software verification, steps leading to a goal in SAT-based planning, or to build candidate conjunctions of theory atoms in SMT solving based on the DPLL(T) approach [6]. The employment of models ranges from giving information to the user—either directly or, more often, after some back-transformation to the application domain—to guiding a search algorithm when a SAT solver is used to iteratively enumerate solutions.

Minimized models try to strip off inessential information from a complete solution produced by a SAT solver. Such reduced models allow, for example, the user to focus on relevant parts of a counterexample trace, or to guide a SAT-based search process more efficiently. E.g., in DPLL(T), smaller SAT models alleviate the work of the theory solver(s), as they get passed smaller conjunctions of theory atoms; by this, the refinement loop typically needs fewer iterations.

In many cases, formulas from the application domain are not in conjunctive normal form (CNF) initially, which is, however, the input format that most SAT solvers require. Thus, they have to be transformed to CNF. A number of efficient procedures for this transformation are available [3,9,12,17]. But this transformation, which typically introduces additional *encoding variables*, increases the gap between the SAT solver's solution and its interpretation in the application domain. The assignment to encoding variables is often not of interest on the application domain level.

M. Järvisalo and A. Van Gelder (Eds.): SAT 2013, LNCS 7962, pp. 224–232, 2013.

To illustrate the problem, consider the formula $F = a \vee (b \wedge c)$, and assume that we are interested in finding a model for F. One such model would be $\{a \to 0,\ b \to 0,\ c \to 1\}$, assigning *true* to a and b, and *false* to c. This model is not minimal though, as setting a to *true* would already be sufficient to make the whole formula F true. So how can we obtain minimal models? Computing them on the CNF level is not sufficient to arrive at minimal models on the level of full propositional logic, as can be seen from our small example. When we convert it to CNF (using the Tseitin transformation), we obtain the clauses $\{\ (\bar{x}\ b)\ (\bar{x}\ c)\ (x\ \bar{b}\ \bar{c})\ (\bar{y}\ a\ x)\ (y\ \bar{a})\ (y\ \bar{x})\ (y)\ \}$, where x represents the subformula $b \wedge c$ and y the complete formula F. A minimal model of this clause set would be $\{x \to 0,\ c \to 0,\ a \to 1,\ y \to 1\}$ and, even after projecting it onto the original problem variables, we would obtain $\{c \to 0,\ a \to 1\}$, which is not the minimal model $\{a \to 1\}$ that we would like to see.

In this paper, we present algorithms that allow to compute minimal models (such as $\{a \to 1\}$ for F) efficiently, using standard SAT solvers to compute an initial (complete) model, which is then minimized. The main contribution of our paper is to also take the CNF encoding into account during minimization.[1]

2   Theoretical Background

We denote the set of propositional formulas by $\mathbb{F}$. Formulas in $\mathbb{F}$ are built from a set of variable symbols $\mathcal{V}$, operators $\{\wedge, \vee, \neg\}$, and constants $\{\top, \bot\}$. For each $F \in \mathbb{F}$, the set $\mathcal{V}_F \subseteq \mathcal{V}$ denotes the set of variables occurring in F. A *variable assignment* for a given formula F is a (possibly partial) function $\alpha : \mathcal{V}_F \rightsquigarrow \{0, 1\}$ that assigns a constant value to some variable in $\mathcal{V}_F$. We use $\mathrm{dom}(\alpha)$ to denote the set of variables for which α is defined. If $\mathrm{dom}(\alpha) = \mathcal{V}_F$, we say that the assignment is *complete*; otherwise it is *partial*. Dealing with partial assignments imposes the need for a three-valued interpretation. The interpretation of a formula F under a (partial) assignment α is denoted by $\mathcal{I}_\alpha$ and defined in Figure 1. Here, 1, 0, and U stand for *true*, *false*, and *undefined*, respectively.

We now extend the standard definition of a model to partial assignments.

Definition 1 (Model). *Given a formula F, a (partial) assignment α is a model (or a satisfying assignment) for F iff $\mathcal{I}_\alpha(F) = 1$. We use $\alpha \models F$ to denote that α is a model of F.*

In what follows, we use M_α to denote the set of *true* literals in an assignment α for a formula F. That is, $M_\alpha = \{v \mid v \in \mathrm{dom}(F) \wedge \alpha(v) = 1\} \cup \{\neg v \mid v \in \mathrm{dom}(F) \wedge \alpha(v) = 0\}$. Note that M_α uniquely defines α and vice versa.

Definition 2 (Model Minimization). *Given a model $\alpha \models F$, a model $\beta \models F$ is called α-minimized if $M_\beta \subseteq M_\alpha$. An α-minimized model β is α-minimal if no further subset $M_\gamma \subset M_\beta$ is a model of F. An α-minimal model β is α-minimum if for each α-minimal model γ it holds that $|M_\gamma| \geq |M_\beta|$. If α is clear from the*

---

[1] In this paper, we specifically consider the Plaisted-Greenbaum encoding [12], but other encodings, such as the original Tseitin encoding [17], are also supported.

$$\mathcal{I}_\alpha(\bot) = 0 \qquad\qquad\qquad \mathcal{I}_\alpha(\top) = 1$$

$$\mathcal{I}_\alpha(v) = \begin{cases} 1, & \text{if } \alpha(v) = 1 \\ 0, & \text{if } \alpha(v) = 0 \\ \mathsf{U}, & \text{if } v \notin \mathrm{dom}(\alpha) \end{cases} \qquad \mathcal{I}_\alpha(F \wedge G) = \begin{cases} 1, & \text{if } \mathcal{I}_\alpha(F) = 1 \text{ and } \mathcal{I}_\alpha(G) = 1 \\ 0, & \text{if } \mathcal{I}_\alpha(F) = 0 \text{ or } \mathcal{I}_\alpha(G) = 0 \\ \mathsf{U}, & \text{otherwise} \end{cases}$$

$$\mathcal{I}_\alpha(\neg F) = \begin{cases} 1, & \text{if } \mathcal{I}_\alpha(F) = 0 \\ 0, & \text{if } \mathcal{I}_\alpha(F) = 1 \\ \mathsf{U}, & \text{if } \mathcal{I}_\alpha(F) = \mathsf{U} \end{cases} \qquad \mathcal{I}_\alpha(F \vee G) = \begin{cases} 1, & \text{if } \mathcal{I}_\alpha(F) = 1 \text{ or } \mathcal{I}_\alpha(G) = 1 \\ 0, & \text{if } \mathcal{I}_\alpha(F) = 0 \text{ and } \mathcal{I}_\alpha(G) = 0 \\ \mathsf{U}, & \text{otherwise} \end{cases}$$

Fig. 1. Interpretation of a formula under a (partial) assignment α

context we may write minimized instead of α-minimized, and similarly for the other terms.

Now let $\mathbb{F}_{\mathsf{cnf}} \subseteq \mathbb{F}$ denote the set of formulas in conjunctive normal form (CNF). Formulas $F \in \mathbb{F}_{\mathsf{cnf}}$ are usually represented as sets of clauses, where a clause is a set of literals. As is well known, each formula can be converted to a equisatisfiable formula in CNF, e.g., by using Tseitin's encoding.

Definition 3 (Tseitin Encoding). *Given a formula $F \in \mathbb{F}$, its Tseitin encoding, $\mathcal{T}(F) \in \mathbb{F}_{\mathsf{cnf}}$, is defined as below. Our definition uses the well-known optimization of Plaisted and Greenbaum [12].*[2]

$$\mathcal{T}(F) = d_F \wedge \mathcal{T}^1(F)$$

$$\mathcal{T}^p(F) = \begin{cases} \mathcal{T}^p_{\mathsf{def}}(F) \wedge \mathcal{T}^p(G) \wedge \mathcal{T}^p(H), & \text{if } F = G \circ H \text{ and } \circ \in \{\wedge, \vee\} \\ \mathcal{T}^p_{\mathsf{def}}(F) \wedge \mathcal{T}^{p \oplus 1}(G), & \text{if } F = \neg G \\ \top, & \text{if } F \in \mathcal{V} \end{cases}$$

$$\mathcal{T}^1_{\mathsf{def}}(F) = \begin{cases} (\neg d_F \vee d_G) \wedge (\neg d_F \vee d_H), & \text{if } F = G \wedge H \\ (\neg d_F \vee d_G \vee d_H), & \text{if } F = G \vee H \\ (\neg d_F \vee \neg d_G), & \text{if } F = \neg G \end{cases}$$

$$\mathcal{T}^0_{\mathsf{def}}(F) = \begin{cases} (d_F \vee \neg d_G \vee \neg d_H), & \text{if } F = G \wedge H \\ (d_F \vee \neg d_G) \wedge (d_F \vee \neg d_H), & \text{if } F = G \vee H \\ (d_F \vee d_G), & \text{if } F = \neg G \end{cases}$$

The Tseitin encoding works by introducing new propositional variables. In more detail, given a formula F, its Tseitin encoding $G = \mathcal{T}(F)$ introduces a new variable symbol d_f for each sub-formula f of F. We call the variable d_F, which stands for the complete formula, the *root variable*. The set of variables $\mathcal{V}_G$ can be partitioned into *input variables* $\mathcal{V}_G^{\mathsf{inp}}$ that stem from the original formula F and new *encoding variables* $\mathcal{V}_G^{\mathsf{enc}}$.

---

[2] Some modern implementations introduce no additional encoding variables for negated formulas and inline the negation.

3 Approach

The starting point of our approach is a Tseitin-encoded formula $\mathcal{T}(F) \in \mathbb{F}_{cnf}$ and a complete satisfying assignment $\alpha \models \mathcal{T}(F)$ for it, as it can be obtained by a standard SAT solver. It then computes a minimized model α' of the original formula $F \in \mathbb{F}$. To do this, it takes structural information about the partitioning of variables in $\mathcal{T}(F)$ into input variables and encoding variables into account, as well as the structural information from the Tseitin encoding.

Our minimization algorithm consists of two parts. The first works on the CNF level, and is based on a transformation of the model minimization problem to a *hitting set problem*, in which we search for a set $M_{\alpha'}$ that contains at least one literal from each clause that is assigned to true. We solve this hitting set problem by converting it to SAT, and using iterative calls to a SAT solver to obtain a minimal model α'.[3] This part is done by procedures `normalize` and `minimize` in Alg. 1. The second part, which works as a pre-processing step, exploits the structure of a Tseitin-encoded formula to further minimize the model (procedure `prune` in Alg. 1). A minimal model for the pruned formula $P \subseteq \mathcal{T}(F)$ is a *minimized* model for F that is often significantly smaller than a *minimal* model for $\mathcal{T}(F)$.

Algorithm 1. High-Level View of Model Minimization Algorithm

Input: Formula $\mathcal{T}(F)$, complete model α of $\mathcal{T}(F)$, root variable d_F
Output: Minimized model α' for F

1 $P = \texttt{prune}(\mathcal{T}(F), d_F)$
2 $N = \texttt{normalize}(P, \alpha)$
3 $\alpha' = \texttt{minimize}(\alpha, N)$
4 **return** α'

The three main steps of our algorithm are explained in what follows, starting with the `normalize` and `minimize` procedures that do not take information about the initial formula's structure into account.

3.1 Normalization

Given a formula $F \in \mathbb{F}_{cnf}$ and a model $\alpha \models F$, the normalization step generates a problem $F' \in \mathbb{F}_{cnf}$ which is an encoding of the hitting set problem mentioned above. This problem is then solved in the minimization step of the algorithm.

The first step of normalization, called *purification*, consists of removing irrelevant literals from F, i.e. those literals which are assigned to *false* by α.

Definition 4 (Purification). *Given a formula $F \in \mathbb{F}_{cnf}$ and a model $\alpha \models F$, the purified formula $p_\alpha(F)$ is defined as follows.*

---

[3] Our main algorithm only computes an approximative solution for the hitting set problem, but in a variant of it we can also compute minimum models (see Sec. 4).

$$p_\alpha(F) = \{C \cap M_\alpha \mid C \in F\}$$

Lemma 1. *Given a formula $F \in \mathbb{F}_{cnf}$ and a model $\alpha \models F$, for any assignment α' for which $M_{\alpha'} \subseteq M_\alpha$, we have $\alpha' \models F$ iff $\alpha' \models p_\alpha(F)$.*

After purification we eliminate negated literals by flipping their signs. As all literals in $p_\alpha(F)$ are pure (i.e. they occur only in one polarity), no new negations are introduced by this step, and all literals are positive afterwards.

The whole process of purification followed by flipping negated literals we call *normalization*. The formula obtained by normalization is denoted by $\nu_\alpha(F)$ and forms the basis for our minimization strategy.

3.2   Iterative Minimization

Computation of a minimal model for F is equivalent to finding a model for $\nu_\alpha(F)$ with a minimal number of *true* variables. Since we are generally only interested in models with a minimal number of input variables (i.e., from $\mathcal{V}_F^{\mathsf{inp}}$), we directly minimize assignments to these.

Minimization works by adding a version of a *cardinality constraint* to $\nu_\alpha(F)$, which starts with a bound $k = |\mathcal{V}_F^{\mathsf{inp}}|$, iteratively decreasing it, and checking by calling a SAT solver whether still a satisfying assignment with this bound exists.

Algorithm 2. Iterative Minimization

Input: Formula $F \in \mathbb{F}_{cnf}$, complete model α of F, input variables $\mathcal{V}_F^{\mathsf{inp}}$
Output: Minimized model $\alpha_{\min}$ as a set $M_{\alpha_{\min}}$ of literals
1 $F' = \nu_\alpha(F)$, $M' = \mathcal{V}_{F'}$
2 **repeat**
3 $\quad$ $C = \emptyset$, $E = \emptyset$, $M = M'$
4 $\quad$ **for** $v \in \mathcal{V}_F^{\mathsf{inp}}$ **do**
5 $\quad\quad$ **if** $v \in M$ **then** $C = \{\neg v\} \cup C$
6 $\quad\quad$ **else** $E = \{\{\neg v\}\} \cup E$
7 $\quad$ $(r, M') = \mathtt{solve}(F' \cup \{C\} \cup E)$
8 **until** $r = \bot$
9 $M_{\alpha_{\min}}^+ = \{v \mid v \in M$, and v has not been flipped by $\nu_\alpha(F)\}$
10 $M_{\alpha_{\min}}^- = \{\neg v \mid v \in M$, and v has been flipped by $\nu_\alpha(F)\}$
11 **return** $M_{\alpha_{\min}}^+ \cup M_{\alpha_{\min}}^-$

Algorithm 2 outlines the procedure. We use a "cardinality clause" C, which forbids assigning all k variables to true. Moreover, we remember variables already excluded from a minimal model in a set E. The SAT solver call $\mathtt{solve}$ in Line 7 is assumed to return both the satisfiability status r ($\top$ for satisfiable, $\bot$ for unsatisfiable) and a model M, if one exists. Construction of clause C ascertains that the constraint is strengthened in each iteration. Finally, we obtain a minimal model of $\nu_\alpha(F)$, which we then map back to the original problem F by taking back the variable flips that were made by the normalization procedure.

Structural Pruning. Assuming that we know that our formula $\mathcal{T}(F)$ uses an encoding like in Definition 3, and given that we also know the root variable d_F and the input variables $\mathcal{V}_F^{\mathsf{inp}}$, we can reconstruct the structure of the original formula, by recursively following the definitions of the sub-formulas of d_F until we reach a definition that is solely based on input variables.

As of Definition 3, for each subformula S of F there exists a variable d_S that is defined by clauses $\mathcal{T}_{\mathsf{def}}^p(S) \subset \mathcal{T}(F)$. It is easy to see that an encoding variable d_S has the same polarity in all its defining clauses. All literals d_X that are used to define d_S are either input variables or are themselves defined by clauses $\mathcal{T}_{\mathsf{def}}^p(X)$. Now let $\mathsf{Clauses}(l, F) = \{C \in F \mid l \in C\}$ denote all clauses in F containing the literal l. If F is clear from the context, we may simply write $\mathsf{Clauses}[l]$.

Lemma 2 (Opposite Polarity). *For all* $C \in \mathcal{T}_{\mathsf{def}}^p(S)$ *and all direct sub-formulas* $d_X \notin \mathcal{V}_{\mathcal{T}(F)}^{\mathsf{inp}}$ *of* S *it holds that*

$$d_X \in C \implies \mathcal{T}_{\mathsf{def}}^p(X) = \mathsf{Clauses}(\neg d_X, \mathcal{T}(F))$$
$$\neg d_X \in C \implies \mathcal{T}_{\mathsf{def}}^p(X) = \mathsf{Clauses}(d_X, \mathcal{T}(F))$$

It follows that by parsing the defining clauses of any formula S we can recursively discover the defining clauses of its sub-formulas. Starting with the top-level Tseitin literal d_F we can thus reconstruct the syntax tree of the original formula.

The idea of structural pruning is to create a new formula $F' \subseteq \mathcal{T}(F)$ by purging all clauses that belong to definitions of sub-formulas that are not satisfied by α. Algorithm 3 outlines the procedure. We start with an empty formula (Line 1) and prepare the set of all satisfied encoding literals (Line 2). We reconstruct parts of the structure of F by following only the definitions of satisfied sub-formulas (Line 6), thus building a new formula that contains only the clauses belonging to the satisfied sub-formulas of F (Line 5).

After pruning we can normalize the new formula $F' \subseteq \mathcal{T}(F)$ and minimize α with respect to the pruned formula as shown above.

4   Experimental Results

We implemented our approach as a patch on top of MiniSAT 2.2.0 and performed the experiments on a PC (3.40GHz × 8 CPU, 8 GB Memory) running Linux (Ubuntu 12.04). Our evaluation benchmarks consist of a collection of satisfiable problems from (1) software checking problems that are shipped with the Alloy Analyzer 4 [16], (2) AIG benchmarks from SAT-Race 2010 [1], and (3) program verification problems generated by JForge [4]. In order to perform minimization, one needs to know the set of input variables of a given CNF formula, which usually occupy the first consecutive block of variable identifiers. Furthermore, in order to perform structural pruning, one needs to know the identifier of the root variable. We modified the CNF generators to produce this information as additional CNF comments ("c input $n") and ("c output $i"), respectively.

Algorithm 3. Structural Pruning

Input: Formula $F \in \mathbb{F}_{cnf}$, model α, input variables $\mathcal{V}_F^{inp}$, root encoding var. d_F
Output: Pruned formula $F' \subseteq F$

1 $F' = \emptyset$
2 $L = \{l \in M_\alpha \mid var(l) \in \mathcal{V}_F^{enc}\}$
3 Stack.push(Clauses$[\neg d_F]$)
4 **while** $C =$ Stack.pop **do**
5 $\quad F' = F' \cup C$
6 $\quad$ **for** $l \in C \cap L$ **do**
7 $\quad\quad L = L \setminus \{l\}$
8 $\quad\quad$ Stack.push(Clauses$[\neg l]$)

Table 1. Experimental results with and without structural pruning

problem	nInput	w/o Pruning			w/ Pruning		
		nInput	%	time (sec)	nInput	%	time (sec)
ibm18-len29-sat	983	764	22.3	0.03	575	41.5	0.04
ibm20-len44-sat	1493	1277	14.5	0.06	991	33.6	0.09
ibm22-len52-sat	2245	1932	13.9	0.11	1664	25.9	0.16
ibm23-len36-sat	1515	1308	13.7	0.06	1083	28.5	0.08
ibm29-len26-sat	362	211	41.7	0.01	134	63.0	0.02
intel-003-k-ind-30	1489	1477	0.8	0.05	1441	3.2	0.08
intel-016.aig.smv.kind-b20	27970	27833	0.5	0.49	26917	3.8	0.46
intel-019-k-ind-10	3786	3729	1.5	0.06	3587	5.3	0.10
intel-025.aig.smv.kind-b30	20399	20357	0.2	0.41	19939	2.3	0.40
intel-025-k-ind-20	13939	13895	0.3	0.50	13504	3.1	0.76
intel-032-k-ind-10	6521	6488	0.5	0.16	6188	5.1	0.24
intel-033.aig.smv.kind-b10	28428	28294	0.5	0.46	26376	7.2	0.42
itox-vc1033	57775	57040	1.3	0.37	56870	1.6	0.28
itox-vc1044	58776	58009	1.3	0.42	57822	1.6	0.29
opt-spantree Closure	673	668	0.7	0.00	201	70.1	0.00
opt-spantree SuccessfulRun	2664	2559	3.9	0.06	2559	3.9	0.07
peterson NotStuck	835	835	0.0	0.01	54	93.5	0.00
set.intersect.cegar	29497	29432	0.2	0.05	4290	85.5	0.03

In this section, we report on those benchmarks where at least 1% of their input variables are *don't care*. We present the quality and performance of our approach with and without structural pruning. The results are given in Table 1. The first column gives the problem name and the second column gives the number of input variables. The next three columns give the final number of input variables, percentage of reduction (of input variables), and the runtime of our minimization approach without structural pruning. The last three columns give the results for our approach with structural pruning.

As can be seen in the table, both approaches (with and without pruning) run quickly; they actually take less than a second to perform minimization even for large CNF formulas. The quality of the results, however, differs substantially. In many cases, pruning can eliminate many more input variables without introducing much runtime overhead. This is because pruning takes advantage of the structure of the Tseitin-encoded formulas and avoids all sub-formulas whose encoding variable is a don't-care.

Optimal Minimization. Since our iterative minimization approach does not guarantee to find a minimum assignment, we performed a second set of experiments in which we compared the outcome of iterative minimization to that of an optimal algorithm. We computed the optimal minimization using a cardinality encoding based on parallel counters [15], and iteratively calling a SAT solver to check whether the number of input variables can be reduced.

In our experiments the simpler iterative minimization approach we presented above was always able to find a minimum assignment. Moreover, its runtime turned out to be much better than the optimal algorithm (up to a factor of 168 in our tests).

5 Related Work and Conclusion

Minimization of SAT models has been a research topic for many years. In literature the minimization goal usually is to reduce the number of positive literals in a model (e.g. [2,10]). Thus the `minimize` function of Koshimura et al. [10] and ours are almost identical. However their algorithm omits the `normalization` step, which means that satisfiability according to their notion of minimality still depends on the negative literals, while our approach guarantees that all negative literals belong to don't care variables.

Other work on model minimization is often directed towards a particular application, such as model checking [11], bounded model checking [8,13,14], SMT solving [5] or QBF solving [7]. None of the approaches seems to work on general formulas, taking only the structural information of the CNF encoding into account as in our approach.

This paper introduced an algorithm for minimizing a given model of a CNF formula with respect to the original input variables (as opposed to the intermediate encoding variables that are introduced during CNF conversion). We transform the model minimization problem to a hitting set problem (also presented as a SAT problem), and solve it by iteratively calling a SAT solver. An optional `pruning` preprocess can be applied when structural information about the CNF encoding is available. Our experiments show that the algorithm performs well with respect to both quality and runtime. Future work that we could envisage is using our minimization approach in model counting.

References

1. SAT-Race 2010 (2010), http://baldur.iti.uka.de/sat-race-2010/ (accessed February 14, 2013)
2. Ben-Eliyahu, R., Dechter, R.: On computing minimal models. Ann. Math. Artif. Intell. 18(1), 3–27 (1996)

3. Boy de la Tour, T.: An optimality result for clause form translation. J. Symb. Comput. 14(4), 283–301 (1992)
4. Dennis, G., Chang, F.S.-H., Jackson, D.: Modular verification of code with SAT. In: International Symposium in Software Testing and Analysis (ISSTA 2006), pp. 109–120 (2006)
5. Dillig, I., Dillig, T., McMillan, K.L., Aiken, A.: Minimum satisfying assignments for SMT. In: Madhusudan, P., Seshia, S.A. (eds.) CAV 2012. LNCS, vol. 7358, pp. 394–409. Springer, Heidelberg (2012)
6. Ganzinger, H., Hagen, G., Nieuwenhuis, R., Oliveras, A., Tinelli, C.: DPLL(T): Fast decision procedures. In: Alur, R., Peled, D.A. (eds.) CAV 2004. LNCS, vol. 3114, pp. 175–188. Springer, Heidelberg (2004)
7. Giunchiglia, E., Narizzano, M., Tacchella, A.: Clause/term resolution and learning in the evaluation of quantified boolean formulas. J. Artif. Intell. Res. (JAIR 2006) 26, 371–416 (2006)
8. Groce, A., Kroening, D.: Making the most of BMC counterexamples. Electr. Notes Theor. Comput. Sci. 119(2), 67–81 (2005)
9. Jackson, P., Sheridan, D.: Clause form conversions for boolean circuits. In: Hoos, H.H., Mitchell, D.G. (eds.) SAT 2004. LNCS, vol. 3542, pp. 183–198. Springer, Heidelberg (2005)
10. Miyuki Koshimura, H.F., Nabeshima, H., Hasegawa, R.: Minimal model generation with respect to an atom set. In: International Workshop on First-Order Theorem Proving (FTP 2009), pp. 49–59 (2009)
11. Nopper, T., Scholl, C., Becker, B.: Computation of minimal counterexamples by using black box techniques and symbolic methods. In: Intl. Conf. on Computer-Aided Design (ICCAD 2007), pp. 273–280 (2007)
12. Plaisted, D.A., Greenbaum, S.: A structure-preserving clause form translation. J. Symb. Comput. 2(3), 293–304 (1986)
13. Ravi, K., Somenzi, F.: Minimal assignments for bounded model checking. In: Jensen, K., Podelski, A. (eds.) TACAS 2004. LNCS, vol. 2988, pp. 31–45. Springer, Heidelberg (2004)
14. Shen, S., Qin, Y., Li, S.: Minimizing counterexample with unit core extraction and incremental SAT. In: Cousot, R. (ed.) VMCAI 2005. LNCS, vol. 3385, pp. 298–312. Springer, Heidelberg (2005)
15. Sinz, C.: Towards an optimal CNF encoding of boolean cardinality constraints. In: van Beek, P. (ed.) CP 2005. LNCS, vol. 3709, pp. 827–831. Springer, Heidelberg (2005)
16. Torlak, E., Jackson, D.: Kodkod: A relational model finder. In: Grumberg, O., Huth, M. (eds.) TACAS 2007. LNCS, vol. 4424, pp. 632–647. Springer, Heidelberg (2007)
17. Tseitin, G.S.: On the complexity of derivation in propositional calculus. In: Slisenko, A.O. (ed.) Studies in Constructive Mathematics and Mathematical Logic, pp. 115–125 (1970)

Solutions for Hard and Soft Constraints Using Optimized Probabilistic Satisfiability

Marcelo Finger*, Ronan Le Bras, Carla P. Gomes, and Bart Selman

Department of Computer Science
Cornell University

Abstract. Practical problems often combine real-world hard constraints with soft constraints involving preferences, uncertainties or flexible requirements. A probability distribution over the models that meet the hard constraints is an answer to such problems that is in the spirit of incorporating soft constraints.

We propose a method using SAT-based reasoning, probabilistic reasoning and linear programming that computes such a distribution when soft constraints are interpreted as constraints whose violation is bound by a given probability. The method, called Optimized Probabilistic Satisfiability (oPSAT), consists of a two-phase computation of a probability distribution over the set of valuations of a SAT formula. Algorithms for both phases are presented and their complexity is discussed.

We also describe an application of the oPSAT technique to the problem of combinatorial materials discovery.

1 Introduction

There are many proposals in the literature that combine logical and probabilistic reasoning, e.g. [23,22,5]. Perhaps the earliest such proposal was made by Boole himself, as a natural extension of Boolean satisfiability [1]. This framework is now called *probabilistic satisfiability (PSAT)*. The semantics is given by assigning a probability distribution π over the set of all 2^n truth assignments of n variables. Given π, one can now assign a probability P to each compound formula by considering the sum of the probabilities of all truth assignments (or models) that satisfy the formula. It has been shown that such a formalization has a number of desirable properties, such as the fact that it satisfies Kolmogorov's probability axioms [19,11].

A set of logical formulas, each assigned some probability value or a probability bound (e.g., $P(A \wedge B) \geq 0.6$), can be viewed as a set of probabilistic constraints. A natural question is whether there exist any probability distribution over all truth assignments that satisfies the probabilistic constraints. This is the consistency problem for probabilistic satisfiability. Note that by assigning probability 0 or 1 to some of the logical formulas, they effectively act as standard logical constraints. So, we can have a mix of logical and probabilistic constraints.

---

* On leave from Department of Computer Science, University of Sao Paulo.

M. Järvisalo and A. Van Gelder (Eds.): SAT 2013, LNCS 7962, pp. 233–249, 2013.

In the mid eighties and early nineties, the consistency problem for PSAT became the focus of much attention because, in principle, it could be used to determine whether expert system sets of rules (hard and soft constraints) were consistent [19,7,12]. Unfortunately, the consistency problem for PSAT appeared to be extremely hard [20]. In particular, since the probability distribution ranges over all truth assignments, it was not even clear how to get a polynomial size witness for a consistent set of PSAT formulas. However, there have been several major breakthroughs in dealing with the complexity of this problem, such as polynomial size witness [7], linear programming algorithms [11] and SAT-based algorithms and the detection of PSAT phase-transition [4].

So, the recent progress has made PSAT a potentially relevant formalism for practical applications, providing an alternative to other approaches. One advantage of the PSAT framework is that its foundations are quite natural and well-grounded.

The goals of this work are three-fold:

(a) To enhance PSAT and introduce a method, called oPSAT, as a modeling framework to deal with mixtures of (hard) logical and soft probabilistic constraints.
(b) To propose a practical algorithmic strategy for solving oPSAT problems.
(c) To demonstrate the practical effectiveness of our proposed approach on a real-world reasoning task in the domain of Materials Discovery.

In this approach, formulas that encode the existence of a *soft violation* in the solution (sometimes called a *defect*) will be modeled by probabilistic constraints. Consider the following underspecified example.

Example 1. There are m students and k summer courses. Each student has a set of potential teammates, with whom coursework will be developed. We want the course enrollment to respect the following constraints:

Hard. For each course, students must decide to develop their coursework alone or in teams of 2 (pairs). A student may have different teammates in different courses, and may work alone in some course and have a teammate in others. Students must enroll in at least one and at most three courses. There is a limit of ℓ students per course.

Soft. Avoid having students with no teammate. A student's enrollment in a course with no team mate is seen as a "soft violation."   □

Example 1 clearly shows the presence of hard and soft constraints. That problem also has some other implicit, data-dependent hard constraints, such as the number of students, courses and list of possible pairs of team mates. An important implicit hard constraint is the definition of a soft violation ("student in a course with no teammates") in terms of the variables present in the hard constraints. There may be no solutions to the hard constraints; or there may be several ones, in which case we are interested in computing a probability distribution over them, which will allow one to answer questions such as "what is the expected number of enrollments?" or "what is the probability that two students will be teammates?".

Yet, Example 1 is underspecified, as no clear way to deal with the soft constraints has been provided. In our method, this additional specification will correspond to a set of probabilistic constraints of the form $P(softViolation_i) \leq p_i$, $1 \leq i \leq m$, where for each student i there is a maximum probability p_i that i enrolls in some course with no teammate. Where do these probabilities come from? There are three main sources:

(a) The probabilities are stipulated or given. In Example 1, the student may be asked with which probability he or she accepts to be with no teammate. In this work, we will assume that this is the case.
(b) The probabilities are learned. For instance, compute p_i from previous editions of the summer course.
(c) The probabilities are minimal. Assume that all p_i are the same and compute the minimal value for which the hard and soft probabilistic constraints are satisfiable. This topic remains for further investigation.

Our method, called *Optimized Probabilistic Satisfiability (oPSAT)* consists of two phases[1]. The first phase is the PSAT problem, which determines if the hard constraints and probability constraints can be jointly satisfied; Section 2 will formalize PSAT and briefly describe a solver method. The output of such a problem, when satisfiable, is a probability distribution over a (small) class of models of the hard constraints. As this solution may not be unique, a second phase is needed to find a "reasonable" or "balanced" solution. By that we mean a distribution with minimal variance of soft violation occurrences. In Section 3, a novel SAT-based column generation method to compute such a distribution is presented.

Then in Section 4 we demonstrate the effectiveness of this approach on a complex real-world application involving the identification of crystallographic structures from high-intensity X-ray diffraction patterns [16,3,15]. The problem arises in the area of so-called combinatorial materials discovery [18].

2   Probabilistic Satisfiability

The PSAT problem is formalized as follows. Let $\mathcal{L}$ be the language of classical propositional formulas. A *PSAT instance* is a set $\Sigma = \{P(\alpha_i) \bowtie_i p_i | 1 \leq i \leq k\}$, where $\alpha_1, \ldots, \alpha_k \in \mathcal{L}$ are classical propositional formulas defined on n logical variables $\mathcal{P} = \{x_1, \ldots, x_n\}$, which are restricted by probability assignments $P(\alpha_i) \bowtie_i p_i$, $\bowtie_i \in \{=, \leq, \geq\}$ and $1 \leq i \leq k$.

There are 2^n possible propositional valuations v over the logical variables, $v : \mathcal{P} \to \{0, 1\}$; each such valuation is extended, as usual, to all formulas, $v : \mathcal{L} \to \{0, 1\}$. A *probability distribution over propositional valuations* $\pi : V \to [0, 1]$, is a function that maps every valuation to a value in the real interval $[0, 1]$ such

---

[1] This method should not be confused with OPTSAT [8].

that $\sum_{i=1}^{2^n} \pi(v_i) = 1$. The probability of a formula α according to distribution π is given by $P_\pi(\alpha) = \sum \{\pi(v_i)|v_i(\alpha) = 1\}$. We simply write $P(\alpha)$ when the distribution is clear from the context.

The definition of PSAT involves linear algebraic notation. We assume a vector b to be a column-vector and b' its transpose, a row-vector. If A is an $m \times n$ matrix, A^j represents its j-th column, and if b is an m-dimensional column, $A[j := b]$ represents the matrix obtained by substituting b for A^j; if A is square matrix, $|A|$ is A's determinant.

From a PSAT instance, construct a $k \times 2^n$ matrix $A = [a_{ij}]$ such that $a_{ij} = v_j(\alpha_i)$. The *probabilistic satisfiability problem* is to decide if there is a probability vector $\pi_{2^n \times 1}$ subject to:

$$A\pi \bowtie p$$
$$\sum \pi_i = 1 \tag{1}$$
$$\pi \geq 0$$

A PSAT instance Σ is *satisfiable* if its associated PSAT restriction (1) has a solution π; in that case, we say that π satisfies Σ. The last two conditions of (1) force π to be a probability distribution. Usually the first two conditions of (1) are combined: A is a $(k+1) \times 2^n$ $\{0,1\}$-matrix with 1's at its first line, p_1 is set to 1 in vector $p_{(k+1)\times 1}$, and the $\bowtie_1$-relation corresponds to "=". In this case, each column A^j, excluding its first position that is always 1, can be seen as a Boolean valuation.

Example 2. We continue Example 1 and for simplicity assume that there is only one course, three students whose enrollment is represented by variables x, y and z, and two potential partnerships of the first student with either of the others, represented by p_{xy} and p_{xz}. These partnerships are mutually exclusive, as x can only have one partner. So we have the hard constraint

$$P(x \wedge y \wedge z \wedge \neg(p_{xy} \wedge p_{xz})) = 1$$

and the soft constraints are probability restriction on the enrollment of a student without partners, set for this example as:

$$P(x \wedge \neg p_{xy} \wedge \neg p_{xz}) \leq 0.25, \quad P(y \wedge \neg p_{xy}) \leq 0.6, \quad P(z \wedge \neg p_{xz}) \leq 0.6.$$

Of all the 2^5 valuations, we consider π such that $\pi(x, y, z, \neg p_{xy}, \neg p_{xz}) = 0.1$, $\pi(x, y, z, p_{xy}, \neg p_{xz}) = 0.4$, $\pi(x, y, z, \neg p_{xy}, p_{xz}) = 0.5$ and $\pi(v) = 0$ for the remaining 29 valuations. This distribution satisfies the PSAT instance. □

It is no coincidence that only a small number of valuations in Example 2 receive non-zero probability. In fact, satisfiable PSAT instances always have a "small" witness.

Proposition 1 ([7]). *If an instance $\Sigma = \{P(\alpha_i) = p_i | 1 \leq i \leq k\}$ has a solution $\pi \geq 0$, then there is a solution $\pi' \geq 0$ with at most $k+1$ non-zero elements.*

From Proposition 1, it follows that PSAT is in NP. As SAT is a special case of PSAT when all $p_i = 1$, PSAT is NP-hard. As a result, PSAT is NP-complete.

There have been several proposed algorithms for PSAT [14,11,13], but its general applicability in practice has only been established with the demonstration that PSAT presents a phase transition [4], just like the SAT problem [17,6].

As in SAT, to display a phase transition the problem must be brought to a normal form. A PSAT instance is in *normal form* if it is partitioned in two sets, $\langle \Gamma, \Psi \rangle$, where $\Gamma = \{P(\alpha_i) = 1 | 1 \le i \le m\}$ and $\Psi = \{P(y_i) = p_i | y_i$ is a variable, $1 \le i \le k\}$. Every PSAT instance can be transformed in a normal form instance $\langle \Gamma, \Psi \rangle$ in polynomial time, such that satisfiability is preserved [4]. The set $\Psi(y_1, \ldots, y_k)$ contains probabilistic restrictions over variables $y_1, \ldots, y_k$ only, and the set $\Gamma(y_1, \ldots, y_k; x_1, \ldots, x_n)$ is a SAT formulas. A valuation v over $y_1, \ldots, y_k$ is Γ-*consistent* if there is an extension of v over $y_1, \ldots, y_k, x_1, \ldots, x_n$ such that $v(\Gamma) = 1$. The following refines Proposition 1.

Proposition 2 ([4]). *A normal form instance $\langle \Gamma, \Psi \rangle$ is satisfiable iff there is a $(k + 1) \times (k + 1)$-matrix A and $\pi \ge 0$ such that $A\pi = p$ and whenever $\pi_j > 0$ then column j of A is Γ-consistent.*

In this work, we will always consider instances to be in normal form. Based on Proposition 2, two algorithms for PSAT solving were proposed in [4], and here we are interested in the one that solves the following optimization problem

$$
\begin{array}{ll}
\text{minimize} & c'\pi \\
\text{subject to} & A\pi = p \text{ and } \pi \ge 0
\end{array}
\tag{2}
$$

where A is a $(k + 1) \times 2^n$ $\{0,1\}$-matrix in (1), π is the probability distribution and c is a $2^n \times 1$ *cost vector*; $c_j = 1$ if A's column j is a Γ-inconsistent valuation, and $c_j = 0$ otherwise. The PSAT instance is satisfiable iff the optimization leads to a cost $c'\pi = 0$.

As A is exponentially large, we do not generate it explicitly. Instead, we use a SAT-solver to generate Γ-consistent columns as the linear optimization *simplex algorithm* requires [21]. The problem is solved iteratively; at each iteration step i, Proposition 2 allows for storing $A_{(i)}$ with $k + 1$ columns and a *column generation* method is employed in which a SAT-based *auxiliary problem* generates a Γ-consistent column that replaces some column in $A_{(i)}$ and decreases the objective function; this method is detailed in Section 2.1. Accordingly, only the components of c and π corresponding to the columns of $A_{(i)}$ are stored.

A *feasible solution* $A_{(i)}$ is a $\{0,1\}$-matrix for which there exists a $\pi_{(k+1) \times 1} \ge 0$ such that $A_{(i)}\pi = p$. It is shown in [4] that an initial feasible solution $A_{(0)}$ always exists and can be easily computed. The simplex method guarantees that the cost function always decreases at each step, by computing the *reduced cost* $\bar{c}_b$ of inserting a column b in a feasible solution A and forcing it to be non-positive [14]:

$$
\bar{c}_b = c_b - c_A A^{-1} b \le 0
\tag{3}
$$

Algorithm 2.1. PSATsolver($\langle \Gamma, \Psi \rangle$)

Input: A normal form PSAT instance $\langle \Gamma, \Psi \rangle$.
Output: Total solution A; or "No", if unsatisfiable.
1: $A_{(0)} :=$ initial feasible solution; $i := 0$; compute $cost^{(i)}$;
2: **while** $cost^{(i)} > 0$ **do**
3:    $b^{(i)} = GenerateColumn(A_{(i)}, p, \Gamma)$; /* Described in Section 2.1 */
4:    **return** "No" **if** $b_1^{(i)} < 0$; /* PSAT instance is unsat */
5:    $A_{(i+1)} = merge(A_{(i)}, b^{(i)})$;
6:    increment i; compute $cost^{(i)}$;
7: **end while**
8: **return** $A_{(i)}$; /* PSAT instance is satisfiable */

---

where c_b and c_A are, respectively, the component of the cost vector corresponding to the column b and the columns of A. In our case, $c_b = 0$, so the goal is to find a column b such that $c_A A^{-1} b \geq 0$.

Algorithm 2.1 presents a method that decides whether a PSAT instance is satisfiable by solving Problem (2), with a positive answer if minimum cost is 0. Let us see an example of Algorithm 2.1 at work.

Example 3. We express the instance of Example 2 in normal form $\langle \Gamma, \Psi \rangle$ by adding variables for each soft violation: s_x, s_y, s_z. Thus

$$\Gamma = \left\{ \begin{array}{c} x,\ y,\ z,\ \neg p_{xy} \vee \neg p_{xz}, \\ (x \wedge \neg p_{xy} \wedge \neg p_{xz}) \to s_x,\ (y \wedge \neg p_{xy}) \to s_y,\ (z \wedge \neg p_{xz}) \to s_z \end{array} \right\}$$
$$\Psi = \{\ P(s_x) = 0.25,\ P(s_y) = 0.6,\ P(s_z) = 0.6\ \}$$

Note that the existence of a soft violation implies the truth of the corresponding variable in s_x, s_y, s_z, but the truth of some of these variables *does not* necessarily imply the occurrence of a soft violation. We now apply Algorithm 2.1.

$$A_{(0)} = \begin{bmatrix} 1 & 1 & 1 & 1 \\ 0 & 0 & 0 & 1 \\ 0 & 0 & 1 & 1 \\ 0 & 1 & 1 & 1 \end{bmatrix} \qquad A_{(1)} = \begin{bmatrix} 1 & 1 & 1 & 1 \\ 0 & 0 & 0 & 1 \\ 0 & 0 & 1 & 1 \\ 0 & 1 & 0 & 1 \end{bmatrix} \qquad A_{(2)} = \begin{bmatrix} 1 & 1 & 1 & 1 \\ 1 & 0 & 0 & 1 \\ 0 & 0 & 1 & 1 \\ 1 & 1 & 0 & 1 \end{bmatrix}$$

$\pi^{(0)} = [0.4\ 0\ 0.35\ 0.25]'$ $\pi^{(1)} = [0.05\ 0.35\ 0.35\ 0.25]'$ $\pi^{(2)} = [0.05\ 0.35\ 0.4\ 0.2]'$
$cost^{(0)} = 0.4$ $cost^{(1)} = 0.05$ $cost^{(2)} = 0$
$b^{(0)} = [1\ 0\ 1\ 0]'$: col 3 $b^{(1)} = [1\ 1\ 0\ 1]'$: col 1

The initial feasible solution $A_{(0)}$ is a line permutation of an upper 1-triangular matrix, has all but its first column Γ-consistent, with lines 2,3,4 corresponding to Ψ-variables s_x, s_y, s_z and leads to $\pi^{(0)}$ and cost 0.4. The first line is always 1 to force the probabilities to add up to 1. Column generation (Section 2.1) produces $b^{(0)}$ which the simplex merging determines to substitute $A_{(0)}$'s third column. This generates $A_{(1)}, \pi^{(1)}$ and decreasing cost 0.05; column generation yields $b^{(1)}$

that substitutes $A_{(1)}$'s first column. In $A_{(2)}$ there are no Γ-inconsistent columns and the cost is 0, so the problem is satisfiable. At each step i, $A_{(i)} \cdot \pi^{(i)} = p$.

The distribution here is distinct from that in Example 2, as here we consider only the variables in Ψ; this also illustrates that the satisfying distribution is not unique. □

2.1 SAT-Based Column Generation

The following describes procedure $GenerateColumn(A_{(i)}, p, \Gamma)$ used in Algorithm 2.1 and adapted for optimization Algorithm 3.1.

A Γ-consistent column b that never increases the value of the objective function is obtained by solving a SAT problem as follows. Consider $x_1, \ldots, x_k$ taking values in $\{0, 1\}$, $a_1, \ldots, a_k, c \in \mathbb{Q}$ and

$$a_1 \cdot x_1 + \cdots a_k \cdot x_k \bowtie c \qquad \bowtie \in \{<, \leq, >, \geq, =, \neq\} \qquad (LR)$$

Linear restriction (LR) can be seen as a propositional formula Δ_{LR}, in the sense that a valuation $v : x_i \mapsto \{0, 1\}$ *satisfies* Δ_{LR} iff v makes (LR) a true condition. Δ_{LR} can be obtained from (LR) in time $O(k)$ [25].

Suppose $1, \ldots, q \leq k + 1$ are the Γ-inconsistent columns of feasible solution A. By (3), a column $b = [1\ y_1 \ldots y_k]'$ that substitutes some A^j and enforces a decreasing *cost* satisfies

$$\sum_{i=1}^{q} A_i^{-1} \cdot [1\ y_1\ \cdots\ y_k]' \geq 0 \qquad (LR_{cost})$$

A valuation that satisfies $\Gamma \wedge \Delta_{LR_{cost}}$ instantiates b. If that formula is satisfiable, $A[j := b]$ is a feasible solution and *cost* never increases.

With respect to the termination of the simplex method, one must ensure that Bland's rule for fixed order of insertion/removal of columns is respected, and thus termination of the simplex optimization is guaranteed [21].[2]

2.2 The Practical Feasibility of PSAT

Prior to the development of very efficient SAT solvers, PSAT was considered "completely impractical" [20]. But the work of [4] has shown that PSAT presents, in practice, the hard/easy phase transition behavior similar to that of SAT [17,6]. Among other things, this means that there are predominantly "easy" cases of satisfiable and unsatisfiable PSAT instances. Of course, PSAT is still several times slower than SAT due to the fact that a PSAT solver invokes a SAT solver several times.

With the current technology of SAT solvers, an auxiliary formula $\Gamma \wedge \Delta_{LR_{cost}}$ with tens or even hundreds of thousands of variables can be mostly dealt without problems. To keep the number of iterations of Algorithm 2.1 under control, it is advisable to keep a small number k of probability restrictions. Several dimensionality reduction techniques may be employed, such as the one described in Example 4.

---

[2] In practice, some SAT solvers, such as zchaff, have an internal behavior that obeys Bland's rule; others, such as minisat, need extra coding precautions to avoid loops.

Example 4. Reconsider Example 1, assuming there are $k > 1$ courses for m students to enroll, but with a limit of ℓ students per course. Consider as a soft violation now a course having any students with no partners, reducing the number of probabilistic constraints from m to $k \ll m$. The probability of a violating course, p_c, can be obtained from the previous one, adopting a simplifying assumption of independence between soft violations, thus obtaining the probability $p_c = 1 - (1 - p_i)^{\ell}$. $\qquad\square$

3 Optimizing Probability Distributions with oPSAT

Solutions to a PSAT problem are not unique, and a second phase is needed to obtain a distribution with desirable properties. This, in some sense, mirrors the two steps of a linear optimization problem using the simplex algorithm. The first phase searches for a feasible solution for the initial constraints, which is what PSAT does; the second phase produces a solution to the constraints that optimizes an objective function.

A first candidate for this objective function is the minimization of the expected value of S, the number of soft violations:

$$E(S) = \sum_{v_i | v_i(\Gamma)=1} S(v_i)\pi(v_i), \qquad \text{where } S(v_i) = \sum_{j=1}^{k} v_i(y_j)$$

However, due to the following result, this initial idea is not applicable. Define a (PSAT) *model linear function* over Ψ-variables $y_1, \ldots, y_k \in \{0, 1\}$:

$$f(y_1, \ldots, y_k) = a_1 y_1 + \cdots + a_k y_k, \qquad \text{where } a_j \in \mathbb{Q}, 1 \leq j \leq k \qquad (4)$$

It is important that *only* variables in Ψ are arguments of f. Note that $E(S)$ is a model linear function with all $a_j = 1$. Also note that the *expected value* of a linear function f according to a probability distribution π is $E_\pi(f) = \sum_j (a_1 v_j(y_1) + \cdots + a_k v_j(y_k))\pi(v_j)$.

Lemma 1. *Consider a satisfiable normal form PSAT instance $\langle \Gamma(y_1, \ldots, y_k; x_1, \ldots, x_n), \Psi\{P(y_j) = p_j | 1 \leq j \leq k\}\rangle$; let $f(y_1, \ldots, y_k)$ be a model linear function. Then for every satisfying probability distribution π, the expected value of f with respect to π is fixed, $E_\pi(f) = \sum a_j p_j$.*

Proof. Directly from the definition of $E_\pi(f)$ and using linearity of E_π:

$$E_\pi(f) = \sum_v (a_1 v(y_1) + \cdots + a_k v(y_k))\pi(v)$$

$$= a_1 \sum_v v(y_1)\pi(v) + \cdots + a_k \sum_v v(y_k)\pi(v)$$

$$= a_1 P_\pi(y_1) + \cdots + a_k P_k(y_k) = \sum_{j=1}^{k} a_j p_j.$$

Note that the use of normal form helped considerably to obtain this result. $\qquad\square$

Lemma 1 shows that there is no point in minimizing the expected number of soft violations, which is a constant for a given PSAT instance.

3.1   Variance Minimization

Lemma 1 implies that the model function to be minimized to obtain a "balanced" probability distribution must be non-linear. The idea is to choose a function that prioritizes assigning higher probability mass to distributions with smaller number of soft violations.

One possible choice is then to minimize the expected value of the square number of soft violations, $E(S^2)$. The minimal value of the expected value of this function tends to assign more weight, that is, a greater probability, to the models with smaller number of soft violations. It also seems a good choice of function that a distribution must minimize to obtain a "balanced" distribution due to the following property.

Theorem 1 (Minimal Variance). *The probability distribution that minimizes $E(S^2)$ is also the probability distribution that minimizes the variance of the number of soft violations,* $\mathrm{Var}(S)$.

Proof. We know from basic statistics that the variance of a function is given by

$$Var(S) = E\left((S - E(S))^2\right) = E(S^2) - (E(S))^2 \tag{5}$$

But, by Lemma 1, $E(S)$ is fixed, so the distribution that minimizes $E(S^2)$, by (5), is also the distribution that minimizes $Var(S)$.   □

So we take the view that a "balanced" distribution that respects soft constraints is one that minimizes the variance of the number of soft violations.

To implement it, we also use a SAT-based column generation to minimize the objective function. The generation of a column b is based on the encoding of the reduced cost given by (3) as $\bar{c}_b = c_b - c_A A^{-1} b < 0$, where c is the cost vector and A is a feasible solution. In the PSAT case, the cost of the new column is $c_b = 0$, but here we do not know a priori its value.

However, there are only a few possible values of $c_b = (S(b))^2$. Thus we iterate $i = 0$ to k, $c_b = i^2$, at each step generating a Γ-consistent SAT formula encoding of (3) with at most i soft violations. Assume *VarianceDecreasingColumn*(i, A, p, Γ) is a column generation function that performs such encoding and submits it to a SAT-solver, obtaining b; again a value $b_1 < 0$ indicates unsatisfiability.

Algorithm 3.1 implements variance minimization and is a variation of Algorithm 2.1. It takes as input the first phase solution to a satisfiable PSAT instance. It contains two nested loops. The outermost one iterates over the computation step (from 0 to k), to be able to compute columns that generate a reduced cost. The inner loop actually performs the column generation optimization step; this loop stops when it is not possible to further minimize the cost for a given number of soft violations set by the outer loop, which may occur if no satisfiable instance for the column generated is obtained.

Algorithm 3.1. MinimizeVariance(Γ, Ψ, A, π)

Input: A PSAT instance $\langle \Gamma, \Psi \rangle$, satisfied by $A\pi = p$.
Output: $\langle A', \pi' \rangle$ such that π' has minimal variance of all solutions to $\langle \Gamma, \Psi \rangle$.
1: $A_{(0)} := A;\ \pi^{(0)} = \pi;\ cost^{(0)} = E_\pi(S^2)$;
2: **for** $i = 0$ **to** k **do**
3:    **repeat**
4:       $b^{(i)} = VarianceDecreasingColumn(i, A_{(i)}, p, \Gamma)$;
5:       **if** $b_1^{(i)} \geq 0$ **then**
6:          $A_{(i+1)} = merge(A_{(i)}, b^{(i)})$;
7:          compute $\pi^{(i+1)}$ and $cost^{(i+1)} = E_{\pi^{(i+1)}}(S^2)$;
8:       **end if**
9:    **until** $b_1^{(i)} < 0$ /* cost cannot be further minimized */
10: **end for**
11: **return** $\langle A_{(k+1)}, \pi^{(k+1)} \rangle$;

Example 5. We continue Example 3, optimizing its output, which had $E(S) = 2 \cdot 0.05 + 1 \cdot 0.35 + 1 \cdot 0.4 + 3 \cdot 0.2 = 1.45$ and $E(S^2) = 4 \cdot 0.05 + 1 \cdot 0.35 + 1 \cdot 0.4 + 9 \cdot 0.2 = 2.75 = cost^{(0)}$. According to Algorithm 3.1, we iterate over the number of soft violations allowed ($i = 0$ to 3). For $i = 0$ and $i = 1$, the computed SAT formula is unsatisfiable; for $i = 2$, a new column is obtained to substitute the third column:

$$A_{(2)} = \begin{bmatrix} 1\,1\,1\,1 \\ 1\,0\,0\,1 \\ 0\,0\,1\,1 \\ 1\,1\,0\,1 \end{bmatrix} \qquad A_{(2)'} = \begin{bmatrix} 1\,1\,1\,1 \\ 1\,0\,0\,0 \\ 0\,0\,1\,1 \\ 1\,1\,0\,1 \end{bmatrix}$$

$$\pi^{(2)} = [0.05\ 0.35\ 0.4\ 0.2]' \qquad \pi^{(2)'} = [0.25\ 0.15\ 0.4\ 0.2]'$$
$$cost^{(2)} = 2.75 \qquad\qquad cost^{(2)'} = 2.35$$
$$b_{(2)} = [1\ 0\ 1\ 1]' : col\ 3$$

The remaining iterations all generate unsatisfiable formulas, so the minimum variance obtained for $i = 2$ is $Var(S) = 2.35 - 1.45^2 = 0.2475$. □

4   oPSAT and Combinatorial Materials Discovery

In this section, we present an application of the proposed oPSAT approach to a practical problem in materials discovery. We first provide some background on this motivating application, before formally defining the problem. Finally, we present an oPSAT encoding for this problem and the experimental results for it.

4.1   Background

In combinatorial materials discovery, the goal is to find intermetallic compounds with desirable physical properties by obtaining measurements on samples from a *thin film* composition spread. This approach has been successfully applied for example to speed up the discovery of new materials with improved catalytic activity for fuel cell applications [24,9]. Nevertheless, the analysis of these measurements, also called the phase-field identification problem, requires a manual and laborious data interpretation component, and our goal is to automate it and reduce its processing time.

Combinatorial materials discovery, and in particular the problem of ternary phase-field identification addressed in this paper, provides unique computational and modeling challenges. While statistical methods and machine learning are important components to address this challenge, they fail to incorporate relationships that are inherent to the basic physics and chemistry of the underlying materials. In fact, a successful approach to materials discovery requires a tight integration of statistical methods, to deal with noise and uncertainty in the measurement data, and optimization and inference techniques, to incorporate a rich set of constraints arising from the underlying materials physics and chemistry. As a consequence, the proposed oPSAT framework seems particularly suited to address this problem.

4.2   Problem Definition

In the composition spread approach, three metals (or oxides) are sputtered onto a silicon wafer using guns pointed at three distinct locations, resulting in a so-called *thin film* (Fig. 1). Different locations (or samples) on the thin film correspond to different concentrations of the sputtered materials, based on their distance to the gunpoints. X-ray diffraction (XRD) is then used to characterize a number of samples on the thin film. For each sample point, it provides the intensity of the electromagnetic waves as a function of the angle of diffraction. The observed diffraction pattern is closely related to the underlying crystal structure, which provides important insights into chemical and physical properties of the corresponding composite material.

The goal of the phase-field identification problem is to identify regions of the thin film that share the same underlying crystal structure. Intuitively, the XRD patterns observed across the *thin film* can be explained as combinations of a small set of basis patterns called *phases*. Finding the phase field corresponds to identifying these *phases* as well as their concentration on the thin film. The main challenge is to model the complex crystallographic process that these phases are subject to (such as the expansion of the lattice, which results in a 'shift' of the XRD pattern), while taking into account the imperfection of the silicon wafer as well as experimental noise of the data.

While it is natural to study the phase-field identification problem on the basis of full XRD curves, constructive interference of the scattered X-rays occurs, by nature, at *specific* angles and creates spikes (or *peaks*) of intensity. In addition, experimental noise combined with variations of the Silicon substrate make the measured intensity of the beam not reliable. As a result, materials scientists mostly rely on peak angles when tackling the phase-field identification problem. Therefore, we use a specialized peak detection algorithm [10] to extract the set of peak angles $\mathcal{Q}(i)$ in the XRD pattern of a sample point i.

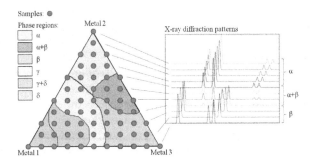

Fig. 1. Example of a *thin film*. Each sample on the silicon wafer corresponds to a different composition, and has an associated measured x-ray diffraction pattern. Colors correspond to different combinations of the basis patterns $\alpha, \beta, \gamma, \delta$. On the right, diffraction patterns of the sample points along the right side of the thin film illustrate how the patterns combine and shift as one moves from one point to a neighboring one.

The goal is then to find a set of peak angles $\mathcal{E}_k$ for each phase k, as well as phase-presence Boolean variables $a_{i,k}$ and scaling factors $s_{i,k} \in \mathbb{R}$ for each sample i and phase k, such that each observed set of peaks $\mathcal{Q}(i)$ is explained. Namely, for each peak $q \in \mathcal{Q}(i)$ we want to have at least one phase k and one peak $e \in \mathcal{E}_k$ of that phase that can explain it, i.e. $\forall q \in \mathcal{Q}(i) \,\exists e \in \mathcal{E}_k$ s.t. $(a_{i,k} \wedge |q - s_{i,k} \cdot e| \leq \epsilon)$ where ϵ is a parameter that depends on the accuracy of the peak detection algorithm.

Moreover, no more than 3 basis patterns can be used to explain the peaks at sample point i, which translates to $|\{k | a_{i,k} = 1\}| \leq 3$. Finally, the sample points are embedded into a graph $\mathcal{G}$, such that there is a vertex for every sample and edges connect samples that are close on the *thin film* (eg. based on the grid). Given this graph, we require that the subgraph induced by $\{i | a_{i,k} = 1\}$ is connected in order for the basis patterns to appear in contiguous locations on the *thin film*. In addition, the scaling factors $s_{i,k}$ should be monotonic along the paths of this graph, and cannot exceed a given value S_{max}.

An analogy with the student enrollment example would be to consider a sample as a student who is enrolling in at most 3 courses (phases assigned to peaks of the sample) and is teaming up with other students (a peak paired with a neighboring peak).

4.3   oPSAT Encoding

We now formulate the phase-field identification problem as an oPSAT encoding. Let K be the set of phases. Also, let G be the set of sample points embedded in a grid, such that each sample has neighbors in one or more of the four directions $\{N, E, S, W\}$. We denote $G(i)$ the peaks of sample point i and l_p the angle of peak $p \in G(i)$. For a peak $p \in G(i)$, we define $N_{p,D} \subseteq G(i')$ the subset of peaks of sample i', where i' is the sample in direction D from i (denoted $i' \in D(i)$), and such that $p' \in N_{p,D}$ if $l_p \leq l'_p \leq l_p.S_{max}$. In other words, $N_{p,D}$ is the set of p's neighbor peaks that can be matched with p according to the direction D and without exceeding the maximum allowed shift (see Fig. 2).

Variables. We define a Boolean variable $x_{p,k}$, for $p \in G(i)$, $i \in G$, $k \in K$, to indicate whether peak p belongs to phase k. Similarly, $z_{i,k}$ indicates whether sample point i contains some peak in phase k, i.e. $z_{i,k} = \bigvee_{p \in G(i)} x_{p,k}$. In addition, a Boolean variable $y_{pp'k}$ indicates that peak p is paired with peak p' for phase k. Therefore, we have $y_{pp'k} \to x_{pk} \wedge x_{p'k}$. Furthermore, we introduce two directions $D_{1k} \in \{N, S\}$ and $D_{2k} \in \{E, W\}$ for each phase k. The direction of a phase is used to impose that any peak of that phase shifts according to that direction. Accordingly, we have: $\bigvee_{p' \notin N_{p, D_{1k}} \cup N_{p, D_{2k}}} y_{pp'k} = 0$ for all $i \in G, p \in G(i), k \in K$. Moreover, in order to introduce probability restrictions on the number of unmatched peaks, we define a Boolean variable d_p that corresponds to whether peak p is paired with a peak of the neighboring samples. Similarly, d_i denotes whether all peaks of sample i are paired, and are channeled to the d_p variables through the following propositional formula: $\neg d_p \vee d_i$ for all $i \in G, p \in G(i)$.

Propositional Formulas. A peak is assigned to at most one phase, i.e. $\sum_k x_{pk} \leq 1$. An unassigned peak is considered unmatched (as illustrated by p_0 in Fig. 2). Namely, $(\bigvee_k x_{pk}) \vee d_p$ for all $i \in G, p \in G(i)$. If a peak is assigned to a phase, then it needs to be paired with a neighboring peak, otherwise it is considered unmatched (see p_1 in Fig. 2). This constraint translates to: $x_{pk} \to \left(\bigvee_{p'} y_{pp'k} \vee d_p \right)$ for all $i \in G, p \in G(i), k \in K$. In addition, a phase should be consistent among the samples in which this phase is involved. Namely, if two adjacent samples share a phase, each peak of one must be paired with a peak of the other, otherwise it is considered unmatched (as illustrated by p_2 in Fig. 2). This translates to: $x_{pk} \wedge z_{i'k} \to \left(\bigvee_{p' \in G(i')} y_{pp'k} \vee d_p \right)$ for all $i \in G, p \in G(i), k \in K, i' \in D_{1k}(i) \cup D_{2k}(i)$. Moreover, we enforce a relaxed form of convex connectivity of a phase on the thin film, requiring that if any two samples that are two or more columns (or rows) apart involve a given phase, then there should be a sample in between them that involves this phase as well. In other words, we require $(x_{pk} \wedge x_{p'k}) \to \bigvee_{i'' \in N_C(i,i'), p'' \in G(i'')} x_{p''k}$, where $N_C(i, i')$ (resp. $N_R(i, i')$) is the set of samples on the grid between the columns (resp. rows) of i and i'. Finally, we impose that a peak cannot be paired with more than one neighboring peak, i.e. $\sum_{k, p' \neq p} y_{pp'k} \leq 1$, for all $i \in G, p \in G(i)$ and $\sum_{k, p' \neq p} y_{p'pk} \leq 1$, for all $i \in G, p \in G(i)$.

Probability Restrictions. We limit the probability that all peaks of a sample i remain unmatched by requiring $P(d_i) \leq p_i$, where p_i is either given or refined by dichotomy search.

Inference Method. For the experimental results described in the following, we computed a probability distribution using oPSAT with variance minimization and used, in order to obtain the accuracy of the computation, the model of the hard (SAT) constraints in that distribution with the highest probability.

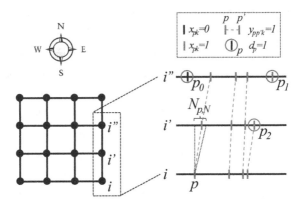

Fig. 2. Examples of soft violations in the oPSAT encoding. Left: Grid of sample points. Right: Pairing the peaks of sample points i, i' and i''. In the case of phase k of direction *North*, the peak p of sample i can only be paired with peaks of i' in $N_{p,N}$. Also, this example illustrates the three possible soft violations for peaks: 1) p_0 is not assigned a phase (assuming one single phase), 2) p_1 is not paired with any other peak, and 3) p_2, assigned to phase k, has no matching peak in i'', although i'' involves phase k.

4.4   Experimental Validation

We evaluate the oPSAT approach on the synthetic data used in [3] and compare with the SMT approach. Note that data from real experiments has to be manually labeled, which unfortunately is not yet available. Data was synthesized based on a known underlying phase map for the Al-Li-Fe system [16], a ternary system composed of 6 phases ($|K| = 6$).

All experiments were conducted on the same machine and using the same C++ implementation of an oPSAT solver, using minisat as the auxiliary SAT solver. The SMT solver used in these experiments was *Z3*[2]. For the oPSAT approach, the model with highest probability in the computed distribution was used to obtain the accuracy results. The maximum probability of a peak to be unmatched, that is, a peak with no phase assigned, was fixed as 2%, and a soft violation was defined as a sample point with some unmatched peak. This soft violation probability was computed over all peaks at that sample point, assuming that the probability of one peak to be unmatched is independent from that of any other peak. Table 1 shows the results of the experiments.

In all cases, the accuracy of the model computed by the oPSAT solver, defined as the percentage of peaks predicted with the same phase as in the synthetic data set, was above 80% (compared to 100% for SMT). On the other hand, the oPSAT implementation presents a dramatic increase of efficiency, of at least two orders of magnitude in all cases, and of about 2,000 times in one case.

Overall, while materials scientists currently proceed to a manual analysis of the high-throughput experimental data, our results provide solutions that are good and useful from the point of view of materials scientists, especially as these solutions are, by design of hard constraints, physically meaningful and

Table 1. Runtime (in seconds) for both SMT and oPSAT approaches on 5 datasets, as well as the accuracy of oPSAT (the accuracy of SMT being 100%). P is the number of sample points, L^* is the average number of peaks per phase, K is the number of basis patterns, and $\#Peaks$ is the overall number of peaks.

Dataset				SMT	oPSAT	
System	P	L^*	K	$\#Peaks$	Time(s)	Time(s)	Accuracy
Al/Li/Fe 28	6	6	170	346	5.3	84.7%
Al/Li/Fe 28	8	6	424	10076	8.8	90.5%
Al/Li/Fe 28	10	6	530	28170	12.6	83.0%
Al/Li/Fe 45	7	6	651	18882	121.1	82.0%
Al/Li/Fe 45	8	6	744	46816	128.0	80.3%

comply with the crystallographic process. In addition, our approach is the first automated method to exhibit short running times, and has great potential to be used within an online setting that guides the data collection itself. Therefore, the acceptable loss of accuracy is made up by a significant gain in speed. Finally, these results advocate the practical feasibility of oPSAT for real applications involving hard and soft constraints.

5   Conclusions

In this work we have described how to use the optimized probabilistic satisfiability (oPSAT) method to deal with problems that combine hard and soft restrictions. We have shown how a probability distribution can be computed to satisfy logic and probabilistic constraints and how it can be optimized to display balanced properties via variance minimization. The technique was then applied to the non-trivial problem of materials discovery with acceptable precision and superior run times than existing methods.

Future work should address the computation of probability constraints that minimize the expected value of soft violations, as well as inference methods that employ the probability distribution computed by the oPSAT method, instead of just considering the model with the largest probability in that distribution. With respect to experimental results, we plan to measure the efficiency of the oPSAT solver on real data, once a manually annotated data set becomes available. The application of oPSAT to other problems combining hard and soft constraints is also a direction to be explored.

Acknowledgments. The first author acknowledges the support of Fapesp-Brazil grant 2011/19860-4 and CNPq grant PQ 302553/2010-0. This work was supported by an NSF Expeditions in Computing Award on Computational Sustainability (Award Number 0832782).

References

1. Boole, G.: An Investigation on the Laws of Thought. Macmillan, London (1854), available on project Gutemberg at http://www.gutenberg.org/etext/15114
2. de Moura, L., Bjørner, N.: Z3: An efficient SMT solver. In: Ramakrishnan, C.R., Rehof, J. (eds.) TACAS 2008. LNCS, vol. 4963, pp. 337–340. Springer, Heidelberg (2008)
3. Ermon, S., Le Bras, R., Gomes, C.P., Selman, B., van Dover, R.B.: SMT-aided combinatorial materials discovery. In: Cimatti, A., Sebastiani, R. (eds.) SAT 2012. LNCS, vol. 7317, pp. 172–185. Springer, Heidelberg (2012)
4. Finger, M., Bona, G.D.: Probabilistic satisfiability: Logic-based algorithms and phase transition. In: Walsh, T. (ed.) IJCAI, pp. 528–533. IJCAI/AAAI (2011)
5. Friedman, N., Getoor, L., Koller, D., Pfeffer, A.: Learning probabilistic relational models. In: IJCAI, pp. 1300–1309. Springer (1999)
6. Gent, I.P., Walsh, T.: The SAT phase transition. In: ECAI 1994 – Proceedings of the Eleventh European Conference on Artificial Intelligence, pp. 105–109. John Wiley & Sons (1994)
7. Georgakopoulos, G., Kavvadias, D., Papadimitriou, C.H.: Probabilistic satisfiability. Journal of Complexity 4(1), 1–11 (1988)
8. Giunchiglia, E., Maratea, M.: Solving optimization problems with dll. In: Brewka, G., Coradeschi, S., Perini, A., Traverso, P. (eds.) ECAI 2006, pp. 377–381. IOS Press (2006)
9. Gregoire, J.M., Tague, M.E., Cahen, S., Khan, S., Abruna, H.D., DiSalvo, F.J., van Dover, R.B.: Improved fuel cell oxidation catalysis in pt1-xtax. Chem. Mater. 22(3), 1080 (2010)
10. Gregoire, J.M., Dale, D., van Dover, R.B.: A wavelet transform algorithm for peak detection and application to powder x-ray diffraction data. Review of Scientific Instruments 82(1), 015105–015105 (2011)
11. Hansen, P., Jaumard, B.: Probabilistic satisfiability. In: Handbook of Defeasible Reasoning and Uncertainty Management Systems, vol. 5, p. 321. Springer, Netherlands (2000)
12. Hansen, P., Jaumard, B., Nguetsé, G.B.D., de Aragão, M.P.: Models and algorithms for probabilistic and bayesian logic. In: IJCAI, pp. 1862–1868 (1995)
13. Hansen, P., Perron, S.: Merging the local and global approaches to probabilistic satisfiability. Int. J. Approx. Reasoning 47(2), 125–140 (2008)
14. Kavvadias, D., Papadimitriou, C.H.: A linear programming approach to reasoning about probabilities. Annals of Mathematics and Artificial Intelligence 1, 189–205 (1990), http://dx.doi.org/10.1007/BF01531078
15. Le Bras, R., Bernstein, R., Gomes, C.P., Selman, B.: Crowdsourcing backdoor identification for combinatorial optimization. In: Proceedings of the 23rd International Joint Conference on Artificial Intelligence, IJCAI 2013 (2013)
16. LeBras, R., Damoulas, T., Gregoire, J.M., Sabharwal, A., Gomes, C.P., van Dover, R.B.: Constraint reasoning and kernel clustering for pattern decomposition with scaling. In: Lee, J. (ed.) CP 2011. LNCS, vol. 6876, pp. 508–522. Springer, Heidelberg (2011), http://dl.acm.org/citation.cfm?id=2041160.2041202
17. Mitchell, D., Selman, B., Levesque, H.: Hard and easy distributions of SAT problems. In: AAAI 1992 – Proceedings of the 10th National Conference on Artificial Intelligence, pp. 459–465 (1992)
18. Narasimhan, B., Mallapragada, S., Porter, M.: Combinatorial Materials Science. Wiley (2007), http://books.google.com/books?id=tRdvxlL7mLOC

19. Nilsson, N.: Probabilistic logic. Artificial Intelligence 28(1), 71–87 (1986)
20. Nilsson, N.: Probabilistic logic revisited. Artificial Intelligence 59(1-2), 39–42 (1993)
21. Papadimitriou, C., Steiglitz, K.: Combinatorial Optimization: Algorithms and Complexity. Dover (1998)
22. De Raedt, L., Frasconi, P., Kersting, K., Muggleton, S. (eds.): Probabilistic Inductive Logic Programming. LNCS (LNAI), vol. 4911. Springer, Heidelberg (2008)
23. Richardson, M., Domingos, P.: Markov logic networks. Machine Learning 62(1-2), 107–136 (2006)
24. Van Dover, R.B., Schneemeyer, L., Fleming, R.: Discovery of a useful thin-film dielectric using a composition-spread approach. Nature 392(6672), 162–164 (1998)
25. Warners, J.P.: A linear-time transformation of linear inequalities into conjunctive normal form. Inf. Process. Lett. 68(2), 63–69 (1998)

Quantified Maximum Satisfiability:
A Core-Guided Approach

Alexey Ignatiev[1,3], Mikoláš Janota[1], and Joao Marques-Silva[1,2]

[1] IST/INESC-ID, Lisbon, Portugal
[2] University College Dublin, Ireland
[3] ISDCT SB RAS, Irkutsk, Russia
{aign,mikolas}@sat.inesc-id.pt, jpms@ucd.ie

Abstract. In recent years, there have been significant improvements in algorithms both for Quantified Boolean Formulas (QBF) and for Maximum Satisfiability (MaxSAT). This paper studies the problem of solving quantified formulas subject to a cost function, and considers the problem in a quantified MaxSAT setting. Two approaches are investigated. One is based on relaxing the soft clauses and performing a linear search on the cost function. The other approach, which is the main contribution of the paper, is inspired by recent work on MaxSAT, and exploits the iterative identification of unsatisfiable cores. The paper investigates the application of these approaches to the concrete problem of computing smallest minimal unsatisfiable subformulas (SMUS), a decision version of which is a well-known problem in the second level of the polynomial hierarchy. Experimental results, obtained on representative problem instances, indicate that the core-guided approach for the SMUS problem outperforms the use of linear search over the values of the cost function. More significantly, the core-guided approach also outperforms the state-of-the-art SMUS extractor Digger.

1 Introduction

When reasoning about quantified Boolean formulas (QBF), different optimization problems can be envisioned. MAX-QSAT [16] is a well-known example. Considering a QBF as a game between the existential and universal players, if the existential player can guarantee that k clauses are satisfied independently of the universal player, then k clauses are said to be *simultaneously satisfiable*. The MAX-QSAT problem is to find the maximum number of simultaneously satisfiable clauses. Original interest in MAX-QSAT was motivated by work on non-approximability results for problems in the polynomial hierarchy. A different optimization problem is to select a subset of clauses of a QBF such that the resulting QBF is true. A related optimization problem assumes the first quantifier to be existential, and asks for an assignment to those existential variables such that the QBF is true and a cost function is optimized. Work in quantified CSP involves computing strategies that optimize some cost function or associating costs with strategies [14,8]. Besides the theoretical interest, there are a number of practical settings where quantified optimization problems find application. This is for example the case when the goal is to optimize a cost function subject to a quantified set of constraints (e. g. the iterative use of QBF for optimizing target values in [13]). Many other concrete examples are given by the optimization versions of decision problems in the polynomial hierarchy [35].

M. Järvisalo and A. Van Gelder (Eds.): SAT 2013, LNCS 7962, pp. 250–266, 2013.

This paper addresses the problem of optimizing a cost function subject to a quantified set of constraints. The cost function will be represented as a set of *soft* clauses, and so this problem is referred to as Quantified MaxSAT (QMaxSAT). Inspired by algorithms for the non-quantified MaxSAT problem [19,2,23,9], this paper develops two novel approaches for QMaxSAT. The first one consists of relaxing all clauses and performing a linear (or binary) search over the values of the cost function. The linear search can either refine upper or lower bounds on the number of falsified soft clauses [19,9]. In contrast, binary search refines both lower and upper bounds [19,23]. The second approach represents the main contribution of this paper, and is inspired by recent work on core-guided MaxSAT, i.e. solving MaxSAT by iteratively computing unsatisfiable subformulas [19]. Thus, this new approach requires QBF solvers to be able to produce unsatisfiable cores. As a result, the second contribution of this paper is to show how recent 2QBF solvers based on abstraction refinement [25,24] can be modified to produce unsatisfiable cores.

The new algorithms for QMaxSAT are evaluated on the problem of computing the *smallest minimally unsatisfiable subformula* (SMUS) [29,32]. The SMUS decision problem is well-known to be in the second level of the polynomial hierarchy (e. g. [20]) and studied in the context of formal verification. Computing SMUSes is also relevant for assessing the quality of computed MUSes in practice. The third contribution of the paper is a novel QMaxSAT formulation for the SMUS problem, and QMaxSAT-based algorithm. Experimental results, obtained on representative problem instances, show that the core-guided QMaxSAT algorithm outperforms Digger, a state-of-the-art algorithm for the SMUS problem [26]. More importantly, these results validate the use of core-guided approaches for QMaxSAT.

The paper is organized as follows. The next section overviews basic definitions on SAT, MaxSAT, and QBF. Section 3 introduces the QMaxSAT problem, and Section 4 proposes several algorithms for QMaxSAT with an arbitrary number of quantification levels. This is completemented by a description in Section 4.1 of how a CEGAR-based 2QBF instrumented to generate unsatisfiable cores, and so used in QMaxSAT algorithms. Section 5 shows the practicality of the framework: it models the SMUS problem as QMaxSAT and describes improvements to the QMaxSAT for the concrete problem of computing an SMUS. Section 6 presents the experimental results on computing SMUSes. Section 7 concludes the paper.

2 Preliminaries

This section provides the notation and basic definitions related to SAT, MaxSAT and QBF.

2.1 Boolean Satisfiability

Let us consider a set of Boolean variables $X = \{x_1, \ldots, x_n\}$, $n \in \mathbb{N}$. A *literal* for variable x_i, $i \in \{1, \ldots, n\}$, is an atomic formula, denoted by l_i, which can be either a *positive* literal x_i, or its negation $\neg x_i$. A set of literals connected by a disjunction is called a *clause*. A conjunction of clauses $\varphi = c_1 \wedge c_2 \wedge \ldots \wedge c_m$, $m \in \mathbb{N}$, is called a formula in *conjunctive normal form* (CNF formula). Whenever convenient, a CNF formula is treated as a set of sets of literals $\varphi = \{c_1, c_2, \ldots, c_m\}$.

An assignment is a total mapping $\mathcal{A}_X : X \to \{0, 1\}$ defined on set X of variables. The notion of assignment $\mathcal{A}_X$ can be extended to literals by setting $\mathcal{A}_X(\neg x_i) = 1 - \mathcal{A}_X(x_i)$ for $x_i \in X$. Hereinafter, expression $\varphi|_{\mathcal{A}_X}$ denotes a formula obtained from a Boolean formula φ by replacing each variable x_i of X with its value $\mathcal{A}_X(x_i)$. The same *restriction* notation $c|_{\mathcal{A}_X}$ is used with regard to a clause c of a CNF formula. Since formula φ expresses some Boolean function $f(x_1, \ldots, x_n)$, $\varphi|_{\mathcal{A}_X}$ defines the value of function f, which in what follows is denoted by $f(\mathcal{A}_X)$. The same notation is used for denoting values of pseudo-Boolean functions.

If $\mathcal{A}_X(l_i) = 1$ then literal l_i is said to be *satisfied* by assignment $\mathcal{A}_X$; if $\mathcal{A}_X(l_i) = 0$ then l_i is *falsified* by $\mathcal{A}_X$. Assignment $\mathcal{A}_X$ satisfies a clause c, i.e. $c|_{\mathcal{A}_X} = 1$, if it satisfies at least one of its literals; otherwise the clause is said to be falsified by $\mathcal{A}_X$ ($c|_{\mathcal{A}_X} = 0$). If for a given CNF formula φ there is an assignment $\mathcal{A}_X$ such that $\varphi|_{\mathcal{A}_X} = 1$, then formula φ is called *satisfiable* and $\mathcal{A}_X$ is its *satisfying assignment*, or *model*. In the remainder of the paper, the set of all models of a CNF formula φ is denoted by $\mathcal{M}(\varphi)$.

2.2 Maximum Satisfiability

The *Maximum Satisfiability* (MaxSAT) is an optimization generalization of SAT formulated as follows: for a given CNF formula $\varphi = \{c_1, c_2, \ldots, c_m\}$, $m \in \mathbb{N}$, find such an assignment $\mathcal{A}_X$ that satisfies the maximum number of clauses of φ. A standard way of dealing with MaxSAT problems is to introduce a set $R = \{r_1, r_2, \ldots, r_m\}$ of additional variables (called *relaxation variables*) and consider a relaxed CNF formula $\varphi_R = \{c_1^R, \ldots, c_m^R\}$, where $c_i^R = c_i \vee r_i$. Observe that φ_R is satisfiable. The MaxSAT problem for φ can be now formulated as follows: given a cost function $f(r_1, \ldots, r_m) = \sum_{i=1}^{m} r_i$, find an assignment $\mathcal{A}_{X \cup R} \in \mathcal{M}(\varphi_R)$ such that for any other assignment $\mathcal{B}_{X \cup R} \in \mathcal{M}(\varphi_R)$

$$f(\mathcal{A}_{X \cup R}) \leq f(\mathcal{B}_{X \cup R})$$

The *partial* MaxSAT problem generalizes MaxSAT and deals with CNF formulas of the form $\varphi = \varphi_S \cup \varphi_H$, where all the clauses of φ_S are declared to be *relaxable* or *soft* while the rest (clauses of φ_H) are declared to be *hard*. The problem is to find an assignment $\mathcal{A}_X$ that satisfies all the hard clauses and maximizes the number of the soft clauses that are satisfied. Analogously to the MaxSAT formulation given above, one can formulate the partial MaxSAT problem by relaxing only the soft clauses and considering a cost function using the corresponding relaxation variables.

2.3 Quantified Boolean Formula

Quantified Boolean formulas (QBFs) are an extension of propositional logic with *existential* and *universal* quantifiers ($\forall$, $\exists$) [11].

In this paper QBFs are assumed to be in *prenex closed* form $Q_1 x_1 \ldots Q_n x_n. \varphi$, where $Q_i \in \{\forall, \exists\}$, x_i are distinct Boolean variables, and φ is a Boolean formula using only the variables x_i and the constants 0 (false), 1 (true). The sequence of quantifiers in a QBF is called the *prefix* and the Boolean formula the *matrix*. The semantics of QBF is defined recursively. A QBF $\exists x_1 Q_2 x_2 \ldots Q_n x_n. \varphi$ is true if and only if

$Q_2x_2 \ldots Q_nx_n. \varphi|_{x_1=1}$ or $Q_2x_2 \ldots Q_nx_n. \varphi|_{x_1=0}$ is true. A QBF $\forall x_1 Q_2x_2 \ldots Q_nx_n. \varphi$ is true iff $Q_2x_2 \ldots Q_nx_n. \varphi|_{x_1=1}$ and $Q_2x_2 \ldots Q_nx_n. \varphi|_{x_1=0}$ are true. To decide whether a given QBF is true or not, is PSPACE-complete [11].

Within a prefix, two adjacent quantifiers of different type, namely $\forall x_i \exists x_{i+1}$ and $\exists x_i \forall x_{i+1}$, are called a *quantifier alternation*. A QBF with k alternations has $k + 1$ *quantification levels*. Whenever possible, for variables $x_1, \ldots, x_n$, $x_i \in X_j$, under the same quantifier Q_j we write $Q_j X_j$ instead of $Q_j x_1 \ldots Q_j x_n$. Therefore, a formula with k quantification levels can be denoted by $Q_1 X_1 \ldots Q_k X_k. \varphi$. Moreover, the prefix $Q_1 X_1 \ldots Q_k X_k$ of a QBF with k quantification levels is usually denoted by $\overrightarrow{Q}$.

In Section 5, devoted to the SMUS problem, we focus on QBFs with 2 levels of quantification, i.e. formulas of the form $\forall X \exists Y. \varphi$ or $\exists X \forall Y. \varphi$, commonly denoted by 2QBF. Deciding whether a formula in 2QBF is true is complete for the second level of the polynomial hierarchy [11].

Section 3 uses the notion of *solution* of QBFs of the form $\psi = \exists X_0 \overrightarrow{Q}. \varphi$. An assignment $\mathcal{A}_{X_0}$ is a solution of ψ iff $\overrightarrow{Q}. \varphi|_{\mathcal{A}_{X_0}}$ is true[1]. Analogously to the set of all models of a CNF formula, the set of all solutions of a QBF ψ, where the first quantifier is $\exists$, is denoted by $\mathcal{M}(\psi)$.

3   Quantified MaxSAT

In this section we consider an optimization formulation of the QBF problem, when one should choose such a solution of the problem (among all solutions), that is optimal with respect to some given criterion. This kind of problems is a natural generalization of MaxSAT: instead of CNF formulas, we consider quantified formulas specified in a general form. Moreover, the optimization criterion in this problem is generalized as well. For example, it is possible to specify it as a minimization problem for some pseudo-Boolean cost function (see [21]).

Consider sets of Boolean variables $X_1, X_2, \ldots, X_k$ and a set of additional variables $E = \{e_1, \ldots, e_l\}$. Let

$$\psi = \exists E \, \overrightarrow{Q}. \varphi \tag{1}$$

be a quantified Boolean formula, where its matrix φ is a propositional formula over the set $(\bigcup_{i=1}^{k} X_i) \cup E$ given in a potentially non-CNF form. Consider a linear[2] pseudo-Boolean function $f(e_1, \ldots, e_l) = \sum_{i=1}^{l} a_i \cdot e_i$ as a cost function. Then the quantified MaxSAT (QMaxSAT) problem can be formulated as the problem of finding an assignment $\mathcal{A}_E \in \mathcal{M}(\psi)$ such that for any other assignment $\mathcal{B}_E \in \mathcal{M}(\psi)$

$$f(\mathcal{A}_E) \leq f(\mathcal{B}_E)$$

Example 1. Consider a 2QBF formula

$$\xi = \exists e_1, e_2 \, \forall x_1, x_2. \varphi,$$

---

[1] Note that solution of a quantified formula defined in this way represents a "portion" of the formula's *model*, which is defined, for example, in [12].

[2] Non-linear pseudo-Boolean formulas can be linearized with the use of auxiliary variables. Some linearization techniques are described in [18,6].

where $\varphi = (\neg e_1 \wedge \neg e_2) \rightarrow (x_1 \vee x_2)$; and a cost function $f(e_1, e_2) = 2 \cdot e_1 + 3 \cdot e_2$. Formula ξ has three possible solutions: $e_1 = 0, e_2 = 1; e_1 = 1, e_2 = 0; e_1 = 1, e_2 = 1$. However, the optimal solution which minimizes the cost function is $e_1 = 1, e_2 = 0$, i. e. $f(1, 0) = \min_{\mathcal{M}(\xi)} f(e_1, e_2) = 2$.

Let us show how the formulated QMaxSAT problem relates to classical (quantifier-free) MaxSAT. First, we define another problem, which is a special case of QMaxSAT. Consider a QBF $\vec{Q}. \varphi$, which is false, and let its matrix φ be in CNF. Then the problem of finding a maximal subset $\varphi' \subset \varphi$ such that $\vec{Q}. \varphi'$ is true, can be easily expressed in terms of QMaxSAT described above. To do this, one should consider a set R of relaxation variables and a CNF formula $\varphi_R = \{c_1 \vee r_1, \ldots, c_m \vee r_m\}, c_i \in \varphi, r_i \in R$, and choose $f(r_1, \ldots, r_m) = \sum_{i=1}^{m} r_i$ as the cost function. Then the problem is to find the *best* solution of QBF $\exists R \, \vec{Q}. \varphi_R$ subject to the cost function f. This problem is obviously a generalization of classical MaxSAT but also a special case of QMaxSAT. Note that although variables E from the QMaxSAT formulation are replaced by relaxation variables R here, they do not play the role of relaxation variables in general (e. g., see the matrix of formula ξ in Example 1).

Due to the close relationship of the QMaxSAT problem to its classical version, an interesting line of work is to apply to this problem the ideas and algorithms developed for non-quantified MaxSAT. The next section gives an explanation of how MaxSAT algorithms can be adapted to the QMaxSAT problem.

Related Work. Optimization problems subject to quantified constraints have been studied elsewhere [16,14,8], but address more general formulations than QMaxSAT. The Max-QSAT problem [16] can be viewed as computing a strategy that maximizes the number of simultaneously satisfiable clauses. Other optimization problems have been studied in the recent past [14,8]. The focus of [14] is approximation algorithms for computing a winning strategy that minimizes some cost function, whereas [8] studies preferences over strategies. To our best knowledge, and besides our work, [8] is the only other reference that proposes an exact algorithm for solving optimization problems over quantified constraints.

4   QMaxSAT Algorithms

One of the simplest approaches to the QMaxSAT problem is to iteratively decide the following formula with a QBF oracle:

$$\exists E \, \vec{Q}. \varphi \wedge (f(e_1, \ldots, e_l) \leq k) \tag{2}$$

Here one can start from a lower bound (e. g. $k = 0$) and increase k until formula (2) becomes true, or decrease it from some upper bound (e. g. $k = \max_{\{0,1\}^l} f$) value while (2) is true. This is analogous to the linear search for non-quantified MaxSAT [9], which refines lower and upper bounds on the value of the cost function[3]. Although these

---

[3] Instead of the linear search algorithms, one can use binary search [19,23]. Binary search algorithms are not covered by this paper.

algorithms are not the main contribution of the paper, we implemented and compared them to our main algorithm for the concrete case of the SMUS problem (see Section 6).

The main goal of this paper is to construct an algorithm which is based on the use of *unsatisfiable cores* (or simply *cores*) similar to the Fu&Malik's algorithm for MaxSAT [19]. Similarly to the linear search that refines lower bounds on the value of the cost function, Fu&Malik's algorithm (we refer to its original version as MSU1 [30]; some authors refer to this algorithm as WPM1 [1]) tests a series of unsatisfiable instances until a satisfiable instance is found. However, instead of dealing with the constraint $f(e_1, \ldots, e_l) \leq k$ and increasing k with each call to a SAT solver, it identifies a small unsatisfiable part of the current formula, which is called an unsatisfiable core. Sequential core computation in MSU1 increases the current cost value with each iteration, i.e. with every new core computed. Thus, each unsatisfiable core increments a possible minimum cost of an assignment that satisfies the constraints.

Recall that function f is linear, i.e. $f(e_1, \ldots, e_l) = \sum_{i=1}^{l} a_i \cdot e_i$. Assume[4], that $a_i = 1, \forall i \in \{1, \ldots, l\}$. For each term e_i of formula $f(e_1, \ldots, e_l) = \sum_{i=1}^{l} e_i$ create a unit clause $\neg e_i$. Denote CNF formula $\{\neg e_1, \neg e_2, \ldots, \neg e_l\}$ by φ_S. Observe that each term e_i of f incrementing its value (i.e. $e_i = 1$) corresponds to a falsified clause $\neg e_i$ of φ_S. This means that an assignment evaluates function f to some value y, $0 \leq y \leq l$, if and only if it satisfies $l - y$ clauses of φ_S. Therefore, function f is evaluated to its minimum value by such an assignment $\mathcal{A}_E$, that maximizes the number of satisfied clauses of φ_S. Let $\#(\varphi_S|_{\mathcal{A}_E})$ be a function that outputs the number of clauses of φ_S that are satisfied by some assignment $\mathcal{A}_E$, i.e. $\#(\varphi_S|_{\mathcal{A}_E}) = \sum_{c \in \varphi_S} c|_{\mathcal{A}_E}$. Instead of QBF ψ (see (1)), consider the formula

$$\psi' = \exists E \overrightarrow{Q}. \varphi \wedge \varphi_S \tag{3}$$

Now we can formulate another way to solve the QMaxSAT problem for QBF ψ subject to the cost function f. It consists in finding an assignment $\mathcal{A}_E \in \mathcal{M}(\psi)$ such that for any other assignment $\mathcal{B}_E \in \mathcal{M}(\psi)$ the following holds: $\#(\varphi_S|_{\mathcal{A}_E}) \geq \#(\varphi_S|_{\mathcal{B}_E})$. On the analogy of partial MaxSAT, CNF φ_S can be treated as a set of *soft* clauses while the original QBF matrix φ is a *hard* part given in a potentially non-CNF form. Let us define a core of formula ψ'. This will enable us to extend the MSU1 algorithm to the case of QMaxSAT.

Definition 1. *A Boolean formula $\varphi_C = \varphi \wedge \varphi'_S$, $\varphi'_S \subseteq \varphi_S$, is called an unsatisfiable core of formula ψ', if and only if the following is false*

$$\exists E \overrightarrow{Q}. \varphi_C$$

According to Definition 1, the hard part of formula ψ' is included into any unsatisfiable core φ_C of ψ'. However, similarly to the core-guided algorithms for the non-quantified MaxSAT, in the algorithm described below we will need only the soft part φ'_S of the core. The algorithm selects soft clauses from the core by calling a function $\text{Soft}(\varphi_C)$.

Algorithm 1 shows the pseudo-code of the MSU1 algorithm adapted to QMaxSAT (we refer to this algorithm as *QMSU1*). For a formula ψ' given in the form (3), which is

---

[4] Otherwise we have a *weighted* version of the problem, and all the ideas described in this section, can be extended as is done for weighted MaxSAT algorithms [30,1].

Algorithm 1. QMSU1 Algorithm

---

input : A QBF $\psi = \exists E\, \vec{Q}.\, \varphi$ s.t. $\mathcal{M}(\psi) \neq \emptyset$, and a CNF φ_S
output : $\mathcal{A}_E \in \mathcal{M}(\psi)$, s.t. $\forall \mathcal{B}_E \in \mathcal{M}(\psi)$: $\#(\varphi_S|_{\mathcal{A}_E}) \geq \#(\varphi_S|_{\mathcal{B}_E})$

1   $R_{all} \leftarrow \emptyset$ // set of all relaxation variables
2   **while** true **do**
3     $\psi'_R = \exists E\, \exists R_{all}\, \vec{Q}.\, \varphi \wedge \varphi_S$
4     $(\text{st}, \varphi_C, \mathcal{A}_E) \leftarrow \text{QBF}(\psi'_R)$ // calling a QBF oracle
5   **if** st = true **then**
6   **return** $\mathcal{A}_E$
7     $R \leftarrow \emptyset$ // set of relaxation variables
8   **foreach** $c \in \text{Soft}(\varphi_C)$ **do**
9   **let** r be a new relaxation variable
10       $R \leftarrow R \cup \{r\}$
11       $\varphi_S \leftarrow \varphi_S \setminus \{c\} \cup \{c \vee r\}$
12     $\varphi \leftarrow \varphi \wedge \text{CNF}(\sum_{r \in R} r \leq 1)$
13     $R_{all} \leftarrow R_{all} \cup R$

---

implicitly defined by a QBF ψ from (1) and a CNF formula φ_S, the QMSU1 algorithm outputs such a solution $\mathcal{A}_E$ of ψ that maximizes the number of satisfied clauses of φ_S over the set $\mathcal{M}(\psi)$. One important pre-condition of the algorithm is that formula ψ must have at least one solution, i.e. $\mathcal{M}(\psi) \neq \emptyset$. The set of all relaxation variables used by the algorithm is denoted by R_{all} and initialized by $\emptyset$ (line 1). At each iteration of the loop the algorithm constructs a relaxed copy ψ'_R of formula ψ' (line 3) and asks a QBF oracle to decide whether it is true or false (line 4). As an answer the oracle returns a 3-tuple $(\text{st}, \varphi_C, \mathcal{A}_E)$. If st = false, then the algorithm considers a set of relaxation variables R (initially set to $\emptyset$) and processes the unsatisfiable core φ_C returned by the QBF oracle. This step consists of relaxing soft clauses of the core, i.e. the algorithm introduces a new relaxation variable $r \in R$ for each soft clause c of the core φ_C, and replaces original clause c with its relaxed copy $c \vee r$ in φ_S. At the end of the iteration QMSU1 adds a CNF encoding of a new cardinality constraint $\sum_{r \in R} r \leq 1$ to the hard part φ of ψ'. Note that since each relaxation variable $r \in R_{all}$ is added only to a clause of the form $\neg e_j$, all of them can be quantified by the same $\exists$-quantifier as variables $e_j \in E$ (see line 3). The algorithm iterates until formula ψ'_R is true and $\mathcal{A}_E \in \mathcal{M}(\psi'_R)$ (line 6). By construction, $\mathcal{A}_E$ maximizes the number of satisfied clauses of φ_S over the set $\mathcal{M}(\psi)$, i.e. it is the solution of a QMaxSAT problem. Note that the algorithm is analogous to the MSU1 algorithm for non-quantified MaxSAT. The only difference is that QMSU1 uses not a SAT solver as an oracle but a QBF solver, and the hard part of the formula can be in a non-CNF form. Thus, the correctness of the algorithm relies on the corresponding result for the MSU1 algorithm [19].

Note that the only requirement imposed by the QMSU1 algorithm on the QBF oracle is the ability to produce a certificate that could validate the answer *true* or *false*. While providing a solution of a formula seems straightforward to implement, the oracle must also be able to explain why the input formula is false, i.e. to extract an unsatisfiable core

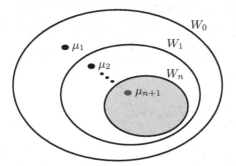

Fig. 1. Gradual strengthening of abstractions until a solution is found

from the formula. A simple solution is to include all the soft clauses in the core. However, the efficiency of the algorithm relies on producing small cores. There exist methods for extracting unsatisfiable cores from unsatisfiable QBF instances for DPLL-based QBF solvers (e. g., see [37]). While the QMSU1 algorithm can use any QBF solver as long as it produces cores, this work uses a CEGAR-based 2QBF solver [25,24] as an underlying QBF oracle for a particular problem (the smallest MUS problem). Section 4.1 describes a method using a CEGAR-based 2QBF solver for extracting unsatisfiable cores. Following the ideas of [24], the method can be easily extended to the case of formulas with an arbitrary number of quantification levels. The task is further simplified by the fact that the QMSU1 algorithm requires only soft part of the core, which depends on variables quantified at the outermost level.

4.1 Extracting Cores in CEGAR-Based 2QBF

Among the many practical uses of the *counterexample guided abstraction refinement* (CEGAR) [15], it can also be applied for solving 2QBF [25]. The key idea of CEGAR is to consider an approximate representation of a problem (called the abstraction) instead of its explicit representation that could be too large to construct or unknown. This section provides a basic overview of the algorithm[5] and describes its modification, which is able to extract an unsatisfiable core of a formula if the formula is false.

For the sake of succintness, in this section we denote assignments to variables of X and Y by μ and ν, respectively. We also assume, that the matrix of the 2QBF is presented as $\varphi_H \wedge \varphi_S$, where φ_S represents a set of soft clauses, and φ_H is a hard part given in a possibly non-CNF form. The algorithm hinges on the idea that the problem $\exists X \forall Y. \varphi_H \wedge \varphi_S$ can be equivalently represented as

$$\exists X. \bigwedge_{\nu \in \{0,1\}^{|Y|}} (\varphi_H \wedge \varphi_S)|_\nu \qquad (4)$$

where the universal quantifier is *expanded* using a conjunction. Since the full expansion (4) of the problem can be exponentially large with respect to the original problem,

---

[5] The reader is referred to [25] for further details and properties of the algorithm.

Algorithm 2. CEGAR loop for 2QBF

> **input** : $\exists X \forall Y. \varphi_H \wedge \varphi_S$
> **output** : (true, μ) if there exists μ s.t. $\forall Y. (\varphi_H \wedge \varphi_S)|_\mu$,
>   (false, $\varphi_H \wedge \varphi'_S$) s.t. $\varphi'_S \subseteq \varphi_S$ otherwise

1 $\omega \leftarrow \emptyset$
2 **while** true **do**
3 $\quad$ $\varphi \leftarrow \mathrm{CNF}(\bigwedge_{\nu \in \omega} \varphi_H|_\nu) \cup \bigwedge_{\nu \in \omega} \varphi_S|_\nu$
4 $\quad$ $(\mathsf{st}_1, \mu, \varphi_C) \leftarrow \mathrm{SAT}(\varphi)$ $\qquad\qquad\qquad\qquad\qquad$ // candidate
5 $\quad$ **if** $\mathsf{st}_1 = $ false **then**
6 $\quad\quad$ $\varphi'_S \leftarrow \{c \in \varphi_S \mid c' \in \varphi_C, \nu \in \omega, c' = c|_\nu\}$
7 $\quad\quad$ **return** (false, $\varphi_H \wedge \varphi'_S$) $\qquad\qquad$ // no candidate found
8 $\quad$ $(\mathsf{st}_2, \nu) \leftarrow \mathrm{SAT}(\neg(\varphi_H \wedge \varphi_S)|_\mu)$ $\qquad\qquad$ // counterexample
9 $\quad$ **if** $\mathsf{st}_2 = $ false **then**
10 $\quad\quad$ **return** (true, μ) $\qquad\qquad\qquad\qquad\qquad$ // solution found
11 $\quad$ $\omega \leftarrow \omega \cup \{\nu\}$ $\qquad\qquad\qquad\qquad\qquad\qquad\qquad$ // refine

it is infeasible to construct such representation in practice. Instead of constructing the full expansion (4), CEGAR constructs a *partial expansion* of the given problem, i. e.

$$\exists X. \bigwedge_{\nu \in W} (\varphi_H \wedge \varphi_S)|_\nu \qquad\qquad (5)$$

where $W \subseteq \{0, 1\}^{|Y|}$. We refer to formula (5) as W-*abstraction*. Observe that for any W, the corresponding W-abstraction is weaker than the full expansion (4). This means that the set of the W-abstraction's solutions is a superset over the set of solutions of the original problem, i. e. some of the W-abstraction's solutions may *not* satisfy (4). The idea of the CEGAR-based algorithm described below is to gradually strengthen the abstraction until a solution of the original problem is found, or the abstraction is proved to be false (see Figure 1).

Algorithm 2 shows the pseudocode of the algorithm. The algorithm maintains a set of assignments W in the variable ω. We start with the abstraction equal to the formula 1, any assignment μ to the variable of X satisfies the abstraction. Assume, that the algorithm encodes the hard part φ_H into a CNF formula by calling a function $\mathrm{CNF}(\varphi_H)$.

In each iteration of the loop, the algorithm first computes a solution to the abstraction, which is maintained in φ and constructed at line 3. We refer to this solution as a *candidate solution*, because it is *not* guaranteed that it is indeed a solution to the original problem. If a SAT oracle says (see line 4) that there is no candidate solution, i. e. the abstraction has no solutions, the original problem does not have any solutions either (recall that the abstraction is always weaker than the problem). In this case the algorithm returns an unsatisfiable core of the input formula in the form $\varphi_H \wedge \varphi'_S$ (line 7). Observe that the soft part φ'_S of the QBF core can be easily extracted from the core φ_C returned by the SAT oracle: φ'_S should include a clause $c \in \varphi_S$ if there is a clause $c' \in \varphi_C$ and an assignment $\nu \in \omega$ such that $c' = c|_\nu$. In other words, the unsatisfiable core φ_C shows the falsity of the W-abstraction even if we consider the abstraction's soft part to be φ'_S instead of φ_S, i.e. $\exists X. \bigwedge_{\nu \in W} (\varphi_H \wedge \varphi'_S)|_\nu$ is false. Note that there can be several clauses $c_i \in \varphi_S$ such that $c' = c_i|_\nu$. However, it is sufficient to include just one of these clauses into φ'_S — by doing this one can get smaller QBF cores.

If the SAT oracle says that there is a candidate solution, then the algorithm checks whether it is indeed a solution of the problem or not. This is done by computing a *counterexample*. For a candidate μ, a counterexample ν is an assignment to the variables of Y such that $\neg\varphi|_{\mu,\nu}$. A counterexample ν serves as a witness that μ is not a solution, i.e. it is not the case that $\forall Y. \varphi|_\mu$ because φ is false when y has the value ν. If no counterexample is found, the current candidate is indeed a solution and can be returned. If a counterexample *is* found, it is added to the set ω which effectively strengthens the abstraction.

5   Smallest MUS Problem

This section considers a concrete application of the QMaxSAT problem — the problem of finding a smallest MUS of a CNF formula. Let $X = \{x_1, \ldots, x_n\}$ be a set of Boolean variables and $\varphi = \{c_1, \ldots, c_m\}$ be a CNF formula. Formula $\psi \subseteq \varphi$ is called a *minimal unsatisfiable subformula* (MUS) of φ, if ψ is unsatisfiable and $\forall c_i \in \psi$ formula $\psi \setminus \{c_i\}$ is satisfiable. The MUS problem is a subject of active research (e. g. [31]).

Definition 2. *Formula ψ^*, $\psi^* \subseteq \varphi$, is called a smallest MUS of φ if*
1. *ψ^* is unsatisfiable;*
2. *for any MUS ψ, $\psi \subseteq \varphi$, the following holds $|\psi^*| \leq |\psi|$.*

The *smallest MUS* problem (SMUS) consists in finding a smallest MUS of a CNF formula. An algorithm that computes an SMUS by searching the space of all unsatisfiable subformulas was presented in [29]. A greedy genetic algorithm that finds approximate solutions of the SMUS problem was proposed in [38]. A branch and bound algorithm for computing SMUSes was described in [32,26]. The decision version of the SMUS problem, i. e. the problem of determining whether a given formula has a smallest MUS of size k, is known to be Σ_2^P-complete (e. g., see [20]). The Digger algorithm, which is a state-of-the-art algorithm for computing an SMUS of a CNF formula, was proposed in [32,26].

Let us formulate an optimization extension of SMUS in terms of QMaxSAT defined in Section 3. First, we consider a set of *selection variables* $S = \{s_1, \ldots, s_m\}$. Instead of formula φ, we consider a formula $\varphi_R = \{c_1 \vee \neg s_1, \ldots, c_m \vee \neg s_m\}$, $c_i \in \varphi$. Let us introduce a function $f : \{0,1\}^m \rightarrow \mathbb{N}$:

$$f(s_1, \ldots, s_m) = \sum_{i=1}^m s_i.$$

Then the problem of finding a smallest MUS of φ consists in finding such an assignment $\mathcal{A}_S \in \mathcal{M}(\neg\varphi_R)$ that for any other assignment $\mathcal{B}_S \in \mathcal{M}(\neg\varphi_R)$ the following holds: $f(\mathcal{A}_S) \leq f(\mathcal{B}_S)$.

As shown in Section 4, to solve this problem, one can use an iterative approach calling a 2QBF oracle which decides whether the following quantified formula is true or false:

$$\exists S \, \forall X. \, \neg \varphi_R \wedge (f(s_1, \ldots, s_m) \leq k) \tag{6}$$

However, one can apply algorithm QMSU1 to this problem as well. Similarly to Section 4, we introduce a set $\varphi_s = \{\neg s_1, \neg s_2, \ldots, \neg s_m\}$ of soft clauses instead of considering constraint $f(s_1, \ldots, s_m) \leq k$ and iteratively ask the QBF oracle to decide the following QBF:

$$\psi = \exists S \, \forall X. \, \neg \varphi_R \wedge \varphi_S \tag{7}$$

Observe that CNF formula φ_S is the set of soft clauses of ψ while $\neg \varphi_R$ is the hard part presented in a non-clausal form. Thus, the QMSU1 algorithm iteratively extracts unsatisfiable cores of formula ψ and relaxes their soft parts, which are some subsets of φ_S, until it finds an assignment $\mathcal{A}_S \in \mathcal{M}(\neg \varphi_R)$ that maximizes the number of satisfied clauses of φ_S. Assignment $\mathcal{A}_S$ corresponds to an SMUS ψ^*, $\psi^* \subseteq \varphi$, such that a clause $c_i \in \psi^*$ iff $\mathcal{A}_S(s_i) = 1$.

5.1   Improvements to the Algorithm

To increase the performance of the Digger algorithm, the authors of [26] use a preprocessing technique — computing a set of disjoint MCSes. An MCS (or *minimal correction set*) of an unsatisfiable CNF formula φ is a subset of clauses $\delta \subset \varphi$ such that $\varphi \setminus \delta$ is satisfiable while $\varphi \setminus \delta \cup c$ is unsatisfiable for any clause $c \in \delta$. There is an important connection between MCSes and MUSes of CNF formulas (see [34,22,10,3,27,28]): any MUS of formula φ is a *minimal hitting set* of the complete set of MCSes of φ. Therefore, the enumeration of disjoint MCSes gives a lower bound of the size of an SMUS, thus, reducing the search space of the Digger algorithm.

Due to the fact that the QMSU1 algorithm does not handle constraints $\leq k$ directly, lower bounds for SMUS themselves cannot be directly used in QMSU1. However, the enumeration of disjoint MCSes can be still helpful while solving SMUS by QMSU1. For example, if a CNF formula φ has an MCS $C = \{c\}$, where c is a clause (so called *unit* MCS), then each MUS of φ contains clause c. Therefore, one of the improvements of QMSU1 for computing an SMUS of formula φ can be enumeration of all the unit MCSes of φ.

Another technique we exploit in our approach is the use of MCSes, found during the preprocessing stage, as unsatisfiable cores of formula (7). Let δ be an MCS of φ. By φ_S^δ we denote a subformula of φ_S containing only clauses of φ_S that correspond to clauses of δ, i.e. $(\neg s_i) \in \varphi_S^\delta$ if $c_i \in \delta$. By definition of an MCS, formula $\varphi \setminus \delta$ is satisfiable. This means that $\varphi_R \wedge \varphi_S^\delta$ is also satisfiable. Then formula

$$\exists S \, \forall X. \, \neg \varphi_R \wedge \varphi_S^\delta$$

is false. Given Definition 1, this means that $\neg \varphi_R \wedge \varphi_S^\delta$ is a core of (7). Therefore, k MCSes computed by preprocessing give us k unsatisfiable cores of (7). Moreover, since all the computed MCSes are disjoint, the cores are disjoint. In practice, the use of this preprocessing technique significantly increases the performance of the QMSU1 algorithm.

6 Experimental Results

A prototype of a solver for the SMUS problem implementing the QMSU1 algorithm was developed with the use of a CEGAR-based 2QBF oracle described in Section 4.1. The underlying SAT solver of our 2QBF oracle implementation is MiniSat 2.2 [17]. We refer to this prototype as *MinUC* (**Min**imum **U**nsatisfiable **C**ore finder). Three versions of this solver were developed. The default one is the core-guided version. The other two include *MinUC-LB* and *MinUC-UB* and implement iterative linear lower and upper bound approaches respectively.

During the course of this research, we implemented a number of efficient algorithms to enumerate disjoint MCSes of CNF formulas. These are beyond the scope of this paper. However, to do a more comprehensive comparison between QMSU1 and Digger, we ran MinUC in three different modes (the corresponding names of the tools are presented in the parentheses):

- without enumerating disjoint MCSes (*MinUC-w*);
- with the use of the Digger's disjoint MCS enumerator (*MinUC-d*);
- with the use of the default built-in disjoint MCS enumerator (*MinUC*).

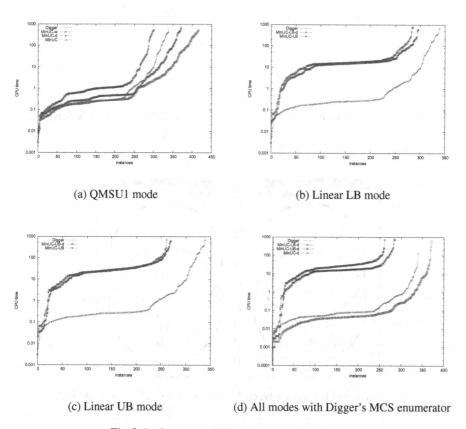

(a) QMSU1 mode   (b) Linear LB mode

(c) Linear UB mode   (d) All modes with Digger's MCS enumerator

Fig. 2. Performance of MinUC compared to Digger

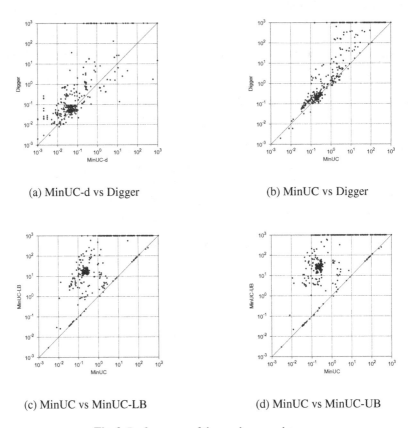

(a) MinUC-d vs Digger

(b) MinUC vs Digger

(c) MinUC vs MinUC-LB

(d) MinUC vs MinUC-UB

Fig. 3. Performance of the used approaches

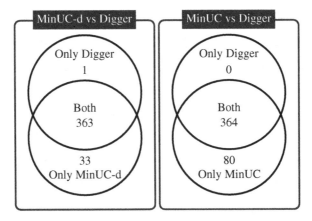

Fig. 4. Number of instances solvable by different solvers: Digger vs MinUC

The set of instances considered includes several sets of benchmarks described below. The first set consists of automotive product configuration benchmarks [36]. Two other sets of benchmarks come from circuit diagnosis. Additionally, we selected instances from the complete set of the MUS competitions benchmarks[6] as follows. Since the SMUS problem is computationally harder than problem of extracting an MUS of formula, we picked all the instances from the MUS competitions that can be solved by muser-2 (see [7]) in 10 seconds. The total number of instances in the set is 682. All experimental results were obtained on an Intel Xeon 5160 3GHz, with 4GB of memory, and running Fedora Linux operating system. The experiments were made with a 800 seconds time limit and a 2GB memory limit. The detailed overview of the results is presented in the following plots.

Figure 2a shows a cactus plot illustrating the performance of the core-guided version of MinUC compared to Digger. The version of MinUC without enumerating disjoint MCSes (MinUC-w) can solve 325 instances only. Digger solves 364 instances while MinUC with the same MCS enumerator (MinUC-d) is able to solve 396 instances. This is 8.8% more than by Digger's result (4.7% of all the 682 instances). In the case of using its own MCS enumerator MinUC demonstrates the best performance with 444 instances solved, having 22% advantage over Digger (11.7% of the total 682 instances). Figure 2b and Figure 2c show similar plots for linear search LB and UB modes respectively. Even with the use of its own MCS enumerator, linear search modes of MinUC perform worse than Digger: MinUC-LB solves 322 while MinUC-UB solves 294 instances. Figure 2d gives a more graphic comparison between Digger and all the versions of MinUC using Digger's MCS enumerator. In this case, the time required to enumerate disjoint MCSes is not taken into account (because it is the same for all the solvers) while in all the other cases it is included in the runtime.

Figure 3 shows scatter plots comparing the QMSU1 versions of MinUC to Digger (see Figure 3a and Figure 3b) and to its linear search versions (Figure 3c and Figure 3d). Figure 4 gives an overview on how many instances are solvable either by Digger or by core-guided MinUC only.

The results indicate that the core-guided version of MinUC has an advantage over other approaches. Digger comes second. MinUC-LB and MinUC-UB have the worst performance. Although the experiment results are quite positive for the current version of the core-guided version of MinUC comparing to Digger, it is still has a potential of possible improvements.

7 Conclusions

This paper studies optimization problems over quantified sets of constraints, and focuses on the concrete case of quantified MaxSAT (QMaxSAT). The main contributions of the paper are: (i) a novel core-guided algorithm for QMaxSAT; (ii) generation of unsatisfiable cores with CEGAR-based QBF solvers; (iii) a QMaxSAT-based approach for solving the smallest MUS (SMUS) problem; and (iv) new pruning techniques for solving the SMUS problem. The novel core-guided algorithm for QMaxSAT is based on

---

[6] http://www.satcompetition.org/2011/

the original work of Fu&Malik [19]. Nevertheless, other algorithms for non-quantified MaxSAT can also be adapted to the quantified case (e. g. [2,23]).

Experimental results on representative problem instances demonstrate that the novel approach for computing SMUSes, based on core-guided QMaxSAT algorithms, significantly outperforms Digger, a state-of-the-art algorithm for computing an SMUS. These results motivate applying core-guided QMaxSAT algorithms to other optimization problems with quantified constraints.

A number of future research directions can be envisioned. Investigating additional optimization problems with quantified constraints will provide a larger set of problem instances. Motivated by a larger universe of problems and problem instances, additional core-guided algorithms can be developed for QMaxSAT. Finally, it will be important to investigate how to extend the algorithms developed in this paper to settings more general than QMaxSAT. For example, MAX-QSAT [16] among others [14,8]. Any QBF solver can be integrated into the QMSU1 algorithm as an oracle as long as it produces unsatisfiable cores. There are known techniques for extracting unsatisfiable cores from unsatisfiable QBF instances for DPLL-based QBF solvers. One of these techniques is proposed in [37] and then followed by recent works on certificate generation for resolution-based QBF solvers (e. g. [4,5,33]). Thus, an interesting subject of future work is integration of a DPLL-based QBF solver into the QMSU1 algorithm and comparison of its performance (in terms of speed and a core size) with performance of the currently implemented CEGAR-based core-producing QBF oracle.

For the concrete application of QMaxSAT, the SMUS problem, several optimizations can be considered. Modern (and efficient) MUS solvers [7] can be used for computing an upper bound on the size of the SMUS. If the lower bound (e. g. due to disjoint cores, or by iterative core extraction) equals the upper bound, then an SMUS will by given by any minimal hitting set of all the disjoint MCSes. Moreover, several preprocessing approaches can be used, several of which are more efficient than the one used in Digger.

Acknowledgments. This work is partially supported by SFI PI grant BEACON (09-/IN.1/I2618), FCT grants ATTEST (CMU-PT/ELE/0009/2009) and POLARIS (PTDC-/EIA-CCO/123051/2010), and multiannual PIDDAC program funds (PEst-OE/EEI/LA-0021/2011).

References

1. Ansótegui, C., Bonet, M.L., Levy, J.: Solving (Weighted) Partial MaxSAT through Satisfiability Testing. In: Kullmann, O. (ed.) SAT 2009. LNCS, vol. 5584, pp. 427–440. Springer, Heidelberg (2009)
2. Ansótegui, C., Bonet, M.L., Levy, J.: A new algorithm for weighted partial MaxSAT. In: AAAI Conference on Artificial Intelligence. AAAI (2010)
3. Bailey, J., Stuckey, P.J.: Discovery of Minimal Unsatisfiable Subsets of Constraints Using Hitting Set Dualization. In: Hermenegildo, M.V., Cabeza, D. (eds.) PADL 2004. LNCS, vol. 3350, pp. 174–186. Springer, Heidelberg (2005)
4. Balabanov, V., Jiang, J.H.R.: Resolution proofs and skolem functions in QBF evaluation and applications. In: Gopalakrishnan, G., Qadeer, S. (eds.) CAV 2011. LNCS, vol. 6806, pp. 149–164. Springer, Heidelberg (2011)

5. Balabanov, V., Jiang, J.H.R.: Unified qbf certification and its applications. Formal Methods in System Design 41(1), 45–65 (2012)
6. Balas, E., Mazzola, J.: Nonlinear 0–1 programming: I. Linearization techniques. Mathematical Programming 30, 1–21 (1984), http://dx.doi.org/10.1007/BF02591796, 10.1007, doi:10.1007/BF02591796
7. Belov, A., Lynce, I., Marques-Silva, J.: Towards efficient MUS extraction. AI Commun. 25(2), 97–116 (2012)
8. Benedetti, M., Lallouet, A., Vautard, J.: Quantified constraint optimization. In: Stuckey, P.J. (ed.) CP 2008. LNCS, vol. 5202, pp. 463–477. Springer, Heidelberg (2008)
9. Berre, D.L., Parrain, A.: The Sat4j library, release 2.2. JSAT 7(2-3), 59–64 (2010)
10. Birnbaum, E., Lozinskii, E.L.: Consistent subsets of inconsistent systems: structure and behaviour. J. Exp. Theor. Artif. Intell. 15(1), 25–46 (2003)
11. Büning, H.K., Bubeck, U.: Theory of quantified boolean formulas. In: Biere, A., Heule, M., van Maaren, H., Walsh, T. (eds.) Handbook of Satisfiability, Frontiers in Artificial Intelligence and Applications, vol. 185, pp. 735–760. IOS Press (2009)
12. Büning, H.K., Subramani, K., Zhao, X.: Boolean Functions as Models for Quantified Boolean Formulas. J. Autom. Reasoning 39(1), 49–75 (2007)
13. Chen, H., Janota, M., Marques-Silva, J.: QBF-based Boolean function bi-decomposition. In: Design, Automation & Test in Europe Conference, pp. 816–819 (2012)
14. Chen, H., Pál, M.: Optimization, games, and quantified constraint satisfaction. In: Fiala, J., Koubek, V., Kratochvíl, J. (eds.) MFCS 2004. LNCS, vol. 3153, pp. 239–250. Springer, Heidelberg (2004)
15. Clarke, E.M., Grumberg, O., Jha, S., Lu, Y., Veith, H.: Counterexample-guided abstraction refinement for symbolic model checking. J. ACM 50(5), 752–794 (2003)
16. Condon, A., Feigenbaum, J., Lund, C., Shor, P.W.: Probabilistically checkable debate systems and nonapproximability of PSPACE-hard functions. Chicago J. Theor. Comput. Sci. (1995)
17. Eén, N., Sörensson, N.: An Extensible SAT-solver. In: Giunchiglia, E., Tacchella, A. (eds.) SAT 2003. LNCS, vol. 2919, pp. 502–518. Springer, Heidelberg (2004)
18. Fortet, R.: Applications de l'algèbre de Boole en recherche opérationnelle. Revue Française d'Informatique et de Recherche Opérationnelle 4, 17–26 (1960)
19. Fu, Z., Malik, S.: On solving the partial MAX-SAT problem. In: Biere, A., Gomes, C.P. (eds.) SAT 2006. LNCS, vol. 4121, pp. 252–265. Springer, Heidelberg (2006)
20. Gupta, A.: Learning Abstractions for Model Checking. Ph.D. thesis, Carnegie Mellon University (June 2006)
21. Hammer, P., Rudeanu, S.: Boolean Methods in Operations Research and Related Areas. Springer (1968)
22. Han, B., Lee, S.J.: Deriving minimal conflict sets by CS-trees with mark set in diagnosis from first principles. IEEE Transactions on Systems, Man, and Cybernetics, Part B 29(2), 281–286 (1999)
23. Heras, F., Morgado, A., Marques-Silva, J.: Core-guided binary search for maximum satisfiability. In: AAAI Conference on Artificial Intelligence. AAAI (2011)
24. Janota, M., Klieber, W., Marques-Silva, J., Clarke, E.: Solving QBF with counterexample guided refinement. In: Cimatti, A., Sebastiani, R. (eds.) SAT 2012. LNCS, vol. 7317, pp. 114–128. Springer, Heidelberg (2012)
25. Janota, M., Marques-Silva, J.: Abstraction-based algorithm for 2QBF. In: Sakallah, K.A., Simon, L. (eds.) SAT 2011. LNCS, vol. 6695, pp. 230–244. Springer, Heidelberg (2011)
26. Liffiton, M.H., Mneimneh, M.N., Lynce, I., Andraus, Z.S., Marques-Silva, J., Sakallah, K.A.: A branch and bound algorithm for extracting smallest minimal unsatisfiable subformulas. Constraints 14(4), 415–442 (2009)

27. Liffiton, M.H., Sakallah, K.A.: On Finding All Minimally Unsatisfiable Subformulas. In: Bacchus, F., Walsh, T. (eds.) SAT 2005. LNCS, vol. 3569, pp. 173–186. Springer, Heidelberg (2005)
28. Liffiton, M.H., Sakallah, K.A.: Algorithms for Computing Minimal Unsatisfiable Subsets of Constraints. J. Autom. Reasoning 40(1), 1–33 (2008)
29. Lynce, I., Marques-Silva, J.P.: On Computing Minimum Unsatisfiable Cores. In: SAT (2004)
30. Manquinho, V., Marques-Silva, J., Planes, J.: Algorithms for weighted boolean optimization. In: Kullmann, O. (ed.) SAT 2009. LNCS, vol. 5584, pp. 495–508. Springer, Heidelberg (2009)
31. Marques-Silva, J.: Minimal unsatisfiability: Models, algorithms and applications (invited paper). In: ISMVL, pp. 9–14. IEEE Computer Society (2010)
32. Mneimneh, M., Lynce, I., Andraus, Z., Marques-Silva, J., Sakallah, K.: A Branch-and-Bound Algorithm for Extracting Smallest Minimal Unsatisfiable Formulas. In: Bacchus, F., Walsh, T. (eds.) SAT 2005. LNCS, vol. 3569, pp. 467–474. Springer, Heidelberg (2005)
33. Niemetz, A., Preiner, M., Lonsing, F., Seidl, M., Biere, A.: Resolution-Based Certificate Extraction for QBF - (Tool Presentation). In: Cimatti, A., Sebastiani, R. (eds.) SAT 2012. LNCS, vol. 7317, pp. 430–435. Springer, Heidelberg (2012)
34. Reiter, R.: A theory of diagnosis from first principles. Artif. Intell. 32(1), 57–95 (1987)
35. Schaefer, M., Umans, C.: Completeness in the polynomial-time hierarchy: a compendium. SIGACT News 33(3), 32–49 (2002)
36. Sinz, C., Kaiser, A., Küchlin, W.: Formal methods for the validation of automotive product configuration data. AI EDAM 17(1), 75–97 (2003)
37. Yu, Y., Malik, S.: Validating the result of a Quantified Boolean Formula (QBF) solver: theory and practice. In: Asia and South Pacific Design Automation Conference, pp. 1047–1051 (2005)
38. Zhang, J., Li, S., Shen, S.: Extracting minimum unsatisfiable cores with a greedy genetic algorithm. In: Australian Conference on Artificial Intelligence, pp. 847–856 (2006)

Nested Boolean Functions as Models
for Quantified Boolean Formulas

Uwe Bubeck and Hans Kleine Büning

Computer Science Institute, University of Paderborn, Germany
{bubeck,kbcsl}@upb.de

Abstract. Nested Boolean functions or Boolean programs are an alternative to the quantified Boolean formula (QBF) characterization of polynomial space. The idea is to start with a set of Boolean functions given as propositional formulas and to define new functions as compositions or instantiations of previously defined ones. We investigate the relationship between function instantiation and quantification and present a compact representation of models and countermodels of QBFs with and without free variables as nested Boolean functions. The representation is symmetric with respect to Skolem models and Herbrand countermodels. For a formula with free variables, it can describe both kinds of models simultaneously in one complete equivalence model which can be Skolem or Herbrand depending on actual assignments to the free variables.

1 Introduction

The satisfiability problem for quantified Boolean formulas (QBF) is the canonical PSPACE-complete problem. As an extension of propositional logic, QBF draws its expressive power from the ability to have quantifiers over propositional variables, where universal quantification $\forall x \, \Phi(x)$ for a variable x and a propositional or quantified Boolean formula Φ is defined to be true if and only if $\Phi(0)$ is true *and* $\Phi(1)$ is true, and $\exists y \, \Phi(y)$ means that $\Phi(0)$ *or* $\Phi(1)$ is true.

Nested Boolean functions (NBF) or Boolean programs have been introduced by Cook and Soltys [6] as an alternative characterization of polynomial space. They extend propositional logic with the ability to define Boolean functions as compositions or instantiations of previously defined functions, starting with a set of initial functions given as propositional formulas. For example, let

$$f_0(p_1, p_2) := (\neg p_1 \wedge p_2) \vee (p_1 \wedge \neg p_2)$$

be an initial function which computes the parity of two binary variables. Then the parity of four variables can be computed by reusing f_0:

$$f_1(p_1, p_2, p_3, p_4) := f_0(f_0(p_1, p_2), f_0(p_3, p_4))$$

The parity of 16 variables can be expressed compactly by reusing f_1, and so on [6]. By replacing all occurrences of f_0 with its definition, we can expand f_1:

$$def(f_1)(p_1, ..., p_4) := (\neg((\neg p_1 \wedge p_2) \vee (p_1 \wedge \neg p_2)) \wedge ((\neg p_3 \wedge p_4) \vee (p_3 \wedge \neg p_4))) \vee$$
$$(((\neg p_1 \wedge p_2) \vee (p_1 \wedge \neg p_2)) \wedge \neg((\neg p_3 \wedge p_4) \vee (p_3 \wedge \neg p_4)))$$

M. Järvisalo and A. Van Gelder (Eds.): SAT 2013, LNCS 7962, pp. 267–275, 2013.
© Springer-Verlag Berlin Heidelberg 2013

The semantics of nested Boolean functions can be defined by this fallback to propositional formulas. But in order to evaluate a NBF sequence $(f_0, ..., f_k)$, i.e. to check whether $f_k(a_1, ..., a_n) = 1$ for given arguments $a_1, ..., a_n \in \{0, 1\}$, it is not necessary to actually create the propositional expansion $def(f_k)$. By immediately replacing subterms whenever their values are known, the formula can be simplified on-the-fly, and polynomial space is sufficient. It can be shown that the evaluation and the satisfiability problem for NBF are PSPACE-complete [6]. Not surprisingly, there exist efficient transformations between QBF and NBF: any NBF can be transformed in linear time into an equivalent QBF [4], while the best known transformation in the other direction needs quadratic time [4].

In this paper, we want to further clarify the relationship between function instantiation and quantification. It is well known that in a closed prenex QBF, e.g. $\Phi = \forall x_n \exists y_n ... \forall x_1 \exists y_1 \; \phi(x_1, ..., x_n, y_1, ..., y_n)$, every existentially quantified variable y_i can be associated with a Boolean function f_i, called *Skolem function* [13], which depends on the values of those universally quantified variables whose quantifiers are outer to $\exists y_i$. For the given Φ, these are $x_i, ..., x_n$. Then Φ is true if and only if there exist $f_1, ..., f_n$ so that $\forall x_n ... \forall x_1 \; \phi(x_1, ..., x_n, f_1(x_1, ..., x_n), ..., f_n(x_n))$ is true. That means each occurrence of an existential variable is replaced with the corresponding Skolem function, and the resulting matrix must be true for all values of the universal variables. If that is the case, we call $f_1, ..., f_n$ a *Skolem model*. Analogously, the universally quantified variables can be mapped to *Herbrand functions* [8]. Φ is false if and only if there is a *Herbrand countermodel* $g_1, ..., g_n$ so that $\exists y_n ... \exists y_1 \; \phi(g_1(y_2, ..., y_n), ..., g_{n-1}(y_n), g_n(), y_1, .., y_n)$ is false. Recently, Balabanov and Jiang have shown how to extract a Skolem model of a true closed QBF from its cube-resolution proof and a Herbrand countermodel from the clause-resolution proof of a false QBF [1]. Earlier approaches use explicit skolemization techniques [2, 3, 9] and do not directly address Herbrand countermodels, or the generated strategies are not explicitly represented as functions [7]. Both Skolem and Herbrand (counter)models are of great practical importance when applications require solutions or explanations of unsatisfiability in addition to a mere decision of satisfiability. But the compact representation of these (counter)models remains a great problem for practical applications [12].

We consider Skolem and Herbrand (counter)models from a more theoretical viewpoint and show that we can compactly encode them by polynomial-size NBFs. More importantly, we further study the duality between Skolem models and Herbrand countermodels by considering QBFs with free variables. While a closed QBF is either true or false, the valuation of a QBF $\Phi(\mathbf{z})$ with free variables $\mathbf{z} = z_1, ..., z_r$ depends on the values of the free variables [10]. Consider the example $\Phi(z) = \forall x \exists y \; (x \vee y) \wedge (\neg x \vee \neg y) \wedge (\neg z \vee \neg y)$. If z is 0, we have $\Phi(0) = \forall x \exists y \; (x \vee y) \wedge (\neg x \vee \neg y) \wedge (\neg 0 \vee \neg y)$, which is a closed QBF and is true with Skolem model $f_y(x) = \neg x$. If z is 1, $\Phi(1) = \forall x \exists y \; (x \vee y) \wedge (\neg x \vee \neg y) \wedge (\neg 1 \vee \neg y)$ is false with Herbrand countermodel $f_x() = 0$. The interesting observation here is that QBFs with free variables can have Skolem models and Herbrand countermodels for different values of the free variables. How are both kinds related when the formula remains the same and only the free variables change? We propose a unified *complete equivalence model*, in which all variables are mapped to functions and show how Skolem and Herbrand (counter)models are embedded in it and how these functions can be computed from the formula matrix by NBFs.

For space considerations, we rely only on the above informal introduction of NBF syntax and semantics and refer the reader to [4] for formal definitions. We furthermore assume in the following that all QBFs are in prenex normal form, i.e. all quantifiers appear in front of a purely propositional matrix. Two QBFs $\Phi(z_1, ..., z_r)$ and $\Psi(z_1, ..., z_r)$ with free variables $z_1, ..., z_r$ are *logically equivalent*, written $\Phi(z_1, ..., z_r) \approx \Psi(z_1, ..., z_r)$ or simply $\Phi \approx \Psi$, if and only if for every truth assignment τ to the free variables $z_1, ..., z_r$ both formulas evaluate to the same truth value [10]. We consider propositional formulas as QBFs in which all variables are free. Furthermore, given a NBF $(f_0, ..., f_k)$ where f_k has arguments $a_1, ..., a_n$, we say $f_k(a_1, ..., a_n)$ is logically equivalent to a QBF $\Phi(a_1, ..., a_n)$, written $f_k(a_1, ..., a_n) \approx \Phi(a_1, ..., a_n)$, if and only if the propositional expansion $def(f_k)(a_1, ..., a_n)$ is equivalent to $\Phi(a_1, ..., a_n)$. When we evaluate a NBF, we use "$=$" instead of "$\approx$" for a more lightweight notation. The length of a QBF is the number of variable occurrences, including the prefix. The length of a NBF $(f_0, ..., f_k)$ is $|f_0| + ... + |f_k|$, where $|f_i|$ is the total number of occurrences of constants, variables and function symbols on the right-hand side of the defining equation of f_i . The parity example above has length $4 + 7$.

2   Quantification as Iterated Function Composition

Definition 1. *Let* $\Phi(\mathbf{z}) = Q_n v_n ... Q_1 v_1 \; \phi(v_1, ..., v_n, \mathbf{z})$ *with* $Q_i \in \{\forall, \exists\}$ *be a QBF with bound variables* $v_1, ..., v_n$ *(*$n \geq 1$ *w.l.o.g), free variables* $\mathbf{z} = z_1, ..., z_r$ *and propositional matrix* ϕ.
Then we iteratively define Boolean functions $F_0, ..., F_n$ *as follows:*

1. $F_0(v_1, ..., v_n, \mathbf{z}) := \phi(v_1, ..., v_n, \mathbf{z})$
2. *For* $i = 1, ..., n$:

$$F_i(v_{i+1}, ..., v_n, \mathbf{z}) := \begin{cases} F_{i-1}(F_{i-1}(0, v_{i+1}, ..., v_n, \mathbf{z}), v_{i+1}, ..., v_n, \mathbf{z}) & \text{, if } Q_i = \forall \\ F_{i-1}(F_{i-1}(1, v_{i+1}, ..., v_n, \mathbf{z}), v_{i+1}, ..., v_n, \mathbf{z}) & \text{, if } Q_i = \exists \end{cases}$$

For the example $\Phi(z) = \forall v_2 \exists v_1 \; (v_1 \vee v_2) \wedge (\neg v_1 \vee \neg v_2) \wedge (\neg z \vee \neg v_1)$ we obtain:

$$F_0(v_1, v_2, z) := (v_1 \vee v_2) \wedge (\neg v_1 \vee \neg v_2) \wedge (\neg z \vee \neg v_1)$$
$$F_1(v_2, z) \quad := F_0(F_0(1, v_2, z), v_2, z) = F_0((\neg v_2 \wedge \neg z), v_2, z) = \neg z \vee v_2$$
$$F_2(z) \quad := F_1(F_1(0, z), z) = F_1(\neg z, z) = \neg z$$

The main idea behind this definition is that we try to assign values to the quantified variables, going from the outermost to the innermost quantifier (NBFs are evaluated by recursion from the last function in the sequence back to the initial functions). Similar to the DPLL algorithm for QBF (QDPLL) [5], we might have to branch in the worst case for $x = 0$ and $x = 1$ on each variable x. If x is universally quantified and the formula is false for $x = 0$, there is no need to try $x = 1$. Similarly, if x is existential and the formula is true for $x = 1$, we do not try $x = 0$. In our NBF encoding, the result of the first branch determines where the second branch is going. If the formula is false for $x = 0$, we stay with $x = 0$, and if it is true for $x = 1$, we stay with $x = 1$. In either case, the arguments to the inner call of F_{i-1} are the same as for the outer call of F_{i-1}, which suggests that it would be an important optimization for a real NBF solver to recognize duplicate instantiations for the same arguments.

Lemma 1. *For $\Phi(\mathbf{z}) = Q_n v_n ... Q_1 v_1 \ \phi(v_1, ..., v_n, \mathbf{z})$ with associated functions $F_0, ..., F_n$ as in Definition 1, it holds for all $i = 1, ..., n$ that*

$$F_i(v_{i+1}, ..., v_n, \mathbf{z}) \approx Q_i v_i ... Q_1 v_1 \ \phi(v_1, ..., v_n, \mathbf{z}) \ .$$

Proof. Let $i = 1$, and consider the case that $Q_1 = \forall$. Then we must show that $\phi(\phi(0, v_2, ..., v_n, \mathbf{z}), v_2, ..., v_n, \mathbf{z}) \approx \forall v_1 \ \phi(v_1, ..., v_n, \mathbf{z})$. Assume that the left-hand side is true for some truth value assignment τ to $v_2, ..., v_n$ and $\mathbf{z}$, that means $\phi(\phi(0, \tau(v_2), ..., \tau(v_n), \tau(\mathbf{z})), \tau(v_2), ..., \tau(v_n), \tau(\mathbf{z})) = 1$. Then it is not possible that the inner instantiation of ϕ is false. Because if we did substitute 0 for the inner instantiation $\phi(0, \tau(v_2), ..., \tau(v_n), \tau(\mathbf{z}))$, the outer instantiation of ϕ would become the same and would thus also be false, which would contradict our assumption that the whole left-hand side is true. If, on the other hand, $\phi(0, \tau(v_2), ..., \tau(v_n), \tau(\mathbf{z})) = 1$ for the inner instantiation, the outer becomes $\phi(1, \tau(v_2), ..., \tau(v_n), \tau(\mathbf{z}))$. With this being true by the initial assumption, we know that $\phi(v_1, \tau(v_2), ..., \tau(v_n), \tau(\mathbf{z}))$ is true for $v_1 = 0$ and $v_1 = 1$, and thus also $\forall v_1 \ \phi(v_1, \tau(v_2), ..., \tau(v_n), \tau(\mathbf{z})) = 1$.

From right to left, $\forall v_1 \ \phi(v_1, \tau(v_2), ..., \tau(v_n), \tau(\mathbf{z})) = 1$ for some τ implies that $\phi(0, \tau(v_2), ..., \tau(v_n), \tau(\mathbf{z})) = 1$ for the inner instantiation of ϕ on the left-hand side, so the outer instantiation on the left becomes $\phi(1, \tau(v_2), ..., \tau(v_n), \tau(\mathbf{z}))$, which is also true by the universal quantification on v_1.

If $Q_1 = \exists$, we must show $\phi(\phi(1, v_2, ..., v_n, \mathbf{z}), v_2, ..., v_n, \mathbf{z}) \approx \exists v_1 \ \phi(v_1, ..., v_n, \mathbf{z})$. For a truth assignment τ which satisfies the left-hand side, the inner instantiation of ϕ on the left is either false or true. Accordingly, $\phi(0, \tau(v_2), ..., \tau(v_n), \tau(\mathbf{z})) = 1$ or $\phi(1, \tau(v_2), ..., \tau(v_n), \tau(\mathbf{z})) = 1$ on the left, and that implies the right-hand side. In the other direction, let $\exists v_1 \ \phi(v_1, \tau(v_2), ..., \tau(v_n), \tau(\mathbf{z})) = 1$ for some τ. Then $\phi(0, \tau(v_2), ..., \tau(v_n), \tau(\mathbf{z})) = 1$ or $\phi(1, \tau(v_2), ..., \tau(v_n), \tau(\mathbf{z})) = 1$. If the latter holds, the inner instantiation on the left-hand side is also true, and the left-hand side becomes $\phi(1, \tau(v_2), ..., \tau(v_n), \tau(\mathbf{z}))$ and is thus true. On the other hand, if $\phi(1, \tau(v_2), ..., \tau(v_n), \tau(\mathbf{z})) = 0$, the left-hand side becomes $\phi(0, \tau(v_2), ..., \tau(v_n), \tau(\mathbf{z}))$, which is true in this case.

For the induction step, assume $F_i(v_{i+1}, ..., v_n, \mathbf{z}) \approx Q_i v_i ... Q_1 v_1 \ \phi(v_1, ..., v_n, \mathbf{z})$ for $i \geq 1$. Then $Q_{i+1} v_{i+1} ... Q_1 v_1 \ \phi(v_1, ..., v_n, \mathbf{z}) \approx Q_{i+1} v_{i+1} F_i(v_{i+1}, ..., v_n, \mathbf{z})$, and we must show $F_i(F_i(0, v_{i+2}, ..., v_n, \mathbf{z}), v_{i+2}, ..., v_n, \mathbf{z}) \approx \forall v_{i+1} F_i(v_{i+1}, ..., v_n, \mathbf{z})$ resp. $F_i(F_i(1, v_{i+2}, ..., v_n, \mathbf{z}), v_{i+2}, ..., v_n, \mathbf{z}) \approx \exists v_{i+1} F_i(v_{i+1}, ..., v_n, \mathbf{z})$. When we let $\phi'(v_{i+1}, ..., v_n, \mathbf{z}) := F_i(v_{i+1}, ..., v_n, \mathbf{z})$, the proof can be obtained in complete analogy to the induction base when substituting ϕ' for ϕ. $\square$

Corollary 1. $F_n(\mathbf{z}) \approx Q_n v_n ... Q_1 v_1 \ \phi(v_1, ..., v_n, \mathbf{z})$
for $\Phi(\mathbf{z}) = Q_n v_n ... Q_1 v_1 \ \phi(v_1, ..., v_n, \mathbf{z})$ with $F_0, ..., F_n$ as in Definition 1.

The length of the NBF $(F_0, ..., F_n)$ is $|\phi| + \Sigma_{i=1...n}(2i+1+2|\phi|)$, which is quadratic in $|\Phi|$. This is the same as for the existing transformation from QBF to NBF in [4] which simulates quantifier expansion, while we now simulate QDPLL. An important difference is that the expansion approach leads to definitions of the form $f_{i+1}(...) := f_i(f_{j_1}(...), ..., f_{j_r}(...))$ and needs multiple initial functions, while we now have definitions $f_{i+1}(...) := f_i(f_i(...), x_2, ..., x_r)$ where recursion occurs only in the first argument of the outer f_i, and we now only need one initial function, which is exactly the matrix of the QBF. This is important for our equivalence models. The following small technical lemma will be helpful later.

Lemma 2. *For $\Phi(\mathbf{z}) = Q_n v_n ... Q_1 v_1 \, \phi(v_1, ..., v_n, \mathbf{z})$ with $F_0, ..., F_n$ as in Def. 1, we have $F_i(v_{i+1}, ..., v_n, \mathbf{z}) \approx Q_i v_i F_{i-1}(v_i, ..., v_n, \mathbf{z})$ for all $i = 1, ..., n$.*

Proof. If $i = 1$, $F_0(v_1, ..., v_n, \mathbf{z})$ is $\phi(v_1, ..., v_n, \mathbf{z})$, and the claim follows immediately from Lemma 1.

For $i > 1$, $Q_i v_i(F_{i-1}(v_i, ..., v_n, \mathbf{z})) \approx Q_i v_i(Q_{i-1}v_{i-1}...Q_1 v_1 \, \phi(v_1, ..., v_n, \mathbf{z}))$ by Lemma 1. Again by Lemma 1, the latter is equivalent to $F_i(v_{i+1}, ..., v_n, \mathbf{z})$. $\square$

3   Equivalence Models for Quantified Boolean Formulas

Equivalence models for QBFs $\Phi(\mathbf{z}) = \forall x_n \exists y_n ... \forall x_1 \exists y_1 \, \phi(x_1, ..., x_n, y_1, ..., y_n, \mathbf{z})$ with free variables $\mathbf{z}$ have been defined in [11] by an equivalence-preserving mapping of existential variables $y_1, ..., y_n$ to functions $h_1(x_1, ..., x_n, \mathbf{z}), ..., h_n(x_n, \mathbf{z})$ with $\Phi(\mathbf{z}) \approx \forall x_n ... \forall x_1 \, \phi(x_1, ..., x_n, h_1(x_1, ..., x_n, \mathbf{z}), ..., h_n(x_n, \mathbf{z}), \mathbf{z})$. For a symmetric treatment of the quantifiers, we will now define the notion of *complete* equivalence models where *all* quantified variables, including the universal ones, are mapped to functions over the free variables.

Definition 2. *Let $\Phi(\mathbf{z}) = Q_n v_n ... Q_1 v_1 \, \phi(v_1, ..., v_n, \mathbf{z})$ with $Q_i \in \{\forall, \exists\}$ be a QBF with bound variables $v_1, ..., v_n$, free variables $\mathbf{z} = z_1, ..., z_r$ and propositional matrix ϕ. A sequence of Boolean functions $h_1(\mathbf{z}), ..., h_n(\mathbf{z})$ is called a* complete equivalence model *if and only if $\Phi(\mathbf{z}) \approx \phi(h_1(\mathbf{z}), ..., h_n(\mathbf{z}), \mathbf{z})$.*

It is easy to see that every QBF has a complete equivalence model: For formulas without free variables, the complete equivalence model consists of constants $h_1, ..., h_n \in \{0, 1\}$ such that Φ is true if and only if $\phi(h_1, ..., h_n)$ is true. Clearly, every true (false, respectively) closed QBF has a satisfying (falsifying, respectively) truth assignment to the matrix, and has thus a complete equivalence model. For a QBF with free variables, a complete equivalence model could always be constructed in a naive way by considering all assignments $\tau(\mathbf{z})$ to the free variables and choosing $h_i(\tau(\mathbf{z})) := \epsilon_i \in \{0, 1\}$ for all $i = 1, ..., n$ such that $\phi(\epsilon_1, ..., \epsilon_n, \tau(\mathbf{z}))$ is true if and only if $\Phi(\tau(\mathbf{z}))$ is true.

The problem of deciding whether a sequence $h_1, ..., h_n \in \{0, 1\}$ is a complete equivalence model for a closed or open QBF is PSPACE-complete. It is in PSPACE, since the equivalence problem for QBF is in PSPACE, and the hardness follows from a similar reduction as in [11]: given a closed QBF $\Phi = Q_n v_n ... Q_1 v_1 \, \phi(v_1, ..., v_n)$, let $\Phi' = \exists v_{n+1} Q_n v_n ... Q_1 v_1 \, (\phi(v_1, ..., v_n) \wedge v_{n+1})$. Then $h_1 = ... = h_{n+1} = 0$ is a complete equivalence model for Φ' iff Φ is false.

Lemma 3. *If $\Sigma_2^P \neq \Pi_2^P$ in the polynomial-time hierarchy, there must exist quantified Boolean formulas with free variables for which every propositional representation of the complete equivalence model requires super-polynomial length.*

Proof. Consider a closed QBF $\Phi = \forall x_n ... \forall x_1 \exists y_m ... \exists y_1\ \phi(x_1, ..., x_n, y_1, ..., y_m)$ for which the satisfiability problem is Π_2^P-complete. Then let $\Phi'(x_1, ..., x_n) :=$ $\exists y_m ... \exists y_1\ \phi(x_1, ..., x_n, y_1, ..., y_m)$. If Φ' had a complete equivalence model with polynomial-size propositional encoding, we could guess in polynomial time propositional formulas $h_1(x_1, ..., x_n), ..., h_m(x_1, ..., x_n)$ and insert them into Φ. If a Π_1^P-oracle accepts $\forall x_n ... \forall x_1\ \phi(x_1, ..., x_n, h_1(x_1, ..., x_n), ..., h_m(x_1, ..., x_n))$, we know that Φ is true and that $h_1, ..., h_m$ have been guessed correctly. In total, we would be able to solve the formula in Σ_2^P, and thus $\Sigma_2^P = \Pi_2^P$. □

Lemma 3 holds analogously for the non-complete equivalence models from [11] and for Skolem/Herbrand (counter)models, even for closed QBFs with only two levels of quantification. Also in practical QBF applications, Skolem/Herbrand (counter)models are often infeasibly large when represented as propositional formulas. We are now going to represent complete equivalence models as NBFs instead, allowing us to place a polynomial bound on the size of these models.

Definition 3. *Let $\Phi(\mathbf{z}) = Q_n v_n ... Q_1 v_1\ \phi(v_1, ..., v_n, \mathbf{z})$ with $Q_i \in \{\forall, \exists\}$ be a QBF with bound variables $v_1, ..., v_n$ ($n \geq 1$ w.l.o.g), free variables $\mathbf{z} = z_1, ..., z_r$ and matrix ϕ. Using the function representation from Definition 1, we map each variable v_k to a model function h_k as follows:*

1. $h_n(\mathbf{z}) := \begin{cases} F_{n-1}(0, \mathbf{z}) & \text{, if } Q_n = \forall \\ F_{n-1}(1, \mathbf{z}) & \text{, if } Q_n = \exists \end{cases}$

2. For $i = n-1, ..., 1$:

$h_i(\mathbf{z}) := \begin{cases} F_{i-1}(0, h_{i+1}(\mathbf{z}), ..., h_n(\mathbf{z}), \mathbf{z}) & \text{, if } Q_i = \forall \\ F_{i-1}(1, h_{i+1}(\mathbf{z}), ..., h_n(\mathbf{z}), \mathbf{z}) & \text{, if } Q_i = \exists \end{cases}$

The intuition here is as follows: according to Definition 1, $F_1, ..., F_n$ are defined by $F_i(v_{i+1}, ..., v_n, \mathbf{z}) := F_{i-1}(F_{i-1}(\sigma_i, v_{i+1}, ..., v_n, \mathbf{z}), v_{i+1}, ..., v_n, \mathbf{z})$ with $\sigma_i \in \{0, 1\}$ according to the quantifier type of Q_i. If we had already found model functions $h_{i+1}, ..., h_n$ (we omit their arguments $\mathbf{z}$ for simplicity), we could substitute them for $v_{i+1}, ..., v_n$: $F_i(h_{i+1}, ..., h_n, \mathbf{z}) \approx F_{i-1}(F_{i-1}(\sigma_i, h_{i+1}, ..., h_n, \mathbf{z}), h_{i+1}, ..., h_n, \mathbf{z})$. To write the latter as $F_{i-1}(h_i, ..., h_n, \mathbf{z})$, we choose $h_i := F_{i-1}(\sigma_i, h_{i+1}, ..., h_n, \mathbf{z})$.

Consider again the example $\Phi(z) = \forall v_2 \exists v_1\ (v_1 \lor v_2) \land (\neg v_1 \lor \neg v_2) \land (\neg z \lor \neg v_1)$ with $F_1(v_2, z) = \neg z \lor v_2$ and $F_2(z) = \neg z$ (Section 2, p. 269). Then $h_2(z) = F_1(0, z) = \neg z$ and $h_1(z) = F_0(1, h_2(z), z) = \neg h_2(z) \land \neg z = z \land \neg z = 0$.

Lemma 4. *For $i = 1, ..., n$:*

$$F_n(\mathbf{z}) \approx F_{i-1}(h_i(\mathbf{z}), ..., h_n(\mathbf{z}), \mathbf{z})$$

Proof. The proof is by backward induction on i. For $i = n$, the right-hand side is $F_{n-1}(h_n(\mathbf{z}), \mathbf{z})$, which is $F_{n-1}(F_{n-1}(0, \mathbf{z}), \mathbf{z})$ if $Q_n = \forall$ and $F_{n-1}(F_{n-1}(1, \mathbf{z}), \mathbf{z})$ otherwise. By Definition 1 (from right to left), this is $F_n(\mathbf{z})$.

For the induction step, let $i \in \{1, ..., n-1\}$. Assume the above equivalence holds for $i+1$, that means $F_n(\mathbf{z}) \approx F_i(h_{i+1}(\mathbf{z}), ..., h_n(\mathbf{z}), \mathbf{z})$. By Definition 1, the right-hand side is $F_{i-1}(F_{i-1}(\sigma_i, h_{i+1}(\mathbf{z}), ..., h_n(\mathbf{z}), \mathbf{z}), h_{i+1}(\mathbf{z}), ..., h_n(\mathbf{z}), \mathbf{z})$ with $\sigma_i \in \{0, 1\}$ according to the quantifier type of Q_i. By the above Definition 3, $h_i(\mathbf{z}) := F_{i-1}(\sigma_i, h_{i+1}(\mathbf{z}), ..., h_n(\mathbf{z}), \mathbf{z})$, so the last expression can be written as $F_{i-1}(h_i(\mathbf{z}), ..., h_n(\mathbf{z}), \mathbf{z})$, i.e. the right-hand side of the statement to be proven.

$\square$

Theorem 1. *For $\Phi(\mathbf{z}) = Q_n v_n ... Q_1 v_1 \ \phi(v_1, ..., v_n, \mathbf{z})$, model functions $h_1, .., h_n$ constructed according to Definition 3 are a complete equivalence model.*

Proof. We have to show that $\phi(h_1(\mathbf{z}), ..., h_n(\mathbf{z}), \mathbf{z}) \approx Q_n v_n ... Q_1 v_1 \ \phi(v_1, ..., v_n, \mathbf{z})$. With $F_0 := \phi$, $\phi(h_1(\mathbf{z}), ..., h_n(\mathbf{z}), \mathbf{z})$ is $F_0(h_1(\mathbf{z}), ..., h_n(\mathbf{z}), \mathbf{z})$. Using Lemma 4 (for $i = 1$), the latter is equivalent to $F_n(\mathbf{z})$, which in turn is equivalent to $Q_n v_n ... Q_1 v_1 \ \phi(v_1, ..., v_n, \mathbf{z})$ by Corollary 1. $\square$

The size of the complete equivalence model constructed according to Definition 3 is $|F_0| + ... + |F_{n-1}| + |h_1| + ... + |h_n|$. The first part is quadratic in $|\Phi|$ as observed in Section 2, and $|h_1| + ... + |h_n| \leq \Sigma_{i=1...n}(i(1 + |\phi|) + 1 + |\phi|)$ (notice that $\mathbf{z}$ stands for at most $|\phi|$ free variable symbols), so the whole model has cubic size when it is written as a sequence of nested Boolean functions.

In [4], a linear-time transformation from NBF to QBF is presented. By applying this transformation to the above NBF representation of the complete equivalence model, we obtain the following corollary:

Corollary 2. *Every QBF $\Phi(\mathbf{z}) = Q_n v_n ... Q_1 v_1 \ \phi(v_1, ..., v_n, \mathbf{z})$ with $Q_i \in \{\forall, \exists\}$ has a complete equivalence model $h_1(\mathbf{z}), ..., h_n(\mathbf{z})$ where each $h_i(\mathbf{z})$ can again be represented as a QBF with free variables $\mathbf{z}$ of size cubic in $|\Phi|$.*

By Lemma 3, if $h_1(\mathbf{z}), ..., h_n(\mathbf{z})$ are represented as propositional formulas, their size cannot be bounded by a polynomial in $|\Phi|$ if $\Sigma_2^P \neq \Pi_2^P$. Since we do have short QBF representations, a further consequence is that there must exist QBFs $\Psi(\mathbf{z})$ for which there are no logically equivalent propositional formulas $\psi(\mathbf{z})$ of length polynomial in $|\Psi(\mathbf{z})|$, unless $\Sigma_2^P = \Pi_2^P$.

Consider again the example $\Phi(z) = \forall v_2 \exists v_1 \ (v_1 \vee v_2) \wedge (\neg v_1 \vee \neg v_2) \wedge (\neg z \vee \neg v_1)$ with $F_1(v_2, z) = \neg z \vee v_2$, $F_2(z) = \neg z$, $h_2(z) = \neg z$ and $h_1(z) = F_0(1, h_2(z), z) = \neg h_2(z) \wedge \neg z = 0$. If $z = 0$, $\Phi(0)$ is true and has a Skolem model. How is this Skolem model embedded in the complete equivalence model (h_1, h_2)? Notice that $h_1(0) = \neg h_2(0)$. Assume we had not mapped the universal variable v_2 (whose quantifier is outer to that of v_1) to h_2 and instead kept v_2 as a parameter to h_1. Then we would have $h_1(v_2, 0) = \neg v_2$. Indeed, the Skolem model function for v_1 is $f_1(v_2) = \neg v_2$. This suggests that we obtain a Skolem model if we apply the mapping from variables v_i to functions h_i only to existential variables and leave the universals as parameters to the functions of existential variables that are quantified further inside. So we modify Definition 3 to leave universal variables untouched. W.l.o.g., we consider only QBFs with alternating quantifiers.

Definition 4. *Let $\Phi(\mathbf{z}) = \exists v_n \forall v_{n-1} ... \exists v_2 \forall v_1 \ \phi(v_1, ..., v_n, \mathbf{z})$ with even $n \geq 2$ be a QBF with alternating quantifiers and free variables $\mathbf{z}$. Using the function* *representation from Definition 1, we map each existentially quantified variable* $v_n, v_{n-2}, ..., v_2$ *to a Skolem function as follows:*

1. $f_n(\mathbf{z}) := F_{n-1}(1, \mathbf{z})$
2. For $i = n - 2, ..., 2$ *(if $n > 2$):*
   $$f_i(v_{i+1}, ..., v_{n-1}, \mathbf{z}) := F_{i-1}(1, v_{i+1}, f_{i+2}(v_{i+3}, ..., v_{n-1}, \mathbf{z}), ..., v_{n-1}, f_n(\mathbf{z}), \mathbf{z})$$

Lemma 5. *For all even i with $0 \leq i \leq n - 2$:*

$$F_n(\mathbf{z}) \approx \forall v_{n-1}...\forall v_{i+1} \; F_i(v_{i+1}, f_{i+2}(v_{i+3}, ..., v_{n-1}, \mathbf{z}), ..., v_{n-1}, f_n(\mathbf{z}), \mathbf{z})$$

Proof. The proof is by backward induction on i. For $i = n - 2$, the right-hand side is $\forall v_{n-1} \; F_{n-2}(v_{n-1}, f_n(\mathbf{z}), \mathbf{z})$. By Lemma 2, this is equivalent to $F_{n-1}(f_n(\mathbf{z}), \mathbf{z}) := F_{n-1}(F_{n-1}(1, \mathbf{z}))$, and that is $F_n(\mathbf{z})$ by Definition 1.

For the induction step, let i be even with $i \in \{0, ..., n - 4\}$ and assume the above equivalence holds for $i + 2$. We must show:

$$F_n(\mathbf{z}) \approx \forall v_{n-1}...\forall v_{i+1} \; F_i(v_{i+1}, f_{i+2}(v_{i+3}, ..., v_{n-1}, \mathbf{z}), ..., v_{n-1}, f_n(\mathbf{z}), \mathbf{z})$$

By Lemma 2:

$$\forall v_{i+1} F_i(v_{i+1}, f_{i+2}(v_{i+3}, ..., v_{n-1}, \mathbf{z}), ..., v_{n-1}, f_n(\mathbf{z}), \mathbf{z})$$
$$\approx F_{i+1}(f_{i+2}(v_{i+3}, ..., v_{n-1}, \mathbf{z}), ..., v_{n-1}, f_n(\mathbf{z}), \mathbf{z})$$

The first argument of F_{i+1} is

$$f_{i+2}(v_{i+3}, ..., v_{n-1}, \mathbf{z}) := F_{i+1}(1, v_{i+3}, f_{i+4}(v_{i+5}, ..., v_{n-1}, \mathbf{z}), ..., v_{n-1}, f_n(\mathbf{z}), \mathbf{z})$$

by Definition 4. If we substitute this into the right-hand side of the previous equivalence, we obtain $F_{i+1}(F_{i+1}(1, v_{i+3}, ..., f_n(\mathbf{z}), \mathbf{z}), v_{i+3}, ..., f_n(\mathbf{z}), \mathbf{z})$, and that is $F_{i+2}(v_{i+3}, ..., f_n(\mathbf{z}), \mathbf{z})$ by Definition 1, because v_{i+2} is existentially quantified. By the induction hypothesis, $\forall v_{n-1}...\forall v_{i+3} \; F_{i+2}(v_{i+3}, ..., f_n(\mathbf{z}), \mathbf{z}) \approx F_n(\mathbf{z})$. $\quad\square$

Corollary 3. *Let $\Phi(\mathbf{z}) = \exists v_n \forall v_{n-1}...\exists v_2 \forall v_1 \; \phi(v_1, ..., v_n, \mathbf{z})$ with even $n \geq 2$ be a QBF with alternating quantifiers and free variables $\mathbf{z}$. Then*

$$\Phi(\mathbf{z}) \approx \forall v_{n-1}...\forall v_1 \; \phi(v_1, f_2(v_3, ..., v_{n-1}, \mathbf{z}), ..., v_{n-1}, f_n(\mathbf{z}), \mathbf{z})$$

for functions $f_2, .., f_n$ as in Definition 4. That means $f_2, ..., f_n$ are a non-complete equivalence model in the sense of [11] and a Skolem model if Φ is closed and true.

Analogously, it is possible to show that Herbrand countermodels can be obtained when omitting the existential variables from the complete equivalence models.

4 Conclusion and Future Work

We have introduced complete equivalence models for QBFs as a generalization of Skolem and Herbrand (counter)models by mapping all quantified variables to Boolean functions, which we can compactly encode by NBFs. These NBFs are essentially recursive instantiations of the propositional matrix of the QBF, which raises the question for future work how restrictions on the matrix, e.g. 2-CNF or Horn, affect the structure of the complete equivalence models. It would also be interesting to investigate whether this recursive computation can be related to the resolution-based (counter)model construction in [1].

References

[1] Balabanov, V., Jiang, J.-H.: Unified QBF Certification and its Applications. Formal Methods in System Design 41, 45–65 (2012)

[2] Benedetti, M.: Evaluating QBFs via Symbolic Skolemization. In: Baader, F., Voronkov, A. (eds.) LPAR 2004. LNCS (LNAI), vol. 3452, pp. 285–300. Springer, Heidelberg (2005)

[3] Benedetti, M.: Extracting Certificates from Quantified Boolean Formulas. In: Proc. 19th Intl. Joint Conf. on Artificial Intelligence (IJCAI 2005), pp. 47–53. Morgan Kaufmann Publishers (2005)

[4] Bubeck, U., Kleine Büning, H.: Encoding Nested Boolean Functions as Quantified Boolean Formulas. Journal on Satisfiability, Boolean Modeling and Computation (JSAT) 8, 101–116 (2012)

[5] Cadoli, M., Schaerf, M., Giovanardi, A., Giovanardi, M.: An Algorithm to Evaluate Quantified Boolean Formulae and its Experimental Evaluation. Journal of Automated Reasoning 28(2), 101–142 (2002)

[6] Cook, S., Soltys, M.: Boolean Programs and Quantified Propositional Proof Systems. The Bulletin of the Section of Logic 28(3), 119–129 (1999)

[7] Goultiaeva, A., Van Gelder, A., Bacchus, F.: A Uniform Approach for Generating Proofs and Strategies for Both True and False QBF Formulas. In: Proc. 22th Intl. Joint Conf. on Artificial Intelligence (IJCAI 2011), pp. 546–553. AAAI Press (2011)

[8] Herbrand, J.: Recherches sur la Théorie de la Demonstration. PhD Thesis, Université de Paris (1930)

[9] Jussila, T., Biere, A., Sinz, C., Kroning, D., Wintersteiger, C.M.: A First Step Towards a Unified Proof Checker for QBF. In: Marques-Silva, J., Sakallah, K.A. (eds.) SAT 2007. LNCS, vol. 4501, pp. 201–214. Springer, Heidelberg (2007)

[10] Kleine Büning, H., Bubeck, U.: Theory of Quantified Boolean Formulas. In: Biere, A., Heule, M., van Maaren, H., Walsh, T. (eds.) Handbook of Satisfiability, pp. 735–760. IOS Press (2009)

[11] Kleine Büning, H., Zhao, X.: Equivalence Models for Quantified Boolean Formulas. In: Hoos, H.H., Mitchell, D.G. (eds.) SAT 2004. LNCS, vol. 3542, pp. 224–234. Springer, Heidelberg (2005)

[12] Niemetz, A., Preiner, M., Lonsing, F., Seidl, M., Biere, A.: Resolution-Based Certificate Extraction for QBF (Tool Presentation). In: Cimatti, A., Sebastiani, R. (eds.) SAT 2012. LNCS, vol. 7317, pp. 430–435. Springer, Heidelberg (2012)

[13] Skolem, T.: Über die Mathematische Logik. Norsk Matematisk Tidsskrift 10, 125–142 (1928)

Factoring Out Assumptions
to Speed Up MUS Extraction*

Jean-Marie Lagniez and Armin Biere

Institute for Formal Models and Verification
Johannes Kepler University, Linz, Austria

Abstract. In earlier work on a limited form of extended resolution for
CDCL based SAT solving, new literals were introduced to factor out
parts of learned clauses. The main goal was to shorten clauses, reduce
proof size and memory usage and thus speed up propagation and conflict
analysis. Even though some reduction was achieved, the effectiveness of
this technique was rather modest for generic SAT solving. In this paper
we show that factoring out literals is particularly useful for incremental SAT solving, based on assumptions. This is the most common approach for incremental SAT solving and was pioneered by the authors
of MINISAT. Our first contribution is to focus on factoring out only
assumptions, and actually all eagerly. This enables the use of compact
dedicated data structures, and naturally suggests a new form of clause
minimization, our second contribution. As last main contribution, we
propose to use these data structures to maintain a partial proof trace for
learned clauses with assumptions, which gives us a cheap way to flush
useless learned clauses. In order to evaluate the effectiveness of our techniques we implemented them within the version of MINISAT used in the
publically available state-of-the-art MUS extractor MUSer. An extensive
experimental evaluation shows that factoring out assumptions in combination with our novel clause minimization procedure and eager clause
removal is particularly effective in reducing average clause size, improves
running time and in general the state-of-the-art in MUS extraction.

1 Introduction

The currently most widespread approach for *incremental* SAT was pioneered by
the authors of MINISAT [1] in context of bounded model checking [2] and finite
model finding [3], and has seen many other important practical applications
since then. It can easily be implemented on top of a standard SAT solver based
on the *conflict driven clause learning* (CDCL) paradigm [4], as described in [1],
by modifying the heuristics for picking decisions, to branch on literals assumed
to be true first. In this paper we refer with *assumptions* to this set of literals
assumed to be true.

Another important application, which makes use of incremental SAT, is the
extraction of a *minimal unsatisfiable set* (MUS) of clauses from a propositional

---

* Supported by FWF, NFN Grant S11408-N23 (RiSE).

M. Järvisalo and A. Van Gelder (Eds.): SAT 2013, LNCS 7962, pp. 276–292, 2013.

formula in *conjunctive normal form* (CNF). The current state-of-the-art in MUS extraction [5] is based on incremental SAT. In the context of MUS extraction [6,7,8,9,10], the focus of this paper, and in similar or related applications of incremental SAT [3,11,12,13,14,15,16], an additional analysis is required, which learns sub-sets of assumptions, under which the formula is proven to be unsatisfiable. In these applications, the number of assumptions is usually not only quite large, e.g. similar in size to the number of original variables and clauses in the CNF, but also the SAT solver is called many times, while the set of assumptions almost stays the same.

As it turns out, current SAT solvers have not been optimized for this actually rather common usage scenario. We propose a new technique for compressing incremental proofs for problems with many assumptions. Our technique is based on the idea of factoring out literals of learned clauses by extended resolution steps, which also forms the basis of related work on speeding up SAT solving in general [17,18]. Clauses *learned* in those applications we are interested in typically contain many literals which are the negation of original assumptions. We call these negations of originally assumed literals also assumptions or more precisely *assumption literals*, if the context requires to distinguish between originally assumed literals ("assumptions") used as decisions and their negations occurring in learned clauses ("assumption literals").

In our approach we factor out these assumption literals in order to shrink learned clauses and reduce the number of literals traversed, particularly during boolean constraint propagation (BCP). This idea, if implemented correctly, does not change the search at all, but it is still quite effective in reducing the time needed for MUS extraction. Further, factoring out assumptions enables the use of compact dedicated data structures, and naturally suggests a new form of clause minimization, which gives another substantial improvement. Recording factored out assumptions explicitly, also gives us simple way to maintain a partial proof trace for learned clauses with assumptions. The trace can be used to compute an approximation of a "clausal core". We can then discard learned clauses out-side this clausal core eagerly, which empirically seems to be a useful strategy.

The authors of [19] observed a similar deficiency when using the assumption based approach for incremental SAT solving in the context of bounded model checking. They propose to use an additional SAT solver, to which assumptions are added as unit clauses. This in turn allows to improve efficiency of preprocessing and inprocessing under assumptions, but prohibits to reuse in the main solver clauses learned by the additional solver. However, according to [19] it is possible, by selectively adding assumptions, to extract "pervasive clauses" from the resolution proofs of clauses learned in the additional solver, with the objective that adding these "pervasive clauses" to the main solver is sound.

While in [19] as in our approach some sort of resolution proof has to be maintained, the main solver in [19] still uses the classical assumption based approach and thus will benefit from our proposed techniques. Finally, the motivations as well as the application characteristics considered in the experimental part differ.

2   Factoring Out Assumptions

In incremental SAT with *many* assumptions, learned clauses contain many assumption literals too (see previous section for the definition of this terminology). Accordingly the average size of learned clauses can become very large (as we will see in Fig. 6). This effect increases the size of the working set (used memory), or more specifically, the average number of traversed literals per visited clause during BCP. The same argument applies to visited clauses during conflict analysis. As a consequence, SAT solver performance degrades.

For every learned clause we propose to replace the "assumption part" by a new fresh literal, called *abbreviation* literal. The replaced part consists of all assumptions and previously added abbreviations. The connection between the abbreviation and the replaced literals is stored in a *definition map* as follows.

$$(p_1 \vee \cdots \vee p_n \vee a_1 \vee \cdots \vee a_m)$$

is factored out into

$$(p_1 \vee \cdots \vee p_n \vee \ell) \quad \text{and} \quad \ell \mapsto \underbrace{a_1 \vee \cdots \vee a_m}_{\mathcal{G}[\ell]}$$

Fig. 1. Factoring out assumptions by introducing a new abbreviation literal ℓ

Let $p_1 \vee \cdots \vee p_n \vee a_1 \vee \cdots \vee a_m$ be a new learned clause, where $p_1, \ldots, p_n$ are original literals and $a_1, \ldots, a_m$ are either assumptions or abbreviations. We pick a fresh abbreviation literal ℓ and instead of the originally learned clause add the clause $p_1 \vee \cdots \vee p_n \vee \ell$ to the clause data base. Then we record $a_1 \vee \cdots \vee a_m$ as the *definition* $\mathcal{G}[\ell]$ of ℓ in the definition map $\mathcal{G}$ (see Fig. 1). For $m \leq 1$ this replacement does not make sense and the original learned clause is kept instead.

Consider the example in Fig. 2 for an (incremental) SAT run under the assumptions $\overline{a_1}, \ldots, \overline{a_6}$. Conflict analysis might learn clauses $\alpha_1, \ldots, \alpha_7$ depicted on the left of Fig. 2(a), where $p_1, \ldots, p_7$ are original literals and $a_1, \ldots, a_6$ assumption literals.[1] Note that the run is not supposed to be complete. Only some clauses are shown together with their antecedent clauses, and original clauses are ignored too (to simplify the example). For instance α_3 is derived through resolution from α_1 and from some other original clauses not shown (the "...").

The result of introducing abbreviations to factor out assumptions is shown on the right. The first clause α_1 is factored into α_1' and the definition $a_1 \vee a_2$ of the new abbreviation literal ℓ_1. The definition is recorded in the definition map, as shown in Fig. 2(b), where ℓ_1 has two incoming arcs, one from a_1 and one from a_2. Further let us point out, that $\alpha_5' = \alpha_5$, because it *keeps* a_2 as single non-original literal, which (as discussed above) reduces the overall number of introduced abbreviations. Finally, note how definitions might recursively depend on other definitions as for ℓ_3, ℓ_4 or ℓ_5, while factoring α_3, α_4, and α_6 respectively.

---

[1] Assumption literals are literals made of a variable which is currently used or was used in an assumption. See again the introduction section for a precise definition.

learned clauses	antecedents		factored clauses
$\alpha_1 : p_2 \vee p_7 \vee a_1 \vee a_2$	$\{\ldots\}$		$\alpha_1' : p_2 \vee p_7 \vee \ell_1$
$\alpha_2 : p_2 \vee a_2 \vee a_3$	$\{\ldots\}$		$\alpha_2' : p_2 \vee \ell_2$
$\alpha_3 : p_7 \vee p_4 \vee \overline{p_6} \vee a_1 \vee a_2 \vee a_4$	$\{\alpha_1, \ldots\}$	factoring	$\alpha_3' : p_7 \vee p_4 \vee \overline{p_6} \vee \ell_3$
$\alpha_4 : p_6 \vee p_8 \vee a_3 \vee a_2 \vee a_5$	$\{\alpha_2, \ldots\}$	$\Longrightarrow$	$\alpha_4' : p_6 \vee p_8 \vee \ell_4$
$\alpha_5 : p_2 \vee p_5 \vee a_2$	$\{\ldots\}$		$\alpha_5' : p_2 \vee p_5 \vee a_2$
$\alpha_6 : p_7 \vee p_4 \vee a_1 \vee a_2 \vee a_4 \vee a_5$	$\{\alpha_3, \alpha_4, \ldots\}$		$\alpha_6' : p_7 \vee p_4 \vee \ell_5$
$\alpha_7 : \overline{p_2} \vee a_6 \vee a_5$	$\{\ldots\}$		$\alpha_7' : \overline{p_2} \vee \ell_6$

(a) Learned clauses (original version left, factored version right)

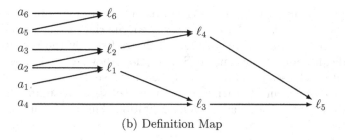

(b) Definition Map

Fig. 2. Factoring out assumptions

As briefly discussed above, assumptions are always assigned first and thus assigning them can actually be seen as a preprocessing resp. initialization step before the actual solving starts. Furthermore, the algorithm for MUS extraction, as implemented in MUSer [9], to which we applied our technique, has the following property: *the set of underlinevariables used in assumptions stays the same over all incremental calls*, with the exception of variables assigned at the top-level. The techniques presented in this paper are sound, even if this property does not hold, i.e. the set of assumptions changes (substantially) from one incremental call to the next. However, if the property does not hold they are probably less effective. We focus on the important problem of MUS extraction here and leave it to future work to apply our techniques to other scenarios of incremental SAT.

Assigning in every incremental call the current set of assumptions during an initialization phase, will imply a unique value for all the (previously introduced) abbreviation literals, unless the set of assumptions turns out to be inconsistent, in which case the solver returns immediately. For that reason we do not have to encode definitions as part of the CNF. Abbreviations are assigned during an initialization phase, as described in the next Section (see also Alg. 2).

2.1   Initialization

After factoring out assumptions and adding abbreviations instead, every learned clause α contains *at most one* assumption or abbreviation. In this case we denote by $r(\alpha)$ this *replacement* literal. For other clauses we assume $r(\alpha)$ to be undefined. The graph represented by the definition map $\mathcal{G}$ can be interpreted as

---

Algorithm 1. assignAbbreviation

Input: ℓ: literal; **var** $\mathcal{I}$: interpretation; $\mathcal{G}$: definition map
1 $removeUnit(\mathcal{G}[\ell])$;
2 **while** $\mathcal{I}(\mathcal{G}[\ell])$ *unassigned* **do**
3 $\quad$ pick *unassigned* $\ell' \in \mathcal{G}[\ell]$;
4 $\quad$ assignAbbreviation$(\mathcal{G}, \ell', \mathcal{I})$;
5 **if** $\mathcal{I}(\mathcal{G}[\ell]) = \bot$ **then** $\mathcal{I} \leftarrow \mathcal{I} \cup \{\neg\ell\}$ **else** $\mathcal{I} \leftarrow \mathcal{I} \cup \{\ell\}$;

---

a (non-cyclic) circuit, which computes consistent values for abbreviations after all the assumption variables have been assigned. Special care has to be taken to handle assumptions and abbreviations, which are fixed by the user in between incremental calls. For instance, in MUS extraction, they are used to permanently select transition clauses [9] to be part of the extracted MUS.

In order to assign an abbreviation, we need to assign assumption variables and, recursively, every abbreviation in its definition. This is formulated in Alg. 1, which has the following arguments: the literal ℓ to be assigned, and (by reference) the current interpretation $\mathcal{I}$ and the definition map $\mathcal{G}$. First, literals assigned at the top-level (units), are removed from $\mathcal{G}[\ell]$. Next, while there is an unassigned literal ℓ' in $\mathcal{G}[\ell]$ and $\mathcal{G}[\ell]$ is itself unassigned by the current interpretation $\mathcal{I}$, we assign ℓ', using the same algorithm recursively. As soon as the value of $\mathcal{G}[\ell]$ under $\mathcal{I}$ is determined, we can also assign ℓ to $\mathcal{I}(\mathcal{G}[\ell])$.

By construction the definitions in the definition map $\mathcal{G}$ are non-cyclic. Further, we assume that every assumption is assigned by $\mathcal{I}$, as discussed in the previous section. Then this algorithm terminates and consistently assigns the value of each abbreviation ℓ to the value of its definition $\mathcal{G}[\ell]$.

2.2 Assigning the Set of Necessary Abbreviations

In the worst case, every learned clause resp. conflict requires a new abbreviation to be added. Therefore, in principle, the definition map grows linearly in the number of conflicts. This not only requires a huge amount of memory, but also needs substantial running time to initialize all the abbreviations of the definition map during incremental SAT calls.

However, since inactive [20] resp. less useful learned [21,22] clauses are frequently collected during the main CDCL loop of the SAT solver anyhow, many abbreviations turn out not to be referenced anymore after a certain point. They become *garbage abbreviations* and could be collected too. Actually, only the assignments of those abbreviations have to be initialized, which are still referenced in learned clauses (recursively). Assigning additional abbreviations is not harmful, but useless.

Algorithm 2 implements an initialization of abbreviations taking this argument into account. It returns an interpretation $\mathcal{I}$, which assigns all abbreviations recursively reachable from the clauses in the CNF Σ (which includes learned clauses). First, the algorithm initializes $\mathcal{I}$ by assigning all assumptions. Next,

---

Algorithm 2. initialization

Input: Σ: CNF formula; $\mathcal{A}$: assumptions; $\mathcal{G}$: a definition map
Result: $\mathcal{I}$ a partial interpretation

1 $\mathcal{I} \leftarrow \mathcal{A} \cup \{\text{top-level units}\}$;
2 **foreach** $\alpha \in \Sigma$ *with* $r(\alpha)$ *defined* **do**
3   **if** $r(\alpha)$ *is unassigned by* $\mathcal{I}$ **then**
4   ⌊ assignAbbreviation($\mathcal{G}, r(\alpha), \mathcal{I}$);
5   ⌊ **if** $\mathcal{I}(\alpha) = \bot$ **then break**;
6 **return** $\mathcal{I}$;

---

it traverses all clauses α to which a replacement $r(\alpha)$ has been added and then calls Alg. 1 to assign the replacement literal. The resulting $\mathcal{I}$ consistently assigns reachable abbreviations to the value of their definition in the definition map $\mathcal{G}$, unless a clause is found that has all its literals assigned to false.

2.3   Assumption Core Analysis

As discussed in the introduction, applications of incremental SAT with assumptions often make use of the SAT solver's ability to return an *assumption core*, i.e., a subset of the given assumptions, which in combination with the given CNF can not to be satisfied. Intuitively, the assumption core exactly contains the assumptions "used" by the SAT solver to derive the inconsistency. In contrast to the concept of MUS, these assumption cores are typically not required to be minimal. As implemented in MINISAT [1] such an assumption core can be computed by a separate conflict analysis routine called "analyzeFinal", which recursively goes through the implication graph to only collect assumptions in contrast to the usual analysis routine of CDCL solvers which cuts off the search for a learned clause as soon as possible, e.g., following the 1st UIP scheme [23].

After factoring out assumptions and adding abbreviations the "analyzeFinal" procedure has to be adapted to care for abbreviations, which is described in Alg. 3. The algorithm takes as input a CNF formula Σ, the current unsatisfiable trail[2] $\mathcal{I}$, a clause α falsified under $\mathcal{I}$, the definition map $\mathcal{G}$, and returns the set of assumptions $\mathcal{C}$ "used" to establish the unsatisfiability proof. It starts by initializing $\mathcal{C}$ and the literals $\mathcal{V}$ already visited with the empty set. Next, the stack $\mathcal{T}$, containing the set of literals that must be further visited, is initialized with the conflict clause α. Then, while there is still an unvisited literal $\ell \in \mathcal{T}$, it is marked. Depending on its type three different cases have to be distinguished. First in line 5, if ℓ is an assumption, then ℓ is added to the conflict clause $\mathcal{C}$. Second in line 6, if ℓ is an abbreviation its definition $\mathcal{G}[\ell]$ is added to $\mathcal{T}$. This is actually the only part where the algorithm has to be adapted to recursively

---

[2] Every literal assigned to true, particularly those found during BCP, are added to a stack, called *trail*, to record the order of assignments. The reason, also called antecedent, of a forced assignment is saved too. Please refer to [1] for more details.

Algorithm 3. analyzeFinal

Input: Σ: CNF; $\mathcal{I}$: trail; α: clause; $\mathcal{G}$: a definition map
Result: $\mathcal{C}$, a subset of the assumptions
1   $\mathcal{C} = \emptyset$; $\mathcal{V} = \emptyset$;
2   $\mathcal{T} \leftarrow \alpha$;
3   **while** $\exists \ell \in \mathcal{T} \setminus \mathcal{V}$ **do**
4     $\mathcal{V} \leftarrow \mathcal{V} \cup \{\ell\}$;
5   **if** ℓ *is an assumption* **then** $\mathcal{C} \leftarrow \mathcal{C} \cup \{\ell\}$;
6   **else if** ℓ *is an abbreviation* **then** $\mathcal{T} \leftarrow \mathcal{T} \cup \mathcal{G}[\ell]$;
7   **else** $\mathcal{T} \leftarrow \mathcal{T} \cup reason(\ell, \mathcal{I})$;
8   **return** $\mathcal{C}$;

---

explore the definition map. Third in line 7, ℓ is neither an assumption nor an abbreviation and the reason of its propagation is added to $\mathcal{T}$ (implication graph exploration). Decision literals are assumed to have an empty set of antecedents.

Example 1. Consider again the example in Fig. 2. Given $\{\overline{a_1}, \overline{a_2}, \overline{a_3}, \overline{a_4}, \overline{a_5}, \overline{a_6}\}$, learning α'_7 allows to conclude that the formula is unsatisfiable. Alg. 3 produces:

$\mathcal{T}$	$\mathcal{V}$	$\mathcal{C}$	ℓ
$\overline{p_2}, \ell_6$	$\emptyset$	$\emptyset$	$undef$
ℓ_6, ℓ_2	$\emptyset$	$\emptyset$	$\overline{p_2}$
ℓ_2, a_5, a_6	ℓ_5	$\emptyset$	ℓ_6
a_5, a_6, a_2, a_3	ℓ_5, ℓ_6	$\emptyset$	ℓ_2
a_6, a_2, a_3	ℓ_5, ℓ_6, a_5	a_5	a_5
a_2, a_3	ℓ_5, ℓ_6, a_5, a_6	a_5, a_6	a_6
a_3	$\ell_5, \ell_6, a_5, a_6, a_2$	a_5, a_6, a_2	a_2
$\emptyset$	$\ell_5, \ell_6, a_5, a_6, a_2, a_3$	a_5, a_6, a_2, a_3	a_3

The resulting learned clause is $(a_5 \vee a_6 \vee a_2 \vee a_3)$. Note, neither a_1 nor a_4 were actually "used" in deriving it. In the next section will make use of such an analysis to eagerly reduce the learned clause·data base.

2.4 Reduce Learned Clause Database

Keeping all learned clauses slows down the SAT solver considerably. Thus heuristics to determine which learned clauses to keep resp. how and when to reduce the learned clause database are an essential part of state-of-the-art SAT solvers [20,21,22]. After an incremental SAT call returned "unsatisfiable", we propose to only keep those learned clauses, which were used to show that the assumed assumptions in this SAT call are inconsistent and discard all others.

Experiments in Sect. 3.2 will give empirical evidence for the effectiveness of these heuristics. Even though it is not a solid argument, an intuitive explanation could be that learned clauses are removed quite frequently anyhow. Further, most likely exactly those learned clauses related to the last set of assumptions are useful in the next SAT call too. This particularly applies to MUS extraction where the assumptions do not change much.

Algorithm 4. eagerLearnedClauseDatabaseReduction

Input: var Δ: set of learned clauses; **var** $\mathcal{G}$: a definition map; $\mathcal{V}$: literals;

1 **foreach** $\alpha \in \Delta$ **do**
2   **if** $r(\alpha)$ *is an abbreviation and* $r(\alpha) \notin \mathcal{V}$ **then**
3        $\Delta \leftarrow \Delta \setminus \alpha$;
4   remove $r(\alpha)$ and its definition from $\mathcal{G}$;

However, in order to apply these heuristics we need to be able to determine whether a certain clause was used in deriving the inconsistency. As it turns out, our definition map can be interpreted as partial proof trace for learned clauses (with assumptions) and thus gives us a cheap way to flush learned clauses and definitions not required to show that the given set of assumptions is inconsistent. Focusing on the remaining relevant learned clauses and definitions in this "core" reduces run time, as our experiments in Sect. 3.2 will show.

Let us continue with Example 1 after learning α'_7. Only α'_2 and α'_7 are required to show unsatisfiability under the given set of assumptions, while α'_4 is not required and thus according to our heuristic should be removed. This eager reduction of the learned clause database can be easily implemented as a post-processing phase using $\mathcal{V}$ computed by analyzeFinal, which is shown in Alg. 4.

2.5 Assumption Aware Clause Minimization

New learned clauses can often be minimized by applying additional resolution steps with antecedent clauses in the implication graph. Two approaches are currently used to achieve this minimization: applying self-subsuming resolution, also called local minimization, or applying recursive minimization[24]. In recursive minimization several resolution steps are tried to determine whether a literal can be removed from the learned clause. In both cases resolutions are only applied if the resulting clause is a strict sub-clause. Sörensson and Biere [24] demonstrated that clause minimization usually improves SAT solver performance. In the following we will either apply this classical recursive minimization, no minimization at all, or a new form of recursive minimization, and thus do not consider local minimization further.

In the incremental setting with many assumptions, our preliminary experiments showed that classical clause minimization is not very effective. Usually the number of literals deleted in classical clause minimizations is rather small. As reason we identified the fact that assumptions are not obtained by unit propagation, and thus cannot be removed from learned clauses through additional resolution steps. Furthermore, non-assumption literals are often blocked by at least one assumption pulled in by resolution steps. The classical minimization algorithm requires that the resulting clause is a strict sub-clause. It is not allowed to contain more assumptions.

This situation is not optimal since assumptions, during one call of the incremental SAT algorithm, are assigned to false and can thus be considered to be irrelevant, at least for this call. Our new minimization procedure makes use of this observation and simply ignores additionally pulled in assumptions during minimization. The resulting "minimized" clause might even increase in size. However, it will never have more non-assumption literals than the original clause.

3   Experiments

The algorithms described above have been implemented within the SAT solver MINISAT [1], starting from the original version, used in the current version of the state-of-the-art MUS extractor MUSer [9]. It heavily makes use of incremental SAT solving with many assumptions following the selector variable-based approach [25]. Our modified version of [1] is called MINISAT$_{abb}$ (MINISAT with abbreviation). We focus on MUS extraction and compare the performance of MUSer for different versions of MINISAT.

For our experiments we used all 295 benchmarks from the MUS track of the SAT Competition 2011 [3] after removing 5 duplicates from the original 300 benchmarks. These benchmarks[4] have their origin in various industrial applications of SAT, including hardware bounded model checking, hardware and software verification, FPGA routing, equivalence checking, abstraction refinement, design debugging, functional decomposition, and bioinformatics. The experiments were performed on machines with Intel® Core^{TM}2 Quad Processor Q9550 with 2.83 GHz CPU frequency with 8 GB memory and running Ubuntu 12.04. Resource limits are the same as in the competition: time limit of 1800 seconds, memory limit of 7680 MB.

In the first experiment we apply our new approach of factoring out assumptions *without* changing clause learning. We then evaluate the impact of our new learned clause reduction scheme and our new clause minimization procedure. The experimental part concludes with more details on memory consumption.

3.1   Factoring Out Assumptions

Fig. 3 shows a comparison between MUSer with our new approach based on factoring out assumptions, called MINISAT$_{abb}$, and the original version of MINISAT. First, in Fig. 3(a) the average size of learned clauses is compared. For many problems, adding clause abbreviations reduces the average size of learned clauses by an order of magnitude.

The main effect of our new technique is to reduce the size of learned clauses. This should also decrease the number of literals traversed while visiting learned clauses during BCP. In the scatter plot in Fig. 3(b) we focus on this metric and compare the average number of traversed literals while running both versions

---

[3] http://www.satcompetition.org/2011

[4] The set of benchmarks is available at http://www.cril.univ-artois.fr/SAT11/

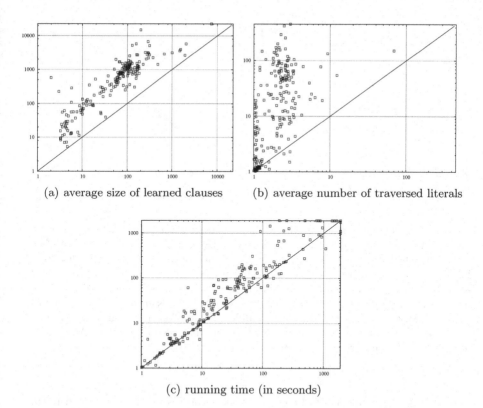

(a) average size of learned clauses   (b) average number of traversed literals

(c) running time (in seconds)

Fig. 3. Comparing MUSer on the 2011 competition instances from the MUS track, using the original MINISAT without abbreviations (y axis) vs. using our new version MINISAT$_{abb}$ with abbreviations (x axis) w.r.t. three different criteria

on the same instance. This includes the literals traversed in clauses visited during BCP, also including original clauses, but of course ignores clauses that are skipped due to satisfied blocking literals [26]. As the plot shows, the reduction in terms of the number of traversed literals is even more than the reduction of the average size of learned clauses. Consequently also the running time reduces considerably, see Fig. 3(c), but of course not in the same scale as in the previous plots. Note that in essence the "same clauses" are learned and thus the number of conflicts and learned clauses does not change.

The net effect of using abbreviations to factor out assumptions is that MUSer based on MINISAT$_{abb}$ solves 272 out of the 295 instances, and runs out of memory on 3 instances, whereas the version with the original MINISAT solves only 261 instances and runs out of memory in 13 cases. Our approach solves more instances, but not, at least primarily, because it runs out of memory less often.

As it turns out in the context of MUS extraction, definition clauses actually do not have to be watched. Further, abbreviation literals never have to be considered as decision and thus also do not have to be added to the priority queue

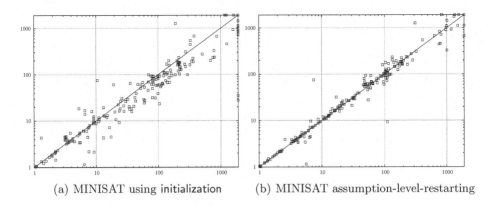

(a) MINISAT using initialization   (b) MINISAT assumption-level-restarting

Fig. 4. On the left we show the running time of MUSer using MINISAT+init, a version of MINISAT, which initializes assumptions explicitly (x axis) vs. the original MINISAT version, which does not initialize them explicitly before search (y axis), both *without* abbreviations. Visiting each learned clause during initialization is time consuming without abbreviations. In the experiment shown on the right we only modified the restart mechanism to backtrack to the decision level of the last assigned assumption instead of backtracking to the root level. The modified version MINISAT+assumption-level-restarting (x axis) performs equally well as the original version of MINISAT (y axis). Running time is measured in seconds with a time limit of 1800 seconds as always.

(implemented as heap in MINISAT) for picking decisions. Thus we need initialization, by assigning all assumptions and abbreviations, the latter in incremental calls only, at the first decision level.

In order to make sure that the improvement observed in the previous experiment is independent from using our new optimized initialization phase, we report in Fig. 4(a) the run times of MUSer using the original version of MINISAT compared to the run times using a modified version of MINISAT, in which the assumption variables are assigned up-front and removed from the priority queue initially too, called MINISAT+init. The results show that using this modified initialization scheme in the original version of MINISAT actually has a negative effect on the performance of MUSer (MUSer using MINISAT solves <u>261</u> instances whereas MUSer using MINISAT+init solves 257 instances) and thus can not be considered to be the main reason for the witnessed improvements in the first experiment. Our explanation for this effect is, that our initialization algorithm in essence needs only one pass over the learned clauses, even just a subset of all learned clauses, while initializing up-front BCP in MINISAT+init needs to visit lots of clauses during initialization. Note, again, that initialization has to be performed at the start of every incremental SAT call and might contribute a substantial part to the overall running time.

Modern SAT solvers based on the CDCL paradigm restart often by frequently backtracking to the root-level (also called top-level) [27,28,29,30] using a specific restart schedule [31,32,33,34]. With assumptions it seems however to be more

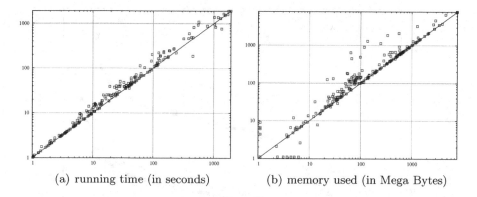

(a) running time (in seconds)   (b) memory used (in Mega Bytes)

Fig. 5. Comparison MUSer using MINISAT_{abb} (y axis) vs. $\text{MINISAT}_{abb}+$g (x axis).

natural to backtrack to the highest decision level, where the last assumption was assigned, which we call *assumption-level*. This technique is implemented in Lingeling [35], since it can naturally be combined with the technique of reusing the trail [36], but is not part of MINISAT. It might be conceivable, that forcing MINISAT to backtrack to the assumption-level during restarts can give the same improvement as initialization up-front. However, the experiment reported in Fig. 4(b) shows, that at least in MUS extraction, this "optimization" is useless.

3.2 Learned Clauses Database Reduction

In this section, we study the impact on the performance of MINISAT_{abb} w.r.t our new reduction algorithm for the learned clause database presented in Sect. 2.4. Fig. 5 compares MUSer using MINISAT_{abb} with and without this more "eager garbage collector", which we denote by $\text{MINISAT}_{abb}+$g resp. MINISAT_{abb}. According to Fig. 5(b) eager garbage collection reduces memory consumption. Moreover, as shown in Fig. 5(a), this memory reduction does not hurt performance, since three more instances are solved (275 vs. 272) and only 1 instance (instead of 3) runs out of memory (see also Tab. 1).

3.3 Minimization of the Learned Clauses

In this section, we compare our new clause minimization procedure to existing variants of clause minimization. We consider three versions of MINISAT and MINISAT_{abb} as back-end in MUSer [9]:

- without clause minimization (called *without*);
- the classical recursive clause minimization (*classic*) [24];
- our new clause minimization procedure (*full*) described in Sect. 2.5.

From the cactus plot in Fig. 6, which compares average size of learned clauses, we can draw the following conclusions. First, classical minimization is not effective in

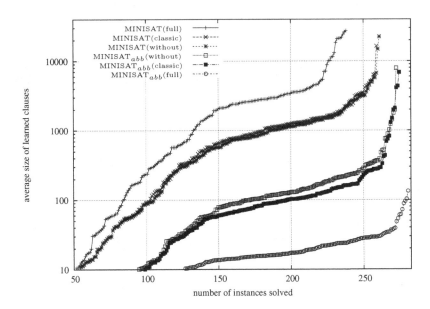

Fig. 6. Cactus plot reporting the average size of learned clauses produced by MUSer using MINISAT without abbreviations and MINISAT$_{abb}$ with abbreviations, and different clause minimization approaches. Factoring out assumptions (lower three curves) is always better than the original scheme (upper three curves). Further, full minimization gives a large improvement but only if assumptions are factored out. Without abbreviations full minimization actually turns out to be detrimental. Classic minimization only gives a small advantage over not using any minimization.

terms of reducing the average size of learned clauses, neither for MINISAT nor for MINISAT$_{abb}$, because it cannot remove assumptions during clause minimization. Classical minimization is slightly more effective with abbreviations than without. However, abbreviations might block self-subsumption during recursive resolution steps and thus prevent further minimization.

Next, we study the impact of our new full clause minimization described in Sect. 2.5 with and without using abbreviations. As reported in [24] for SAT solving *without* assumptions, recursive clause minimization typically is able to reduce the average size of learned clauses by one third. In the SAT solving *with* assumptions, as previously noted, assumptions prevent this reduction. With full clause minimization, however, we get back to the same reduction ratio of around 30% considering only literals that are neither assumptions nor abbreviations. Nevertheless, since deleting one literal is often necessary to apply additional resolutions, many new assumptions are added to the minimized clause. Using full minimization in MINISAT without abbreviations increases the average size of learned clauses by an order of magnitude, whereas MINISAT$_{abb}$ does not have this problem, since assumptions and abbreviations are factored out.

Table 1. The table shows the number of solved instances by MUSer within a time limit of 1800 seconds and a memory limit of 7680 MB, for different back-end SAT solver: the original MINISAT, then MINISAT$_{abb}$ with abbreviations, and finally MINISAT$_{abb}$+g with abbreviations and eager learned clause garbage collection. For each version of these three SAT solvers we further use three variants of learned clause minimization. The approach with abbreviations, eager garbage collection and full learned clause minimization, e.g. using all of our suggested techniques, works best and improves the state-of-the-art in MUS extraction from 261 solved instances to **281**.

	MINISAT	MINISAT$_{abb}$	MINISAT$_{abb}$+g
	#solved(MO)	#solved(MO)	#solved(MO)
without minimization	259(15)	272(3)	273(3)
classic minimization	261(13)	272(3)	275(1)
full minimization	238(25)	276(0)	**281(0)**

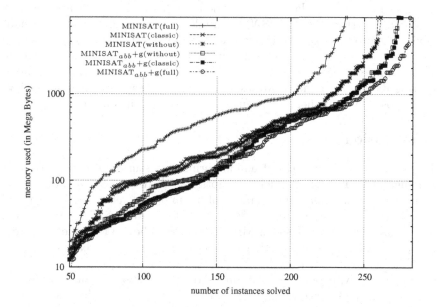

Fig. 7. Memory usage of MUSer based on the original MINISAT without abbreviations and MINISAT$_{abb}$+g with both abbreviations and eager garbage collection. Both versions of the MINISAT are combined with three different clause minimization strategies. Note, that even with eager garbage collection, which reduces memory consumption, e.g., see Fig. 5(b), the effect of our techniques on overall memory usage is not particularly impressive and leaves room for further optimization.

Actually, our new full clause minimization procedure in combination with MINISAT$_{abb}$ is able to reduce the average size of learned clauses by two orders of magnitude w.r.t the best version of MINISAT without abbreviations, while already one order of magnitude is obtained by MINISAT$_{abb}$ just by using abbreviations alone (with or without using classical clause minimization procedure).

In another experiment we measured the effect of our new garbage collection procedure Alg. 4. As it turns out, the average size is not influenced by adding this procedure, but as Tab. 1 shows, it has a positive impact on the number solved instances independent from the minimization algorithm used. Finally, this tables also shows that the reduction of the average size of learned clauses directly translates into an increase of the number of solved instances. The combination of our new techniques improves the state-of-the-art of MUS extraction considerably.

3.4   Memory Usage

We conclude the experiments with a more detailed analysis of memory usage for the various considered versions of MUSer. As expected, Fig. 7 shows that shorter clauses need less memory. However, the effect in using our new techniques on overall memory usage is less pronounced than their effect w.r.t. to reducing average learned clause length. The main reason is that definitions have to be stored too. However, MINISAT with full clause minimization but without abbreviations produces a huge increase in memory consumption by an order of magnitude. This shows that factoring out assumptions is the key to make full clause minimization actually work. Also note, that our current implementation for storing definitions is not optimized for memory usage yet, and we believe that it is possible to further reduce memory consumption considerably.

4   Conclusion

In this paper we introduced the idea of factoring out assumptions, in the context of incremental SAT solving under assumptions. We developed techniques that work particularly well for large numbers of assumptions and many incremental SAT calls, as it is common, for instance, in MUS extraction. We implemented these techniques in the SAT solver MINISAT$_{abb}$ and showed that they lead to a substantial reduction in solving time if used in the SAT solver back-end of the state-of-the-art MUS extractor MUSer [9].

More specifically, experimental results show that factoring out assumptions by introducing abbreviations is particularly effective in reducing the average learned clause length, which in turn improves BCP speed. Even though memory usage is not reduced at the same level as average learned clause lengths, using abbreviations leads to shorter running time. Furthermore, the ability to factor out assumptions is crucial for a new form of clause minimization, which gave another substantial improvement. In general, we improved the state-of-the-art in MUS extraction considerably.

Our prototype MINISAT$_{abb}$ uses rather basic data structures, which can be improved in several ways. Memory usage could be reduced by a more sophisticated implementation of managing abbreviations. Further, in the current implementation, identical definitions are not shared. A hashing scheme could cheaply detect this situation and would allow to reuse already existing definitions instead of introducing new ones. This should reduce memory usage further and also speed up the initialization phase.

Finally, it would be interesting to combine the techniques presented in this paper with more recent results on MUS preprocessing [10] and preprocessing under assumptions [19,37] resp. inprocessing [38]. We also want to apply our approach to high-level MUS extraction [7,8,16].

Software and more details about the experiments including log files are available at http://fmv.jku.at/musaddlit.

References

1. Eén, N., Sörensson, N.: An extensible SAT-solver. In: Giunchiglia, E., Tacchella, A. (eds.) SAT 2003. LNCS, vol. 2919, pp. 502–518. Springer, Heidelberg (2004)
2. Eén, N., Sörensson, N.: Temporal induction by incremental SAT solving. ENTCS 89(4), 543–560 (2003)
3. Claessen, K., Sörensson, N.: New techniques that improve MACE-style finite model finding. In: Proc. CADE-19 Workshop: Model Computation - Principles, Algorithms, Applications (2003)
4. Marques-Silva, J., Lynce, I., Malik, S.: 4. In: Conflict-Driven Clause Learning SAT Solvers. Frontiers in Artificial Intel. and Applications, vol. 185, pp. 131–153. IOS Press (February 2009)
5. Belov, A., Lynce, I., Marques-Silva, J.: Towards efficient MUS extraction. AI Commun. 25(2), 97–116 (2012)
6. Grégoire, É., Mazure, B., Piette, C.: Extracting MUSes. In: Proc. ECAI 2006. Frontiers in Artificial Intel. and Applications, vol. 141, pp. 387–391. IOS Press (2006)
7. Nadel, A.: Boosting minimal unsatisfiable core extraction. In: Proc. FMCAD 2010, pp. 221–229. IEEE (2010)
8. Ryvchin, V., Strichman, O.: Faster extraction of high-level minimal unsatisfiable cores. In: Sakallah, K.A., Simon, L. (eds.) SAT 2011. LNCS, vol. 6695, pp. 174–187. Springer, Heidelberg (2011)
9. Belov, A., Marques-Silva, J.: Accelerating MUS extraction with recursive model rotation. In: Proc. FMCAD 2011, pp. 37–40. FMCAD (2011)
10. Belov, A., Järvisalo, M., Marques-Silva, J.: Formula preprocessing in MUS extraction. In: Piterman, N., Smolka, S.A. (eds.) TACAS 2013. LNCS, vol. 7795, pp. 108–123. Springer, Heidelberg (2013)
11. Fu, Z., Malik, S.: On solving the partial MAX-SAT problem. In: Biere, A., Gomes, C.P. (eds.) SAT 2006. LNCS, vol. 4121, pp. 252–265. Springer, Heidelberg (2006)
12. Andraus, Z.S., Liffiton, M.H., Sakallah, K.A.: Refinement strategies for verification methods based on datapath abstraction. In: Proc. ASP-DAC 2006, pp. 19–24. IEEE (2006)
13. Marques-Silva, J., Planes, J.: On using unsatisfiability for solving maximum satisfiability. CoRR abs/0712.1097 (2007)
14. Brummayer, R., Biere, A.: Effective bit-width and under-approximation. In: Moreno-Díaz, R., Pichler, F., Quesada-Arencibia, A. (eds.) EUROCAST 2009. LNCS, vol. 5717, pp. 304–311. Springer, Heidelberg (2009)
15. Eén, N., Mishchenko, A., Amla, N.: A single-instance incremental SAT formulation of proof - and counterexample - based abstraction. In: Proc. FMCAD 2010, pp. 181–188 (2010)
16. Nöhrer, A., Biere, A., Egyed, A.: Managing SAT inconsistencies with HUMUS. In: Proc. VaMoS 2012, pp. 83–91. ACM (2012)

17. Huang, J.: Extended clause learning. AI 174(15), 1277–1284 (2010)
18. Audemard, G., Katsirelos, G., Simon, L.: A restriction of extended resolution for clause learning SAT solvers. In: Proc. AAAI 2010 (2010)
19. Nadel, A., Ryvchin, V.: Efficient SAT solving under assumptions. In: Cimatti, A., Sebastiani, R. (eds.) SAT 2012. LNCS, vol. 7317, pp. 242–255. Springer, Heidelberg (2012)
20. Goldberg, E.I., Novikov, Y.: BerkMin: A fast and robust Sat-solver. In: Proc. DATE 2002, pp. 142–149. IEEE (2002)
21. Audemard, G., Simon, L.: Predicting learnt clauses quality in modern SAT solvers. In: Proc. IJCAI 2009, pp. 399–404. Morgan Kaufmann (2009)
22. Audemard, G., Lagniez, J.-M., Mazure, B., Saïs, L.: On freezing and reactivating learnt clauses. In: Sakallah, K.A., Simon, L. (eds.) SAT 2011. LNCS, vol. 6695, pp. 188–200. Springer, Heidelberg (2011)
23. Zhang, L., Madigan, C.F., Moskewicz, M.W., Malik, S.: Efficient conflict driven learning in boolean satisfiability solver. In: Proc. ICCAD 2001, pp. 279–285 (2001)
24. Sörensson, N., Biere, A.: Minimizing learned clauses. In: Kullmann, O. (ed.) SAT 2009. LNCS, vol. 5584, pp. 237–243. Springer, Heidelberg (2009)
25. Oh, Y., Mneimneh, M.N., Andraus, Z.S., Sakallah, K.A., Markov, I.L.: AMUSE: a minimally-unsatisfiable subformula extractor. In: Proc. DAC 2004, pp. 518–523. ACM (2004)
26. Chu, G., Harwood, A., Stuckey, P.J.: Cache conscious data structures for boolean satisfiability solvers. JSAT 6(1-3), 99–120 (2009)
27. Gomes, C.P., Selman, B., Kautz, H.A.: Boosting combinatorial search through randomization. In: Proc. AAAI/IAAI 1998, pp. 431–437 (1998)
28. Huang, J.: The effect of restarts on the effectiveness of clause learning. In: Proc. IJCAI 2007 (2007)
29. Pipatsrisawat, K., Darwiche, A.: RSat!2.0: SAT solver description. Technical Report Technical Report D–153, Automated Reasoning Group, Comp. Scienc. Dept., UCLA (2007)
30. Biere, A.: Picosat essentials. JSAT 4(2-4), 75–97 (2008)
31. Luby, M., Sinclair, A., Zuckerman, D.: Optimal speedup of Las Vegas algorithms. Information Processing Letters 47 (1993)
32. Biere, A.: Adaptive restart strategies for conflict driven SAT solvers. In: Kleine Büning, H., Zhao, X. (eds.) SAT 2008. LNCS, vol. 4996, pp. 28–33. Springer, Heidelberg (2008)
33. Ryvchin, V., Strichman, O.: Local restarts. In: Kleine Büning, H., Zhao, X. (eds.) SAT 2008. LNCS, vol. 4996, pp. 271–276. Springer, Heidelberg (2008)
34. Audemard, G., Simon, L.: Refining restarts strategies for SAT and UNSAT. In: Milano, M. (ed.) CP 2012. LNCS, vol. 7514, pp. 118–126. Springer, Heidelberg (2012)
35. Biere, A.: Lingeling and friends at the SAT Competition 2011. FMV Report Series Technical Report 11/1, Johannes Kepler University, Linz, Austria (2011)
36. van der Tak, P., Ramos, A., Heule, M.: Reusing the assignment trail in CDCL solvers. JSAT 7(4), 133–138 (2011)
37. Nadel, A., Ryvchin, V., Strichman, O.: Preprocessing in incremental SAT. In: Cimatti, A., Sebastiani, R. (eds.) SAT 2012. LNCS, vol. 7317, pp. 256–269. Springer, Heidelberg (2012)
38. Järvisalo, M., Heule, M.J.H., Biere, A.: Inprocessing rules. In: Gramlich, B., Miller, D., Sattler, U. (eds.) IJCAR 2012. LNCS, vol. 7364, pp. 355–370. Springer, Heidelberg (2012)

On the Interpolation between Product-Based Message Passing Heuristics for SAT

Oliver Gableske

Ulm University, Theoretical Computer Science, 89081 Ulm, Germany
oliver.gableske@uni-ulm.de
https://www.gableske.net

Abstract. This paper introduces a notational frame to characterize the four basic product-based Message Passing (MP) heuristics currently available for SAT: Belief Propagation (BP), Survey Propagation (SP), Expectation Maximization BP Global (EMBPG) and Expectation Maximization SP Global (EMSPG). Using this framework, the paper introduces *indirect structural interpolation* (ISI). Using this technique, we create a hierarchy of heuristics – each new level in this hierarchy consists of heuristics strictly more general than their predecessors. The final result is the $\rho\sigma\text{PMP}^i$ heuristic, which is able to mimic all product-based MP heuristics and is hence a generalization for all them.

1 Introduction

SAT is one of the most studied combinatorial problems and the interest in practically applicable algorithms to solve this problem (called SAT solvers) has increased during the past decades. Several classes of SAT solvers are available, like CDCL, SLS, and look-ahead solvers. A similarity between all of them is their application of variable and value selection heuristics to advance the search. Well known examples for such heuristics are VSIDS and phase-saving [16, 17].

Message Passing (MP) is a comparatively new class of such heuristics [8, 2]. An MP heuristic creates estimations of variable assignments, called biases. An MP-based SAT solver then uses these biases for assigning variables and thus can simplify the formula under the assumption that these biases capture a partial satisfying assignment. We call the simplification using biases *MP-Inspired Decimation* (MID). Estimating biases and performing MID is repeated until the formula is solved by the set of given assignments or a conflict occurs. If a conflict occurs, the solver either reports the failure of the MID approach or, in conjunction with a complete search strategy like CDCL, can learn from the conflict, back-jump, and repeat the bias computation along with MID. SAT solvers that apply an MP heuristic and MID are henceforth called *MID solvers*.

Two well known MP heuristics to solve large uniform random k-SAT formulas are Belief Propagation (BP) and Survey Propagation (SP) [13, 4]. The computational cost of BP is comparatively low, but its ability to create biases that lead to a satisfying assignment (called the *success rate*) is comparatively low as well. An increased focus on MP heuristics came with SP which has a higher success

M. Järvisalo and A. Van Gelder (Eds.): SAT 2013, LNCS 7962, pp. 293–308, 2013.

rate than BP. Both BP and SP do, however, suffer from the fact that they might not converge towards a stable configuration during their message passing (called the *convergence rate*). Hence, BP and SP might fail to provide biases, and a MID solver relying on them to advance the search would therefore fail as well.

The low convergence rates of BP and SP are a major issue when trying to solve crafted formulas. Here, BP and SP fail to converge and are therefore unable to provide biases at all. This inspired the development of two Expectation Maximization variants that are guaranteed to converge: EMBPG and EMSPG [9–11]. The guaranteed convergence in the EM variants allows them to always provide biases, but their success rates are worse than those of the non-EM variants when solving uniform random k-SAT formulas. All together, the non-EM variants are helpful when solving random formulas, but they are not helpful when solving crafted formulas. In contrast, the EM variants are helpful when solving crafted formulas, but they are not helpful when solving random formulas.

While the main difference between the non-EM and the EM variants is their convergence behavior, the main difference between the BP and SP variants is how careful they provide biases. Intuitively, the SP variants favor variables that are more *unlikely* to be assigned wrong, resulting in an increased success rate. However, when decreasing the ratio of random formulas, the SP variants might converge into a *paramagnetic state* where all variables appear unbiased. Here, the SP variants converge but nonetheless fail to provide biases, and the less careful BP behavior would be preferable. All together, the lack of flexibility of the MP heuristics renders it difficult to design a robust MID solver. The situation can be improved by designing MP heuristics with an increased flexibility as follows.

First, we propose the interpolation of the non-EM and the EM variants aiming for an improved flexibility regarding convergence. We introduce σEMBPG^i (interpolating BP and EMBPG) and σEMSPG^i (interpolating SP and EMSPG). Using the *interpolation parameter* $\sigma \in [0, 1]$, these heuristics can mimic the convergence behavior of the non-EM variants ($\sigma = 0$), the EM variants ($\sigma = 1$), and anything in between ($\sigma \in (0, 1)$). The σ heuristics can be seen as generalizations regarding convergence of the non-EM and EM variants they interpolate.

Second, we propose the interpolation of the BP and SP variants aiming for an improved flexibility regarding carefulness. We introduce ρSP^i (interpolating BP and SP) and ρEMSPG^i (interpolating EMBPG and EMSPG). Using the interpolation parameter $\rho \in [0, 1]$, these heuristics can mimic the carefulness of the BP variants ($\rho = 0$), the SP variants ($\rho = 1$), and anything in between ($\rho \in (0, 1)$). The ρ heuristics can be seen as generalizations regarding carefulness of the BP and SP variants they interpolate.

Finally, one can ask for an MP heuristic that has the flexibility to *simultaneously* adapt its convergence and carefulness behavior. We introduce $\rho\sigma\text{PMP}^i$, which uses two interpolation parameters $\rho, \sigma \in [0, 1]$. It interpolates between all the above heuristics and is therefore the most flexible one. It can be seen as a generalization of all the other MP heuristics presented in this paper. Hence, it should be the most useful one to solve random and crafted formulas alike. The empirical results presented in Section 5 do indeed support this intuition.

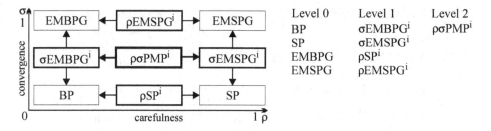

Fig. 1. A conceptual overview of the MP heuristics presented in this paper. The arrows indicate which heuristics are interpolated using *indirect structural interpolation* (ISI) in order to derive the respective interpolation.

We call BP, SP, EMBPG, and EMSPG the level 0 (L0) heuristics. We introduce *indirect structural interpolation* (ISI) to derive from the L0 heuristics the L1 heuristics: ρSP^i, $\rho EMSPG^i$, $\sigma EMBPG^i$, and $\sigma EMSPG^i$. From the L1 heuristics, we derive the L2 heuristic $\rho\sigma PMP^i$. The levels L0 to L2 define a *hierarchy of generality* with $\rho\sigma PMP^i$ being the most general heuristic. See Fig. 1.

The contributions of this paper are summarized as follows. First, a notational frame to present product-based MP heuristics is introduced and used to explain the basic L0 heuristics along with a generic MID solver genMID. Second, a technique to interpolate two given MP heuristics into a new and more general one is introduced. This technique, called ISI, is used to derive the L1 and L2 heuristics mentioned above. The final result is the L2 heuristic $\rho\sigma PMP^i$. The strength of $\rho\sigma PMP^i$ is twofold. First, it can mimic the behavior of the L0 and L1 heuristics which makes their separate implementation unnecessary. Second, $\rho\sigma PMP^i$ has the ability to achieve an MP behavior that cannot be achieved using the L0 or L1 heuristics. Using $\rho\sigma PMP^i$ in a MID solver increases its flexibility and allows to improve its robustness using the new possibilities of parameter tuning.

2 A Notational Frame and a Generic MID Solver

In this section we first repeat some definitions regarding SAT. Afterwards, we explain MP heuristics on a conceptual level and present a notational frame to explain them in more detail. Furthermore, we explain MID, and how MP heuristics and MID are combined into a generic MID solver, called genMID.

Definition 1. *Let V be a set of n Boolean variables. Let $v \in V$. A CNF formula F is a set of m clauses. Each clause $c \in F$ is a disjunction of literals, where a literal is a variable v or its negation $\bar{v}$. The sets $C_v^+, C_v^-, C_v = C_v^+ \cup C_v^-$ comprise the sets of clauses containing v as a positive, negative, or arbitrary literal. Let $\Box$ denote the empty clause. We call $\alpha : V \to \{0,1\}$ an assignment.*

We continue by explaining the conceptual functioning of product-based MP heuristics. Let F be a CNF formula. For convenience, assume that F does not contain unit clauses, tautological clauses, or pure literals.

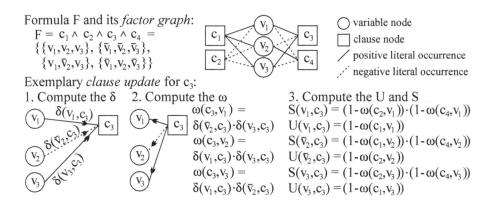

Formula F and its *factor graph*:

$F = c_1 \wedge c_2 \wedge c_3 \wedge c_4 =$
$\{\{v_1,v_2,v_3\}, \{\bar{v}_1,\bar{v}_2,\bar{v}_3\},$
$\{v_1,\bar{v}_2,v_3\}, \{\bar{v}_1,v_2,\bar{v}_3\}\}$

○ variable node
□ clause node
╱ positive literal occurrence
⋰ negative literal occurrence

Exemplary *clause update* for c_3:

1. Compute the δ 2. Compute the ω

$\omega(c_3,v_1) =$
$\delta(\bar{v}_2,c_3) \cdot \delta(v_3,c_3)$

$\omega(c_3,v_2) =$
$\delta(v_1,c_3) \cdot \delta(v_3,c_3)$

$\omega(c_3,v_3) =$
$\delta(v_1,c_3) \cdot \delta(\bar{v}_2,c_3)$

3. Compute the U and S

$S(v_1,c_3) = (1-\omega(c_2,v_1)) \cdot (1-\omega(c_4,v_1))$
$U(v_1,c_3) = (1-\omega(c_1,v_1))$
$S(\bar{v}_2,c_3) = (1-\omega(c_1,v_2)) \cdot (1-\omega(c_4,v_2))$
$U(\bar{v}_2,c_3) = (1-\omega(c_2,v_2))$
$S(v_3,c_3) = (1-\omega(c_2,v_3)) \cdot (1-\omega(c_4,v_3))$
$U(v_3,c_3) = (1-\omega(c_1,v_3))$

Fig. 2. Top: A CNF formula and its *factor graph* representation. Bottom: The order of computations that are performed for a *clause update*.

Definition 2 (Factor Graph). *A* factor graph *[13] of a CNF F is an undirected graph that contains two types of nodes and two types of edges. The nodes are partitioned into* variable nodes *(one for each $v \in \mathcal{V}$) and* clause nodes *(one for each $c \in F$). The edges are partitioned into* positive *(solid) edges and* negative *(dotted) edges. A* positive *(neg.) edge between a variable and a clause node exists, if and only if the variable appears as positive (neg.) literal in the clause.*

See Fig. 2 (top) for an example of a factor graph. Conceptually, an MP heuristic H sends messages along the edges of the factor graph. The goal is to reach a stable configuration of the messages. These messages are then used to derive biases for the variables, which can be seen as suggestions for assignments. We distinguish between *warning messages* and *disrespect messages* as follows.

Warning messages, $\omega_H \in [0,1]$, are send from clause nodes to variable nodes. Intuitively, a warning message from clause c to variable v, denoted $\omega_H(c,v)$, tells variable v how much clause c is in need of this variable to satisfy it. Intuitively, we understand a ω_H close to one as a strong warning, which means that clause c wants v to be assigned in such a way that it satisfies its literal occurrence in c.

Disrespect messages, $\delta_H \in [0,1]$, are send from variable nodes to clause nodes. Intuitively, a disrespect message from variable v along a solid line to a clause c, denoted $\delta_H(v,c)$, tells clause c how likely it is that v will be assigned to false (and hence *not* satisfy c). A disrespect message from variable v along a dotted line to a clause c, denoted $\delta_H(\bar{v},c)$, tells clause c how likely it is that v will be assigned to true (and hence *not* satisfy c). We combine both cases for δ_H messages (positive or negative literal $l \in c$) by writing $\delta_H(l,c)$. Intuitively, we understand a δ_H close to one as a strong disrespect, which means that variable v will most likely *not* be assigned such that $l \in c$ becomes satisfied. The exact computation of δ_H depends on H, and will be introduced for each of these heuristics later. For a conceptual overview it suffices to understand that $\forall c \in F : \forall l \in c : \delta_H(l,c) \in [0,1]$.

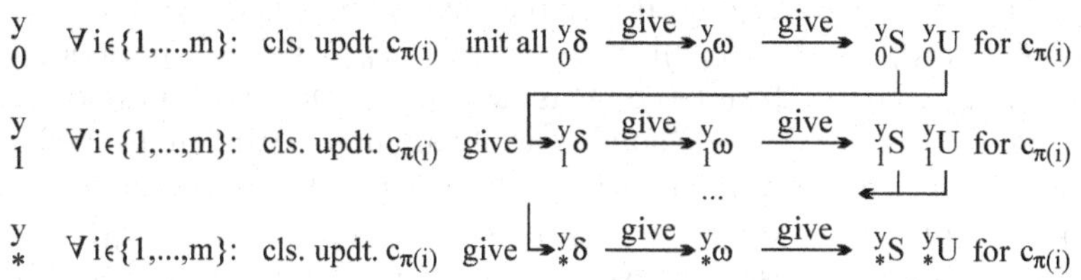

$$\begin{array}{l} {}^y_0 \quad \forall i \in \{1,...,m\}: \text{ cls. updt. } c_{\pi(i)} \quad \text{init all } {}^y_0\delta \xrightarrow{\text{give}} {}^y_0\omega \xrightarrow{\text{give}} {}^y_0S \; {}^y_0U \text{ for } c_{\pi(i)} \\[2ex] {}^y_1 \quad \forall i \in \{1,...,m\}: \text{ cls. updt. } c_{\pi(i)} \quad \text{give} \hookrightarrow {}^y_1\delta \xrightarrow{\text{give}} {}^y_1\omega \xrightarrow{\text{give}} {}^y_1S \; {}^y_1U \text{ for } c_{\pi(i)} \\[1ex] \qquad\qquad\qquad\qquad\qquad\qquad\qquad\qquad ... \\[1ex] {}^y_* \quad \forall i \in \{1,...,m\}: \text{ cls. updt. } c_{\pi(i)} \quad \text{give} \hookrightarrow {}^y_*\delta \xrightarrow{\text{give}} {}^y_*\omega \xrightarrow{\text{give}} {}^y_*S \; {}^y_*U \text{ for } c_{\pi(i)} \end{array}$$

Fig. 3. The iterations performed in a cycle y with an arbitrary permutation $\pi \in S_m$

For all *product-based* MP heuristics it is $\omega_H(c, v) = \prod_{l \in c \setminus \{v, \bar{v}\}} \delta_H(l, c)$ and

$$S_H(l,c) = \begin{cases} \prod_{d \in C_v^-} 1 - \omega_H(d,v), l = v \\ \prod_{d \in C_v^+} 1 - \omega_H(d,v), l = \bar{v} \end{cases} \qquad U_H(l,c) = \begin{cases} \prod_{d \in C_v^+ \setminus \{c\}} 1 - \omega_H(d,v), l = v \\ \prod_{d \in C_v^- \setminus \{c\}} 1 - \omega_H(d,v), l = \bar{v} \end{cases}$$

H performs a *clause update* for clause c by computing the current $\delta_H(l, c)$ messages for all $l \in c$ and, using these values, updates all $\omega_H(c, v)$. Then, it updates the $S_H(l, c)$ and $U_H(l, c)$ values using the updated $\omega_H(c, v)$. See Fig. 2 (bottom).

Intuitively, the $S_H(l, c) \in [0, 1]$ value captures the freedom of variable v to be assigned in such a way that it *satisfies* l (and hence *satisfies* c). Likewise, the $U_H(l, c) \in [0, 1]$ value captures the freedom of variable v to be assigned in such a way that it *falsifies* l (and hence does *not* satisfy c). We call the S_H and U_H values the *cavity freedom values*. Clause c itself is called the *cavity clause*, since it is ignored during the computation of S_H and U_H. Intuitively, a clause update captures the tendency for an assignment to v in S_H and U_H, while assuming $c \notin F$. This is known as the *cavity method* (see [15] for details).

Performing a clause update with the above equations for all clauses $c \in F$ exactly once is called an *iteration*. The ordering of the clauses in which the updates are performed is arbitrary (assume a random clause permutation $\pi \in S_m$). A cycle y is a finite tuple of iterations. Intuitively, the terms *cycle and iteration* capture the notion of *passing time* during the computations of H, where iteration z of cycle y is understood as a single point in time. In the following, we denote by ${}^y_z\delta_H(l, c)$ the disrespect message send by H from l to c in iteration z of cycle y. Likewise we denote the warning and cavity freedom values of a specific iteration z in cycle y with ${}^y_z\omega_H(c, v), {}^y_zS_H(l, c)$, and ${}^y_zS_H(l, c)$.

Conceptually, MP heuristics perform cycles of iterations starting in iteration $z = 0$ with arbitrarily initialized ${}^y_0\delta_H$ (assume random values in $(0, 1)$). The ${}^y_0\omega_H$, ${}^y_0S_H$, and ${}^y_0U_H$ values for iteration 0 then follow with the above equations.

In order to compute ${}^y_z\delta_H$ with $z > 0$, H relies on the ${}_{z-1}{}^yS_H$, and ${}_{z-1}{}^yU_H$ values of iteration $z - 1$. See Fig. 3. The equation δ_H used for this computation is defined by H. For example in Belief Propagation, the equation looks as follows.

$$ {}^y_z\delta_{BP}(l,c) = \frac{{}_{z-1}{}^yU_{BP}(l,c)}{{}_{z-1}{}^yU_{BP}(l,c) + {}_{z-1}{}^yS_{BP}(l,c)} $$

Iterating in cycle y continues until the following *abort condition* holds: $\forall c \in F$: $\forall v \in c$: $\left| {}_z^y \omega_H(c,v) - {}_{z-1}^y \omega_H(c,v) \right| < \omega_{max}$ (with ω_{max} being a parameter of H, commonly $\omega_{max} = 0.01$). Intuitively, H is said to be *converged* in cycle y as soon as all warning messages do not change notably anymore. We denote the iteration of cycle y in which H converged with $*$. The corresponding messages from this iteration are called *equilibrium messages*, denoted ${}_*^y \delta_H(l,c)$ and ${}_*^y \omega_H(c,v)$.

It is possible that H does not converge in a cycle y. In case the above abort condition never holds, H would not terminate. A straightforward way to circumvent this is to use another parameter for H called $z_{max} > 1$ (commonly $z_{max} = 1000$), such that H declares a failure and stops if iteration z_{max} is reached. We formally define the terms introduced above as follows.

Definition 3 (Messages, Cavity Freedom, Cavity Clause). *Let H be an MP heuristic. Let $v \in \mathcal{V}, c \in F$. Let literal $l \in c$ correspond to variable v. With iteration z in cycle y, we define the* disrespect message *of heuristic H as a function ${}_z^y \delta_H(l,c) \in [0,1]$. We define the* warning message *of heuristic H as*

$$ {}_z^y \omega_H(c,v) = \prod_{l \in c \setminus \{v, \bar{v}\}} {}_z^y \delta_H(l,c) $$

We define the cavity freedom *of l in* cavity clause c *in iteration z of cycle y as*

$$ {}_z^y S_H(l,c) = \begin{cases} \prod_{d \in C_v^-} 1 - {}_z^y \omega_H(d,v), l = v \\ \prod_{d \in C_v^+} 1 - {}_z^y \omega_H(d,v), l = \bar{v} \end{cases} \qquad {}_z^y U_H(l,c) = \begin{cases} \prod_{d \in C_v^+ \setminus \{c\}} 1 - {}_z^y \omega_H(d,v), l = v \\ \prod_{d \in C_v^- \setminus \{c\}} 1 - {}_z^y \omega_H(d,v), l = \bar{v} \end{cases} $$

Assume H converged in cycle y. The equilibrium messages ${}_*^y \omega_H(c,v)$ are used to derive variable biases. These biases are then used to make assignments and simplify F (MID). The derivation of variable biases works as follows.

Definition 4 (Variable Freedom). *Let H be converged in cycle y. We define the* freedom *of v in cycle y to be assigned to true ($\mathcal{T}$) or false ($\mathcal{F}$) as*

$$ {}^y \mathcal{T}_H(v) = \prod_{c \in C_v^-} 1 - {}_*^y \omega_H(c,v) \qquad and \qquad {}^y \mathcal{F}_H(v) = \prod_{c \in C_v^+} 1 - {}_*^y \omega_H(c,v) $$

An intuitive interpretation of a large ${}^y \mathcal{T}_H(v)$ is that the equilibrium messages indicate that none of the clauses in which v appears as *negative* literal $\bar{v}$ need $\bar{v}$ to become satisfied. Hence, v can be assigned to *true* (similar for ${}^y \mathcal{F}_H(v)$ and assignment *false*). A large value for ${}^y \mathcal{T}_H(v) \cdot {}^y \mathcal{F}_H(v)$ means that the variable has a strong global freedom, which indicates that v can be assigned arbitrarily.

Even though a variable can have large $\mathcal{T}$ and $\mathcal{F}$ values, it cannot be assigned to both true and false simultaneously. The variable's tendency to either true or false is captured in the *variable magnetization* as follows.

Definition 5 (Variable Magnetization). *We call ${}^y \mu_H^+(v), {}^y \mu_H^-(v), {}^y \mu_H^\pm(v) \in [0,1]$ the* positive, negative, *and* total *magnetization of v in cycle y. Additionally, we call ${}^y \mu_H(v) = {}^y \mu_H^+(v) + {}^y \mu_H^-(v) + {}^y \mu_H^\pm(v)$ its* global *magnetization in cycle y.*

The μ functions are defined separately for each H. Intuitively, a large $^y\mu_H^+(v)$ value indicates that *both* $^y\mathcal{T}_H(v)$ and $^y\mathcal{F}_H(v)$ suggest an assignment to true. In other words, the $^y\mu_H^+(v)$ value will be large if $^y\mathcal{T}_H(v)$ is large and $^y\mathcal{F}_H(v)$ is small. The *bias* of a variable is computed using the magnetization values as follows.

Definition 6 (Variable Bias). *We call* $^y\beta_H^+(v), ^y\beta_H^-(v) \in [0,1]$, *and* $^y\beta_H(v) \in [-1,1]$ *the* positive, negative, and total bias *of variable* v *in cycle* y. *It is*

$$^y\beta_H^+(v) = \frac{^y\mu_H^+(v)}{^y\mu_H(v)} \qquad ^y\beta_H^-(v) = \frac{^y\mu_H^-(v)}{^y\mu_H(v)} \qquad ^y\beta_H(v) = {}^y\beta_H^+(v) - {}^y\beta_H^-(v)$$

If $^y\beta_H(v)$ has a value close to -1 ($+1$) the variable is biased towards the assignment false (true). Finally, we define the clause count of variables and literals.

Definition 7 (Clause Count). *With* $v \in \mathcal{V}$ *let* l *correspond to* v.

$$t(v) = |C_v^-| \qquad f(v) = |C_v^+| \qquad s(l) = \begin{cases} |C_v^-|, l = v \\ |C_v^+|, l = \bar{v} \end{cases} \qquad u(l) = \begin{cases} |C_v^+|, l = v \\ |C_v^-|, l = \bar{v} \end{cases}$$

Assume that we are applying H in a MID solver `genMID`. Assume that H converged in cycle y and has computed $^y\beta_H(v)$ for all variables. The solver `genMID` will use two parameters to decide what to do with those biases. First, $\beta_{min} \in (0,1)$ (commonly $\beta_{min} = 0.01$) that defines when the set of biases is strong enough to be understood as meaningful. Second, $p \in (0,1)$ (commonly $p = 0.05$) which defines the fractional amount of variables that will be assigned according to their bias in case they are meaningful.

In case $\sum_{i=1}^n \max(^y\beta_H^-(v_i), ^y\beta_H^+(v_i)) \geq \beta_{min}$, solver `genMID` will assign the $p \cdot n$ highest biased variables according to their bias and simplifies the formula using unit propagation and pure literal elimination, resulting in F' (MID). It then checks if $F' \in$ SAT or if $\square \in F'$. In case $F' \in$ SAT, `genMID` outputs the assignment to all variables, declares satisfiability and stops. In case $\square \in F'$, `genMID` either reports a failure and stops, or learns from the conflict and backjumps. If `genMID` did not stop, it continues by calling H in cycle $y + 1$ on F'.

In case $\sum_{i=1}^n \max(^y\beta_H^-(v_i), ^y\beta_H^+(v_i)) < \beta_{min}$, `genMID` declares a paramagnetic state. All variables appear unbiased and no further meaningful MID is possible. The remaining F must be solved by a different algorithm (e.g. SLS or CDCL).

Refer to Fig. 4 for the pseudo-code of a generic MID solver. The type of search that `genMID` performs is solely based on the equations for heuristic H, which are based on $\delta_H, \mu_H^+, \mu_H^-$, and $\mu_H^\pm$. In other words, if a MID solver is to be derived, it suffices to define these equations and include them in the call to `genMID`. For example, assume we set $z_{max}, \omega_{max}, \beta_{min}, p$ and the equations for Belief Propagation are given: $\delta_{BP}, \mu_{BP}^+, \mu_{BP}^-$, and $\mu_{BP}^\pm$. A BP based MID solver search on F is performed by `genMID`$(F, z_{max}, \omega_{max}, \beta_{min}, p, \delta_{BP}, \mu_{BP}^+, \mu_{BP}^-, \mu_{BP}^\pm)$.

The following section will provide the δ and μ equations for various product-based MP heuristics. In conjunction with the `genMID` algorithm they explain how the corresponding MID solvers work.

genMID(CNF formula F, z_{max}, ω_{max}, β_{min}, p, δ_H, μ_H^+, μ_H^-, $\mu_H^\pm$)

    $\alpha = \{\}$; $y := 0$; //Initialize the empty assignment and the cycle

    $\forall c \in F : \forall l \in c : {}_*^0\delta_H(l,c) \in_R (0,1)$;//Initialize δ messages randomly

    REPEAT

        $y := y+1$; $z := -1$; $\pi \in_R \mathcal{S}_m$; //Where π is a random clause permutation

        REPEAT //Perform cycle y

            $\omega' := 0$; $z := z+1$;

            $\forall i \in \{1, \ldots, m\}$://Perform iteration z

                $c := c_{\pi(i)}$;//Get next clause according to clause permutation π

                IF $z = 0$ //Initialize cycle y with previous equilibrium messages

                    $\forall l \in c : {}_0^y\delta_H(l,c) := {}_*^{y-1}\delta_H(l,c)$;

                    $\forall v \in c : {}_0^y\omega_H(c,v) := \prod_{l \in c \setminus \{v,\bar{v}\}} {}_0^y\delta_H(l,c)$;

                    $\forall l \in c :$ compute ${}_0^y S_H(l,c)$ and ${}_0^y U_H(l,c)$ using ${}_0^y\omega_H(c,v)$;

                ELSE //Iteration $z > 0$ relying on values from iteration $z - 1$

                    $\forall l \in c :$ compute ${}_z^y\delta_H(l,c)$ using ${}_{z-1}^y S_H(l,c)$, ${}_{z-1}^y U_H(l,c)$;

                    $\forall v \in c : {}_z^y\omega_H(c,v) := \prod_{l \in c \setminus \{v,\bar{v}\}} {}_z^y\delta_H(l,c)$;

                    IF $\omega' < \left| {}_z^y\omega_H(c,v) - {}_{z-1}^y\omega_H(c,v) \right|$ //Retain largest

                      $\omega' := \left| {}_z^y\omega_H(c,v) - {}_{z-1}^y\omega_H(c,v) \right|$;//message difference

                    $\forall l \in c :$ compute ${}_z^y S_H(l,c)$ and ${}_z^y U_H(l,c)$ using ${}_z^y\omega_H(c,v)$;

        UNTIL ($z > 0$ **AND** $\omega' < \omega_{max}$) **OR** $z > z_{max}$//Concludes cycle y

        IF $\omega' < \omega_{max}$ //Check if convergence has been reached in cycle y

            $\forall v \in \mathcal{V} :$ compute ${}^y\beta_H(v)$;//Uses the μ_H and the cycle equilibrium msgs

            IF $\sum_{i=1}^n \max({}^y\beta_H^-(v_i), {}^y\beta_H^+(v_i)) < \beta_{min}$//Check if paramagnetic

                output PARAMAG,α; **STOP**;//Might call SLS($\alpha(F)$) or CDCL($\alpha(F)$)

            sort($\mathcal{V}$); //Sort variables, highest bias first

            $\alpha := \alpha \cup \{p \cdot n$ highest biased variables according to bias$\}$; //Extend α

            $F := \alpha(F)$; //Perform MID, with unit propagation and pure literal removal

            IF $\square \in F$: output UNKNOWN; **STOP**;//Or learn and backjump like CDCL

            ELSE IF $F \in$ SAT: output SATISFIABLE,α; **STOP**;

        ELSE output UNKNOWN; **STOP**;//Unconverged after z_{max} iterations

    UNTIL TRUE //Perform MP and MID as long as possible

Fig. 4. The generic MID solver as conceptually introduced in [4]. Independent of $z_{max}, \omega_{max}, \beta_{min}$, and p, the type of search that genMID performs depends only on H.

2.1 A Summary of the Equations for the L0 Heuristics

In this section we briefly present the δ and μ equations for the L0 heuristics. For the δ equations, we assume an arbitrary iteration z of cycle y ($y, z > 0$). The cases $y = 0$ or $z = 0$ belong to the initialization (see the previous section). For the μ (and β) equations, we assume convergence in iteration $*$ of cycle y. The β equations follow from the μ equations with Definition 6. For brevity, we use a *simplified notation* and write $U = {}_{z-1}^y U_H(l,c)$, $S = {}_{z-1}^y S_H(l,c)$, $\mathcal{T} = {}^y\mathcal{T}_H(v)$, $\mathcal{F} = {}^y\mathcal{F}_H(v)$, as well as $u = u(l)$, $s = s(l)$, $t = t(v)$, and $f = f(v)$.

The equations for Belief Propagation (BP) [13], Survey Propagation (SP) [4, 5], Expectation Maximization BP Global (EMBPG) [9, 10], and Expectation

Maximization SP Global (EMSPG) [9, 10] are as follows.

$$
{}^y_z\delta_{\mathrm{BP}}(l,c) = \frac{U}{U+S}, \quad {}^y\beta_{\mathrm{BP}}(v) = \frac{\mathcal{T}-\mathcal{F}}{\mathcal{T}+\mathcal{F}}, \quad \begin{array}{ll} {}^y\mu^{\pm}_{\mathrm{BP}}(v) = 0, & {}^y\mu^{+}_{\mathrm{BP}}(v) = \mathcal{T}, \\ {}^y\mu^{-}_{\mathrm{BP}}(v) = \mathcal{F}, & {}^y\mu_{\mathrm{BP}}(v) = \mathcal{T}+\mathcal{F} \end{array}
$$

$$
{}^y_z\delta_{\mathrm{SP}}(l,c) = \frac{U(1-S)}{U(1-S)+S}, \quad {}^y\beta_{\mathrm{SP}}(v) = \frac{\mathcal{T}-\mathcal{F}}{\mathcal{T}+\mathcal{F}-\mathcal{T}\mathcal{F}}, \quad {}^y\mu^{\pm}_{\mathrm{SP}}(v) = \mathcal{T}\mathcal{F},
$$
$$
{}^y\mu^{+}_{\mathrm{SP}}(v) = \mathcal{T}(1-\mathcal{F}), \quad {}^y\mu^{-}_{\mathrm{SP}}(v) = \mathcal{F}(1-\mathcal{T}), \quad {}^y\mu_{\mathrm{SP}}(v) = \mathcal{T}+\mathcal{F}-\mathcal{T}\mathcal{F}
$$

$$
{}^y_z\delta_{\mathrm{EMBPG}}(l,c) = \frac{uU+s}{u(1+U)+s(1+S)}, \quad {}^y\beta_{\mathrm{EMBPG}}(v) = \frac{f(1-\mathcal{F})-t(1-\mathcal{T})}{t(1+\mathcal{T})+f(1+\mathcal{F})}
$$
$$
{}^y\mu^{\pm}_{\mathrm{EMBPG}}(v) = 0, \quad\quad\quad {}^y\mu^{+}_{\mathrm{EMBPG}}(v) = t\mathcal{T}+f,
$$
$$
{}^y\mu^{-}_{\mathrm{EMBPG}}(v) = f\mathcal{F}+t, \quad\quad {}^y\mu_{\mathrm{EMBPG}}(v) = t(1+\mathcal{T})+f(1+\mathcal{F})
$$

$$
{}^y_z\delta_{\mathrm{EMSPG}}(l,c) = \frac{uU+s(1-S)}{(u+s)(1+US)}, \quad {}^y\beta_{\mathrm{EMSPG}}(v) = \frac{f(1-2\mathcal{F})-t(1-2\mathcal{T})}{(t+f)(1+\mathcal{T}\mathcal{F})}
$$
$$
{}^y\mu^{\pm}_{\mathrm{EMSPG}}(v) = (t+f)\mathcal{T}\mathcal{F}, \quad\quad {}^y\mu^{+}_{\mathrm{EMSPG}}(v) = t\mathcal{T}+f(1-\mathcal{F}),
$$
$$
{}^y\mu^{-}_{\mathrm{EMSPG}}(v) = f\mathcal{F}+t(1-\mathcal{T}), \quad\quad {}^y\mu_{\mathrm{EMSPG}}(v) = (t+f)(1+\mathcal{T}\mathcal{F})
$$

Given the above equations, we derive the L1 heuristics using ISI as follows.

3 Indirect Structural Interpolation (ISI) – L1 Heuristics

In this section we explain the idea behind ISI and extend the notational frame to define the term *L1 heuristic*. Furthermore, we present a detailed derivation of $\rho\mathrm{SP}^i$ to show how ISI is applied in practice. We conclude this section by summarizing the equations for all L1 heuristics as shown in Fig. 1.

3.1 The Idea behind Indirect Structural Interpolation

Given two L0 heuristics H_1 and H_2, we aim for an interpolation τH^i using an interpolation parameter τ, such that τH^i is able to mimic H_1 (with $\tau = 0$) and H_2 (with $\tau = 1$), and interpolate in between (with $\tau \in (0,1)$). The goal is to be able to control a specific behavior in which H_1 and H_2 differ by adjusting τ. For example, given BP and SP along with interpolation parameter ρ, heuristic $\rho\mathrm{SP}^i$ is supposed to be able to adjust its *carefulness* for presenting biases. Here, $\rho = 0$ means least careful and mimics BP, and $\rho = 1$ means most careful and mimics SP. Setting $\rho \in (0,1)$ is supposed to yield an intermediate carefulness, expressed in biases that are between those computed by BP and SP.

In order to derive τH^i, we need to derive the equations $\delta^i_{\tau\mathrm{H}}$ (which then implicitly gives $\omega^i_{\tau\mathrm{H}}$), as well as $\mu^{i+}_{\tau\mathrm{H}}, \mu^{i-}_{\tau\mathrm{H}}$, and $\mu^{i\pm}_{\tau\mathrm{H}}$ (which then implicitly give $\beta^{i+}_{\tau\mathrm{H}}, \beta^{i-}_{\tau\mathrm{H}}$, and $\beta^{i\pm}_{\tau\mathrm{H}}$). ISI is a technique to derive these equations. In order to explain ISI, we extend our notational frame for L1 heuristics to capture an application of τ. Compare the following definition to Definition 3.

Definition 8 (L1 Messages, Cavity Values). *Let τH^i be an L1 MP heuristic using interpolation parameter $\tau \in [0,1]$. Let $v \in \mathcal{V}, c \in F$. Let literal $l \in c$ correspond to v. With iteration z in cycle y, we define the disrespect message of τH^i as $^y_z\delta^i_H(l,c,\tau) \in [0,1]$, and its warning message as*

$$^y_z\omega^i_{\tau H}(c,v,\tau) = \prod_{l\in c\setminus\{v,\bar{v}\}} {}^y_z\delta^i_{\tau H}(l,c,\tau)$$

We define the cavity freedom of l in cavity clause c similar to those given in Definition 3, but with an extended signature to capture the application of τ. Additionally, these functions rely on $^y_z\omega^i_{\tau H}(c,v,\tau)$ instead of $^y_z\omega_H(c,v)$.

Similarly, we extend the definitions for variable freedom, variable magnetization, and variable bias. Compare the following definition to Definitions 4, 5 and 6.

Definition 9 (L1 Variable Freedom, Magnetization, Bias). *Let τH^i be converged in cycle y. We define the freedom of v in cycle y to be assigned to true ($\mathcal{T}$) or false ($\mathcal{F}$) as*

$$^y\mathcal{T}^i_{\tau H}(v,\tau) = \prod_{c\in C_v^-} 1 - {}^y_*\omega^i_{\tau H}(c,v,\tau) \quad and \quad ^y\mathcal{F}^i_{\tau H}(v,\tau) = \prod_{c\in C_v^+} 1 - {}^y_*\omega^i_{\tau H}(c,v,\tau)$$

Additionally, we define the positive, negative, and total magnetization of v as presented in Definition 5, but with an extended signature of the functions to capture the application of τ. The positive, negative, and total bias are then

$$^y\beta^{i+}_{\tau H}(v,\tau) = \frac{^y\mu^{i+}_{\tau H}(v,\tau)}{^y\mu^i_H(v,\tau)} \qquad ^y\beta^{i-}_{\tau H}(v,\tau) = \frac{^y\mu^{i-}_{\tau H}(v,\tau)}{^y\mu^i_{\tau H}(v,\tau)} \qquad ^y\beta^i_{\tau H}(v,\tau) = \frac{^y\beta^+_{\tau H}(v,\tau)}{-{}^y\beta^-_{\tau H}(v,\tau)}$$

For brevity we extend the *simplified notation* and write $U^i = {}_{z-1}^{\;\;\;y}U^i_{\tau H}(l,c,\tau)$, $S^i = {}_{z-1}^{\;\;\;y}S^i_{\tau H}(l,c,\tau)$, $T^i = {}^y\mathcal{T}^i_{\tau H}(v,\tau)$, and $\mathcal{F}^i = {}^y\mathcal{F}^i_{\tau H}(v,\tau)$.

ISI is a technique to derive the equations for τH^i. For the $\delta_{\tau H}$ equation, it *linearly interpolates the structure* of the numerators and denominators of δ_{H_1} and δ_{H_2} separately, and combines both terms into a quotient. Additionally, it replaces U with U^i and S with S^i to capture the application of τ. Regarding ρSP^i, the derivation of $\delta^i_{\rho SP}$ using δ_{BP} and δ_{SP} from Section 2.1 works as follows.

$$^y_z\delta^i_{\rho SP}(l,c,\rho) = \frac{(1-\rho)\{U^i\} + \rho\{U^i(1-S^i)\}}{(1-\rho)\{U^i + S^i\} + \rho\{U^i(1-S^i) + S^i\}} = \frac{U^i(1-\rho S^i)}{U^i(1-\rho S^i) + S^i}.$$

For the $\mu^i_{\tau H}$ equations, ISI *linearly interpolates the structure* of the μ_{H_1} and μ_{H_2} equations. Additionally, it replaces $\mathcal{T}$ with $\mathcal{T}^i$ and $\mathcal{F}$ with $\mathcal{F}^i$ to capture the application of τ. The resulting equations for ρSP^i are as follows.

$$\left.\begin{array}{l} ^y\mu^{i\pm}_{\rho SP}(v,\rho) = (1-\rho)\{0\} + \rho\{\mathcal{T}^i\mathcal{F}^i\} = \rho\mathcal{T}^i\mathcal{F}^i \\[4pt] ^y\mu^{i+}_{\rho SP}(v,\rho) = (1-\rho)\{\mathcal{T}^i\} + \rho\{\mathcal{T}^i(1-\mathcal{F}^i)\} = \mathcal{T}^i(1-\rho\mathcal{F}^i) \\[4pt] ^y\mu^{i-}_{\rho SP}(v,\rho) = (1-\rho)\{\mathcal{F}^i\} + \rho\{\mathcal{F}^i(1-\mathcal{T}^i)\} = \mathcal{F}^i(1-\rho\mathcal{T}^i) \end{array}\right\} \begin{array}{l} \Rightarrow {}^y\mu^i_{\rho SP}(v,\rho) \\[4pt] = \mathcal{T}^i + \mathcal{F}^i \\[4pt] -\rho\mathcal{T}^i\mathcal{F}^i \end{array}$$

The $\beta^i_{\tau H}$ equations then follow from the $\mu^i_{\tau H}$ equations (see Definition 9). For ρSP^i, it is $^y\beta^i_{\rho SP}(v, \rho) = (\mathcal{T}^i - \mathcal{F}^i)/(\mathcal{T}^i + \mathcal{F}^i - \rho \mathcal{T}^i \mathcal{F}^i)$. Compare the equations for ρSP^i with the those of BP and SP in Section 2.1. The following definition formalizes the term of an L1 heuristic.

Definition 10 (Level 1 Heuristic). *Given two L0 heuristics H_1 and H_2 and $\tau \in [0, 1]$. We call τH^i a level 1 heuristic of H_1 and H_2 using τ, if and only if*

1. *τH^i is an MP heuristic that properly defines $\delta^i_{\tau H}(l, c, \tau)$, $\mu^{i+}_{\tau H}(v, \tau)$, $\mu^{i-}_{\tau H}(v, \tau)$, $\mu^{i\pm}_{\tau H}(v, \tau) \in [0, 1]$; the latter then give $\beta^i_{\tau H}(v, \tau) \in [-1, 1]$ via Definition 9.*
2. *If $\tau = 0$ and $\forall c \in F, l \in c : {}^0_*\delta_{H_1}(l, c) = {}^0_*\delta^i_{\tau H}(l, c, 0)$ (equal initialization), then $\forall v \in V : \beta_{H_1}(v) = \beta^i_{\tau H}(v, 0)$ (τH^i computes the same biases as H_1)*
3. *If $\tau = 1$ and $\forall c \in F, l \in c : {}^0_*\delta_{H_2}(l, c) = {}^0_*\delta^i_{\tau H}(l, c, 1)$ (equal initialization), then $\forall v \in V : \beta_{H_2}(v) = \beta^i_{\tau H}(v, 1)$ (τH^i computes the same biases as H_2)*

In [14], an interpolation between BP and SP is given, called ρ-SP. On the first glance, this interpolation looks similar to ρSP^i, however, there is a subtle difference. While the computation of the magnetization function $^y\mu^{i\pm}_{\rho SP}$ (and thereby the computation of the bias $^y\beta^i_{\rho SP}$) applies ρ in ρSP^i, algorithm ρ-SP from [14] ignores ρ in the computation for $\mu(*)$ (see [14], Section 2.2.2, page 10). Hence, ρ-SP always computes biases using the global freedom of the variables. In case $\rho = 0$, this does *not* result in the same biases as BP would compute. Hence, ρ-SP is missing Property 2 of Def. 10 and is therefore no L1 interpolation.

In [1], an interpolation is derived that is meant to be an interpolation between BP and SP. However, the very first equation regarding SP in that paper (Equation 6) is wrong. It is claimed to be the transport-message of SP, which should then be equal to the "interior of the product" of Equation 28 from [4]. However, in [4], Equation 27 clearly states how $\Pi^u_{j \to a}$ looks like when applied in the numerator of Equation 28. In [1], the numerator of Equation 6 is missing one of the two products – it would need to apply both A^1_i and A^0_i. Hence, whatever the resulting interpolation is, it is no interpolation between BP and SP.

Applying ISI as shown above in relation to Fig. 1 gives the L1 heuristics.

3.2 A Summary of the Equations for the L1 Heuristics

Similar to Section 2.1 we present the equations for the L1 heuristics. Applying ISI on BP and SP using interpolation parameter ρ results in ρSP^i:

$$^y_z\delta^i_{\rho SP}(l, c, \rho) = \frac{U^i(1 - \rho S^i)}{U^i(1 - \rho S^i) + S^i}, \qquad ^y\beta^i_{\rho SP}(v, \rho) = \frac{\mathcal{T}^i - \mathcal{F}^i}{\mathcal{T}^i + \mathcal{F}^i - \rho \mathcal{T}^i \mathcal{F}^i}$$

$$^y\mu^{i\pm}_{\rho SP}(v, \rho) = \rho \mathcal{T}^i \mathcal{F}^i, \qquad\qquad ^y\mu^{i+}_{\rho SP}(v, \rho) = \mathcal{T}^i(1 - \rho \mathcal{F}^i),$$

$$^y\mu^{i-}_{\rho SP}(v, \rho) = \mathcal{F}^i(1 - \rho \mathcal{T}^i), \qquad ^y\mu^i_{\rho SP}(v, \rho) = \mathcal{T}^i + \mathcal{F}^i - \rho \mathcal{T}^i \mathcal{F}^i$$

Applying ISI on EMBPG and EMSPG using parameter ρ results in ρEMSPG^i:

$$\begin{aligned}
{}^y_z\delta^i_{\rho\text{EMSPG}}(l,c,\rho) &= \frac{uU^i + s(1 - \rho S^i)}{u(1 + U^i(1 - \rho(1 - S^i))) + s(1 + S^i(1 - \rho(1 - U^i)))} \\
{}^y\beta^i_{\rho\text{EMSPG}}(v,\rho) &= \frac{f(1 - (1 + \rho)\mathcal{F}^i) - t(1 - (1 + \rho)\mathcal{T}^i)}{t(1 + \mathcal{T}^i(1 - \rho(1 - \mathcal{F}^i))) + f(1 + \mathcal{F}^i(1 - \rho(1 - \mathcal{T}^i)))} \\
{}^y\mu^{i\pm}_{\rho\text{EMSPG}}(v,\rho) &= \rho(t + f)\mathcal{T}^i\mathcal{F}^i, \quad {}^y\mu^{i+}_{\rho\text{EMSPG}}(v,\rho) = t\mathcal{T}^i + f(1 - \rho\mathcal{F}^i), \\
{}^y\mu^{i-}_{\rho\text{EMSPG}}(v,\rho) &= f\mathcal{F}^i + t(1 - \rho\mathcal{T}^i), \\
{}^y\mu^{i}_{\rho\text{EMSPG}}(v,\rho) &= t(1 + \mathcal{T}^i(1 - \rho(1 - \mathcal{F}^i))) + f(1 + \mathcal{F}^i(1 - \rho(1 - \mathcal{T}^i)))
\end{aligned}$$

Applying ISI on BP and EMBPG using parameter σ results in σEMBPG^i:

$$\begin{aligned}
{}^y_z\delta^i_{\sigma\text{EMBPG}}(l,c,\sigma) &= \frac{U^i + \sigma\big[(u - 1)U^i + s\big]}{U^i + S^i + \sigma\big[(u - 1)U^i + (s - 1)S^i + u + s\big]} \\
{}^y\beta^i_{\sigma\text{EMBPG}}(v,\sigma) &= \frac{\mathcal{T}^i - \mathcal{F}^i + \sigma\big[(t - 1)\mathcal{T}^i - (f - 1)\mathcal{F}^i + f - t\big]}{\mathcal{T}^i + \mathcal{F}^i + \sigma\big[(t - 1)\mathcal{T}^i + (f - 1)\mathcal{F}^i + f + t\big]} \\
{}^y\mu^{i\pm}_{\sigma\text{EMBPG}}(v,\sigma) &= 0, \quad {}^y\mu^{i+}_{\sigma\text{EMBPG}}(v,\sigma) = \mathcal{T}^i + \sigma\big[(t - 1)\mathcal{T}^i + f\big], \\
{}^y\mu^{i-}_{\sigma\text{EMBPG}}(v,\sigma) &= \mathcal{F}^i + \sigma\big[(f - 1)\mathcal{F}^i + t\big], \\
{}^y\mu^{i}_{\sigma\text{EMBPG}}(v,\sigma) &= \mathcal{T}^i + \mathcal{F}^i + \sigma\big[(t - 1)\mathcal{T}^i + (f - 1)\mathcal{F}^i + f + t\big]
\end{aligned}$$

Applying ISI on SP and EMSPG using parameter σ results in σEMSPG^i:

$$\begin{aligned}
{}^y_z\delta^i_{\sigma\text{EMSPG}}(l,c,\sigma) &= \frac{U^i(1 - S^i) + \sigma\big[uU^i + (s - U^i)(1 - S^i)\big]}{U^i(1 - S^i) + S^i + \sigma\big[(s + u + 1)U^iS^i + u + s - U^i - S^i\big]} \\
{}^y\beta^i_{\sigma\text{EMSPG}}(v,\sigma) &= \frac{\mathcal{T}^i - \mathcal{F}^i + \sigma\big[\mathcal{F}^i - \mathcal{T}^i + f(1 - 2\mathcal{F}^i) - t(1 - 2\mathcal{T}^i)\big]}{\mathcal{T}^i + \mathcal{F}^i - \mathcal{T}^i\mathcal{F}^i + \sigma\big[-\mathcal{T}^i - \mathcal{F}^i + \mathcal{T}^i\mathcal{F}^i + (t + f)(1 + \mathcal{T}^i\mathcal{F}^i)\big]} \\
{}^y\mu^{i\pm}_{\sigma\text{EMSPG}}(v,\sigma) &= \mathcal{T}^i\mathcal{F}^i + \sigma\big[\mathcal{T}^i\mathcal{F}^i(f + t - 1)\big] \\
{}^y\mu^{i+}_{\sigma\text{EMSPG}}(v,\sigma) &= \mathcal{T}^i(1 - \mathcal{F}^i) + \sigma\big[\mathcal{T}^i(t + \mathcal{F}^i - 1) + f(1 - \mathcal{F}^i)\big] \\
{}^y\mu^{i-}_{\sigma\text{EMSPG}}(v,\sigma) &= \mathcal{F}^i(1 - \mathcal{T}^i) + \sigma\big[\mathcal{F}^i(f + \mathcal{T}^i - 1) + t(1 - \mathcal{T}^i)\big] \\
{}^y\mu^{i}_{\sigma\text{EMSPG}}(v,\sigma) &= \mathcal{T}^i + \mathcal{F}^i - \mathcal{T}^i\mathcal{F}^i + \sigma\big[-\mathcal{T}^i - \mathcal{F}^i + \mathcal{T}^i\mathcal{F}^i + (t + f)(1 + \mathcal{T}^i\mathcal{F}^i)\big]
\end{aligned}$$

In reference to Fig. 1, parameter ρ can be understood as horizontal adaption that controls the *carefulness* of the ρ-heuristics, while parameter σ can be understood as vertical adaption that controls the *convergence* of the σ-heuristics. We derive a heuristic that allows for both adaptations *simultaneously* in the next section.

4   Deriving the L2 Heuristic $\rho\sigma$PMP^i

Given the L1 heuristics σEMBPG^i and σEMSPG^i, we will once more apply ISI using a second interpolation parameter ρ to derive the L2 heuristic $\rho\sigma$PMP^i. Similar to the extension of the notational frame in the previous section, we will extend the notation once again in a similar way to capture the application of ρ.

Definition 11 (L2 Messages, Cavity Values). *Let $\tau_1\tau_2H^i$ be an L2 MP heuristic using interpolation parameters $\tau_1,\tau_2 \in [0,1]$. Let $v \in \mathcal{V}, c \in F$. Let literal $l \in c$ correspond to v. With iteration z in cycle y, we define the disrespect message of $\tau_1\tau_2H^i$ as ${}^y_z\delta^i_H(l,c,\tau_1,\tau_2) \in [0,1]$, and its warning message as*

$$
{}^y_z\omega^i_{\tau_1\tau_2H}(c,v,\tau_1,\tau_2) = \prod_{l\in c\backslash\{v,\bar v\}} {}^y_z\delta^i_{\tau_1\tau_2H}(l,c,\tau_1,\tau_2)
$$

We define the cavity freedom *of l in cavity clause c similar to those given in Definition 3, but with an extended signature to capture the application of τ_1,τ_2. Additionally, these functions rely on ${}^y_z\omega^i_{\tau_1\tau_2H}(c,v,\tau_1,\tau_2)$ instead of ${}^y_z\omega_H(c,v)$.*

The extension of the notation to capture the L2 variable freedom, magnetization and biases is done in a similar way, but we refrain from presenting this here. The definition for an L2 heuristic is similar to the one for L1 heuristics, except for the usage of the extended functions above that capture the application of τ_1,τ_2. Again, we extend the *simplified notation* and write $U^I = {}_{z-1}U^i_{\tau_1\tau_2H}(l,c,\tau_1,\tau_2)$, $S^I = {}^y_zS^i_{\tau_1\tau_2H}(l,c,\tau_1,\tau_2)$, $\mathcal{T}^I = {}^y\mathcal{T}^i_{\tau_1\tau_2H}(v,\tau_1,\tau_2)$, and $\mathcal{F}^I = {}^y\mathcal{F}^i_{\tau_1\tau_2H}(v,\tau,\tau_2)$.

Applying ISI once more using interpolation parameter ρ on the L1 heuristics σEMBPG^i and σEMSPG^i results in the following equations for $\rho\sigma\text{PMP}^i$.

$$
{}^y_z\delta^i_{\rho\sigma\text{PMP}}(l,c,\rho,\sigma) =
$$
$$
\frac{U^I + \sigma\left[(u-1)U^I+s\right] - \rho S^I\left\{U^I+\sigma(s-U^I)\right\}}{(1-\sigma)\left[U^I+S^I-\rho U^IS^I\right]+\sigma\left[s(1+S^I(1-\rho+\rho U^I))+u(1+U^I(1-\rho+\rho S^I))\right]}
$$

$$
{}^y\beta^i_{\rho\sigma\text{PMP}}(v,\rho,\sigma) =
$$
$$
\frac{(1-\sigma)\left[\mathcal{T}^I-\mathcal{F}^I\right]+\sigma\left[f(1-(1+\rho)\mathcal{F}^I)-t(1-(1+\rho)\mathcal{T}^I)\right]}{(1-\sigma)\left[\mathcal{T}^I+\mathcal{F}^I-\rho\mathcal{T}^I\mathcal{F}^I\right]+\sigma\left[f(1+\mathcal{F}^I(1-\rho+\rho\mathcal{T}^I))+t(1+\mathcal{T}^I(1-\rho+\rho\mathcal{F}^I))\right]}
$$

$$
{}^y\mu^{i\pm}_{\rho\sigma\text{PMP}}(v,\rho,\sigma) = \rho\left\{\mathcal{T}^I\mathcal{F}^I + \sigma\left[(t+f-1)\mathcal{T}^I\mathcal{F}^I\right]\right\}
$$

$$
{}^y\mu^{i+}_{\rho\sigma\text{PMP}}(v,\rho,\sigma) = \mathcal{T}^I + \sigma\left[(t-1)\mathcal{T}^I+f\right] - \rho\mathcal{F}^I\left\{\mathcal{T}^I+\sigma(f-\mathcal{T}^I)\right\}
$$

$$
{}^y\mu^{i-}_{\rho\sigma\text{PMP}}(v,\rho,\sigma) = \mathcal{F}^I + \sigma\left[(f-1)\mathcal{F}^I+t\right] - \rho\mathcal{T}^I\left\{\mathcal{F}^I+\sigma(t-\mathcal{F}^I)\right\}
$$

$$
{}^y\mu^i_{\rho\sigma\text{PMP}}(v,\rho,\sigma) = \begin{cases} (1-\sigma)\left[\mathcal{T}^I+\mathcal{F}^I-\rho\mathcal{T}^I\mathcal{F}^I\right] \\ \quad +\sigma\left[f(1+\mathcal{F}^I(1-\rho+\rho\mathcal{T}^I))+t(1+\mathcal{T}^I(1-\rho+\rho\mathcal{F}^I))\right]\end{cases}
$$

This heuristic is therefore able to *simultaneously* adapt its carefulness and convergence behavior with ρ and σ. The next section covers how the MP heuristics mentioned in this paper are connected and how $\rho\sigma\text{PMP}^i$ performs in practice.

5   An MP Hierarchy and Practical Aspects of $\rho\sigma\text{PMP}^i$

The main difference of heuristics from different levels L0 to L2 is their varying application of interpolation parameters $\rho,\sigma \in [0,1]$. Consider the parameter space $(\rho,\sigma) \in [0,1]^2$ as a plane. Each point in this plane characterizes exactly one MP behavior, even though this point might be covered by several heuristics.

Benchmark	S/U	Solver Performance							
		Lingeling		DimetheusJW		DimetheusMP			
		%	PAR10	%	PAR10	%	PAR10	ρ	σ
battleship	S	42.1	11641.9	47.4	10627.2	89.5	2130.1	0.5002	0.0025
battleship	U	77.8	4645.4	55.6	8919.7	55.6	8890.4	0.4463	1.0000
em-all	S	100.0	39.9	75.0	5263.7	100.0	75.4	0.8606	0.1295
em-compact	S	25.0	15228.1	0.0	20000.0	37.5	12728.5	0.9229	0.7946
em-explicit	S	100.0	127.3	75.0	5473.3	100.0	157.1	0.2932	0.2698
em-fbcolors	S	12.5	17509.9	12.5	17723.3	37.5	12662.9	0.0000	0.1731
grid-pebbling	S	100.0	5.7	100.0	16.5	100.0	8.0	0.9931	0.3890
grid-pebbling	U	100.0	4.6	88.9	2226.9	100.0	4.7	0.5884	0.0035
sgen1	S	16.7	16667.4	16.7	16677.7	27.8	14460.9	0.0937	0.6563
k3-r4.200	S	0.0	20000.0	0.0	20000.0	100.0	22.7	0.9929	0.0004
k3-r4.237	S	0.0	20000.0	0.0	20000.0	75.0	5026.8	0.9961	0.0000
k4-r9.000	S	0.0	20000.0	0.0	20000.0	100.0	10.0	0.8592	0.0000
k4-r9.526	S	0.0	20000.0	0.0	20000.0	100.0	5.2	0.9530	0.0000

Fig. 5. The table shows an excerpt of results of the empirical study given in [7]. The benchmark column indicates the types of formulas. The S/U column indicates if the formulas are satisfiable. The % column shows the success rate for these formulas. The PAR10 column gives the penalized avg. runtime. The timeout was set to 2000 seconds.

As an example, the point $p_0 = (\rho, \sigma) = (0, 0)$ is covered by BP. Hence, BP is able to perform the MP behavior characterized by this point. However, p_0 is also covered by ρSP^i when setting $\rho = 0$. This means, that the same MP behavior can also be achieved with ρSP^i. Furthermore, the heuristic $\sigma EMBPG^i$ covers p_0 with $\sigma = 0$, and $\rho\sigma PMP^i$ covers p_0 with $\rho = \sigma = 0$. All together, these four heuristics are able to perform MP in a way commonly known as "Belief Propagation" which is characterized by p_0.

Other points in the parameter plane are, however, covered by different heuristics. For example the point $(0.5, 0)$ is *exclusively* covered by ρSP^i and $\rho\sigma PMP^i$. Furthermore, the point $(0.5, 0.5)$ is *solely* covered by $\rho\sigma PMP^i$ and the MP behavior it characterizes can only be achieved by this heuristic. The ability to cover different types of MP behavior implies a hierarchy of generality for the L0 (least general) to L2 (most general) product-based MP heuristics.

Even though $\rho\sigma PMP^i$ can mimic the behavior of all L0/L1 heuristics, a comparison of the update functions for the heuristics gives the impression that this might result in a serious performance drawback due to the higher arithmetical complexity. Empirical tests show that this is not as immense as initially expected. Even in the most extreme example of computing BP with $\rho\sigma PMP^i$, the additional CPU time required is less than 10%. We believe that this is due to the fact that modern CPUs handle arithmetic operations very efficiently. A separate implementation of the L0 and L1 heuristics in a MID solver is therefore unnecessary. The L2 heuristic can mimic them with a negligible overhead.

Using $\rho\sigma PMP^i$ in a MID solver increases the solver's flexibility and allows for a more sophisticated parameter tuning in order to adapt the solver to a wide range of formulas. Fig. 5 shows results of an empirical study that was conducted on crafted and random formulas from the SAT Challenge 2012 to determine the optimal settings to ρ and σ on these classes. The compared solvers are `Lingeling` [3] for reference, and `Dimetheus` [6] which is a CDCL-based MID solver following `genMID`. `Dimetheus` comes in two versions. The `MP` version uses $\rho\sigma\mathrm{PMP}^i$ in order to guide the decision making as shown in `genMID`. The `JW` version replaces MP with the Jeroslow-Wang heuristic [12]. Both versions rely on VSIDS and phase-saving to advance the search and merely use either MP or JW to *initialize* these two heuristics. To state it clearly: both versions perform similar CDCL search and only differ in the initialization of VSIDS and phase-saving.

Note the initially weak performance of `DimetheusJW`, and how `DimetheusMP` improves this performance all across the board, sometimes being able to outperform even `Lingeling`. It is also important to note that the best performance in the crafted domain requires $\rho, \sigma \notin \{0,1\}$, which implies that $\rho\sigma\mathrm{PMP}^i$ gives a better performance than the L0 and L1 heuristics. The parameter tuning on the random domain clearly prefers the L1 interpolation $\rho\mathrm{SP}^i$. Even though $\rho\sigma\mathrm{PMP}^i$ can gradually enforce convergence, the results show that it is not helpful to do so – which is an insight that has not yet been provided. We refer the reader to [7] for more details, additional results, and a discussion of the empirical study.

6   Conclusions and Future Work

This paper introduced a *notational frame* to present product-based Message Passing heuristics, and explained four basic (L0) variants currently available for SAT: BP, SP, EMBPG, and EMSPG. Furthermore, this paper explained how *MID solvers* work and provided a generic implementation in algorithm `genMID`.

Using this framework, the paper introduced *indirect structural interpolation* (ISI), which is a technique to derive an interpolation of two MP heuristics. Applying ISI on the L0 heuristics resulted in the set of L1 heuristics. These heuristics are strictly more general than the heuristics they interpolate since they cannot only mimic their behavior but are also able to gradually adapt between them. Depending on the interpolated heuristics, this adaption influences either the carefulness of the L1 heuristics to present biases or how much they enforce convergence. A second application of ISI on two of the L1 heuristics resulted in the L2 heuristic $\rho\sigma\mathrm{PMP}^i$ that can *simultaneously* adapt its carefulness and convergence behavior. An important insight is that $\rho\sigma\mathrm{PMP}^i$ can mimic the L0 and L1 heuristics but it is not confined to them, which renders the separate implementation of the L0 and L1 heuristics unnecessary. Furthermore, as shown in the empirical study given in [7], it increases the flexibility of MID solvers and extends the possibilities for parameter tuning, which leads to an increased robustness of CDCL-based MID solvers on various types of CNF formulas.

So far, the integration of MP and CDCL merely affects the way CDCL initializes VSIDS and phase-saving. A tighter integration of MP biases might result in new clause database maintenance schemes as well as new restart schedules, but this remains a matter of future work.

References

1. Aurell, E., Gordon, U., Kirkpatrick, S.: Comparing Beliefs, Surveys and Random Walks. In: Adv. in Neural Information Processing Sys. 17, p. 49. MIT Press (2005)
2. Battaglia, D., Kolář, M., Zecchina, R.: Minimizing energy below the glass thresholds. Physical Review E 70, 036107 (2004)
3. Biere, A.: Lingeling SAT solver, Version ala-b02aala,
   `http://fmv.jku.at/lingeling/lingeling-ala-b02aa1a-121013.tar.gz`
4. Braunstein, A., Mézard, M., Zecchina, R.: Survey Propagation: An Algorithm for Satisfiability. Journal of Rand. Struct. and Algo., 201 (2005)
5. Chavas, J., Furtlehner, C., Mézard, M., Zecchina, R.: Survey-propagation decimation through distributed local computations. Journal of Statistical Mechanics: Theory and Experiment (2005) 1742-5468 / 05 / P11016
6. Gableske, O.: Dimetheus SAT solver, Version 1.6,
   `https://www.gableske.net/dimetheus`
7. Gableske, O., Müelich, S., Diepold, D.: On the Performance of CDCL based Message Passing Inspired Decimation using $\rho\sigma\text{PMP}^i$. Submitted to the Proceedings of the Pragmatics of SAT Workshop, POS 2013 (2013)
8. Hsu, E.I., McIlraith, S.A.: VARSAT: Integrating Novel Probabilistic Inference Techniques with DPLL Search. In: Kullmann, O. (ed.) SAT 2009. LNCS, vol. 5584, pp. 377–390. Springer, Heidelberg (2009)
9. Hsu, E.I., McIlraith, S.A.: Characterizing Propagation Methods for Boolean Satisfiability. In: Biere, A., Gomes, C.P. (eds.) SAT 2006. LNCS, vol. 4121, pp. 325–338. Springer, Heidelberg (2006)
10. Hsu, E.I., Muise, C.J., Beck, J.C., McIlraith, S.A.: Probabilistically Estimating Backbones and Variable Bias: Experimental Overview. In: Stuckey, P.J. (ed.) CP 2008. LNCS, vol. 5202, pp. 613–617. Springer, Heidelberg (2008)
11. Hsu, E., Kitching, M., Bacchus, F., McIlraith, S.: Using Expectation Maximization to Find Likely Assignments for Solving CSPs. In: AAAI 2007, p. 224. AAAI Press (2007)
12. Jeroslow, R., Wang, J.: Solving Propositional Satisfiability Problems. Annals of Mathematics and Artificial Intelligence 1, 167 (1990)
13. Schischang, K., Frey, B., Loeliger, H.: Factor Graphs and the sum-product algorithm. IEEE Trans. Inform. Theory 47, 498 (2002)
14. Maneva, E., Mossel, E., Wainwright, M.J.: A New Look at Survey Propagation and its Generalizations, arXiv:cs/0409012v3, 01 (February 2008)
15. Mézard, M., Parisi, G.: The cavity method at zero temperature. Journal of Statistical Physics 111(1/2) (April 2003)
16. Moskewicz, M., Madigan, C., Zhao, Y., Zhang, L., Malik, S.: Chaff: Engineering an Efficient SAT Solver. In: DAC 2001, p. 530. ACM (2001)
17. Pipatsrisawat, K., Darwiche, A.: A lightweight component caching scheme for Satisfiability Solvers. In: Marques-Silva, J., Sakallah, K.A. (eds.) SAT 2007. LNCS, vol. 4501, pp. 294–299. Springer, Heidelberg (2007)

Improving Glucose for Incremental SAT Solving with Assumptions: Application to MUS Extraction

Gilles Audemard[1,*], Jean-Marie Lagniez[2,**], and Laurent Simon[3]

[1] CRIL, Artois University, France
audemard@cril.fr
[2] FMV, JKU University, Austria
Jean-Marie.Lagniez@jku.at
[3] LRI, University Paris Sud, France
simon@lri.fr

Abstract. Beside the important progresses observed in SAT solving, a number of applications explicitly rely on incremental SAT solving only. In this paper, we focus on refining the incremental SAT Solver Glucose, from the SAT engine perspective, and address a number of unseen problems this new use of SAT solvers opened. By playing on clause database cleaning, assumptions managements and other classical parameters, we show that our approach immediately and significantly improves an intensive assumption-based incremental SAT solving task: Minimal Unsatisfiable Set. We believe this work could bring immediate benefits in a number of other applications relying on incremental SAT.

1 Introduction

The Satisfiability (SAT) status of a propositional formula is one of the most fundamental question in computer science, heavily studied since the 70's, both theoretically and practically. This problem captures the hardness of a large set of difficult – but practically interesting –problems that could arise in many applications. Following the famous proof of SAT NP-Completeness [1], a number of work have reduced many other problems to SAT. In 2001 [2], based on the ideas of Marques-Silva and Sakallah [3], a new era of SAT solvers was born, called "Modern" SAT Solvers [4]. The tremendous progresses observed in SAT had an important impact in many other problems of computer science: the State of the Art in solving a number of NP-Complete problems is now to use a SAT solver engine. The efficiency of data structures used in SAT solvers also allows to work with problems of millions of variables and clauses, which typically correspond to SAT encodings of problems above NP [5,6].

However, since a few years, a new use of SAT solvers has emerged, called "Incremental SAT Solving". Even if this mechanism was already proposed in earlier versions of Minisat [4], its importance has grown in the very last years. It is now the state of the art to rely on this new paradigm in many applications, like in bounded model checking [7], extraction of a Minimal Unsatisfiable Set (MUS) [8], Maximum Satisfiability (MaxSAT) [9] or even inductive verification [10]. In this context, SAT solvers

---

* Partially supported by CNRS and OSEO, under the ISI project "Pajero".
** Partially supported by FWF, NFN Grant S11408-N23 (RiSE), an EC FEDER grant.

M. Järvisalo and A. Van Gelder (Eds.): SAT 2013, LNCS 7962, pp. 309–317, 2013.

are not run on a single, potentially huge, SAT problem. They are rather called thousands times on a number of instances close to each other (with removed/added clauses), which allows to reuse as much informations as possible between successive SAT calls. However, it is clear that, for some applications (when constraints can be removed), the information held by the learned clauses can not be directly reused. For allowing the removal of clauses, it is necessary to add "assumptions" to the SAT solver. Assumptions are literals that are assumed to be true and which are always picked first for decisions during a single run.

In this paper, we take a typical use of incremental SAT solving, heavily relying on assumptions, hoping that our results will be immediately useful for other incremental uses. We base our work on the highly tunable and open source Muser [11] (see also [12,8,13] for MUS extraction). We focus here on the SAT engine and demonstrate how we can improve the result of the overall incremental SAT solving by carefully taking into account the essential ingredients of SAT solvers. In particular, we are interested in pushing to the incremental case the solver glucose [14]. Memory consumption is indeed a strong limitation of SAT solving with assumptions, because each clause may contain hundreds of literals. Being able to remove useless clauses is essential and successfully applying glucose strategies may be crucial.

2   SAT and Incremental SAT

So-called "modern SAT solvers" are based on the *conflict driven clause learning* paradigm (CDCL) [2]. If they were initially introduced as an extension of the DPLL algorithm [15], with a powerful conflict clause learning [3,16] scheme, it is acknowledged that they must be described as a mix between backtrack search and plain resolution engines. They integrate a number of effective techniques including clause learning, highly dynamic variable ordering heuristic [2], polarity heuristic [17], clause deletion strategy [18,14,19] and search restarts [20,21] (see [22] for a detailed survey).

2.1   Incremental SAT Solving with Selectors

In this paper, we focus on improving SAT engines for *incremental SAT Solving* (see [23] for a detailed introduction). In incremental SAT solving, the same solver is ran on a number of instances close to each others and the solver state is memorized between each call. For instance, the final state of variable ordering [2], polarity cache [17] and clause deletion/restarts strategies [18,14,19] can be easily saved for the next call. However, in most of the cases, *i.e.* as soon as initial constraints can be removed, learned clauses can not be directly reused. This is why it is necessary to add "assumptions" to the SAT solver. Assumptions in SAT solvers were already proposed in early versions of Minisat [4] and are daily used in many distinct applications (*s.t.* bounded model checking [7], extraction of a minimal unsatisfiable set [8] ...). A set of assumptions $\mathcal{A}$ is defined as a set of literals that are assumed to be true and which are picked for decisions first, always in the top of the search tree ([23] proposed another way of handling them, which can be easily used with our approach). Then, if during the search, it is needed to flip the assignment of one of these assumptions to false, the problem is unsatisfiable under the initial assumptions.

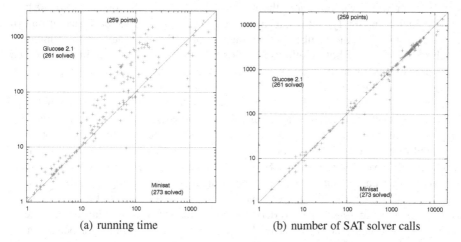

(a) running time   (b) number of SAT solver calls

Fig. 1. Comparison of `Glucose` 2.1 performances against `Minisat` in `Muser`. Figure (Left) us the loglog plot of `Minisat` (x-axis) against `Glucose` (y-axis) in CPU time (seconds). Figure (Right) is the loglog plot of `Minisat` (x-axis) agains `Glucose` (y-axis) in number of calls to the SAT solver by `Muser`.

When an assumption is used for activating/deactivating a clause (one fresh assumption for each clause), it is called a *selector* for this clause: if the literal is false (resp. true) under the current assumptions, then the clause is activated (resp. deactivated). Therefore, since these additional literals appear only positively in the formula, the learned clauses obtained during the search process keep track of all the initial clauses used to produce them. Then, removing the set of clauses which are derived from an initial clause with selector a can be easily performed by simply assuming a to true.

2.2 Using Glucose in Minimal UNSAT Set Extraction Problem

As a typical application of incremental SAT solving with selectors, we focus on the *minimal unsatisfiable set* (MUS) extraction problem [24]. In this approach, a unique selector is added to *each clause* and the SAT solver is incrementally called with most of these additional literals as assumptions. This application is quite typical and challenging for incremental SAT solvers: the number of assumptions is usually quite large (equal to the number of clauses in the formula) and the SAT solver is called many times. We rely our work on the highly tunable and open source `Muser` [11] but we focus only on the SAT engine. Let us precise that we use `Muser` with default options, *i.e.* the hybrid algorithm is used (essentially deletion-based) with clause-set refinement and model rotation (see [11] for more details). Note also that we use the 300 benchmarks from the MUS track of the SAT 2011 competition. Experiments are done on Intel XEON X5550 quad-cores 2.66 GHz with 32Gb RAM. CPU time limit is 2400s.

If we want to be able to efficiently reuse learned clauses, it is essential to be able to distinguish which clauses may be useful for futures calls. In this context, it may be a good choice to use `Glucose` [14], a variant of `Minisat` [4] based on an efficient scoring mechanism of clauses, called Literal Block Distance (LBD). So, as a starting point for our work, we compared the integration of Glucose against Minisat in Muser. Our surprisingly bad results are summarized Fig. 2.2 (for each scatter plot of the paper, we provide the number of benchmarks solved by each method, the number of benchmarks solved by both methods (259 for Fig 1(a)) and the less points are located in the area of a given method, the best this last one is). The two figures are quite interesting, however. On one side, the CPU time comparison, Figure 1(a), is in clear favor of Minisat. On the other side, Figure 1(b) shows this comparison from a number of SAT calls point of view. On this last figure, the comparison is not so clear, and some interesting points show that Glucose can sometimes require less SAT calls. Moreover, we found that, over the 259 run both solved by Glucose and Minisat, 103 runs of Glucose have fewer calls than Minisat, 68 runs of Glucose have more calls than Minisat and 88 runs have the same number of calls. Using Glucose instead of Minisat seems a good option if we can fix the CPU time problem. Intuitively, we think that Glucose is able to derive proofs with fewer initial clauses, thus helping Muser to quickly converge. Together with the idea of proposing a better clause database management, this was our initial motivations for this work.

How can we explain such a disappointing result? Glucose is updating the scores of clauses during unit-propagation. Thus, when clauses tend to have thousands of literals, it adds a prohibitive penalty. However, it is also important to notice that, on hard MUS problems, Minisat showed (not reported here) some limitations for scaling-up, mainly due to memory consumption problems (6 Memory out). On hard MUS problems, one may have to handle hundreds of hard UNSAT calls, and the ability of Glucose to handle long runs with many learned clauses may be crucial here. In the following, we will show how it is possible to adapt Glucose mechanisms to incremental SAT solving.

3 Improving Incremental SAT Engines

3.1 Assumptions and LBD

In Minisat, each assumption has its own decision level, except when it was propagated by other assumptions assignments. Thus, in most of the cases, the LBD score will be dominated by the number of assumptions in the clause. For most of the learned clauses, this value will be quickly approximating its size. Given the fact that, in addition, the LBD score is very discriminating (clauses of LBD $n + 1$ are significantly less important than clauses of LBD n), we must get rid of assumption literals during LBD scoring. We propose to adapt the LBD by simply skipping assumption literals during its computation. This new LBD scoring will be called hereafter "New LBD".

Tab. 1 shows some statistics about the above remark. On four representative instances, we detail the initial LBD score obtained and the new LBD score as previously described. As we can see, it is clear that using the initial LBD score is not meaningful in the context of SAT solving with selectors. The second part of Tab. 1 shows the same statistics with the New LBD definition. As one may observe, the CPU time is significantly improved. Moreover, LBD scores are smaller and are no more related to the size of learned clauses (when counting assumptions). Furthermore, average size of learned

Table 1. For some characteristics instances, we provide the number of clauses and, for each LBD score, we also report the time to compute the MUS, the maximum and average size, and the LBD of learned clauses

Instance	#C	LBD					New LBD				
			size		LBD			size		LBD	
		time	avg	max	avg	max	time	avg	max	avg	max
fdmus_b21_96	8541	29	1145	5980	1095	5945	11	972	6391	8	71
longmult6	8853	46	694	3104	672	3013	14	627	2997	11	61
dump_vc950	360419	110	522	36309	498	35873	67	1048	36491	8	307
g7n	15110	190	1098	16338	1049	16268	75	1729	17840	27	160

clauses may be larger for the new LBD score than for the initial one: good LBD clauses are not necessarily short ones.

However, as we can see Fig. 2, even if we increase Glucose performances, this new version is very close to Minisat (in number of solved instances), but slower. At this stage, such results are relatively disappointing: Glucose is supposed to be more efficient than Minisat in non incremental SAT solving. In order to improve its overall performances, we propose hereafter some very important improvements.

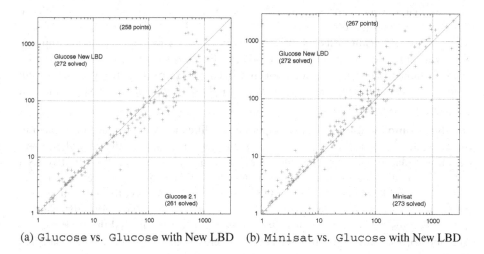

(a) Glucose vs. Glucose with New LBD   (b) Minisat vs. Glucose with New LBD

Fig. 2. Comparison of Glucose 2.1 performances against Glucose with a new LBD scoring in Muser. Fig. 2(a) is the loglog plot of Glucose 2.1 (x-axis) against Glucose with New LBD (y-axis) in CPU time (seconds). Fig. 2(b) is the loglog plot of Minisat (x-axis) against Glucose with New LBD (y-axis) in CPU time (seconds).

3.2 Improving Performances

As pointed out Tab 1, the produced learned clauses can be quite large. For example, the average size of learned clauses on instance g7n is 1729 (with clauses with more than 10,000 literals). Thus, even a simple operation like clause traversal may quickly induce a prohibitive overhead when applied with selectors. This is very problematic as fundamental operations of SAT solvers are based on clauses traversals, *i.e.* BCP, updates of LBD scores or even simplification of clauses database (when removing satisfied clauses). Considering this operation as one of the main bottleneck for SAT solving with selectors, we propose several improvements specialized for these operations.

Efficient Traversal of Clauses with Many Selectors. As previously noticed, in most of the cases, learned clauses will be dominated by selector literals but, as they are not taken into account in the new LBD scoring, we do not have to check them during clauses traversals. Moreover, since these literals can be used as watchers during the BCP process, it is impossible to split the set of clause's literals into two independent sets (selectors/non selectors): we must be able to visit selector literals when needed. However, it is possible to store, in each learned clause, the number of initial literals and the total size of the clause (obviously, the number of selectors is redundant). In addition, we propose to push all the selectors at the end of the learned clauses. As a first, immediate, impact of this new clause arrangement, we can hope to speed up LBD score updates: we can stop the traversal as soon as all initial literals have been checked (number of initial literals in the clause is known).

Improving BCP. If the above data structure modification allows to speed up LBD updates, the BCP engine also needs to traverse clauses when looking for new watchers. Suppose that, during the propagation of an assumption, the new watcher a for a clause c is chosen among the other selectors of the clause. Then, if a is assigned to false, c will be traversed again. This can be easily avoided: when propagating an assumption, we always traverse the entire clause in order to find a new watcher which is true (the clause is satisfied) or which is not a selector. Because Glucose is firing very frequent restarts, we also limit any restart to backtrack only until the decision level of the first non-selector decision level.

Database Simplification. The last important modification we introduced is related to the removal of satisfied learned clauses. This is very important in the case of Muser: each time a clause is known to be out of a MUS, its associated selector is definitively set to true. Then, all learned clauses where it appears will be satisfied. However, scanning all literals of all learned clauses to find these true literals can break the potential benefit of this technique. In the new version of Glucose, we limit the search for satisfied learned clauses to the two watched literals only. Since we changed the BCP process (see above), we expect that this will be sufficient to find and remove most satisfied clauses.

As a last small modification of Glucose, we keep current indicators for clause database cleanings and restarts from one run to the other, allowing the behavior of the solver to reflect the very high frequency of runs in most incremental SAT solving.

3.3   Overall Evaluation

We implemented all the above described techniques and called it GlucoseInc. Fig. 3 compares GlucoseInc against Minisat and Glucose. Results are clear: this new version outperforms the other methods (see also Fig. 4).

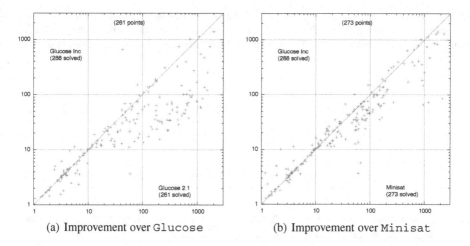

(a) Improvement over Glucose   (b) Improvement over Minisat

Fig. 3. Comparison of Glucose 2.1 (resp. Minisat) performances against GlucoseInc in Muser. Fig. 3(a) (resp. 3(b)) use the loglog plot of Glucose (resp. Minisat), x-axis, against GlucoseInc (y-axis) in CPU time (seconds) in Muser.

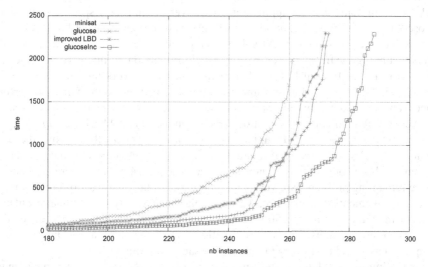

Fig. 4. Cactus plot. x-axis represents the number of instances and y-axis represents the time needed to solve them if they were ran in parallel

As we wrote in section 2.2, our goal was to improve performances of incremental SAT solving, and thus, performances for each call to the SAT solver. Tab. 2 shows that this goal is reached. It provides for Minisat, Glucose and GlucoseInc and for the same previous 4 instances, the number of SAT calls and the average time needed for each of them. Interestingly enough, one can notice that for each instance the total time needed to extract a MUS is improved, even if more SAT calls may be necessary. This is unfortunately a general remark: by dropping the quality of the LBD measure, we

Table 2. For representative instances, we provide the number of clauses, and for each incremental SAT engine, we provide the number of calls to the SAT solver and the average time needed for each call. For GlucoseInc, we also provide the total time (for other methods, this is done in Tab. 1)

Instance	#C	Minisat		Glucose		GlucoseInc		
		#SAT calls	avg	#SAT calls	avg	time	#SAT calls	avg
fdmus_b21_96	8541	2103	0.009	2134	0.02	11	2153	0.004
longmult6	8853	706	0.01	1027	0.03	13	748	0.01
dump_vc950	360419	7	135	11	11.5	65	9	7.2
g7n	70492	4791	0.02	4393	0.08	67	4779	0.01

loose the first observation we made. Now, using Glucose instead of Minisat does not show any improvements in the number of SAT calls by Muser.

4   Conclusion

One of the reasons for the success of SAT solvers is certainly their ability to be used as Black Boxes in many distinct applications. However, when a very specific new use is proposed, it may be important to adapt SAT solvers technologies accordingly, directly at the engine level. Incremental SAT solving is typically a new field of research for SAT solvers technologies. In this paper, we proposed to focus on an intensive incremental SAT solving task, Minimal UNSAT Set extraction, but by working only on the SAT engine level. This study led to an impressive improvement in MUS extraction performances, and we believe that our approach can also lead to substantial – and direct – improvements in many other applications relying on incremental SAT solving.

By playing only on the SAT engine, we were able to drastically improve the performances of glucose on MUS extraction. It has to be pointed out that the set of problems we based our study on is now solved at 96% (on 2400s). So, there is a chance that the remaining problems are very hard, and may not be a good set of problems for further improvements. For instance, we were able to solve only 4 additional problems by increasing the CPU time to 15000 seconds on the unsolved problems. Moreover, the efficient handling of clauses by glucose allows it to have no Memory Out problems where Minisat had 6 Memory Out under the same conditions.

This preliminary work has a lot of promising perspectives. For example, we plan to study dependencies beween selectors and to take them into account in order to improve performances. An other interesting future work is related to the adaptation of the SAT solver (in term of heuristics, restarts, ...) with respect to previous calls.

References

1. Cook, S.A.: The complexity of theorem-proving procedures. In: Proc. STOC 1971, pp. 151–158 ACM (1971)
2. Moskewicz, M., Conor, C., Zhao, Y., Zhang, L., Malik, S.: Chaff: Engineering an Efficient SAT Solver. In: Proc. DAC 2001 (2001)

3. Silva, J.P.M., Sakallah, K.: Grasp – a new search algorithm for satisfiability. In: Proc. CAD 1996, pp. 220–227 (1996)
4. Eén, N., Sörensson, N.: An Extensible SAT-solver. In: Giunchiglia, E., Tacchella, A. (eds.) SAT 2003. LNCS, vol. 2919, pp. 502–518. Springer, Heidelberg (2004)
5. Biere, A., Cimatti, A., Clarke, E.M., Fujita, M., Zhu, Y.: Symbolic model checking using sat procedures instead of bdds. In: Proc. DAC 1999, pp. 317–320 (1999)
6. Velev, M.N., Bryant, R.E.: Effective use of boolean satisfiability procedures in the formal verification of superscalar and vliw microprocessors. Journal of Symbolic Computation 35(2), 73–106 (2003)
7. Eén, N., Sörensson, N.: Temporal induction by incremental SAT solving. Electronic Notes in Theoretical Computer Science 89(4), 543–560 (2003)
8. Nadel, A.: Boosting minimal unsatisfiable core extraction. In: Proc. FMCAD 2010, pp. 221–229 (2010)
9. Fu, Z., Malik, S.: On solving the partial MAX-SAT problem. In: Biere, A., Gomes, C.P. (eds.) SAT 2006. LNCS, vol. 4121, pp. 252–265. Springer, Heidelberg (2006)
10. Bradley, A.R.: IC3 and beyond: Incremental, inductive verification. In: Madhusudan, P., Seshia, S.A. (eds.) CAV 2012. LNCS, vol. 7358, p. 4. Springer, Heidelberg (2012)
11. Belov, A., Marques-Silva, J.: Accelerating MUS extraction with recursive model rotation. In: Proc. FMCAD 2011, pp. 37–40 (2011)
12. Grégoire, É., Mazure, B., Piette, C.: Extracting muses. In: Proc. ECAI 2006. Frontiers in Artificial Intel. and Applications, vol. 141, pp. 387–391. IOS Press (2006)
13. Ryvchin, V., Strichman, O.: Faster extraction of high-level minimal unsatisfiable cores. In: Sakallah, K.A., Simon, L. (eds.) SAT 2011. LNCS, vol. 6695, pp. 174–187. Springer, Heidelberg (2011)
14. Audemard, G., Simon, L.: Predicting learnt clauses quality in modern SAT solvers. In: Proc. IJCAI 2009, pp. 399–404 (2009)
15. Davis, M., Logemann, G., Loveland, D.: A machine program for theorem-proving. Commun. ACM (1962)
16. Zhang, L., Madigan, C., Moskewicz, M., Malik, S.: Efficient conflict driven learning in a boolean satisfiability solver. In: Proc. CAD 2001, pp. 279–285 (2001)
17. Pipatsrisawat, K., Darwiche, A.: A lightweight component caching scheme for satisfiability solvers. In: Marques-Silva, J., Sakallah, K.A. (eds.) SAT 2007. LNCS, vol. 4501, pp. 294–299. Springer, Heidelberg (2007)
18. Goldberg, E.I., Novikov, Y.: BerkMin: A fast and robust SAT-solver. In: Proc. DATE 2002, pp. 142–149 (2002)
19. Audemard, G., Lagniez, J.-M., Mazure, B., Saïs, L.: On freezing and reactivating learnt clauses. In: Sakallah, K.A., Simon, L. (eds.) SAT 2011. LNCS, vol. 6695, pp. 188–200. Springer, Heidelberg (2011)
20. Biere, A.: Adaptive restart strategies for conflict driven SAT solvers. In: Kleine Büning, H., Zhao, X. (eds.) SAT 2008. LNCS, vol. 4996, pp. 28–33. Springer, Heidelberg (2008)
21. Audemard, G., Simon, L.: Refining restarts strategies for SAT and UNSAT. In: Milano, M. (ed.) CP 2012. LNCS, vol. 7514, pp. 118–126. Springer, Heidelberg (2012)
22. Marques-Silva, J., Lynce, I., Malik, S.: 4. In: Conflict-Driven Clause Learning SAT Solvers. Frontiers in Artificial Intelligence and Applications, vol. 185, p. 980. IOS Press (2009)
23. Nadel, A., Ryvchin, V.: Efficient SAT solving under assumptions. In: Cimatti, A., Sebastiani, R. (eds.) SAT 2012. LNCS, vol. 7317, pp. 242–255. Springer, Heidelberg (2012)
24. Oh, Y., Mneimneh, M., Andraus, Z., Sakallah, K., Markov, I.: AMUSE: a minimally-unsatisfiable subformula extractor. In: Design Automation Conference, pp. 518–523 (2004)

A SAT Approach to Clique-Width

Marijn J.H. Heule[1,*] and Stefan Szeider[2,**]

[1] Department of Computer Sciences, The University of Texas at Austin, USA
[2] Institute of Information Systems, Vienna University of Technology, Vienna, Austria

Abstract. Clique-width is a graph invariant that has been widely studied in combinatorics and computer science. However, computing the clique-width of a graph is an intricate problem, the exact clique-width is not known even for very small graphs. We present a new method for computing the clique-width of graphs based on an encoding to propositional satisfiability (SAT) which is then evaluated by a SAT solver. Our encoding is based on a reformulation of clique-width in terms of partitions that utilizes an efficient encoding of cardinality constraints. Our SAT-based method is the first to discover the exact clique-width of various small graphs, including famous graphs from the literature as well as random graphs of various density. With our method we determined the smallest graphs that require a small pre-described clique-width.

1 Introduction

Clique-width is a fundamental graph invariant that has been widely studied in combinatorics and computer science. Clique-width measures in a certain sense the "complexity" of a graph. It is defined via a graph construction process involving four operations where only a limited number of vertex labels are available; vertices that share the same label at a certain point of the construction process must be treated uniformly in subsequent steps. This graph composition mechanism was first considered by Courcelle, Engelfriet, and Rozenberg [10,11] and has since then been an important topic in combinatorics and computer science.

Graphs of small clique-width have advantageous algorithmic properties. Algorithmic meta-theorems show that large classes of NP-hard optimization problems and #P-hard counting problems can be solved in *linear time* on classes of graphs of bounded clique-width [7,8]. Similar results hold for the graph invariant *treewidth*, however, clique-width is more general in the sense that graphs of small treewidth also have small clique-width, but there are graphs of small clique-width but arbitrarily high treewidth [9,6]. Unlike treewidth, dense graphs (e.g., cliques) can also have small clique-width.

All these algorithms for graphs of small clique-width require that a certificate for the graph having small clique-width is provided. However, it seems that computing the certificate, or just deciding whether the clique-width of a graph

---

* Research supported in part by the National Science Foundation under grant CNS-0910913 and DARPA contract number N66001-10-2-4087.
** Research supported by the ERC, grant reference 239962 (COMPLEX REASON).

M. Järvisalo and A. Van Gelder (Eds.): SAT 2013, LNCS 7962, pp. 318–334, 2013.

is bounded by a given number, is a very intricate combinatorial problem. More precisely, given a graph G and an integer k, deciding whether the clique-width of G is at most k is NP-complete [16]. Even worse, the clique-width of a graph with n vertices of degree greater than 2 cannot be approximated by a polynomial-time algorithm with an absolute error guarantee of n^ϵ unless P = NP, where $0 \leq \epsilon < 1$ [16]. In fact, it is even unknown whether graphs of clique-width at most 4 can be recognized in polynomial time [5]. There are approximation algorithms with an exponential error that, for fixed k, compute $f(k)$-expressions for graphs of clique-width at most k in polynomial time (where $f(k) = (2^{3k+2}-1)$ by [30], and $f(k) = 8^k - 1$ by [29]).

Because of this intricacy of this graph invariant, the exact clique-width is not known even for very small graphs.

Clique-width via SAT. We present a new method for determining the clique-width based on a sophisticated SAT encoding which entails the following ideas:

1. *Reformulation.* The conventional construction method for determining the clique-width of a graph consists of many steps. In the worst case, the number of steps is quadratic in the number of vertices. Translating this construction method into SAT would result in large instances, even for small graphs. We reformulated the problem in such a way that the number of steps is less than the number of vertices. The alternative construction method allows us to compute the clique-width of much larger graphs.

2. *Representative encoding.* Applying the frequently-used direct encoding [35] on the reformulation results in instances that have no arc consistency [18], i.e., unit propagation may find conflicts much later than required. We developed the representative encoding that is compact and realizes arc consistency.

Experimental Results. The implementation of our method allows us for the first time to determine the exact clique-width of various graphs, including famous graphs known from the literature, as well as random graphs of various density.

1. *Clique-width of small Random Graphs.* We determined experimentally how the clique-width of random graphs depends on the density. The clique-width is small for dense and sparse graphs and reaches its maximum for edge-probability 0.5. The larger n, the steeper the increase towards 0.5. These results complement the asymptotic results of Lee et al. [27].

2. *Smallest Graphs of Certain Clique-width.* In general it is not known how many vertices are required to form a graph of a certain clique-width. We provide these numbers for clique-width $k \in \{1,\dots,7\}$. In fact, we could compute the total number of connected graphs (modulo isomorphism) with a certain clique-width with up to 10 vertices. For instance, there are only 7 connected graphs with 8 vertices and clique-width 5 (modulo isomorphism), and no graphs with 9 vertices and clique-width 6. There are 68 graphs with 10 vertices and clique-width 6. The smallest one has 18 edges.

3. *Clique-width of Famous Named Graphs.* Over the last 50 years, researchers in graph theory have considered a large number of special graphs. These special graphs have been used as counterexamples for conjectures or for showing the tightness of combinatorial results. We considered several prominent graphs from the literature and computed their exact clique-width. These results may be of interest for people working in combinatorics and graph theory.

Related Work. We are not aware of any implemented algorithms that compute the clique-width exactly or heuristically. However, algorithms have been implemented that compute upper bounds on other width-based graph invariants, including *treewidth* [14,19,26], *branchwidth* [33], *Boolean-width* [24], and *rank-width* [2]. Samer and Veith [31] proposed a SAT encoding for the exact computation of treewidth. Boolean-width and rank-width can be used to approximate clique-width, however, the error can be exponential in the clique-width; in contrast, treewidth and branchwidth can be arbitrarily far from the clique-width, hence the approximation error is unbounded [4].

Our SAT encoding is based on a new characterization of clique-width that is based on partitions instead of labels. A similar partition-based characterization of clique-width, has been proposed by Heggernes et al. [23]. There are two main differences to our reformulation. Firstly, our characterization of clique-width uses three individual properties that can be easily expressed by clauses. Secondly, our characterization admits the "parallel" processing of several parts of the graph that are later joined together.

Full Version. Because of space constraints some proofs have been omitted or shortened. Detailed proofs can be found in the full version, available at arxiv.org/abs/1304.5498.

2   Preliminaries

2.1   Formulas and Satisfiability

We consider propositional formulas in Conjunctive Normal Form (*CNF formulas*, for short), which are conjunctions of clauses, where a clause is a disjunction of literals, and a literal is a propositional variable or a negated propositional variables. A CNF formula is *satisfiable* if its variables can be assigned true or false, such that each clause contains either a variable set to true or a negated variable set to false. The satisfiability problem (SAT) asks whether a given formula is satisfiable.

2.2   Graphs and Clique-Width

All graphs considered are finite, undirected, and without self-loops. We denote a graph G by an ordered pair $(V(G), E(G))$ of its set of vertices and its set of edges, respectively. An edge between vertices u and v is denoted uv or equivalently vu. For basic terminology on graphs we refer to a standard text book [13].

Let k be a positive integer. A k-*graph* is a graph whose vertices are labeled by integers from $\{1, \ldots, k\}$. We consider an arbitrary graph as a k-graph with all vertices labeled by 1. We call the k-graph consisting of exactly one vertex v (say, labeled by i) an *initial* k-*graph* and denote it by $i(v)$. The *clique-width* of a graph G is the smallest integer k such that G can be constructed from initial k-graphs by means of repeated application of the following three operations.

1. Disjoint union (denoted by $\oplus$);
2. Relabeling: changing all labels i to j (denoted by $\rho_{i \to j}$);
3. Edge insertion: connecting all vertices labeled by i with all vertices labeled by $j, i \neq j$ (denoted by $\eta_{i,j}$ or $\eta_{j,i}$); already existing edges are not doubled.

A construction of a k-graph using the above operations can be represented by an algebraic term composed of $\oplus$, $\rho_{i \to j}$, and $\eta_{i,j}$, $(i, j \in \{1, \ldots, k\}$, and $i \neq j)$. Such a term is called a k-*expression* defining G. Thus, the clique-width of a graph G is the smallest integer k such that G can be defined by a k-expression.

Example 1. The graph $P_4 = (\{a, b, c, d\}, \{ab, bc, cd\})$ is defined by the 3-expression $\eta_{2,3}(\rho_{2 \to 1}(\eta_{2,3}(\eta_{1,2}(1(a) \oplus 2(b)) \oplus 3(c))) \oplus 2(d))$. Hence cwd$(P_4) \leq 3$.   ⊣

2.3   Partitions

As partitions play an important role in our reformulation of clique-width, we recall some basic terminology. A *partition* of a set S is a set P of nonempty subsets of S such that any two sets in P are disjoint and S is the union of all sets in P. The elements of P are called *equivalence classes*. Let P, P' be partitions of S. Then P' is a *refinement* of P if for any two elements $x, y \in S$ that are in the same equivalence class of P' are also in the same equivalence class of P (this entails the case $P = P'$).

3   A Reformulation of Clique-Width without Labels

Initially, we developed a SAT encoding of clique-width based on k-expressions. Even after several optimization steps, this encoding was only able to determine the clique-width of graphs consisting of at most 8 vertices. We therefore developed a new encoding based on a reformulation of clique-width which does not use k-expressions. In this section we explain this reformulation, in the next section we will discuss how it can be encoded into SAT efficiently.

Consider a finite set V, the *universe*. A *template* T consists of two partitions cmp(T) and grp(T) of V. We call the equivalence classes in cmp(T) the *components* of T and the equivalence classes in grp(T) the *groups* of T. For some intuition about these concepts, imagine that components represent induced subgraphs and that groups represent sets of vertices in some component with the same label in a k-expression. A *derivation* of length t is a finite sequence $\mathcal{D} = (T_0, \ldots, T_t)$ satisfying the following conditions.

D1. $|\mathrm{cmp}(T_0)| = |V|$ and $|\mathrm{cmp}(T_t)| = 1$.
D2. $\mathrm{grp}(T_i)$ is a refinement of $\mathrm{cmp}(T_i)$, $0 \le i \le t$.
D3. $\mathrm{cmp}(T_{i-1})$ is a refinement of $\mathrm{cmp}(T_i)$, $1 \le i \le t$.
D4. $\mathrm{grp}(T_{i-1})$ is a refinement of $\mathrm{grp}(T_i)$, $1 \le i \le t$.

We would like to note that D1 and D2 together imply that $|\mathrm{grp}(T_0)| = |V|$. Thus, in the first template T_0 all equivalence classes (groups and components) are singletons, and when we progress through the derivation, some of these sets are merged, until all components are merged into a single component in the last template T_t.

The *width* of a component $C \in \mathrm{cmp}(T)$ is the number of groups $g \in \mathrm{grp}(T)$ such that $g \subseteq C$. The width of a template is the maximum width over its components, and the width of a derivation is the maximum width over its templates. A *k-derivation.* is a derivation of width at most k. A derivation $\mathcal{D} = (T_0, \dots, T_t)$ is a derivation *of* a graph $G = (V, E)$ if V is the universe of the derivation and the following three conditions hold for all $1 \le i \le t$.

Edge Property: For any two vertices $u, v \in V$ such that $uv \in E$, if u, v are in the same group in T_i, then u, v are in the same component in T_{i-1}.
Neighborhood Property: For any three vertices $u, v, w \in V$ such that $uv \in E$ and $uw \notin E$, if v, w are in the same group in T_i, then u, v are in the same component in T_{i-1}.
Path Property: For any four vertices $u, v, w, x \in V$, such that $uv, uw, vx \in E$ and $wx \notin E$, if u, x are in the same group in T_i and v, w are in the same group in T_i, then u, v are in the same component in T_{i-1}.

The neighborhood property and the path property could be merged into a single property if we do not insist that all mentioned vertices are distinct. However, two separate properties provide a more compact SAT encoding.

The following example illustrates that a derivation can define more than one graph, in contrast to a k-expression, which defines exactly one graph.

Example 2. Consider the derivation $\mathcal{D} = (T_0, \dots, T_3)$ with universe $V = \{a, b, c, d\}$ and

$$\mathrm{cmp}(T_0) = \{\{a\}, \{b\}, \{c\}, \{d\}\}, \quad \mathrm{grp}(T_0) = \{\{a\}, \{b\}, \{c\}, \{d\}\},$$
$$\mathrm{cmp}(T_1) = \{\{a, b\}, \{c\}, \{d\}\}, \quad \mathrm{grp}(T_1) = \{\{a\}, \{b\}, \{c\}, \{d\}\},$$
$$\mathrm{cmp}(T_2) = \{\{a, b, c\}, \{d\}\}, \quad \mathrm{grp}(T_2) = \{\{a\}, \{b\}, \{c\}, \{d\}\},$$
$$\mathrm{cmp}(T_3) = \{\{a, b, c, d\}\}, \quad \mathrm{grp}(T_3) = \{\{a, b\}, \{c\}, \{d\}\}.$$

The width of $\mathcal{D}$ is 3. Consider the graph $G = (V, \{ab, ad, bc, bd\})$. To see that $\mathcal{D}$ is a 3-derivation of G, we need to check the edge, neighborhood, and path properties. We observe that a, b are the only two vertices such that $ab \in E(G)$ and both vertices appear in the same group of some T_i (here, we have $i = 3$). To check the edge property, we only need to verify that a, b are in the same component of T_2, which is true. For the neighborhood property, the only relevant choice of three vertices is a, b, c ($bc \in E(G)$, $ac \notin E(G)$, and a, b in a group of T_3). The neighborhood property requires that b, c are in the same component in T_2, which is the case. The path property is satisfied since there is no template in which two pairs of vertices belong to the same group, respectively.

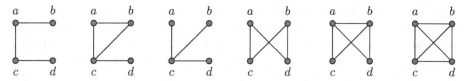

Fig. 1. All connected graphs with four vertices (up to isomorphism). The 3-derivation of Example 2 defines all six graphs. The clique-width for all, but the first graph is 2.

Similarly we can verify that $\mathcal{D}$ is a derivation of the graph $G' = (V, \{ab, bc, cd\})$. In fact, for all connected graphs with four vertices, there exists an isomorphic graph that is defined by $\mathcal{D}$ (see Figure 1). However, $\mathcal{D}$ is not a derivation of the graph $G'' = (V, \{ab, ac, bd, cd\})$ since the neighborhood property is violated: $bd \in E(G'')$ and $ad \notin E(G'')$, a, b belong to the same group in T_3, while a, d do not belong to the same component in T_2. ⊣

We call a derivation $(T_0, \ldots, T_t)$ to be *strict* if $|\text{cmp}(T_{i-1})| > |\text{cmp}(T_i)|$ holds for all $1 \leq i \leq t$. It is easy to see that if two consecutive templates in a derivation of a graph G have the same components, then one of the two templates can be omitted, and we still have a derivation of G. This yields the next lemma, whose detailed proof can be found in the full version of this paper.

Lemma 1. *If G has a k-derivation, it has a strict k-derivation.*

Lemma 2. *Every strict k-derivation of a graph with n vertices has length at most $n - 1$.*

Proof. Let $(T_0, \ldots, T_t)$ be a strict k-derivation of a graph with n vertices. Since $|\text{cmp}(T_0)| = n$ and $|\text{cmp}(T_0)| = 1$, it follows that $t \leq n - 1$. □

In the proofs of the next two lemmas we need the following concept of a k-*expression tree*, which is the parse tree of a k-expression equipped with some additional information. Let ϕ be a k-expression for a graph $G = (V, E)$. Let Q be the parse tree of ϕ with root r. Consider a node x of Q and let ϕ_x be the subexpression of ϕ whose parse tree is the subtree of Q rooted at x. Then x is labeled with the k-graph G_x constructed by the k-expression ϕ_x. Thus the leaves of Q are labeled with initial k-graphs and the root r is labeled with a labeled version of G. One $\oplus$-node of the parse tree can represent several directly subsequent $\oplus$-operations (e.g., the operation $(x \oplus y) \oplus z$ can be represented by a single node with three children). Evidently, k-expressions and their k-expression trees can be effectively transformed into each other.

We introduce some additional terminology on k-expression trees. We call a non-leaf node of Q an $\oplus$-node, η-node, or ρ-node, according to the operation it represents. We define the *rank* $R(x)$ of node x of Q as the largest number of $\oplus$-nodes that appear on a path from a leaf of Q to x. Hence leaves have rank 0. We denote the set of nodes of Q of rank i by $V_i(Q)$.

Lemma 3. *From a k-expression of a graph G we can obtain a k-derivation of G in polynomial time.*

Proof. Let ϕ be a k-expression of $G = (V, E)$ and let Q be the corresponding k-expression tree. We let $t := R(r)$ and define a derivation $\mathcal{D} = (T_0, \ldots, T_t)$ by setting $\mathrm{cmp}(T_i) = \{ V(G_x)) : x \in V_i(Q) \}$ and $\mathrm{grp}(T_i) = \bigcup_{x \in V_i(Q)} \mathrm{grp}(G_x)$ where $\mathrm{grp}(G_x)$ denotes the partition of $V(G_x)$ into sets of vertices that have the same label. By construction, $\mathcal{D}$ is a derivation with universe V. Furthermore, since ϕ is a k-expression, $|\mathrm{grp}(G_x)| \leq k$ for all nodes x of Q. Hence $\mathcal{D}$ is a k-derivation. It is not difficult to verify that $\mathcal{D}$ is a k-derivation of G by checking the edge, neighborhood, and path properties. Due to space limitations, this part of the proof is only in the full version of this paper. The above procedure for generating the k-derivation can clearly be carried out in polynomial time. $\quad\square$

Example 3. Consider the 3-expression ϕ for the graph P_4 of Example 1. Applying the procedure described in the proof of Lemma 3 we obtain the 3-derivation $\mathcal{D}$ of Example 2. $\quad\dashv$

Lemma 4. *From a k-derivation of a graph G we can obtain a k-expression of G in polynomial time.*

Proof. Due to space restrictions we only sketch the proof; a detailed proof can be found in the full version of the paper. Given a k-derivation $\mathcal{D} = (T_0, \ldots, T_t)$ of a graph $G = (V, E)$, we first construct a k-expression $\phi_\oplus$ that only contains $\oplus$-operations and initial graphs with label 1. These $\oplus$-operations reflect how components are merged along the derivation $\mathcal{D}$. Next we insert ρ-operations directly before $\oplus$-operations and obtain a k-expression $\phi_{\oplus,\rho}$. These $\oplus$-operations reflect how groups are merged along the derivation $\mathcal{D}$. In a last step we insert η-operations. For each edge $uv \in E$ we take the smallest subexpression of $\phi_{\oplus,\rho}$ that defines a graph containing both u and v (the outermost operation of this subexpression is clearly $\oplus$). Let a, b be the labels of u, v in this graph, respectively. It follows from the edge property that $a \neq b$. Right after this subexpression we insert the operation $\eta_{a,b}$. It follows from the neighborhood and path properties that this η-operation does not insert any edges not contained in E. $\quad\square$

We note that we could have saved some ρ-operations in the proof of Lemma 4. In particular the k-expression produced may contain ρ-operations where the number of different labels before and after the application of the ρ-operation remains the same. It is easy to see that such a ρ-operations can be omitted if we change labels of some initial k-graphs accordingly.

Example 4. Consider the derivation $\mathcal{D}$ of graph G in Example 2. We construct a 3-expression of G using the procedure as described in the proof of Lemma 4. First we obtain $\phi_\oplus = ((1(a) \oplus 1(b)) \oplus 1(c)) \oplus 1(d)$. Next we insert ρ operations to represent how the groups evolve through the derivation: $\phi_{\oplus,\rho} = \rho_{1\to2}((1(a) \oplus \rho_{1\to2}(1(b)) \oplus \rho_{1\to3}1(c))) \oplus 1(d)$. Finally we add η operations, and obtain $\phi_{\oplus,\rho,\eta} = \eta_{1,2}(\rho_{1\to2}(\eta_{2,3}(\eta_{1,2}(1(a) \oplus \rho_{1\to2}(1(b))) \oplus \rho_{1\to3}1(c))) \oplus 1(d))$. $\quad\dashv$

By Lemma 2 we do not need to search for k-derivations of length $> n-1$ when the graph under consideration has n vertices. The next lemma improves this bound to $n - k + 1$ which provides a significant improvement for our SAT encoding, especially if the graph under consideration has large clique-width.

Lemma 5. *Let $1 \le k \le n$. If a graph with n vertices has a k-derivation, then it has a k-derivation of length $n - k + 1$.*

Proof. Due to space restrictions we sketch the proof only, a detailed proof can be found in the full version of this paper. Let $\mathcal{D} = (T_0, \ldots, T_t)$ be a k-derivation of a graph $G = (V, E)$ with $|V| = n$. By Lemma 1 we may assume that $\mathcal{D}$ is strict. Let i be the largest index $1 \le i \le t$ where all components of T_i have size at most k and put $\ell = t - i$. It is not difficult to see that $\ell \le n - k$. We define a new template T_i' with $\mathrm{cmp}(T_i') = \mathrm{cmp}(T_i)$ and $\mathrm{grp}(T_i') = \mathrm{grp}(T_0)$, and we set $\mathcal{D}' = (T_0, T_i', T_{i+1}, \ldots, T_t)$. It can be verified that $\mathcal{D}'$ is a k-derivation of G. □

Example 5. Again, consider the derivation $\mathcal{D}$ of Example 2. $\mathcal{D}$ defines P_4 which has clique-width 3 [9]. According to Lemma 5, it should have a derivation of length $n - k + 1 = 4 - 3 + 1 = 2$. We can obtain such a derivation by removing T_1 from $\mathcal{D}$, which gives $\mathcal{D}' = (T_0, T_2, T_3)$. ⊣

By combining Lemmas 3, 4, and 5, we arrive at the main result of this section.

Proposition 1. *Let $1 \le k \le n$. A graph G with n nodes has clique-width at most k if and only if G has a k-derivation of length at most $n - k + 1$.*

4   Encoding a Derivation of a Graph

Let $G = (V, E)$ be graph, and $t > 0$ an integer. We are going to construct a CNF formula $F_{\mathrm{der}}(G, t)$ that is satisfiable if and only if G has a derivation of length t. We assume that the vertices of G are given in some arbitrary but fixed linear order $<$.

For any two distinct vertices u and v of G and any $0 \le i \le t$ we introduce a *component variable* $c_{u,v,i}$. Similarly, for any two distinct vertices u and v of G with $u < v$ and any $0 \le i \le t$ we introduce a *group variable* $g_{u,v,i}$. Intuitively, $c_{u,v,i}$ or $g_{u,v,i}$ are true if and only if u and v are in the same component or group, respectively, in the ith template of an implicitly represented derivation of G.

The formula $F_{\mathrm{der}}(G, t)$ is the conjunction of all the clauses described below. The following clauses represent the conditions D1–D4.

$$(\bar{c}_{u,v,0}) \wedge (c_{u,v,t}) \wedge (c_{u,v,i} \vee \bar{g}_{u,v,i}) \wedge (\bar{c}_{u,v,i-1} \vee c_{u,v,i}) \wedge (\bar{g}_{u,v,i-1} \vee g_{u,v,i})$$
$$\text{for } u, v \in V, \ u < v, \ 0 \le i \le t.$$

We further add clauses that ensure that the relations of being in the same group and of being in the same component are transitive.

$$(\bar{c}_{u,v,i} \vee \bar{c}_{v,w,i} \vee c_{u,w,i}) \wedge (\bar{c}_{u,v,i} \vee \bar{c}_{u,w,i} \vee c_{v,w,i}) \wedge (\bar{c}_{u,w,i} \vee \bar{c}_{v,w,i} \vee c_{u,v,i}) \ \wedge$$
$$(\bar{g}_{u,v,i} \vee \bar{g}_{v,w,i} \vee g_{u,w,i}) \wedge (\bar{g}_{u,v,i} \vee \bar{g}_{u,w,i} \vee g_{v,w,i}) \wedge (\bar{g}_{u,w,i} \vee \bar{g}_{v,w,i} \vee g_{u,v,i})$$
$$\text{for } u, v, w \in V, \ u < v < w, \ 0 \le i \le t.$$

In order to enforce the *edge property* we add the following clauses for any two vertices $u, v \in V$ with $u < v$, $uv \in E$ and $1 \le i \le t$:

$$(c_{u,v,i-1} \vee \bar{g}_{u,v,i}).$$

Further, to enforce the *neighborhood property*, we add for any three vertices $u, v, w \in V$ with $uv \in E$ and $uw \notin E$ and $1 \le i \le t$, the following clauses.

$$(c_{\min(u,v),\max(u,v),i-1} \vee \bar{g}_{\min(v,w),\max(v,w),i})$$

Finally, to enforce the *path property* we add for any four vertices u, v, w, x, such that $uv, uw, vx \in E$, and $wx \notin E$, $u < v$ and $1 \leq i \leq t$ the following clauses:

$$(c_{u,v,i-1} \vee \bar{g}_{\min(u,x),\max(u,x),i} \vee \bar{g}_{\min(v,w),\max(v,w),i})$$

The following statement is a direct consequence of the above definitions.

Lemma 6. $F_{\mathrm{der}}(G, t)$ *is satisfiable if and only if G has a derivation of length t.*

5 Encoding a k-Derivation of a Graph

In this section, we describe how the formula $F_{\mathrm{der}}(G, t)$ can be extended to encode a derivation of width at most k. Ideally, one wants to encode that unit propagation results in a conflict on any assignment of component and group variables representing a derivation containing a component with more than k groups. First we will describe the conventional direct encoding [35] followed by our representative encoding. Only the latter encoding realizes arc consistency [18].

5.1 Direct Encoding

We introduce new Boolean variables $l_{v,a,i}$ for $v \in V$, $1 \leq a \leq k$, and $0 \leq i \leq t$. The purpose is to assign each vertex for each template a group number between 1 and k. The intended meaning of a variable $l_{v,a,i}$ is that in T_i, vertex v has group number a. Let $F(G, k, t)$ denote the formula obtained from $F_{\mathrm{der}}(G, t)$ by adding the following three sets of clauses. The first ensures that every vertex has at least one group number, the second ensures that every vertex has at most one group number, and the third ensures that two vertices of the same group share the same group number.

$$(l_{v,1,i} \vee l_{v,2,i} \vee \cdots \vee l_{v,k,i}) \qquad \text{for } v \in V, \, 0 \leq i \leq t,$$
$$(\bar{l}_{v,a,i} \vee \bar{l}_{v,b,i}) \qquad \text{for } v \in V, \, 1 \leq a < b \leq k, \, 0 \leq i \leq t,$$
$$(\bar{l}_{u,a,i} \vee \bar{l}_{v,a,i} \vee \bar{c}_{u,v,i} \vee g_{u,v,i}) \wedge (\bar{l}_{u,a,i} \vee l_{v,a,i} \vee \bar{g}_{v,w,i}) \wedge (l_{v,a,i} \vee \bar{l}_{v,a,i} \vee \bar{g}_{u,v,i})$$
$$\text{for } u, v \in V, \, u < v, \, 1 \leq a \leq k, \, 0 \leq i \leq t.$$

Together with Lemma 6 this construction directly yields the following statement.

Proposition 2. *Let $G = (V, E)$ be graph and $t = |V| - k + 1$. Then $F(G, k, t)$ is satisfiable if and only if $\mathrm{cwd}(G) \leq k$.*

Example 6. Let $G = (V, E)$ and $k = 2$. Vertices $u, v, w \in V$ in template T_i, are in one component, but in different groups. Hence the corresponding component variables are true, and the corresponding group variables are false. The clauses containing the variables $l_{u,a,i}, l_{v,a,i}, l_{w,a,i}$ with $a \in \{1, 2\}$ after removing falsified literals are: $(l_{u,1,i} \vee l_{u,2,i})$, $(l_{v,1,i} \vee l_{v,2,i})$, $(l_{w,1,i} \vee l_{w,2,i})$, $(\bar{l}_{u,1,i} \vee \bar{l}_{v,1,i})$, $(\bar{l}_{u,1,i} \vee \bar{l}_{w,1,i})$, $(\bar{l}_{v,1,i} \vee \bar{l}_{w,1,i})$, $(\bar{l}_{u,2,i} \vee \bar{l}_{v,2,i})$, $(\bar{l}_{u,2,i} \vee \bar{l}_{w,2,i})$, $(\bar{l}_{v,2,i} \vee \bar{l}_{w,2,i})$. These clauses cannot be satisfied, yet unit propagation will not result in a conflict. Therefore, a SAT solver may not be able to cut off the current branch.   ⊣

5.2   The Representative Encoding

To overcome the unit propagation problem of the direct encoding, as described in Example 6, we propose the *representative encoding* which uses two types of variables. First, for each $v \in V$ and $1 \leq i \leq t$ we introduce a representative variable $r_{v,i}$. This variable, if assigned to true, expresses that vertex v is the representative of a group in template T_i. In each group, only one vertex can be the representative and we choose to make the first vertex in the lexicographical ordering the representative. This results in the following clauses:

$$(r_{v,i} \vee \bigvee_{u \in V, u < v} g_{u,v,i}) \wedge \bigwedge_{u \in V, u < v} (\bar{r}_{v,i} \vee \bar{g}_{u,v,i}) \qquad \text{for } v \in V, 0 \leq i \leq t$$

Additionally we introduce auxiliary variables to efficiently encode that the number of representative vertices in a component is at most k. These auxiliary variables are based on the *order encoding* [34]. Consider a (non-Boolean) variable $L_{v,i}$ with domain $D = \{1, \ldots, k\}$, whose elements denote the group number of vertex v in template T_i. In the direct encoding, we used k variables $l_{v,a,i}$ with $a \in D$. Assigning $l_{v,a,i} = 1$ in that encoding means $L_{v,i} = a$. Alternatively, we can use *order variables* $o_{v,a,i}^{>}$ with $v \in V$, $a \in D \setminus \{k\}$, $0 \leq i \leq t$. Assigning $o_{v,a,i}^{>} = 1$ means $L_{v,i} > a$. Consequently, $o_{v,a,i}^{>} = 0$ means $L_{v,i} \leq a$.

Example 7. Given an assignment to the order variables $o_{v,a,i}^{>}$, one can easily construct the equivalent assignment to the variables in the direct encoding (and the other way around). Below is a visualization of the equivalence relation with $k = 5$. In the middle is a binary representation of each of the k labels by concatenating the Boolean values to the order variables.

$$L_v = 1 \leftrightarrow l_{v,1,i} = 1 \leftrightarrow 0000 \leftrightarrow o_{v,1,i}^{>} = o_{v,2,i}^{>} = o_{v,3,i}^{>} = o_{v,4,i}^{>} = 0$$
$$L_v = 2 \leftrightarrow l_{v,2,i} = 1 \leftrightarrow 1000 \leftrightarrow o_{v,1,i}^{>} = 1, o_{v,2,i}^{>} = o_{v,3,i}^{>} = o_{v,4,i}^{>} = 0$$
$$L_v = 3 \leftrightarrow l_{v,3,i} = 1 \leftrightarrow 1100 \leftrightarrow o_{v,1,i}^{>} = o_{v,2,i}^{>} = 1, o_{v,3,i}^{>} = o_{v,4,i}^{>} = 0$$
$$L_v = 4 \leftrightarrow l_{v,4,i} = 1 \leftrightarrow 1110 \leftrightarrow o_{v,1,i}^{>} = o_{v,2,i}^{>} = o_{v,3,i}^{>} = 1, o_{v,4,i}^{>} = 0$$
$$L_v = 5 \leftrightarrow l_{v,5,i} = 1 \leftrightarrow 1111 \leftrightarrow o_{v,1,i}^{>} = o_{v,2,i}^{>} = o_{v,3,i}^{>} = o_{v,4,i}^{>} = 1 \qquad \dashv$$

Although our encoding is based on the variables from the order encoding, we use none of the associated clauses. We implemented the original order [34], which resulted in many long clauses and the performance was comparable to the direct encoding.

Instead, we combined the representative and order variables. Our use of the order variables can be seen as the encoding of a sequential counter [32]. We would like to point out that if u and v are both representative vertices in the same component of template T_i and $u < v$, then $o_{u,a,i}^{>} = 0$ and $o_{v,a,i}^{>} = 1$ must hold for some $1 \leq a < k$. Consequently, $o_{u,k-1,i}^{>} = 0$ (vertex u has not the highest group number in T_i), $o_{v,1,i}^{>} = 1$ (vertex v has not the lowest group number in T_i), and $o_{u,a,i}^{>} \rightarrow o_{v,a+1,i}^{>}$: These constraints can be expressed by the following clauses.

$$(\bar{c}_{u,v,i} \vee \bar{r}_{u,i} \vee \bar{r}_{v,i} \vee \bar{o}_{u,k-1,i}^{>}) \wedge (\bar{c}_{u,v,i} \vee \bar{r}_{u,i} \vee \bar{r}_{v,i} \vee o_{v,1,i}^{>}) \wedge$$
$$\bigwedge_{1 \leq a < k-1} (\bar{c}_{u,v,i} \vee \bar{r}_{u,i} \vee \bar{r}_{v,i} \vee \bar{o}_{u,a,i}^{>} \vee o_{v,a+1,i}^{>}) \qquad \text{for } u, v \in V, u < v, 0 \leq i \leq t.$$

Example 8. Consider a graph $G = (V, E)$ with $u, v, w, x \in V$ and the representative encoding with $k = 3$. We will show that if u,v,w, and x are all in the same component and they are all representatives of their respective group numbers in template T_i, then unit propagation will result in a conflict (because there are four representatives and only three group numbers). Observe that all corresponding component and representative variables are true. This example, with falsified literals removed, contains the clauses $(\bar{o}^>_{u,2,i})$, $(\bar{o}^>_{u,1,i} \vee \boldsymbol{o}^>_{\boldsymbol{v,2,i}})$, $(o^>_{v,1,i})$, $(\bar{o}^>_{u,2,i})$, $(\bar{o}^>_{u,1,i} \vee \boldsymbol{o}^>_{\boldsymbol{w,2,i}})$, $(o^>_{w,1,i})$, $(\bar{o}^>_{u,2,i})$, $(\bar{o}^>_{u,1,i} \vee o^>_{x,2,i})$, $(o^>_{x,1,i})$, $(\bar{o}^>_{v,2,i})$, $(\boldsymbol{\bar{o}}^>_{\boldsymbol{v,1,i}} \vee \boldsymbol{o}^>_{\boldsymbol{w,2,i}})$, $(o^>_{w,1,i})$, $(\bar{o}^>_{v,2,i})$, $(\boldsymbol{\bar{o}}^>_{\boldsymbol{v,1,i}} \vee o^>_{x,2,i})$, $(o^>_{x,1,i})$, $(\bar{o}^>_{w,2,i})$, $(\boldsymbol{\bar{o}}^>_{\boldsymbol{w,1,i}} \vee o^>_{x,2,i})$, $(o^>_{x,1,i})$. Literals that are falsified by unit clauses are shown in bold. Notice that $(\bar{o}^>_{v,1,i} \vee o^>_{w,2,i})$ is falsified, i.e., a conflicting clause.   ⊣

Both the direct and representative encoding require $n(n + k - 1)(n - k + 2)$ variables. The number of clauses depends on the set of edges. In worst case, the number of clauses can be $\mathcal{O}(n^5 - n^4 k)$ due to the path condition.

6   Experimental Results

In this section we report the results we obtained by running our SAT encoding on various classes of graphs. Given a graph $G = (V, E)$, we compute that G has clique-width k by determining for which value of k it holds that $F(G, k, |V| - k + 1)$ is satisfiable and $F(G, k - 1, |V| - k + 2)$ is unsatisfiable. We used the SAT solver Glucose version 2.2 [1] to solve the encoded problems. Glucose solved the hardest instances about twice as fast (or more) as other state-of-the-art solvers such as Lingeling [3], Minisat [15] and Clasp [17]. We used a 4-core Intel Xeon CPU E31280 3.50GHz, 32 Gb RAM machine running Ubuntu 10.04.

Although the direct and representative encodings result in CNF formulas of almost equal size, there is a huge difference in costs to solve these instances. To determine the clique-width of the famous named graphs (see below) using the direct encoding takes about two to three orders of magnitude longer as compared to the representative encoding. For example, we can establish that the Paley graph with 13 vertices has clique-width 9 within a few seconds using the representative encoding, while the solver requires over an hour using the direct encoding. Because of the huge difference in speed, we discarded the use of the direct encoding in the remainder of this section.

We noticed that upper bounds (satisfiable formulas) are obtained much faster than lower bounds (unsatisfiable formulas). The reason is twofold. First, the whole search space needs to be explored for lower bounds, while for upper bounds, one can be "lucky" and find a solution fast. Second, due to our encoding, upper bound formulas are smaller (due to a smaller t) which makes them easier. Table 1 shows this for a random graph with 20 vertices for the direct encoding and the representative encoding.

We examined whether adding symmetry-breaking predicates could improve performance. We used Saucy version 3 for this purpose [25]. After the addition of the clauses with representative variables, the number of symmetries is drastically

Table 1. Runtimes in seconds of the direct and representative encoding on a random graph with 20 vertices and 95 edges for different values of k. Up to $k = 9$ the formulas are unsatisfiable, afterwards they are satisfiable. Timeout (TO) is 20,000 seconds.

k	3	4	5	6	7	8	9	10	11	12	13	14	15	16
direct	1.39	14.25	101.1	638.5	18,337	TO	TO	TO	TO	30.57	0.67	0.50	0.10	0.10
repres	0.62	2.12	8.14	12.14	33.94	102.3	358.6	9.21	0.40	0.35	0.32	0.29	0.29	0.28

reduced. However, one can generate symmetry-breaking predicates for $F_{\mathrm{der}}(G, t)$ and add those instead. Although it is helpful in some cases, the average speed-up was between 5 to 10%.

Our experimental computations are ongoing. Below we report on some of the results we have obtained so far.

6.1   Random Graphs

The asymptotics of the clique-width of random graphs have been studied by Lee et al. [27]. Their results show that for random graphs on n vertices the following holds asymptotically almost surely: If the graphs are very sparse, with an edge probability below $1/n$, then clique-width is at most 5; if the edge probability is larger than $1/n$, then the clique-width grows at least linearly in n. Our first group of experiments complements these asymptotic results and provides a detailed picture on the clique-width of small random graphs. We have used the SAT encoding to compute the clique-width of graphs with 10, 15, and 20 vertices, with the edge probability ranging from 0 to 1. A plot of the distribution is displayed in Figure 2. It is interesting to observe the symmetry at edge probability $1/2$,

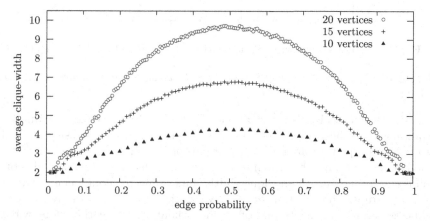

Fig. 2. Average clique-width of random graphs with edge probabilities between 0 and 1. Each dot in the graph represents the average clique-width of 100 graphs.

and the how the steepness of the curve increases with the number of vertices. Note the "shoulders" of the curve for very sparse and very dense graphs.

6.2   The Clique-Width Numbers

For every $k > 0$, let n_k denote the smallest number such that there exists a graph with n_k vertices that has clique-width k. We call n_k the kth *clique-width number*. From the characterizations known for graphs of clique-width 1, 2, and 3, respectively [5], it is easy to determine the first three clique-width numbers (1, 2, and 4). However, determining n_4 is not straightforward, as it requires nontrivial arguments to establish clique-width lower bounds. We would like to point out that a similar sequence for the graph invariant *treewidth* is easy to determine, as the complete graph on n vertices is the smallest graph of treewidth $n - 1$. One of the very few known graph classes of unbounded clique-width for which the exact clique-width can be determined in polynomial time are grids [23]; the $k \times k$ grid with $k \geq 3$ has clique-width $k + 1$ [20]. Hence grids provide the upper bounds $n_4 \leq 9$, $n_5 \leq 16$, $n_6 \leq 25$, and $n_7 \leq 36$. With our experiments we could determine $n_4 = 6$, $n_5 = 8$, $n_6 = 10$, $n_7 = 11$, $n_8 \leq 12$, and $n_9 \leq 13$. It is known that the path on four vertices (P_4) is the unique smallest graph in terms of the number of vertices with clique-width 3. We could determine that the triangular prism (3-Prism) is the unique smallest graph with clique-width 4, and that there are exactly 7 smallest graphs with clique-width 5. There are 68 smallest graphs with clique-width 6 and one of them has only 18 edges. See Figure 3 for an illustration. Additionally, we found several graphs of size 11 with clique-width 7 by extending a graph of size 10 with clique-width 6.

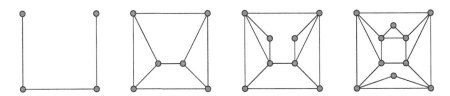

Fig. 3. Smallest graphs with clique-width 3, 4, 5, and 6 (from left to right)

Proposition 3. *The clique-width sequence starts with the numbers 1, 2, 4, 6, 8, 10, 11.*

We used Brendan McKay's software package Nauty [28] to avoid checking isomorphic copies of the same graph. There are several other preprocessing methods that can speed up the search for small graphs of prescribed clique-width $k \geq 2$. Obviously, we can limit the search to *connected* graphs, as the clique-width of a graph is clearly the maximum clique-width of its connected components. We can also ignore graphs that contain *twins*—two vertices that have exactly the same neighbors—as we can delete one of them without changing the clique-width. Similarly, we can ignore graphs with a *universal vertex*, a vertex that is adjacent to all other vertices, as it can be deleted without changing the clique-width.

All these filtering steps are subsumed by the general concept of *prime graphs*. Consider a graph $G = (V, E)$. A vertex $u \in V$ *distinguishes* vertices $v, w \in V$ if $uv \in E$ and $uw \notin E$. A set $M \subseteq V$ is a *module* if no vertex from $V \setminus M$ distinguishes two vertices from M. A module M is *trivial* if $|M| \in \{0, 1, |V|\}$. A graph is *prime* if it contains only trivial modules. It is well-known that the clique-width of a graph is either 2 or the maximum clique-width of its induced prime subgraphs [9]. Hence, in our search, we can ignore all graphs that are not prime. We can efficiently check whether a graph is prime [21]. The larger

Table 2. Number of connected and prime graphs with specified clique-width, modulo isomorphism

			clique-width				
$|V|$	connected	prime	2	3	4	5	6
4	6	1	0	1	0	0	0
5	21	4	0	4	0	0	0
6	112	26	0	25	1	0	0
7	853	260	0	210	50	0	0
8	11,117	4,670	0	1,873	2,790	7	0
9	261,080	145,870	0	16,348	125,364	4,158	0
10	11,716,571	8,110,354	0	142,745	5,520,350	2,447,190	68

Table 3. Clique-width of named graphs. Sizes are reported for the unsatisfiables.

graph	$|V|$	$|E|$	cwd	variables	clauses	UNSAT	SAT
Brinkmann	21	42	10	8,526	163,065	3,932.56	1.79
Chvátal	12	24	5	1,800	21,510	0.40	0.09
Clebsch	16	40	8	3,872	60,520	191.02	0.09
Desargues	20	30	8	7,800	141,410	3,163.70	0.26
Dodecahedron	20	30	8	7,800	141,410	5,310.07	0.33
Errera	17	45	8	4,692	79,311	82.17	0.16
Flower snark	20	30	7	8,000	148,620	276.24	3.9
Folkman	20	40	5	8,280	168,190	11.67	0.36
Franklin	12	18	4	1,848	21,798	0.07	0.04
Frucht	12	18	5	1,800	20,223	0.39	0.02
Hoffman	16	32	6	4,160	64,968	8.95	0.46
Kittell	23	63	8	12,006	281,310	179.62	18.65
McGee	24	36	8	13,680	303,660	8,700.94	59.89
Sousselier	16	27	6	4,160	63,564	3.67	11.75
Paley-13	13	39	9	1,820	22,776	12.73	0.05
Paley-17	17	68	11	3,978	72,896	194.38	0.12
Pappus	18	27	8	5,616	90,315	983.67	0.14
Petersen	10	15	5	1,040	9,550	0.10	0.02
Poussin	15	39	7	3,300	50,145	9.00	0.21
Robertson	19	38	9	6,422	112, 461	478.83	0.76
Shrikhande	16	48	9	3,680	59,688	129.75	0.11

the number of vertices, the larger the fraction of non-prime graphs (considering connected graphs modulo isomorphism). Table 2 gives detailed results.

6.3   Famous Named Graphs

The graph theoretic literature contains several graphs that have names, sometimes inspired by the graph's topology, and sometimes after their discoverer. We have computed the clique-width of several named graphs, the results are given in Table 3 (definitions of all considered graphs can be found in Math-World [37]). The *Paley graphs*, named after the English mathematician Raymond Paley (1907–1933), stick out as having large clique-width. Our results on the clique-width of Paley graphs imply some upper bounds on the 9th and 11th clique-width numbers: $n_9 \leq 13$ and $n_{11} \leq 17$.

7   Conclusion

We have presented a SAT approach to the exact computation of clique-width, based on a reformulation of clique-width and several techniques to speed up the search. This new approach allowed us to systematically compute the exact clique-width of various small graphs. We think that our results could be of relevance for theoretical investigations. For instance, knowing small vertex-minimal graphs of certain clique-width could be helpful for the design of discrete algorithms that recognize graphs of bounded clique-width. Such graphs can also be useful as gadgets for a reduction to show that the recognition of graphs of clique-width 4 is NP-hard, which is still a long-standing open problem [16]. Furthermore, as discussed in Section 1, there are no heuristic algorithms to compute the clique-width directly, but heuristic algorithms for related parameters can be used to obtain upper bounds on the clique-width. Our SAT-based approach can be used to empirically evaluate how far heuristics are from the optimum, at least for small and medium-sized graphs.

So far we have focused in our experiments on the exact clique-width, but for various applications it is sufficient to have good upper bounds. Our results (see Table 1) suggest that our approach can be scaled to medium-sized graphs for the computation of upper bounds. We also observed that for many graphs the upper bound of Lemma 5 is not tight. Thus, we expect that if we search for shorter derivations, which is significantly faster, this will yield optimal or close to optimal solutions in many cases.

Finally, we would like to mention that our SAT-based approach is very flexible and open. It can easily be adapted to variants of clique-width, such as linear clique-width [22,16], m-clique-width [12], or NLC-width [36]. Hence, our approach can be used for an empirical comparison of these parameters.

Acknowledgement. The authors acknowledge the Texas Advanced Computing Center (TACC) at The University of Texas at Austin for providing grid resources that have contributed to the research results reported within this paper.

References

1. Audemard, G., Simon, L.: Predicting learnt clauses quality in modern sat solvers. In: Proceedings of the 21st International Jont Conference on Artifical Intelligence, IJCAI 2009, pp. 399–404. Morgan Kaufmann Publishers Inc., San Francisco (2009)
2. Beyß, M.: Fast algorithm for rank-width. In: Kučera, A., Henzinger, T.A., Nešetřil, J., Vojnar, T., Antoš, D. (eds.) MEMICS 2012. LNCS, vol. 7721, pp. 82–93. Springer, Heidelberg (2013)
3. Biere, A.: Lingeling and friends entering the SAT Challenge 2012. In: Balint, A., Belov, A., Diepold, A., Gerber, S., Järvisalo, M., Sinz, C. (eds.) Solver and Benchmark Descriptions. Department of Computer Science Series of Publications B, vol. B-2012-2, pp. 33–34. University of Helsinki (2012)
4. Bui-Xuan, B.-M., Telle, J.A., Vatshelle, M.: Boolean-width of graphs. Theoretical Computer Science 412(39), 5187–5204 (2011)
5. Corneil, D.G., Habib, M., Lanlignel, J.-M., Reed, B., Rotics, U.: Polynomial-time recognition of clique-width ≤ 3 graphs. Discr. Appl. Math. 160(6), 834–865 (2012)
6. Corneil, D.G., Rotics, U.: On the relationship between clique-width and treewidth. SIAM J. Comput. 34(4), 825–847 (2005)
7. Courcelle, B., Makowsky, J.A., Rotics, U.: Linear time solvable optimization problems on graphs of bounded clique-width. Theory Comput. Syst. 33(2), 125–150 (2000)
8. Courcelle, B., Makowsky, J.A., Rotics, U.: On the fixed parameter complexity of graph enumeration problems definable in monadic second-order logic. Discr. Appl. Math. 108(1-2), 23–52 (2001)
9. Courcelle, B., Olariu, S.: Upper bounds to the clique-width of graphs. Discr. Appl. Math. 101(1-3), 77–114 (2000)
10. Courcelle, B., Engelfriet, J., Rozenberg, G.: Context-free handle-rewriting hypergraph grammars. In: Ehrig, H., Kreowski, H.-J., Rozenberg, G. (eds.) Graph Grammars 1990. LNCS, vol. 532, pp. 253–268. Springer, Heidelberg (1991)
11. Courcelle, B., Engelfriet, J., Rozenberg, G.: Handle-rewriting hypergraph grammars. J. of Computer and System Sciences 46(2), 218–270 (1993)
12. Courcelle, B., Twigg, A.: Constrained-path labellings on graphs of bounded clique-width. Theory Comput. Syst. 47(2), 531–567 (2010)
13. Diestel, R.: Graph Theory, 2nd edn. Graduate Texts in Mathematics, vol. 173. Springer, New York (2000)
14. Alex Dow, P., Korf, R.E.: Best-first search for treewidth. In: Proceedings of the Twenty-Second AAAI Conference on Artificial Intelligence, Vancouver, British Columbia, Canada, July 22-26, pp. 1146–1151. AAAI Press (2007)
15. Eén, N., Sörensson, N.: An extensible SAT-solver. In: Giunchiglia, E., Tacchella, A. (eds.) SAT 2003. LNCS, vol. 2919, pp. 502–518. Springer, Heidelberg (2004)
16. Fellows, M.R., Rosamond, F.A., Rotics, U., Szeider, S.: Clique-width is NP-complete. SIAM J. Discrete Math. 23(2), 909–939 (2009)
17. Gebser, M., Kaufmann, B., Neumann, A., Schaub, T.: clasp: A conflict-driven answer set solver. In: Baral, C., Brewka, G., Schlipf, J. (eds.) LPNMR 2007. LNCS (LNAI), vol. 4483, pp. 260–265. Springer, Heidelberg (2007)
18. Gent, I.P.: Arc consistency in SAT. In: van Harmelen, F. (ed.) 15th European Conference on Artificial Intelligence (ECAI 2002), pp. 121–125. IOS Press (2002)
19. Gogate, V., Dechter, R.: A complete anytime algorithm for treewidth. In: Proceedings of the Proceedings of the Twentieth Conference Annual Conference on Uncertainty in Artificial Intelligence (UAI 2004), Arlington, Virginia, pp. 201–208. AUAI Press (2004)

20. Golumbic, M.C., Rotics, U.: On the clique-width of perfect graph classes extended abstract. Internat. J. Found. Comput. Sci. 11(3), 423–443 (2000); Selected papers from In: Widmayer, P., Neyer, G., Eidenbenz, S. (eds.) WG 1999. LNCS, vol. 1665, pp. 135–443. Springer, Heidelberg (1999)

21. Habib, M., Paul, C.: A survey of the algorithmic aspects of modular decomposition. Computer Science Review 4(1), 41–59 (2010)

22. Heggernes, P., Meister, D., Papadopoulos, C.: Characterising the linear clique-width of a class of graphs by forbidden induced subgraphs. Discr. Appl. Math. 160(6), 888–901 (2012)

23. Heggernes, P., Meister, D., Rotics, U.: Computing the clique-width of large path powers in linear time via a new characterisation of clique-width. In: Kulikov, A., Vereshchagin, N. (eds.) CSR 2011. LNCS, vol. 6651, pp. 233–246. Springer, Heidelberg (2011)

24. Hvidevold, E.M., Sharmin, S., Telle, J.A., Vatshelle, M.: Finding good decompositions for dynamic programming on dense graphs. In: Marx, D., Rossmanith, P. (eds.) IPEC 2011. LNCS, vol. 7112, pp. 219–231. Springer, Heidelberg (2012)

25. Katebi, H., Sakallah, K.A., Markov, I.L.: Conflict anticipation in the search for graph automorphisms. In: Bjørner, N., Voronkov, A. (eds.) LPAR-18 2012. LNCS, vol. 7180, pp. 243–257. Springer, Heidelberg (2012)

26. Koster, A.M.C.A., Bodlaender, H.L., van Hoesel, S.P.M.: Treewidth: Computational experiments. Electronic Notes in Discrete Mathematics 8, 54–57 (2001)

27. Lee, C., Lee, J., Oum, S.-I.: Rank-width of random graphs. J. Graph Theory 70(3), 339–347 (2012)

28. McKay, B.D.: Practical graph isomorphism. In: Proceedings of the Tenth Manitoba Conference on Numerical Mathematics and Computing, Vol. I (Winnipeg, Man., 1980), Winnipeg, Man, vol. 30, pp. 45–87 (1981)

29. Oum, S.-I.: Approximating rank-width and clique-width quickly. ACM Transactions on Algorithms 5(1) (2008)

30. Oum, S.-I., Seymour, P.: Approximating clique-width and branch-width. J. Combin. Theory Ser. B 96(4), 514–528 (2006)

31. Samer, M., Veith, H.: Encoding treewidth into SAT. In: Kullmann, O. (ed.) SAT 2009. LNCS, vol. 5584, pp. 45–50. Springer, Heidelberg (2009)

32. Sinz, C.: Towards an optimal CNF encoding of boolean cardinality constraints. In: van Beek, P. (ed.) CP 2005. LNCS, vol. 3709, pp. 827–831. Springer, Heidelberg (2005)

33. Cole Smith, J., Ulusal, E., Hicks, I.V.: A combinatorial optimization algorithm for solving the branchwidth problem. Comput. Optim. Appl. 51(3), 1211–1229 (2012)

34. Tamura, N., Taga, A., Kitagawa, S., Banbara, M.: Compiling finite linear CSP into SAT. Constraints 14(2), 254–272 (2009)

35. Walsh, T.: SAT v CSP. In: Dechter, R. (ed.) CP 2000. LNCS, vol. 1894, pp. 441–456. Springer, Heidelberg (2000)

36. Wanke, E.: k-NLC graphs and polynomial algorithms. Discr. Appl. Math. 54(2-3), 251–266 (1994); Efficient algorithms and partial k-trees

37. Weisstein, E.: MathWorld online mathematics resource, mathworld.wolfram.com

Cliquewidth and Knowledge Compilation

Igor Razgon[1,*] and Justyna Petke[2]

[1] Department of Computer Science and Information Systems,
Birkbeck, University of London
igor@dcs.bbk.ac.uk
[2] Department of Computer Science,
University College London
J.Petke@cs.ucl.ac.uk

Abstract. In this paper we study the role of cliquewidth in succinct representation of Boolean functions. Our main statement is the following: Let Z be a Boolean circuit having cliquewidth k. Then there is another circuit Z^* computing the same function as Z having treewidth at most $18k+2$ and which has at most $4|Z|$ gates where $|Z|$ is the number of gates of Z. In this sense, cliquewidth is not more 'powerful' than treewidth for the purpose of representation of Boolean functions. We believe this is quite a surprising fact because it contrasts the situation with graphs where an upper bound on the treewidth implies an upper bound on the cliquewidth but not vice versa.

We demonstrate the usefulness of the new theorem for knowledge compilation. In particular, we show that a circuit Z of cliquewidth k can be compiled into a Decomposable Negation Normal Form (DNNF) of size $O(9^{18k}k^2|Z|)$ and the same runtime. To the best of our knowledge, this is the first result on efficient knowledge compilation parameterized by cliquewidth of a Boolean circuit.

1 Introduction

Statement of the Results. Cliquewidth is a graph parameter, probably best known for its role in the design of fixed-parameter algorithms for graph-theoretic problems [2]. In this context the most interesting property of cliquewidth is that it is 'stronger' than treewidth in the following sense: if all graphs in some (infinite) class have treewidth bounded by some constant c, then the cliquewidth of the graphs of this class is also bounded by a constant $O(2^c)$. However, the opposite is not true. Consider, for example, the class of all complete graphs. The treewidth of this class is unbounded while the cliquewidth of any complete graph is 2. Thus, classes of bounded cliquewidth contain dense graphs, unlike the case of bounded treewidth.

In this paper we essentially show that, roughly speaking, cliquewidth of a Boolean function is not a stronger parameter than its treewidth. In particular,

---

* I would like to thank Fedor Fomin for his help in shaping of my understanding of the structural graph parameters.

given a Boolean circuit Z, we define its cliquewidth as the cliquewidth of the DAG of this circuit and the treewidth as the treewidth of the undirected graph underlying this DAG. The main theorem of this paper states that for any circuit Z of cliquewidth k there is another circuit Z^* computing the same function whose treewidth is at most $18k + 2$ and the number of gates is at most 4 times the number of gates of Z. Moreover, if Z is accompanied with the respective clique decomposition then such a circuit Z^* (and the tree decomposition of width $18k + 2$) can be obtained in time $O(k^2 n)$. The definition of circuit treewidth is taken from [14] and the definition of circuit cliquewidth naturally follows from the treewidth definition. In fact, the relationship between circuit treewidth and cliquewidth is put in [14] as an open question.

We demonstrate that the main theorem is useful for knowledge compilation. In particular, we show that any circuit Z of cliquewidth k can be compiled into decomposable negation normal form (DNNF) [3] of size $O(9^{18k} k^2 |Z|)$ (where $|Z|$ is the number of gates) by an algorithm taking the same runtime. To the best of our knowledge, this is the first result on space-efficient knowledge compilation parameterized by cliquewidth. We believe this result is interesting because the parameterization by cliquewidth, compared to treewidth, allows to capture a wider class of inputs including those circuits whose underlying graphs are dense.

This bound is obtained as an immediate corollary of the main theorem and the $O(9^t t^2 |Z|)$ bound on the DNNF size for the given circuit Z, where t is the treewidth of Z. The intermediate step for the latter result is an $O(3^p (|C| + n))$ bound of the DNNF size of the given CNF where C and n are, respectively the number of clauses and variables of this CNF and p is the treewidth of its *incidence* graph. All these 3 bounds significantly extend the currently existing bound $O(2^r n)$ of [3] where r is the treewidth of the *primal* graph of the given CNF. For example, if the given CNF has large clauses (and hence a large treewidth of the primal graph) then the $O(2^r n)$ bound becomes practically infeasible while the $O(3^p (C + n))$ bound may be still feasible provided a small treewidth of the incidence graph and a number of clauses polynomially dependent on n.

Related Work. The algorithmic power of cliquewidth stems from the meta-theorem of [2] stating that any problem definable in Monadic Second Order Logic (MSO$_1$) can be solved in linear time for a class of graphs of fixed cliquewidth k. The cliquewidth of the given graph is NP-hard to compute [8] and it is not known to be FPT. On the other hand, cliquewidth is FPT approximable by an FPT computable parameter called *rankwidth* [13,11]. As said above, there are classes of graphs with unrestricted treewidth and bounded cliquewidth. However, it has been shown in [10] that the only reason for treewidth to be much larger than cliquewidth is the presence of a large complete bipartite graph (biclique) in the considered graph. In fact, we prove the main theorem of this paper by applying a transformation that eliminates all bicliques from the DAG of the given circuit.

DNNFs have been introduced as a knowledge compilation formalism in [3], where it has been shown that any CNF on n variables of treewidth t of the primal graph can be compiled into a DNNF of size $O(2^t tn)$ with the same runtime. A detailed analysis of special cases of DNNF has been provided in [6]. In particular, it has been shown that Free Binary Decision Diagrams (FBDD) and hence Ordered Binary Decision Diagrams (OBDD) can be seen as special cases of DNNF. In fact, there is a separation between DNNF and FBDD [4]. This additional expression power of DNNF has its disadvantages: a number of queries that can be answered in polynomial time (polytime) for FBDD and OBDD are NP-complete for DNNF [6]. This trade-off led to investigation of subclasses of DNNF that, on the one hand, retain the succinctness of DNNF for CNFs of small treewidth and, on the other hand, have an increased set of queries that can be answered in polytime. Probably the most notable result obtained in this direction are Sentential Decision Diagrams (SDD) [5] that, on one hand, can answer in polytime the equivalence query (possibility to answer this query in polytime for OBDDs is probably the main reason why this formalism is very popular in the area of verification) and, on the other hand, retain the same upper bound dependence on treewidth as DNNF.

In fact the size of OBDD can also be efficiently parameterized by the treewidth of the initial representation of the considered function. Indeed, there is an OBDD of size $O(n2^p)$ where p is the pathwidth of the primal graph of the given CNF and of size $(n^{O(t)})$ where t is the treewidth of the graph, see e.g. [9]. It is shown in [14] that similar pattern retains if we consider the pathwidth and treewidth of a circuit but in the former case p is replaced by an exponential function of p and in the latter case, t is replaced by a double exponential function of t.

Important Remark. Due to space constraints, proofs of some statements are either omitted or replaced by sketches. A complete version of this work is available at http://arxiv.org/abs/1303.4081.

2   Preliminaries

A *labeled* graph $G = (V, E, \mathbf{S})$ is defined by the usual set $V(G)$ of vertices and a set $E(G)$ of edges and also by $\mathbf{S}(G)$, a partition of $V(G)$. Each element of $\mathbf{S}(G)$ is called a *label*. A *simplified clique decomposition* (SCD) is a pair $(T, \mathbf{G})$ where T is a rooted tree and $\mathbf{G}$ is a family of labeled graphs. Each node t of T is associated with a graph $G(t)$, which is defined as follows. If t is a leaf node, then $G(t) = (\{v\}, \emptyset, \{\{v\}\})$. Assume that t has two children t_1 and t_2 and let $G_1 = G(t_1)$ and $G_2 = G(t_2)$. Then $V(G_1) \cap V(G_2) = \emptyset$ and $G(t) = (V(G_1) \cup V(G_2), E(G_1) \cup E(G_2), \mathbf{S}(G_1) \cup \mathbf{S}(G_2))$. Finally, assume that t has only one child t_1 and let $G_1 = G(t_1)$. Graph $G(t)$ can be obtained from G_1 by one of the following three operations:

- **Adding a New Vertex.** There is $v \notin V(G_1)$ such that $G(t) = (V(G_1) \cup \{v\}, E(G_1), \mathbf{S}(G_1) \cup \{\{v\}\})$.
- **Union of Labels.** There are $S_1, S_2 \in \mathbf{S}(G_1)$ such that $G(t) = (V(G_1), E(G_1), (\mathbf{S}(G_1) \setminus \{S_1, S_2\}) \cup \{S\})$. We say that S_1 and S_2 are the *children* of S.
- **New Adjacency.** There are $S_1, S_2 \in \mathbf{S}(G_1)$ such that $G(t) = (V(G_1), E(G_1) \cup \{\{u, v\} | u \in S_1, v \in S_2\}, \mathbf{S}(G_1))$. We say that S_1 and S_2 are *adjacent*.

The width of a node t of T is $|\mathbf{S}(G(t))|$. The width of $(T, \mathbf{G})$ is the largest width of a node t of T. Let r be the root of T. Then we say that $(T, \mathbf{G})$ is an SCD of $G(r)$ and of $(V(G(r)), E(G(r)))$ (the unlabeled version of $G(r)$). The *simplified cliquewidth* (SCW) of a graph G is the smallest width among all SCDs of G. The definition of SCD is closely related to the standard notion of clique decomposition. In fact SCW of a graph G is at most twice larger than the cliquewidth of G. The details are provided in the complete version.

Clique decomposition and SCD are easily extended to the directed case. In fact the notion of cliquewidth has been initially proposed for the directed case, as noted in [7]. The only change is that the new adjacency operation adds to $G(t)$ all possible directed arcs from label S_1 to label S_2 instead of undirected edges. In this case we say that there is an arc from S_1 to S_2.

We denote $\bigcup_{t \in V(T)} \mathbf{S}(G(t))$ by $\mathbf{S} = \mathbf{S}(T, \mathbf{G})$ and call it *the set of labels* of $(T, \mathbf{G})$. The set $\mathbf{S}$ is very important for our reasoning. In fact, we use SCD because we believe it allows a more intuitive definition of set $\mathbf{S}$ than the stadard clique deocmposition.

A tree decomposition of a graph G is a pair $(T, \mathbf{B})$ where T is a tree and the elements of $\mathbf{B}$ are subsets of vertices called *bags*. There is a mapping between the nodes of T and elements of $\mathbf{B}$. Let us say a vertex v of G is *contained* in a node t of T if v belongs to the bag $B(t)$ of t. Two properties of a tree decomposition are *connectedness* (all the nodes containing the given vertex v form a subtree of T), *adjacency* (each edge $\{u, v\}$ is a subset of some bag), and *union* (the union of all bags is $V(G)$). In this paper we consider the treewidth of a directed graph as the treewidth of the underlying undirected graph.

Boolean circuits considered in this paper are over the basis $\{\vee, \wedge, \neg\}$ with the unbounded fan-in. In such a circuit there are input gates (having only output wires) corresponding to variables and constants *true* and *false*. The output of each gate of a circuit Z computes a function on the set of input variables. We denote by $functions(Z)$ the set of all functions computed by the gates of Z. The number of gates of Z is denoted by $|Z|$.

A clique or tree decomposition of a circuit Z is the respective decomposition of the DAG of Z. In our discussion, we often associate the vertices of the DAG with the respective gates. *De Morgan circuits* are a subclass of circuits where the inputs of all the NOT gates are variables (i.e. the outputs of NOT gates serve as negative literals). For a gate g of Z, denote by $Var(g)$ the set of variables having a path to g in the DAG of Z. A circuit Z has the *decomposability* property if for any two in-neighbours g_1 and g_2 of an AND gate g, $Var(g_1) \cap Var(g_2) = \emptyset$. DNNF is a decomposable De Morgan circuit. When we consider a general circuit Z, we assume that it does not have constant input gates, since these gates can be propagated by removal of some gates of Z, which in turn does not increase the cliquewidth nor the treewidth of the circuit. However, for convenience of reasoning, we may use constant input gates when we describe construction of a DNNF. If the given circuit Z is a CNF then its variables-clauses relation can be represented by the *incidence graph*, a bipartite graph with parts corresponding to variables and clauses and a variable-clause edge representing occurrence of a variable in a clause.

3   From Small Cliquewidth to Small Treewidth

The central result of this section is the following theorem:

Theorem 1. *Let F be a circuit of cliquewidth k over n variables Then there is a circuit F^* of treewidth at most $18k+2$ and $|F^*| \leq 4|F|$ such that $functions(F) \subseteq functions(F^*)$. Moreover, given F and a clique decomposition of F of width k there is an $O(k^2 n)$ algorithm constructing F^* and a tree decomposition of F^* of width at most $18k + 2$ having at most $2|F|$ bags.*

The rest of this section is the proof of Theorem 1. The main idea of the proof is to replace 'parts' of the given circuit forming large bicliques by circuits computing equivalent functions where such bicliques do not occur. As an example consider a CNF of 3 clauses $C_1 = (a_1 \vee a_2 \vee a_3 \vee b_1)$, $C_2 = (a_1 \vee a_2 \vee a_3 \vee b_2)$ and $C_3 = (a_1 \vee a_2 \vee a_3 \vee b_3)$. The circuit of this graph contains a biclique of order 3 created by C_1, C_2, C_3 on one side and a_1, a_2, a_3 on the other one. This biclique can be eliminated by the introduction of an additional OR gate C_4 having input a_1, a_2, a_3 and output c_4 so that the clauses C_1, C_2, C_3 are transformed into $(b_1 \vee c_4), (b_2 \vee c_4), (b_3 \vee c_4)$, respectively. It is not hard to see that the new circuit computes the same function as the original one. This is the main idea behind the construction of circuit F^*. The formal description of the construction is given below.

For the purpose of construction of F^* we consider a type respecting SCD $(T, \mathbf{G})$ of F where each non-singleton label is one of the following:

- A *unary* label containing input gates and negation gates.
- An AND label containing AND gates.
- An OR label containing OR gates.

The following lemma essentially follows from splitting each label of the given clique decomposition into three type respecting labels.

Lemma 1. *Let k be the cliquewidth of F and let k^* be the smallest width of an SCD of F that respects types. Then $k^* \leq 6k$.*

Given a type respecting SCD $(T, \mathbf{G})$, let us construct the circuit F^*. In the first stage, we associate each label $S \in \mathbf{S}$ with a set of gates as follows:

- If S is non-singleton then it is associated with an AND gate denoted by $oand(S)$ and an OR gate denoted by $oor(S)$.
- If S is non-singleton and does not contain input gates then it is associated with an additional gate called $in(S)$ whose type is determined as follows: If S is an AND or OR label then $in(S)$ is an AND or OR gate, respectively. If S is a unary label then $in(S)$ is a circuit (perceived as a single atomic gate) consisting of two NOT gates, the output of one of them is the input of the other. So, the input of the former and the output of the latter are, respectively, the input and output of $in(S)$.

- Each singleton label $\{g\}$ is associated with the gate g of F. We call the gates associated with singleton labels *original gates* because they are the gates of F^* appearing in F. For the sake of uniformity, for each original gate g associated with a singleton label S, we put $g = oand(S) = oor(S) = in(S)$.

The wires of F^* are described below. When we say that there is a wire from gate g_1 to gate g_2, we mean that the wire is *from the output* of g_1 *to the input* of g_2.

- **Child-Parent Wires.** Let S_1 and S_2 be labels of $(T, \mathbf{G})$ such that S_1 is a child of S_2. Then there is a wire from $oand(S_1)$ to $oand(S_2)$ and a wire from $oor(S_1)$ to $oor(S_2)$.
- **Parent-Child Wires.** Let S_1 and S_2 be as above and assume that S_2 does not contain input gates. Then there is a wire from $in(S_2)$ to $in(S_1)$. That is, the direction of child-parent wires is opposite to the direction of parent-child wires.[1]
- **Adjacency wires.** Assume that in $(T, \mathbf{G})$ there is an arc from S_1 to S_2 (established by the new adjacency node). Then the following cases apply:
  - If S_2 is an AND label then put a wire from $oand(S_1)$ to $in(S_2)$.
  - If S_2 is an OR label then put a wire from $oor(S_1)$ to $in(S_2)$.
  - If S_2 is a unary label consisting of negation gates only then put a wire from an arbitrary one of $oand(S_1)$ or $oor(S_1)$ to $in(S_2)$.

Finally, we remove $in(S)$ gates that have no inputs. This removal may be iterative as removal of one gate may leave without input another one.

It is not hard to see by construction that F and F^* have the same input gates. This gives us possibility to state the following theorem with proof in Section 3.1.

Theorem 2. F^* *is a well formed circuit (that is, F^* satisfies the definition of a Boolean circuit). The output of each original gate g of F^* computes exactly the same function (in terms of input gates) as in F.*

In Section 3.2, we prove that the treewidth of F^* is not much larger than the width of $(T, \mathbf{G})$.

Theorem 3. *There is a tree decomposition of F^* with at most $2|F|$ bags having width at most $3k + 2$, where k is the width of $(T, \mathbf{G})$.*

Proof of Theorem 1. Due to Theorem 2, $functions(F) \subseteq functions(F^*)$. If we take $(T, \mathbf{G})$ to be of the smallest possible type respecting width then the treewidth of F^* is at most $18k + 2$ by combination of Theorem 3 and Lemma 1.

To compute the number of gates of F^*, let n be the number of gates of F, which is also the number of singleton labels of $(T, \mathbf{G})$. Since each non-singleton label has two children (i.e. in the respective tree of labels each non-leaf node is binary), the number of non-singleton labels is at most $n - 1$. By construction, F^* has one gate per singleton label plus at most 3 gates per non-singleton label, which adds up to at most $4n$. The technical details of the runtime derivation are omitted due to space constraints.

---

[1] We would like to thank the anonymous referee, for helping us to identify a typo in this definition that occurred in the first version of the manuscript.

3.1   Proof of Theorem 2

We start with establishing simple combinatorial properties of F^* (Lemmas 2,3, 4,5). A *path* in a circuit is a sequence of gates so that the output of every gate (except the last one) is connected by a wire to the input of its successor. Let us call a path a *connecting* path if it contains exactly one adjacency wire.

Lemma 2. − *Any path P of F^* starting at an original gate and not containing adjacency wires contains child-parent wires only.*
− *Any path P of F^* ending at an original gate and not containing adjacency wires contains parent-child wires only.*

Proof. The only possible wire to leave the original gate is a child-parent wire. Any path starting from an original gate and containing child-parent wires only ends up in an *oand* or *oor* gate. This means that the next wire (if not an adjacency one) can be only another child-parent wire. Thus the correctness of the lemma for all the paths of length i implies its correctness for all such paths of length $i + 1$, confirming the first statement.

For the second statement, we start from an original gate and go back *against* the direction of wires. The reasoning similar to the previous paragraph applies with the *in* gates of non-singleton labels replacing the *oor* and *oand* ones.   ∎

Lemma 3. *Let g_1 and g_2 be gates of F such that g_2 is an AND or an OR gate. Then there is a wire from g_1 to g_2 in F if and only if F^* has a connecting path from g_1 to g_2 such that all the gates of this path except possibly g_1 are of the same type as g_2.*

Proof. We prove only the case where g_2 is an AND gate, the other case is symmetric. Let P be a connecting path of F^* from g_1 to g_2 of the specified kind. Let g_1' and g_2' be, respectively, the tail and the head gates of the adjacency wire. Then either $g_1 = g_1'$ or the suffix of P ending at g_1' consists of child-parent wires only according to Lemma 2. It follows that g_1' corresponds to a label containing g_1. Analogously, we conclude that either $g_2 = g_2'$ or the suffix of P starting at g_2' contains only parent-child labels and hence the label corresponding to g_2' contains g_2. Existence of the adjacency wire from the label of g_1' to the label of g_2' means that the SCD introduces all wires from the gates in the label of g_1' to the gates in the label of g_2'. In particular, there is a wire from g_1 to g_2 in F.

Conversely, assume that there is a wire from g_1 to g_2 in F. Then there are labels S_1 and S_2 containing g_1 and g_2, respectively, such that $(T, \mathbf{G})$ introduces an adjacency arc from S_1 to S_2. By construction of F^* there is a gate g_1' corresponding to S_1 and a gate g_2' corresponding to S_2 such that F^* has an adjacency wire from g_1' to g_2'. Moreover, by the definition of a type respecting SCD, S_2 is an AND label, hence $g_2' = in(S_2)$ is an AND gate. Furthermore, by construction of F^* either $g_2 = g_2'$ or there is a path from g_2' to g_2 consisting of parent-child arcs only and AND gates only. Indeed, if S_2 is not a singleton then there is a wire from $in(S_2)$ to $in(S_3)$ containing g_2 since S_2 is partitioned by its children. Iterative application of this argument produces a path from g_2' to g_2. Since g_2

is an AND gate, all gates in this path are AND gates by construction. Thus the suffix exists. What about the prefix? By construction, $g_1' = oand(S_1)$. Since S_1 contains g_1, either $g_1' = g_1$ or there is a path from g_1 to g_1' involving child-parent wires and AND gates only: just start at g_1 and go every time to the $oand$-gate of the parent until S_1 has been reached. Thus we have established existence of the desired prefix.

It remains to be shown that the prefix and suffix do not intersect. However, this is impossible due to the disjointness of S_1 and S_2.   ∎

Lemma 4. *Let g_1 and g_2 be the gates of F such that g_2 is a* NOT *gate. Then F has a wire from g_1 to g_2 if and only if there is a connecting path P in F^* from g_1 to g_2 with the adjacency wire (g_1', g_2') such that $g_1 = g_1'$ and all the intermediate vertices in the suffix of P starting from g_1' are in-gates of unary labels containing negation gates only.*

Proof. Let P be a connecting path of F^* of the specified form. Then either $g_2' = g_2$ or g_2' corresponds to a label containing g_2. In both cases this means that F has a wire from g_1 to g_2.

Conversely, assume that F has a wire from g_1 to g_2. Then there are labels S_1 and S_2 containing g_1 and g_2 such that $(T, \mathbf{G})$ sets an adjacency wire from S_1 to S_2. Observe that S_1 cannot contain more than one element because in this case g_2, a NOT gate, will have two inputs. Furthermore, either S_2 contains g_2 only or S_2 is a unary label containing negation gates only (because the input gates do not have input wires). In the latter case, the desired suffix from the head of the adjacency arc to g_2 follows by construction.   ∎

Lemma 5. *Any path of F^* between two original gates that does not involve other original gates is a connecting path.*

Proof. First of all, let us show that any path of F^* between original gates involves at least one adjacency wire. Indeed, by Lemma 2, any path leaving an original gate and not having adjacency wires has only child-parent wires. Such wires lead only to bigger and bigger labels and cannot end up at a singleton one. It follows that at least one adjacency wire is needed.

Let us show that additional adjacency wires cannot occur without original gates as intermediate vertices. Indeed, the head of the first adjacency wire is an *in* gate of some label S. Unless S is a singleton, the only wires leaving $in(S)$ are parent-child wires to the *in* gates of the children of S. Applying this argumentation iteratively, we observe that no other wires except parent-child wires are possible until the path meets the *in* gate of a singleton label. However, this is an original gate that cannot be an intermediate node in our path. It follows that any path between two original gates without other original gates cannot involve 2 adjacency wires. Combining with the previous paragraph, it follows that any such path involves exactly one adjacency wire, i.e. it is a connecting path.   ∎

Using the lemmas above, it can be shown that any cycle in F^* involves at least one original gate and that this implies that F contains a cycle as well, a contradiction showing that F^* is acyclic. The technical details of this derivation are omitted due to space constraints. By construction, each wire connects output to input and there are no gates (except the input gates of course) having no input. It follows that F^* is a well formed circuit.

For each gate g of F^* denote by $f(g, F^*)$ the function computed by a subcircuit of F^* rooted by g. We establish properties of these functions from which Theorem 2 will follow by induction. In the following we sometimes refer to $f(g, F^*)$ as the function of g.

Lemma 6. *For each* NOT *gate* g *of* F^*, $f(g, F^*)$ *is the negation of* $f(g', F^*)$, *where* g' *is the input of* g *in* F.

Proof. According to Lemma 4, F^* has a path from g' to g where all vertices except the first one are NOT gates. Since all of them but the last one are doubled, there is an odd number of such NOT gates. Each NOT gate has a single input, hence the function of each gate of the path (except the first one) is the negation of the function of its predecessor. Hence these functions are, alternatively, the negation of the function of g' and the function of g'. Since the number of NOT gates in the path is odd, the function of g is the negation of the function of g', as required. ∎

In order to establish a similar statement regarding AND and OR gates we need two auxiliary lemmas.

Lemma 7. *For each label* S, $f(oand(S), F^*)$ *is the conjunction of* $f(g, F^*)$ *of all original gates* g *contained in* S. *Similarly,* $f(oor(S), F^*)$ *is the disjunction of the functions of such gates.*

Proof. We prove the lemma only for the *oand* gates as for the *oor* gates the proof is symmetric. The proof easily goes by induction. For an original gate this is just a conjunction of a single element, namely itself, and this is clear by construction. For a larger label S, it follows by construction that $f(oand(S), F^*) = f(oand(S_1), F^*) \wedge f(oand(S_2), F^*))$, where S_1 and S_2 are the children of S. For S_1 and S_2 the rule holds by the induction assumption. Hence, $f(oand(S), F^*)$ is the conjunction of all the functions of all the original gates in the union of S_1 and S_2, that is, the conjunction of the functions of all the original gates contained in S, as required. ∎

Let us call a path of F^* *semi-connecting* if it starts with an adjacency wire and the rest of the wires are parent-child ones.

Lemma 8. *Let* S *be an* AND *label. Then* $f(in(S), F^*)$ *is the conjunction of the functions of all gates from which there is a semi-connecting path to* $in(S)$. *For the* OR *label the statement is analogous with the conjunction replaced by disjunction.*

Proof. We provide the proof only for the AND label, for the OR label the proof is analogous with the corresponding replacements of AND by OR and conjunctions by disjunctions.

The proof is by induction on the decreasing size of labels. For the largest AND label S, all the input wires are the adjacency wires. Clearly the considered function is the conjunction of the functions of the gates at the tails of these adjacency wires. It remains to see if there are no more gates to arrive at $in(S)$ by semi-connected paths. But any such gate, after passing through the adjacency wire must meet an ancestor of S and, by the maximality assumption, S has no ancestors.

The same reasoning as above is valid for any label S without ancestors. If S has ancestors, then $f(in(S), F^*)$ is the conjunction of the functions of the gates at the tails of the adjacency wires incident to $in(S)$ and the function of the in gate of the parent of S. By the induction assumption, this function is in fact a conjunction of the gates at the tails of the adjacency wires incident to $in(S)$ plus those connected to $in(S)$ by semi-connected paths through the parent. Since any semi-connected path either directly hits $in(S)$ at the head of an adjacency wire or approaches it through the parent, the statement is proven.   ∎

Lemma 9. *The function of any original AND gate g of F^* is the conjunction of the functions of the singleton gates whose outputs are the inputs of g in F. The same happens for the OR gate and the disjunction.*

Proof. As before, we prove the statement for the AND gate, for the OR gate it is analogous with the respective substitutions. By construction and Lemma 8, $f(g, F^*)$ is the conjunction of functions of all *oand* gates (since there are no other ones) connected to g by semi-connected paths. Let us call the labels of these *oand* gates the *critical labels*. Combining this with Lemma 7, we see that $f(g, F^*)$ is in fact a conjunction of the functions of all original gates contained in the critical labels. It remains to show that these gates are exactly the in-neighbours of g in F. Let us take a particular in-neighbour g'. By Lemma 3, there is a connecting path from g' to g and by Lemma 7, the tail of the adjacency wire of this path is the *oand* gate of a critical label, so g' is in the required set. Conversely, assume that g' is a gate in the required set. Specify a critical label S g' belongs to. Clearly, there is a child-parent path from g' to $oand(S)$ which, together with a semi-connected path from $oand(S)$ to g, makes a connecting path. The latter means that in F there is a wire from g' to g according to Lemma 3, as required.   ∎

Proof of Theorem 2. Let us order the gates of F topologically and do induction on the topological order. The first gate is an input gate and the function of the input is just the corresponding variable both in F and in F^*. Otherwise, the gate is AND or OR or NOT gate. In the former two cases, according to Lemma 9 the function of g in F^* is the conjunction (or disjunction, in case of OR) of the functions of its inputs in F, the same relation as in F. The theorem holds regarding the inputs by the induction assumption, hence the function of g in F^* is the same as in F. Regarding the NOT gate, the argumentation is analogous, employing Lemma 6.   ∎

3.2   Proof of Theorem 3

Let us define the undirected graph $H = H(T, \mathbf{G})$ called the *representation graph* of $(T, \mathbf{G})$ as follows. The vertices of this graph are the labels of $(T, \mathbf{G})$ and two vertices S_1 and S_2 are adjacent if and only if either S_1 is a child of S_2 (or vice versa of course) or S_1 and S_2 are adjacent in $(T, \mathbf{G})$ (meaning that the new adjacency operation is applied on S_1 and S_2). We call the first type of edges *child-parent* edges and the second type *adjacency* edges.

Lemma 10. *Let t be the treewidth of H. Then the treewidth of F^* is at most $3t + 2$.*

Proof (Sketch). Observe that if we contract the gates in F^* of each label into a single vertex, eliminate directions and remove multiple occurrences of edges, we obtain a graph isomorphic to H. The desired tree decompositom is obtained from the tree decomposition of H by replacing the occurrence of each vertex of H in a bag by the gates corresponding to this vertex. Thus, there is a tree decomposition of F^* with at most $3(t + 1)$ elements in each bag, that is the treewidth of F^* is at most $3t + 2$. ∎

Lemma 11. *The treewidth of H is at most k, where k is the width of $(T, \mathbf{G})$.*

Proof. For each node t of T, let $S(t)$ be the set of labels of $G(t)$ and let $B(t)$ be the set of vertices of H corresponding to $S(t)$. Denote the set of all $B(t)$ by $\mathbf{B}$. We are going to show that $(T, \mathbf{B})$ is a tree decomposition of graph H' obtained from H by removal of all child-parent edges.

First of all, observe that for each $v \in V(H)$, the subgraph T_v of T consisting of all nodes containing v is a subtree of T. Indeed, let us consider T as a rooted tree with the root t being the same as in $(T, \mathbf{G})$. Let t_1 and t_2 be two nodes containing v. Then one of them is an ancestor of the other. Indeed, otherwise t_1 and t_2 are nodes of two disjoint subtrees T_1 and T_2 whose roots t_1' and t_2' are children of some node t^*. By the definition of SCD, $G(t_1')$ is disjoint with $G(t_2')$ and it is not hard to conclude from the definition that $V(G(t_1)) \subseteq V(G(t_1'))$ and $V(G(t_2)) \subseteq V(G(t_2'))$ are disjoint. Since any label is a subset of the set of vertices of the graph it belongs to, $S(t_1)$ and $S(t_2)$ cannot have a common label and hence $B(t_1)$ and $B(t_2)$ cannot have a joint node. Furthermore, it is not hard to observe, if t_1 is ancestor of t_2 and $S \in S(t_1) \cap S(t_2)$ then S belongs to $S(t')$ of all nodes t' in the path between t_1 and t_2. It follows that the node of H corresponding to S belongs to $B(t')$ of all these nodes t'. Thus we have shown that if t_1 and t_2 contain v they cannot belong to different connected components of T_v, confirming the connectedness of T_v.

Next, we observe that if v_1 and v_2 are incident to an adjacency edge then there is a node t containing both v_1 and v_2. Indeed, let S_1 and S_2 be the labels corresponding to v_1 and v_2, respectively. Let t be the node where the adjacency operation regarding S_1 and S_2 is applied. Then both S_1 and S_2 belong to $S(t)$ and, consequently, t contains both v_1 and v_2. Finally, by construction, each vertex of H is contained in some node.

To obtain the desired tree decomposition of H, we are going to modify $(T, \mathbf{B})$ to acquire two properties: that the number of nodes of the resulting tree is at most $2|F|$ and that each parent-child pair u, v is contained in some node t. For the former just iteratively remove all nodes whose operations are new adjacency.

If the node t being removed is not the root then make the parent of t to be the parent of the only child of t (since t has only one child the tree remains binary). The latter property can be established by adding at most one vertex to each bag of the resulting structure $(T', \mathbf{B}')$. Indeed, for each non-singleton label S, let $t(S)$ be the node where this label is created by the union operation. Then both children of S belong to the only child of $t(S)$. Let $(T', \mathbf{B}^*)$ be obtained from $(T', \mathbf{B}')$ as follows. For each non-singleton label S, add the vertex corresponding to S to the bag of the child of $t(S)$. Since at most one new label is created per node of T', at most one vertex is added to each bag. It is not hard to see both of the modifications preserve properties stated in the previous paragraphs and achieve the desired properties regarding the child-parent edges. Since each bag of $(T, \mathbf{B}')$ contains at most $k+1$ elements, we conclude that the treewidth of H is at most k. Since the number of bags is at most as the number of labels, we conclude that the number of bags is at most $2|F|$ ∎

Proof of Theorem 3. Immediately follows from the combination of Lemma 10 and Lemma 11. ∎

4   Application to Knowledge Compilation

In this section we demonstrate an application of Theorem 1 to knowledge compilation by showing the existence of an algorithm compiling the given circuit Z into DNNF. Both the time complexity of the algorithm and the space complexity of the resulting DNNF are fixed-parameter linear parameterized by the cliquewidth of Z. More precisely, the statement is the following:

Theorem 4. *Given a single-output circuit Z of cliquewidth k, there is a DNNF of Z having size $O(9^{18k}k^2|Z|)$. Moreover, given a clique decomposition of Z of width k, there is a $O(9^{18k}k^2|Z|)$ algorithm constructing such a DNNF.*

Theorem 4 is an immediate corollary of Theorem 1 and the following one:

Theorem 5. *Given a circuit Z of treewidth p, there is a DNNF of Z having size $O(9^p p^2|Z|)$. Moreover, such a DNNF can be constructed by an algorithm of the same runtime that gets as input the circuit Z and a tree decomposition of Z of width p having $O(Z)$ bags.*

The rest of this section is a proof of Theorem 5. Our first step is Tseitin transformation from circuit Z into a CNF F'. For this purpose we assume that Z does not have paths of 2 or more NOT gates. Depending on whether this path is of odd or even length, it can be replaced by a single NOT gate or by a wire, without treewidth increase. In this case the variables $y_1, \ldots, y_m$ of F' are the variables of Z and the outputs of AND and OR gates of Z. Under this assumption, it is not hard to see that the inputs of each gate are literals of $y_1, \ldots, y_m$. Then the output x of Z is either y_i or $\neg y_i$ for some i. Let us call x the output literal.

The CNF F' is a conjunction of the singleton clause containing the output literal and the CNFs associated with each AND and OR gate. Let C be an AND gate with inputs $t_1, \ldots, t_r$ and output z. Then the resulting CNF is $(t_1 \vee \neg z) \wedge \ldots \wedge (t_r \vee \neg z) \wedge (\neg t_1 \vee \ldots \vee \neg t_r \vee z)$. If C is an OR gate then the resulting CNF is $(\neg t_1 \vee z) \wedge \ldots \wedge (\neg t_r \vee z) \wedge (t_1 \vee \ldots \vee t_r \vee \neg z)$. We call the last clause of the CNF of C the *carrying* clause w.r.t. C and the rest are *auxiliary* ones w.r.t. C and the corresponding input.

To formulate the property of Tseitin transformation that we need for our transformation, let us extend the notation. We consider sets of literals that do not contain a variable and its negation. For a set S of literals, $Var(S)$ is the set of variables of S. Let $V' \subseteq Var(S)$. The *projection* $Pr(S, V')$ of S to V' is the subset S' of S such that $Var(S') = V'$. Let $\mathbf{S}$ be a family of sets of literals over a set V of variables. Then the projection $Proj(\mathbf{S}, V')$ of $\mathbf{S}$ to $V' \subseteq V$ is $\{Proj(S, V') | S \in \mathbf{S}\}$. Denote by $Var(Z)$ and $Var(F')$ the sets of variables of Z and F', respectively. Let us say that a set S of literals with $Var(S) = Var(Z)$ is a *satisfying assignment* of Z if Z is true on the truth assignment on $Var(Z)$ that assigns all the literals of S to true. For a CNF, the definition is analogous. The well known property of Tseitin transformation is the following:

Lemma 12. *Let* $\mathbf{S}_1$ *and* $\mathbf{S}_2$ *be the sets of satisfying assignments of* F' *and* Z, *respectively. Then* $Proj(\mathbf{S}_1, Var(Z)) = \mathbf{S}_2$.

Lemma 12 is useful because of the following nice property of DNNF.

Lemma 13. *(Theorem 9 of [3]). Let* Z *be a DNNF let* $V' \subseteq Var(Z)$ *and let* Z' *be the DNNF obtained from* Z *by replacing the variables of* $Var(Z) \setminus V'$ *with the true constant. Let* $\mathbf{S}$ *and* $\mathbf{S}'$ *be sets of satisfying assignments of* Z *and* Z', *respectively. Then* $\mathbf{S}' = Proj(\mathbf{S}, V')$.

Thus it follows from Lemmas 12 and 13 that having compiled F' into a DNNF D', a DNNF D of Z can be obtained by replacing the variables of $Var(F') \setminus Var(Z)$ with the *true* constant. Clearly, this does not incur any additional gates. In order to obtain a DNNF of F', we observe that the treewidth of the incidence graph of F' is not much larger than the treewidth of Z.

Lemma 14. *Let* $(T, \mathbf{B})$ *be a tree decompositoion of* Z *of width* p. *There is a* $O(p^2|T|)$ *time algorithm (|T| is the number of nodes of* T) *transforming* $(T, \mathbf{B})$ *into a tree decomposition* $(T^*, \mathbf{B}^*)$ *of the incidence graph* G' *of* F' *having width at most* $2p + 1$ *and with* $|T^*| = O(p^2|T|)$.

Proof (Sketch). Let F'' be the CNF obtained from F' by removal of all the clauses but the carrying ones and let G'' be the respective incidence graph. Transform $(T, \mathbf{B})$ into $(T, \mathbf{B}'')$ as follows:

- Replace each occurrence of an AND or OR gate X with the respective carrying clause and the variable corresponding to the output of X.
- Replace each occurrence of a NOT gate with the variable corresponding to the input of the gate (it may either be an input variable of Z or the output variable of some AND or OR gate).

It can be observed by a straightforward inspection that $(T, \mathbf{B}'')$ is indeed a tree decomposition of G'' of width $2p + 1$.

Next, we observe that for each AND or OR gate X of Z and for each variable u of F' corresponding to an input of X and for variable y of F' corresponding to the output of X, there is a node t of $(T, \mathbf{B}'')$ containing both y and u. Indeed, let C be the carrying clause corresponding to X. By construction, whenever t contains C, t also contains y. By the adjacency property, there is at least one t containing C and u. Since this last t contains also y, this is a desired clause. Pick one node with the specified property and denote it by $t(y, u)$. Add to T a new node t' with $t(y, u)$ being its only neighbour. The bag of t' will contain y, u, and $C(y, u)$ the auxiliary clause of X corresponding to the input u. Do so for all the auxiliary clauses. Finally, properly add a node whose bag contains the variable y of the output literal and the singleton clause containing this literal (the neighbour of this new node should be an existing node containing y). Let $(T^*, \mathbf{B}^*)$ be the resulting structure. It is not hard to observe by construction that $(T^*, \mathbf{B}^*)$ satisfies the statement of the lemma.   ■

It remains to show that a space-efficient DNNF can be created parameterized by the treewidth of the incidence graph.

Theorem 6. *Let F be a CNF and let $(T', \mathbf{B}')$ be a tree decomposition of the incidence graph of F. Then F has a DNNF of size $O(3^t |T'|)$ where t is the width of $(T', \mathbf{B}')$. Moreover, given F and $(T', \mathbf{B}')$ such a DNNF can be constructed by an algorithm having the same runtime.*

We omit the proof of Theorem 6 due to space constraints. It is similar to the proof of Theorem 16 of [3], essentially based on dynamic programming. The difference is that in addition to branching on assignments of variables of the given bag, the algorithm also needs to branch on the clauses of that bag that are not satisfied by the currently considered assignment of variables. Three choices need to be considered for each clause: to not satisfy the clause at all (this choice is needed for 'coordination' with the 'parent bag'), to satisfy the clause by the variables of the left child and to satisfy the clause by the variables of the right child. These 3 choices increase the base of the exponent from 2 to 3.

Remark. It is not hard to see that any tree decomposition can be transformed (without the increase of width) into another one whose number of nodes is at most as big as the number of vertices. Having this in mind, Theorem 6 canbe reformulated with $O(3^t (CL + n))$ istead $O(3^t |T'|)$, where CL and n are, respectively the number of clauses and the number of variables of F. With this reformulation, Theorem 6 becomes of an independent interest because it extends the result of Darwiche [3] from the primal to the incidence graph of the given CNF without increasing much the base of the exponent.

Proof of Theorem 5. The construction of a DNNF for Z consists of 4 stages: transform Z into F' by the Tseitin transformation; transform the tree decomposition of Z into a tree decomposition of the incidence graph of F'; obtain a DNNF of F' as specified by Theorem 6 and obtain a DNNF of Z as specified in Lemma

13. The correctness of this procedure follows from the above discussion. The time and space complexities easily follow from the combination of the complexities of intermediate stages.   ∎

5   Discussion

In this paper we presented a theorem that shows that a circuit of cliquewidth k can be transformed into, roughly speaking, an equivalent circuit of treewidth $18k + 2$ with at most 4 times more gates. A consequence of this statement is that any space-efficient knowledge compilation parameterized by the *treewidth* of the input circuit can be transformed into a space efficient knowledge compilation parameterized by the *cliquewidth* of the input circuit. We elaborated this consequence on the example of DNNF. As a result we obtained a theoretically efficient but formidably looking space complexity of $(9^{18k}k^2n)$. Therefore, the first natural question is how to reduce the base of the exponent.

The next question for further investigation is to check if the proposed upper bound can be applied to SDD [5] which is more practical than DNNF in the sense that it allows a larger set of queries to be efficiently handled. To answer this question positively, it will be sufficient to extend Theorem 6 to the case of SDD, the 'upper' levels of the reasoning will be applied analogously to the case of DNNF.

It is important to note that rankwidth is a better parameter for capturing dense graphs than cliquewidth in the sense that rankwidth of a graph does not exceed its treewidth plus one [12] as well as cliquewidth [13], while cliquewidth can be exponentially larger than treewidth (and hence rankwidth) [1]. Also, computing of rankwidth, unlike cliquewidth, is known to be FPT [11]. Therefore, it is interesting to investigate the relationship between the rankwidth and the treewidth of a Boolean function. For this purpose rankwidth has to be extended to directed graphs [15]. It is worth saying that if the question is answered negatively, i.e. that the treewidth of a circuit can be exponentially larger than its rankwidth, it would be an interesting circuit complexity result.

Finally, recall that all the upper bounds on the DNNF size obtained in this paper are polynomial in the *size* of the circuit which can be much larger than the number of variables. On the other hand, the upper bound on the DNNF size parameterized by the treewidth of the primal graph of the given CNF is polynomial in the number of variables [3]. Can we do the same in the circuit case?

References

1. Corneil, D.G., Rotics, U.: On the relationship between clique-width and treewidth. SIAM J. Comput. 34(4), 825–847 (2005)
2. Courcelle, B., Makowsky, J.A., Rotics, U.: Linear time solvable optimization problems on graphs of bounded clique-width. Theory Comput. Syst. 33(2), 125–150 (2000)

3. Darwiche, A.: Decomposable negation normal form. J. ACM 48(4), 608–647 (2001)
4. Darwiche, A.: On the tractable counting of theory models and its application to truth maintenance and belief revision. Journal of Applied Non-Classical Logics 11(1-2), 11–34 (2001)
5. Darwiche, A.: Sdd: A new canonical representation of propositional knowledge bases. In: IJCAI, pp. 819–826 (2011)
6. Darwiche, A., Marquis, P.: A knowledge compilation map. J. Artif. Intell. Res. (JAIR) 17, 229–264 (2002)
7. Dvorák, W., Szeider, S., Woltran, S.: Reasoning in argumentation frameworks of bounded clique-width. In: COMMA, pp. 219–230 (2010)
8. Fellows, M.R., Rosamond, F.A., Rotics, U., Szeider, S.: Clique-width is NP-complete. SIAM J. Discrete Math. 23(2), 909–939 (2009)
9. Ferrara, A., Pan, G., Vardi, M.Y.: Treewidth in verification: Local vs. Global. In: Sutcliffe, G., Voronkov, A. (eds.) LPAR 2005. LNCS (LNAI), vol. 3835, pp. 489–503. Springer, Heidelberg (2005)
10. Gurski, F., Wanke, E.: The tree-width of clique-width bounded graphs without $K_{n,n}$. In: Brandes, U., Wagner, D. (eds.) WG 2000. LNCS, vol. 1928, pp. 196–205. Springer, Heidelberg (2000)
11. Hlinený, P., Oum, S.I.: Finding branch-decompositions and rank-decompositions. SIAM J. Comput. 38(3), 1012–1032 (2008)
12. Oum, S.I.: Rank-width is less than or equal to branch-width. Journal of Graph Theory 57(3), 239–244 (2008)
13. Oum, S.I., Seymour, P.D.: Approximating clique-width and branch-width. J. Comb. Theory, Ser. B 96(4), 514–528 (2006)
14. Jha, A.K., Suciu, D.: On the tractability of query compilation and bounded treewidth. In: ICDT, pp. 249–261 (2012)
15. Kanté, M.M., Rao, M.: F-rank-width of (edge-colored) graphs. In: Winkler, F. (ed.) CAI 2011. LNCS, vol. 6742, pp. 158–173. Springer, Heidelberg (2011)

A Rank Lower Bound for Cutting Planes Proofs of Ramsey's Theorem

Massimo Lauria

KTH Royal Institute of Technology, Stockholm
lauria@kth.se

Abstract. Ramsey's Theorem is a cornerstone of combinatorics and logic. In its simplest formulation it says that there is a function r such that any simple graph with $r(k, s)$ vertices contains either a clique of size k or an independent set of size s. We study the complexity of proving upper bounds for the number $r(k, k)$. In particular we focus on the propositional proof system cutting planes; we prove that the upper bound "$r(k, k) \leq 4^k$" requires cutting planes proof of high rank. In order to do that we show a protection lemma which could be of independent interest.

1 Introduction

The Ramsey's Theorem for simple graphs claims that if a graph is big enough, it has either a clique or an independent set of moderate size. To be more specific, for any k and s there is a number $r(k, s)$ which is the smallest such that *any graph* with at least $r(k, s)$ vertices contains either a clique of size k or an independent set of size s.

Discovering the actual value of r is challenging, and so far only few points have been computed exactly. For this reason there is great interest in asymptotic estimates. Erdős and Szekeres proved in [14] that

$$r(k, s) \leq \binom{k + s - 2}{k - 1}.$$

Erdős [13] proved a lower bound for the diagonal numbers (i.e. $k = s$):

$$r(k, k) \geq (1 + o(1)) \frac{k}{\sqrt{2e}} 2^{k/2},$$

as one of the first applications of his probabilistic method. Of course there have been some improvements since: to the author's knowledge the current state of the art regarding diagonal numbers $r(k, k)$ is represented by a lower bound of Spencer [28] and an upper bound of Conlon [11].

For the off-diagonal Ramsey numbers (i.e. $r(k, s)$ for $k \neq s$) the state of the art is by Bohman and Keevash (lower bound [3]) and Ajtai, Komlós and Szemerédi (upper bound [1]). The maximally unbalanced numbers $r(3, t)$ got further attention (see [22,1]).

M. Järvisalo and A. Van Gelder (Eds.): SAT 2013, LNCS 7962, pp. 351–364, 2013.

The study of Ramsey theorem in proof theory is well established in literature. In bounded arithmetic there are papers attempting to classify the power of a theory in comparison with Ramsey Theorem. It is also considered a good candidate for separating low levels of bounded depth Frege [25].

A propositional statement of the form "$r(k, k) \leq N$" become easier to prove as N increases. In particular if $m = r(k, k)$ then the statement "$r(k, k) \leq m$" is the hardest possible. Krishnamurthy and Moll [24] proposed this statement as a candidate of a hard formula to prove. They also proved a lower bound on the *width* of the clauses appearing in its resolution refutations. Krajíček later proved an exponential lower bound on the length of bounded depth Frege proofs [23], for the same statement.

Proving a weaker bound should be easier. Indeed it is possible to give a short proof that "$r(k, k) \leq 4^k$" in a relatively weak fragment of sequent calculus (namely, any formula in the proof has bounded depth) [25,23]. It is not clear how strong the proof system must be in order to prove efficiently this statement. Recently Pudlák has shown that resolution is not enough, since the length of a resolution proof of "$r(k, k) \leq 4^k$" must be exponential in the length of the formula itself (see [27]). The propositional complexity of off-diagonal Ramsey upper bounds has received less attention, and the only known results are from [8].

In the context of proof complexity research, cutting planes is one of the most studied proof systems after resolution, so it is natural to ask whether Ramsey's Theorem is hard for it. Cutting planes has been originally introduced as a technique to solve integer programs (see [17,9]). The original idea is to do a canonical linear programming optimization. If the optimum is at a fractional point, it is possible to get an valid inequality which can be "rounded" in order to remove that point from the set of feasible solutions.

Cutting planes was later proposed as a proof system [12], indeed it is possible to view the previous process as a sequence of inferences: a new inequality is either as positive combination or as a rounding of previously derived inequalities. Another way to describe the rounding rule is the following: if the inequality $\sum_i a_i x_i \leq A$ is valid and all a_i are integers divisible by c, then any integer solution would also satisfy $\sum_i \frac{a_i}{c} x_i \leq \lfloor \frac{A}{c} \rfloor$, which is not valid for fractional solutions if A is fractional.

Studying the length of proofs in cutting planes is a way to study the running time for integer linear programming solvers based on the rounding rule. Unfortunately this seems to be difficult. The only lower bound known for *unrestricted* cutting planes refutations is due to Pudlák [26], and it deals with a relatively artificial formula. Lower bounds for more natural formulas exist for cutting plane proofs of restricted forms (e.g. when the numeric coefficients are small [6] or the proof is tree-like [19]). Another restricted form of cutting planes is the one where every proof line has small "degree of falsity" (a complexity measure introduced in [16]). If the degree of falsity is sufficiently small, then the proof system has a sub-exponential simulation in resolution [18]. This implies that most strong resolution lower bounds generalize to this limited version of cutting planes. In particular this is true for [27].

Ramsey's Theorem is a natural is probably difficult for cutting planes. Since length lower bound are out of reach with the current techniques, we focus on the "rank" of a refutation: that is the depth (in term of rounding rule applications) of the refutation. The focus on auxiliary complexity measures in not new in proof complexity, and it is not limited to cutting planes. Well known examples are "width" in resolution, "degree" in polynomial calculus, and "rank" in geometric proof systems like Lovász-Schrijver and sum-of-squares. These measures relate with the actual proof length, in the sense that there are proof search algorithms which runs in time $n^{O(r)}$ on formulas with n variables and measure r. Indeed Chvátal et al. [10] prove that under some technical conditions if there is a cutting planes proof of rank r then there is one of size $n^{O(r)}$. For further information about cutting planes refutations and the notion of rank (also called Chvátal rank) we suggest the reader to refer to [21, Chapter 19].

In this paper we are going to prove that Ramsey's Theorem requires rank $\Omega(2^{k/2})$. The result does not follow from the classic protection lemma for cutting planes [7, Lemma 3.1], so we need to prove a different one which could be of independent interest.

The rest of the paper has the following structure. In Section 2 we give necessary preliminaries: we formally introduce the cutting planes proof system in Section 2.1 and we describe the integer inequalities encoding the Ramsey's theorem in Section 2.2. We then define the rank of a cutting planes proof in Section 2.3. In Section 3 we give the proof of the main theorem (Theorem 2), and in Section 4 we discuss about improvements and related open problems.

2   Preliminaries

2.1   Cutting Planes Proof System

Cutting planes is a technique to solve mixed integer linear programs. In this paper we consider an inference system for refuting unsatisfiable CNFs based on the cutting planes technique. We encode propositional clauses as affine inequalities which have 0–1 solutions if and only if the corresponding assignments satisfy the original clauses. A clause $\vee_i l_i$ translates to the inequality $\sum_i f_i \geq 1$ where

$$f_i = \begin{cases} x & \text{if } l_i = x \\ 1 - x & \text{if } l_i = \neg x \end{cases} \tag{1}$$

For example the clause

$$\neg x \vee y \vee \neg z \tag{2}$$

translates as

$$-x + y - z \geq -1 \tag{3}$$

after summing the constant terms.

After such encoding, any proof that there are no integer solutions for the linear program is a refutation of the corresponding CNF, so we can define a *proof system* for the UNSAT language by the means of cutting planes.

A proof system for UNSAT is a polynomial time machine P which has in input a CNF ϕ and a candidate refutation Π. If the formula ϕ is unsatisfiable there must be some refutation Π for which $P(\phi, \Pi)$ accepts. If ϕ is satisfiable then P does not accept any pair (ϕ, Π).

The study of proof systems was initially motivated by the fact that **NP** is the class of languages with short proof of membership. So in order to separate **NP** from **coNP** we may show that all proof systems for UNSAT require superpolynomial length refutations for some formulas.

Nowadays the study of proof systems focuses in large part on those systems which model actual SAT solvers, automatic theorem provers and algorithms for combinatorial optimization. This is because the study of complexity measures of the refutation process usually gives insight about the performance of such algorithms. In particular most of these algorithms use heuristics to solve what computer science considers hard problems; a proof system has nondeterministic nature, so it models the best possible heuristic and any lower bound on (for example) proof length usually translates to a lower bound on the running time of all such algorithms.

A refutation in cutting planes (as defined in [12]) is an inference process which starts with the inequalities encoding the CNF, and ends with a false inequality $1 \leq 0$. Two inference rules are available.

Positive linear combination

$$\frac{a^T \cdot x \leq A \qquad b^T \cdot x \leq B}{(\alpha a + \beta b)^T \cdot x \leq (\alpha A + \beta B)}$$

for any non negative α, β.

Integer division with rounding

$$\frac{(c \cdot a)^T \cdot x \leq A}{a^T \cdot x \leq \lfloor \frac{A}{c} \rfloor} \quad .$$

Positive linear combination is sound in general. Integer division with rounding is only sound on integer solutions. The rule says that if the integer coefficients of the variables have a common factor c, then dividing everything by c keeps the left side of the inequality to be integer. Thus it is possible to strengthen the right side to the closest integer. Such proof system is complete, since it is possible to transform any resolution refutation of a CNF into a cutting planes refutation of the same CNF.

2.2   Ramsey Statement

Informally, the classical "Ramsey's Theorem" claims that any big enough structure, however complicated, contains an instance of a regular substrucure. A specific instance of Ramsey's theorem on graphs claims that for any two numbers k and s there is an integer number $r(k, s)$ such that any graph with $r(k, s)$ vertices has either a clique of size k or an independent set of size s. In [14] it was proved that $r(k,k) \leq 4^k$ or, equivalently, that any graph with n vertices has either a clique or an independent set of size $\lceil \frac{\log n}{2} \rceil$.

Theorem 1 (Erdös, Szekeres 1935 [14]). *Any graph with 4^k vertices has either a clique of size k or an independent set of size k.*

We study cutting planes proofs of this Ramsey statement. Actually we study refutations of its negation, encoded as a CNF. For any unordered pair of vertices we indifferently denote by either $x_{i,j}$ or $x_{j,i}$ the propositional variable whose intended meaning is that an edge in the graph connects vertices i and j. Let U be a set of vertices, we have two types of clauses.

$$\mathsf{NoCli}(U) := \bigvee\nolimits_{\{i,j\} \in \binom{U}{2}} \neg x_{i,j} \tag{4}$$

$$\mathsf{NoInd}(U) := \bigvee\nolimits_{\{i,j\} \in \binom{U}{2}} x_{i,j} \tag{5}$$

We encode "$r(k,k) > 4^k$" as the following CNF, which has $\binom{4^k}{2}$ variables and $2\binom{4^k}{k}$ clauses of width $\binom{k}{2}$.

$$\mathrm{RAM}_k := \left(\bigwedge_{U \in \binom{[4^k]}{k}} \mathsf{NoCli}(U) \right) \wedge \left(\bigwedge_{U \in \binom{[4^k]}{k}} \mathsf{NoInd}(U) \right). \tag{6}$$

In cutting planes refutations the clauses are represented as follows:

$$\mathsf{NoCli}(U) : \sum_{\{i,j\} \in \binom{U}{2}} x_{i,j} \leq \binom{k}{2} - 1 \tag{7}$$

$$\mathsf{NoInd}(U) : \sum_{\{i,j\} \in \binom{U}{2}} x_{i,j} \geq 1 \tag{8}$$

which can be succinctly represented as

$$1 \leq \sum_{\{i,j\} \in \binom{U}{2}} x_{i,j} \leq \binom{k}{2} - 1. \tag{9}$$

In the rest of the paper we keep everything expressed as a function of k. To get a picture on the proof complexity of this formula it is useful to state it at least once in term of the number n of vertices in the graph. This customary for propositional formulas related to graph theory. Here $n = 4^k$: the formula has $\Theta(n^2)$ variables and $n^{\Theta(\log n)}$ clauses of width $\Theta(\log n)$, so it has quasi-polynomial length with respect to the number of variables. In this paper we prove a rank lower bound of roughly $\Omega(\sqrt[4]{n})$.

2.3 The Rank of a Cutting Planes Refutation

One complexity measure for cutting planes is the "rank" of an inference. Other geometric proof systems, with specific inference rules, have similar notions of rank. The rank of cutting planes proof system is also called Chvátal Rank.

The linear program that we use to encode the CNF does not take into account the fact that we care about integer solutions only. Indeed the initial polyhedron contains fractional solutions that we want to ignore. We do that by adding further inequalities which are valid on integer solutions but may remove fractional ones. The "integer division with rounding" inference rule is the way employed by cutting planes to add such inequalities. All initial inequalities have rank 0. A line obtained applying the "positive linear combination" rule from two inequalities of rank r_1 and r_2 has rank $\max\{r_1, r_2\}$. A line obtained from an inequality of rank r using the division rule has rank $r + 1$.

Thus the rank represents the nesting of integer division applications in the refutation. The rank of a refutation is the largest rank among the refutation lines. The rank of an unsatisfiable CNF is the smallest rank among all possible refutations.

The notion of rank has also a geometric interpretation: a point p has rank r if there is an inequality of rank $r + 1$ which is not satisfied by p, and such that p satisfies all inequalities of rank r. More concretely we can think the inequalities to define a chain of polyhedrons $P_0 \supseteq \ldots \supseteq P_i \supseteq \ldots \supseteq P_I$, where P_i contains all points of rank $\geq i$, and P_I is the convex hull of all integer solutions of the linear program. It is a well known fact that there is $r \geq 0$ such that $P_r = P_I$. If the CNF has no solution then $P_I = \emptyset$, and the rank of P_I corresponds to such r.

To show that the rank of a refutation is at least r, is sufficient to show that there is a point in P_r. To do that the only known technique is the use of protection lemmata, which roughly say that if some points in a structured set (called "protection set") have rank i, then another point has rank $i + 1$.

In particular it is possible to define a prover-delayer game as follows: prover challenges the delayer to exhibit a protection set for a point p_0. Delayer either gives up or shows a set S_0. At the next round the prover picks a point $p_1 \in S_0$ and asks again for a protection set. If the Delayer has a strategy to play the game for r rounds, then the point p_0 has rank at least r.

3   A Protection Lemma for Fractional Graphs

The fractional points that we will use in this paper have a peculiar structure. We only use half integral points (i.e. each coordinate is either 0, $\frac{1}{2}$, or 1), which in turn is a natural encoding of partially specified graphs: 0 encodes non-edges, 1 encodes edges, $\frac{1}{2}$ encodes unspecified edges. The points we are interested in have additional structure, as described by the following definition.

Definition 1 (Fractional graph). *A "fractional graph" is a pair $G = (V, E)$ on the vertex set V when E is a function $E : \binom{V}{2} \to \{0, \frac{1}{2}, 1\}$. Consider $U \subseteq V$ such that for all $\{u, v\}$*

$$E(\{u, v\}) = \frac{1}{2} \text{ if and only if } \{u, v\} \nsubseteq U,$$

then we say that G is integral on the vertex set U. U is called the integral part of G.

It is clear that a fractional graph is an half-integral point in the space $[0,1]^{\binom{V}{2}}$, thus the notion of rank applies to fractional graphs. The integral part of a fractional graph is unique.

Remark on Notation: in the following we use $x_{i,j}$ to denote the variables referring to edges in the graph, and we denote an inequality as "$a \cdot x \leq b$" or "$ax \leq b$". We denote as G both the fractional graph and the corresponding point in the space. Indeed for a fractional graph $G = (V, E)$ the notation "$a \cdot G$" indicates the value

$$\sum_{\{u,v\} \in \binom{V}{2}} a_{u,v} E(\{u, v\}).$$

Fractional graphs are actually vectors with coordinates in $[0,1]$, so we can make convex combination of them. For this paper we just need the average between two graphs.

Definition 2 (Graph average). *Given two fractional graphs $G_1 = (V, E_1)$ and $G_2 = (V, E_2)$ we consider the average of them (denoted as $\frac{1}{2}G_1 + \frac{1}{2}G_2$) to be the graph $H = (V, \frac{E_1 + E_2}{2})$.*

The average of two fractional graphs is not necessarily a fractional graph according to our definition. It is in the particular conditions that we enforce in the definition of protection sets and in the rest of the paper.

Definition 3 (Protection set). *Consider a fractional graph G which is integral on the vertices in I and a set of graph pairs $\left(G'_{\{u,v\}}, G''_{\{u,v\}}\right)$, one graph pair for each vertex pair $\{u, v\}$ disjoint from I. The set of graph pairs is a protection set for G if for all pairs it holds that:*
- *both $G'_{\{u,v\}}$ and $G''_{\{u,v\}}$ are integral on $I \cup \{u, v\}$;*
- *$G = \frac{1}{2}G'_{\{u,v\}} + \frac{1}{2}G''_{\{u,v\}}$.*

If p is a point in $[0,1]^{\binom{V}{2}}$ we denote $p_{a,b}$ has the value of the coordinate of p corresponding to edge $\{a, b\}$. In particular if p represents a fractional graph $G = (V, E)$ then $p_{a,b} = E(\{a, b\})$. The following simple lemma highlights the peculiar structure of a protection set for G.

Lemma 1. *Consider a graph G with integral part I and choose a pair $(G'_{\{u,v\}}, G''_{\{u,v\}})$ from some protection set for G. Let p, p', p'' to be the points representing $G, G'_{\{u,v\}}, G''_{\{u,v\}}$, respectively. The following hold:*
1. *for any $\{a, b\} \subseteq I$, $p_{a,b} = p'_{a,b} = p''_{a,b}$;*
2. *for any $\{a, b\} \nsubseteq I$ and $\{a, b\} \subseteq I \cup \{u, v\}$, $p_{a,b} = \frac{1}{2}$ and $p'_{a,b} = 1 - p''_{a,b}$.*

Proof. Point (1) holds because edge $\{a, b\}$ is in the integral part: $p_{a,b}$ must be integer and equal to $\frac{p'_{a,b} + p''_{a,b}}{2}$, so the values of $p'_{a,b}$ and $p''_{a,b}$ must be equal to $p_{a,b}$; to prove (2) notice that $\{a, b\} \nsubseteq I$ immediately implies that $p_{a,b} = \frac{1}{2}$. Both $G'_{\{u,v\}}$ and $G''_{\{u,v\}}$ have integral edge $\{a, b\}$, so the values $p'_{a,b}, p''_{a,b}$ must be opposite in order to average to $\frac{1}{2}$. □

We show a protection lemma for fractional graphs which essentially states that the previous definition of protection set is meaningful, and thus will be useful to get rank lower bounds. This protection lemma is different from the ones already known: every point in a protection set has additional integer values in the coordinates, and in constructions from literature such coordinates must be disjoint and independently settable (see [7]). In our construction this is not needed, which allows us to use protection sets made *by fractional graphs*.

We now focus on the sequence of polytopes $[0, 1]^{\binom{V}{2}} \supseteq P_0 \supseteq P_1 \supseteq \cdots \supseteq P_i \supseteq \cdots$, where P_i is the set of points of rank at least i.

Lemma 2 (Protection Lemma). *Let G be a fractional graph with an even number of vertices and an integral part of even size. If G has a protection set contained in P_i, then $G \in P_{i+1}$.*

Proof. The fractional graph G is the average of two points in P_i by construction, so $G \in P_i$ as well. Assume by contradiction that $G \notin P_{i+1}$, then it holds that $a \cdot G > b$ where $ax \leq b$ is an inequality of rank $i+1$. We can derive such inequality by integer division from an inequality $a'x \leq b'$ of rank i, where

$$a'_{u,v} = q a_{u,v} \qquad b' = qb + r \qquad \text{for some } q, r \in \mathbb{Z} \text{ with } 0 < r < q. \tag{10}$$

Since $G \in P_i$ we have $a' \cdot G \leq b' < q(b + 1)$. Putting all together we have that $b < a \cdot G < b + 1$.

Fix I to be the integral vertices of G, and $J = V(G) \setminus I$. The value of $a \cdot G$ is strictly less than $b + 1$ but it is strictly larger than b, so it must be $b + \frac{1}{2}$. The coefficient vector a is integral, thus it follows that

$$\sum_{\{u,v\} \in J} a_{u,v} + \sum_{u \in J, w \in I} a_{u,w} \equiv 1 \pmod 2 \tag{11}$$

because otherwise $a \cdot G$ would be integral.

We now show that equation (11) implies that we can find at least one pair $\{u, v\} \subseteq J$ for which it holds that:

$$a_{u,v} + \sum_{w \in I} a_{u,w} + \sum_{w \in I} a_{v,w} \equiv 1 \pmod 2. \tag{12}$$

To see this denote $b_u := \sum_{w \in I} a_{u,w}$ for all $u \in J$. Equations (11) and (12) can be rewritten as

$$\sum_{\{u,v\} \in J} a_{u,v} + \sum_{u \in J} b_u \equiv 1 \pmod 2 \tag{13}$$

and

$$a_{u,v} + b_u + b_v \equiv 1 \pmod 2. \tag{14}$$

We partition J in two classes $J_0 = \{u \in J : b_u \equiv_2 0\}$ and $J_1 = \{u \in J : b_u \equiv_2 1\}$. If there is a pair $\{u, v\}$ such that $b_u \equiv b_v \pmod 2$ and $a_{u,v} \equiv 1 \pmod 2$ we are done; if there is a pair $\{u, v\}$ such that $b_u \not\equiv b_v \pmod 2$ and $a_{u,v} \equiv 0 \pmod 2$

we are also done. If neither happens then we can manipulate equation (13) as follows

$$1 \equiv \sum_{\{u,v\}\in J} a_{u,v} + \sum_{u\in J} b_u \equiv \sum_{u\in J_0}\sum_{v\in J_1} a_{u,v} + \sum_{u\in J_1} b_u \equiv |J_0||J_1| + |J_1| \pmod 2,$$

which is a contradiction: $|J_0||J_1| + |J_1| = (|J_0|+1)(|J|-|J_0|)$ and since $|J|$ is even, the right hand side is always even.

Fix any pair $\{u,v\}$ such that equation (12) holds. We consider $a \cdot G$ as the sum of three contributions: the sum over the integral edges of G, the sum over the edges enumerated in equation (12) for the chosen pair $\{u,v\}$, and the sum over the rest of the edges. Let us call these sums A,B and C respectively: clearly $A + B + C = b + \frac{1}{2}$. All edges in G corresponding to the sum B have value $\frac{1}{2}$, so by equation (12) B is half integral, and in particular follows that $A + C$ is integer.

Consider the two graphs $G'_{\{u,v\}}$ and $G''_{\{u,v\}}$ in the protection set. By definition they must differ from G *only* on the edges which coefficients are in the summation (12), thus $a \cdot G'_{\{u,v\}} = A + B' + C$ and $a \cdot G''_{\{u,v\}} = A + B'' + C$ for some B' and B''. On these edges the two graphs have integral values, so B' and B'' are integer numbers.

It follows that numbers $a \cdot G'_{\{u,v\}}$ and $a \cdot G''_{\{u,v\}}$ are integral and (being the two graphs in P_i) they are strictly smaller than $b + 1$. Thus the two values are at most b. G is the average of the two graphs, so it follows that $a \cdot G \le b$, which contradicts the assumption that $G \notin P_{i+1}$.   □

We are now ready to prove the lower bound on rank of cutting planes proof of the Ramsey number upper bound.

Theorem 2. *For all even $k \ge 4$, cutting planes rank of formula* RAM_k *is at least* $2^{k/2-1}$.

Proof. Consider the following Prover-Delayer game:

Initial choice (round 0): let P_0 be the polytope described by the linear system of RAM_k, and let G_0 a fractional graph with empty integral part (i.e. all edges have value $\frac{1}{2}$).

Delayer choice (round $i > 0$): delayer shows a protection set for G_{i-1} contained in P_0.

Prover choice (round $i > 0$): prover sets G_i to be an arbitrary element of the protection set of G_{i-1} shown by delayer.

For $k \ge 4$, fractional graph G_0 satisfies all equations (9), thus it is a point of the initial polytope P_0. Lemma 2 says that if delayer reaches round i, then G_0 has rank at least i. To prove the theorem it is sufficient to show a strategy for Delayer for playing up to round $2^{k/2-1}$.

At each step i in the prover-delayer game G_i is a fractional graph with an integral part of $2i$ vertices, since each application of Lemma 2 adds exactly two vertices. Furthermore at each step we keep a bijection σ_i between the integral part of G_i and $\{1\ldots 2i\}$.

We are going to build the protection sets using a model graph H on vertex set $\{1 \ldots 2^{k/2}\}$. The indicator variable $h_{i,j}$ is either 1 if $\{i, j\} \in E(H)$ or 0 otherwise. We call "diagonal pair" any pair of the form $\{2m-1, 2m\}$, for some $m \in [2^{k/2-1}]$. We need H to have properties in the following claim:

Claim 1. There exists a graph H such that
 - H has neither a clique nor an independent set of size k;
 - for every H' obtained from H by arbitrarily changing the diagonal edges, the previous property holds for H';
 - given any diagonal pair $\{2m - 1, 2m\}$ and any vertex $a < 2m - 1$, it holds that
$$h_{a,2m-1} = 1 - h_{a,2m}.$$

This graph H has $2m = 2^{k/2}$ vertices, so the fact that it has no clique and no independent set of size k does not necessarily violate the Ramsey's theorem. Indeed we will show later that such graph H exists.

Delayer Strategy: delayer uses such H to define its strategy against prover. The main idea is that at each round a new pair of vertices in G_0 is mapped to some diagonal pair of H. Each G_i in the trace of the game is almost a copy of the graph induced by the vertices $\{1 \ldots 2i\}$ on H. We say "almost", because the value on the diagonal pair will be changed arbitrarily. We call σ_i the mapping at round i, and we define σ_0 to be the empty mapping.

At round i we want to show a protection set for G_i, with integral part I. For each u and v not in I, we define the two graphs $G'_{u,v}$ and $G''_{u,v}$ by adding $\{u, v\}$ to the integral part in the following way: for every $a \in I$

$$p'_{a,u} := h_{\sigma_i(a),(2m-1)}$$
$$p'_{a,v} := h_{\sigma_i(a),2m}$$
$$p''_{a,u} := h_{\sigma_i(a),2m}$$
$$p''_{a,v} := h_{\sigma_i(a),(2m-1)}$$
$$p'_{u,v} := 0$$
$$p''_{u,v} := 1,$$

where p, p', p'' are the point representing fractional graphs G_i, $G'_{u,v}$ and $G''_{u,v}$, respectively. The other coordinates of p' and p'' keep the values of p. By construction the defined graphs make a protection set, because they satisfy the conditions of Definition 3.

After Prover Choice: prover can choose either $G'_{u,v}$ or $G''_{u,v}$ for some pair $\{u, v\}$. If prover chooses $G'_{u,v}$ then we extend σ_i to σ_{i+1} by adding the mapping $u \mapsto (2m - 1)$ and $v \mapsto 2m$. Otherwise we add the mapping $u \mapsto 2m$ and $v \mapsto (2m - 1)$.

Finally we show that the player can play for $e = 2^{k/2-1}$ rounds. In order to play that many rounds we need to argue that G_e is contained in P_0, or equivalently that it satisfies equations (9). Take an arbitrary set of vertices $K \subseteq$

$V(G_e)$ of size $k \geq 4$: if there is even a single vertex out of the integral part, then the sum contains at least two half-integral variables. None of the bounds in (9) is violated.

If K is contained in the integral part of G_e, notice that the latter is isomorphic to some H' which is obtained from H by arbitrarily changing the edges on the diagonal pairs. By Claim 1 graph H' does not contain homogeneous vertices of size k. Thus Equation (9) on K is satisfied.

We have proved that $G_e \in P_0$. That means (using Lemma 2) that $G_{e-1} \in P_1$, $G_{e-2} \in P_2$, ..., and so on until $G_0 \in P_e$. This shows that P_e is not the empty polytope, and that inequality $0 \leq -1$ has rank larger than e. This concludes the proof of the theorem. □

Proof (of Claim 1). Consider any $i \leq 2^{k/2-1}$. We determine independently at random the 0–1 values of $h_{v,(2i-1)}$ for all vertices $v < 2i - 1$, and we set $h_{(v,2i)} := 1 - h_{v,(2i-1)}$. This definition immediately enforces the third condition of the claim. We get the first and the second condition by probabilistic method: we show that with positive probability any set of vertices of size k contains both an edge and a non-edge which *are not on diagonal pairs*. This is true by construction for any set K containing a diagonal pair $\{2m - 1, 2m\}$ plus some other vertex $v < 2m - 1$. Let $\mathcal{K}_0$ the family of sets of size k with no diagonal pair, and $\mathcal{K}_1$ the family of sets of size k such that the two smallest elements form a diagonal pair. The size of the families are

$$|\mathcal{K}_0| = 2^k \binom{n/2}{k} \qquad |\mathcal{K}_1| = 2^{k-2} \binom{n/2}{k-1}.$$

Families $\mathcal{K}_0$ and $\mathcal{K}_1$ are empty unless $k \geq 8$, and the graph H has no homogeneous sets of size k by construction. Consider $k \geq 8$. There are $\binom{k}{2}$ independent random edges in sets from $\mathcal{K}_0$, and $\binom{k}{2} - 1$ in sets from $\mathcal{K}_1$. Fix $n = 2^{k/2}$, and notice that n is even. Then

$$\Pr[H \text{ has a homogeneous set of size } k] \leq \sum_{K \in \mathcal{K}} \Pr[K \text{ is homogeneous}] \leq$$

$$\leq |\mathcal{K}_0| \frac{2}{2^{\binom{k}{2}}} + |\mathcal{K}_1| \frac{2}{2^{\binom{k}{2}-1}} \leq \frac{2}{2^{\binom{k}{2}}} \left[2^k \binom{n/2}{k} + 2^{k-1} \binom{n/2}{k-1} \right] < 1, \quad (15)$$

for $n = 2^{k/2}$. □

4   Conclusion

We have seen that Ramsey's Theorem requires refutations of large rank. Of course the actual rank depends on the value of $r(k, k)$ itself: the proof may focus on the first $r(k, k)$ vertices and the corresponding $\binom{r(k,k)}{2}$ variables. Thus in order to improve the rank lower bound it is necessessary to understand better the Ramsey number itself, in particular its lower bounds.

Rank is just an auxiliary complexity measure: the interest of proof complexity revolves around the length of proofs. Unfortunately there is very little understanding about the length of cutting planes refutations: the only lower bound known is based on the interpolation technique [26]. This means that the formula for which the lower bound is proved has ad-hoc structure and is not interesting per se. Such lower bound has been proved by harnessing the connection between cutting planes inferences and monotone computation [26,5]. It is an open problem how to prove length lower bounds for natural formulas, in particular using combinatorial techniques which allow to study more general CNFs.

A natural question is whether the rank has a role here. In other proof systems (e.g. resolution and polynomial calculus) a good lower bound on an auxiliary complexity measure implies proof length lower bounds [2,20]. It is interesting to notice that even if this implication is true then it must have some limitations, since there are formulas with large rank (i.e. the square root of the number of variables) and small refutations [7]. The latter also happens in resolution and polynomial calculus (with width and degree complexity measure, respectively. See [15,4]). Still the study of such auxiliary measures allowed proof size lower bounds.

In order to understand the relation between rank and length of cutting planes proof the following question is unavoidable:

Open Problem 1. *Is there any k-CNF formula on n variables with polynomial length refutations and cutting planes rank $\Omega(n)$?*

As mentioned before there is a formula on n variables, polynomial length refutation and rank $\Omega(\sqrt{n})$ (see [7]). Thus any rank-length connection which holds in general would not be useful to prove a length lower bound for Ramsey's Theorem, given the current knowledge. So even if a rank-length trade-off is proved, that would not solve the following problem:

Open Problem 2. *Does RAM_k have a cutting planes refutation of polynomial length?*

For further open problems about cutting planes refutations we suggest to refer to the book [21, Chapter 19].

Acknowledgment. The author did most of this work while he was at the Math Institute of the Czech Academy of Science, funded by the Eduard Čech Center. While finalizing the paper, the author has been supported by the European Research Council under the European Union's Seventh Framework Programme (FP7/2007–2013) / ERC grant agreement no 279611.

References

1. Ajtai, M., Komlós, J., Szemerédi, E.: A note on Ramsey numbers. Journal of Combinatorial Theory, Series A 29(3), 354–360 (1980)
2. Ben-Sasson, E., Wigderson, A.: Short proofs are narrow - resolution made simple. J. ACM 48(2), 149–169 (2001)

3. Bohman, T., Keevash, P.: The early evolution of the h-free process. Inventiones Mathematicae 181(2), 291–336 (2010)
4. Bonet, M.L., Galesi, N.: Optimality of size-width tradeoffs for resolution. Computational Complexity 10(4), 261–276 (2001)
5. Bonet, M.L., Pitassi, T., Raz, R.: Lower bounds for cutting planes proofs with small coefficients. In: Proceedings of the Twenty-Seventh Annual ACM Symposium on Theory of Computing, pp. 575–584. ACM (1995)
6. Bonet, M.L., Pitassi, T., Raz, R.: Lower bounds for cutting planes proofs with small coefficients. The Journal of Symbolic Logic 62(3), 708–728 (1997)
7. Buresh-Oppenheim, J., Galesi, N., Hoory, S., Magen, A., Pitassi, T.: Rank bounds and integrality gaps for cutting planes procedures. Theory of Computing 2(4), 65–90 (2006)
8. Carlucci, L., Galesi, N., Lauria, M.: Paris-harrington tautologies. In: Proc. of IEEE 26th Conference on Computational Complexity, pp. 93–103 (2011)
9. Chvátal, V.: Edmonds polytopes and a hierarchy of combinatorial problems. Discrete Mathematics 4(4), 305–337 (1973)
10. Chvátal, V., Cook, W., Hartmann, M.: On cutting-plane proofs in combinatorial optimization. Linear Algebra and its Applications 114, 455–499 (1989)
11. Conlon, D.: A new upper bound for diagonal ramsey numbers. Annals of Mathematics 170(2), 941–960 (2009)
12. Cook, W., Coullard, C.R., Turán, G.: On the complexity of cutting-plane proofs. Discrete Applied Mathematics 18(1), 25–38 (1987)
13. Erdös, P.: Some remarks on the theory of graphs. Bull. Amer. Math. Soc 53, 292–294 (1947)
14. Erdős, P., Szekeres, G.: A combinatorial problem in geometry. In: Gessel, I., Rota, G.-C. (eds.) Classic Papers in Combinatorics, Modern Birkhäuser Classics, pp. 49–56. Birkhäuser, Boston (1987)
15. Galesi, N., Lauria, M.: Optimality of size-degree tradeoffs for polynomial calculus. ACM Transaction on Computational Logic 12, 4:1–4:22 (2010)
16. Goerdt, A.: The cutting plane proof system with bounded degree of falsity. In: Börger, E., Jäger, G., Kleine Büning, H., Richter, M.M. (eds.) CSL 1991. LNCS, vol. 626, pp. 119–133. Springer, Heidelberg (1992)
17. Gomory, R.E.: Outline of an algorithm for integer solutions to linear programs. Bulletin of the American Mathematical Society 64(5), 275–278 (1958)
18. Hirsch, E.A., Nikolenko, S.I.: Simulating cutting plane proofs with restricted degree of falsity by resolution. In: Bacchus, F., Walsh, T. (eds.) SAT 2005. LNCS, vol. 3569, pp. 135–142. Springer, Heidelberg (2005)
19. Impagliazzo, R., Pitassi, T., Urquhart, A.: Upper and lower bounds for tree-like cutting planes proofs. In: Proceedings of the Symposium on Logic in Computer Science, LICS 1994, pp. 220–228. IEEE (1994)
20. Impagliazzo, R., Pudlák, P., Sgall, J.: Lower bounds for the polynomial calculus and the gröbner basis algorithm. Computational Complexity 8(2), 127–144 (1999)
21. Jukna, S.: Boolean Function Complexity: Advances and Frontiers. Springer (2012)
22. Kim, J.H.: The Ramsey number $r(3, t)$ has order of magnitude $t^2/\log(t)$. Random Structures and Algorithms 7(3), 173–208 (1995)
23. Krajíček, J.: A note on propositional proof complexity of some Ramsey-type statements. Archive for Mathematical Logic 50, 245–255 (2011), doi:10.1007/s00153-010-0212-9
24. Krishnamurthy, B., Moll, R.N.: Examples of hard tautologies in the propositional calculus. In: 13th ACM Symposium on Th. of Computing, STOC 1981, pp. 28–37 (1981)

25. Pudlák, P.: Ramsey's theorem in Bounded Arithmetic. In: Schönfeld, W., Börger, E., Kleine Büning, H., Richter, M.M. (eds.) CSL 1990. LNCS, vol. 533, pp. 308–317. Springer, Heidelberg (1991)

26. Pudlák, P.: Lower bounds for Resolution and Cutting Plane proofs and monotone computations. Journal of Symbolic Logic 62(3), 981–998 (1997)

27. Pudlák, P.: A lower bound on the size of resolution proofs of the ramsey theorem. Inf. Process. Lett. 112(14-15), 610–611 (2012)

28. Spencer, J.: Asymptotic lower bounds for Ramsey functions. Discrete Mathematics 20, 69–76 (1977)

The Complexity of Theorem Proving
in Autoepistemic Logic*

Olaf Beyersdorff

School of Computing, University of Leeds, UK

Abstract. Autoepistemic logic is one of the most successful formalisms for nonmonotonic reasoning. In this paper we provide a proof-theoretic analysis of sequent calculi for credulous and sceptical reasoning in propositional autoepistemic logic, introduced by Bonatti and Olivetti [5]. We show that the calculus for credulous reasoning obeys almost the same bounds on the proof size as Gentzen's system LK. Hence proving lower bounds for credulous reasoning will be as hard as proving lower bounds for LK. This contrasts with the situation in sceptical autoepistemic reasoning where we obtain an exponential lower bound to the proof length in Bonatti and Olivetti's calculus.

1 Introduction

Autoepistemic logic is one of the most popular nonmonotonic logics which is applied in a diversity of areas as commonsense reasoning, belief revision, planning, and reasoning about action. It was introduced by Moore [19] as a modal logic with a single modal operator L interpreted as "is known". Semantically, autoepistemic logic describes possible views of an ideally rational agent on the grounds of some objective information. Autoepistemic logic has been intensively studied, both in its semantical as well as in its computational aspects (cf. [18]). The main computational problems in autoepistemic logic are the credulous and sceptical reasoning problems, formalising that a given formula holds under some, respectively all, views of the agent. Thus these problems can be understood as generalisations of the classical problems SAT and TAUT. However, in autoepistemic logic, these tasks are presumably harder than their propositional counterparts as they are complete for the second level of the polynomial hierarchy [12].

   In this paper we target at understanding the complexity of autoepistemic logic in terms of theorem proving. Traditionally, the main objective in proof complexity has been the investigation of propositional proofs [7, 16]. During the last decade there has been growing interest in proof complexity of non-classical logics, most notably modal and intuitionistic logics [14, 15], and strong results have been obtained (cf. [1] for an overview and further references). For autoepistemic logic, Bonatti and Olivetti [5] designed elegant sequent calculi for both credulous and sceptical reasoning. In this paper we provide a proof-theoretic analysis of these calculi. Our main results show that *(i)* the calculus for credulous

---

* Research supported by a grant from the John Templeton Foundation.

M. Järvisalo and A. Van Gelder (Eds.): SAT 2013, LNCS 7962, pp. 365–376, 2013.

autoepistemic reasoning obeys almost the same bounds to the proof size as the classical sequent calculus LK and *(ii)* the calculus for sceptical autoepistemic reasoning has exponential lower bounds to the size and length of proofs.

These results are interesting to compare with previous findings for default logic—another principal approach in nonmonotonic logic. In a wider attempt to a proof-theoretic formalisation of nonmonotic logics, Bonatti and Olivetti [5] also devise calculi for default logic which were proof-theoretically analysed in [2, 10]. Default logic is known to admit a very close relation to autoepistemic logic via translations [13], but these are not directly applicable to transfer proof complexity results from default logic to the autoepistemic calculi. Our findings on autoepistemic logic in the present paper confirm results from [2] where the authors establish a similar polynomial dependence between proof lengths in LK and credulous default reasoning. Combining results from [2] with Theorem 4 of this paper, we can infer that credulous reasoning in default and in autoepistemic logic have the same complexity in theorem proving. On the other hand, [2] also provides an unconditional exponential lower bound for sceptical default reasoning. This reveals an interesting general picture for nonmonotonic logics: while credulous reasoning is equivalent to classical reasoning in terms of lengths of proofs, lower bounds are easier to obtain for sceptical reasoning. We comment further on the broader picture in Section 5.

This paper is organised as follows. In Sect. 2 we start with some background information on autoepistemic logic and proof systems. Our results on the proof complexity of credulous and sceptical autoepistemic reasoning follow in Sects. 3 and 4, respectively. In Sect. 5, we conclude with a discussion and open questions.

2   Preliminaries

We assume familiarity with propositional logic and basic notions from complexity theory (cf. [16]). By $\mathcal{L}$ we denote the set of all propositional formulas over some fixed standard set of connectives. By $\top$ and $\bot$ we denote the logical constants true and false, respectively. For formulas φ, σ, θ, the notation $\varphi[\sigma/\theta]$ means that all occurrences of subformulas σ in φ are replaced by θ.

Autoepistemic Logic. Autoepistemic logic is an extension of classical logic that has been proposed by Moore [19]. The logic is non-monotone in the sense that an increase in information may decrease the number of consequences. The language of autoepistemic logic $\mathcal{L}^{ae}$ consists of the language $\mathcal{L}$ of classical propositional logic augmented by an unary modal operator L. Intuitively, for a formula φ, the formula $L\varphi$ means that φ is *believed* by a rational agent. We emphasize that L-operators could be nested. Classical propositional formulas without occurrence of L are called *objective* formulas. A set of *premises* is a finite set of $\mathcal{L}^{ae}$ formulas.

Propositional assignments are extended to assignments for autoepistemic logic by considering all formulas of the form $L\varphi$ as propositional atoms, *i.e.*, in autoepistemic logic an assignment is a mapping from all propositional variables and formulas $L\varphi$ to $\{0,1\}$. This yields an immediate extension of the consequence relation to autoepistemic logic: if $\Phi \subseteq \mathcal{L}^{ae}$ and $\varphi \in \mathcal{L}^{ae}$, then $\Phi \models \varphi$ iff φ is true under every assignment which satisfies all formulas from Φ. As in classical logic we define $Th(\Phi) = \{\varphi \in \mathcal{L}^{ae} \mid \Phi \models \varphi\}$.

The main semantical notion in autoepistemic logic are *stable expansions* which correspond to all possible views an ideally rational agent might adopt on the knowledge of some set of premises $\Sigma \subseteq \mathcal{L}^{ae}$. Formally, a stable expansion of $\Sigma \subseteq \mathcal{L}^{ae}$ was defined by Moore [19] as a set $\Delta \subseteq \mathcal{L}^{ae}$ satisfying the fixed-point equation

$$\Delta = Th\left(\Sigma \cup \{L\varphi \mid \varphi \in \Delta\} \cup \{\neg L\varphi \mid \varphi \notin \Delta\}\right).$$

Informally, a stable expansion corresponds to a possible view of an agent, allowing him to derive all statements of his view from the given premises Σ together with his believes and disbelieves.

A set of premises Σ can have none or several stable expansions. A sentence $\varphi \in \mathcal{L}^{ae}$ is *credulously* entailed by Σ if φ holds in *some* stable expansion of Σ. If φ holds in *every* expansion of Σ, then φ is *sceptically* entailed by Σ. We give some examples which will be important later on.

Example 1. (a) If the premises Σ only consist of objective formulas, then Σ has exactly one stable expansion, namely the deductive closure of Σ (together with closure under L) if Σ is consistent and $\mathcal{L}^{ae}$ if Σ is inconsistent. (b) The set $\{p \leftrightarrow Lp\}$ has two stable expansions, one containing p and Lp and the other containing both $\neg p$ and $\neg Lp$. (c) The set $\{Lp\}$ has no stable expansion.

Proof Systems. Cook and Reckhow [8] defined the notion of a *proof system* for an arbitrary language L as a polynomial-time computable function f with range L. A string w with $f(w) = x$ is called an f-*proof* for $x \in L$. Proof systems for $L = \text{TAUT}$ are called *propositional proof systems*. The sequent calculus LK of Gentzen [11] is one of the most important and best studied propositional proof systems. It is well known that LK and Frege systems mutually p-simulate each other (cf. [16] for background information on proof systems and definitions of LK and Frege).

There are two measures which are of primary interest in proof complexity. The first is the minimal *size* of an f-proof for some given element $x \in L$. To make this precise, let $s_f^*(x) = \min\{|w| \mid f(w) = x\}$ and $s_f(n) = \max\{s_f^*(x) \mid |x| \leq n\}$. We say that the proof system f is t-*bounded* if $s_f(n) \leq t(n)$ for all $n \in \mathbb{N}$. If t is a polynomial, then f is called *polynomially bounded*. Another interesting parameter of a proof is the *length* defined as the number of proof steps. This measure only makes sense for proof systems where proofs consist of lines containing formulas or sequents. This is the case for LK and most systems studied in this paper. For such a system f, we let $t_f^*(\varphi) = \min\{k \mid f(\pi) = \varphi \text{ and } \pi \text{ uses } k \text{ steps}\}$ and $t_f(n) = \max\{t_f^*(\varphi) \mid |\varphi| \leq n\}$. Obviously, it holds that $t_f(n) \leq s_f(n)$, but the two measures are even polynomially related for a number of natural systems as extended Frege (cf. [16]).

The Antisequent Calculus. Bonatti and Olivetti's calculi for autoepistemic logic use three main ingredients: classical propositional sequents and rules of LK, antisequents to refute formulas, and autoepistemic rules. In this section we introduce Bonatti's *antisequent calculus AC* from [4]. In AC we use *antisequents* $\Gamma \nvdash \Delta$, where $\Gamma, \Delta \subseteq \mathcal{L}$. Semantically, $\Gamma \nvdash \Delta$ is true if there exists an assignment satisfying $\bigwedge \Gamma$ and falsifying $\bigvee \Delta$. Axioms of AC are all sequents $\Gamma \nvdash \Delta$, where Γ and Δ are disjoint sets of propositional variables. The inference rules of AC are shown in Fig. 1. Bonatti [4] shows soundness and completeness of the calculus

$$\frac{\Gamma \nvdash \Sigma, \alpha}{\Gamma, \neg\alpha \nvdash \Sigma} \; (\neg\nvdash) \qquad\qquad \frac{\Gamma, \alpha \nvdash \Sigma}{\Gamma \nvdash \Sigma, \neg\alpha} \; (\nvdash \neg)$$

$$\frac{\Gamma, \alpha, \beta \nvdash \Sigma}{\Gamma, \alpha \wedge \beta \nvdash \Sigma} \; (\wedge\nvdash) \qquad \frac{\Gamma \nvdash \Sigma, \alpha}{\Gamma \nvdash \Sigma, \alpha \wedge \beta} \; (\nvdash \bullet\wedge) \qquad \frac{\Gamma \nvdash \Sigma, \beta}{\Gamma \nvdash \Sigma, \alpha \wedge \beta} \; (\nvdash \wedge\bullet)$$

$$\frac{\Gamma \nvdash \Sigma, \alpha, \beta}{\Gamma \nvdash \Sigma, \alpha \vee \beta} \; (\nvdash \vee) \qquad \frac{\Gamma, \alpha \nvdash \Sigma}{\Gamma, \alpha \vee \beta \nvdash \Sigma} \; (\bullet\vee \nvdash) \qquad \frac{\Gamma, \beta \nvdash \Sigma}{\Gamma, \alpha \vee \beta \nvdash \Sigma} \; (\vee\bullet \nvdash)$$

$$\frac{\Gamma, \alpha \nvdash \Sigma, \beta}{\Gamma \nvdash \Sigma, \alpha \to \beta} \; (\nvdash\to) \qquad \frac{\Gamma \nvdash \Sigma, \alpha}{\Gamma, \alpha \to \beta \nvdash \Sigma} \; (\bullet\to\nvdash) \qquad \frac{\Gamma, \beta \nvdash \Sigma}{\Gamma, \alpha \to \beta \nvdash \Sigma} \; (\to\bullet\nvdash)$$

Fig. 1. Inference rules of the antisequent calculus AC

AC. Proofs in the antisequent calculus are always short as observed in [2] (the bounds are not stated explicitly, but are implicit in the proof):

Proposition 2 (contained in [2]). $s_{AC}(n) \leq n^2$ and $t_{AC}(n) \leq n$.

The polynomial upper bounds on the complexity of AC are not surprising, since, to prove $\Gamma \nvdash \Delta$ we could alternatively guess assignments to the propositional variables in Γ and Δ and thereby verify antisequents in NP.

3 Proof Complexity of Credulous Autoepistemic Reasoning

We can now describe the calculus *CAEL* of Bonatti and Olivetti [5] for credulous autoepistemic reasoning. A *credulous autoepistemic sequent* is a 3-tuple $\langle \Sigma, \Gamma, \Delta \rangle$, denoted by $\Sigma; \Gamma \hspace{-0.3em}\sim\hspace{-0.3em} \Delta$, where Σ, Γ, and Δ are sets of $\mathcal{L}^{ae}$-formulas. Moreover, all formulas of Σ are of the form $L\alpha$ or $\neg L\alpha$ and are called *provability constraints*. Semantically, the sequent $\Sigma; \Gamma \hspace{-0.3em}\sim\hspace{-0.3em} \Delta$ is true, if there exists a stable expansion E of Γ which satisfies all of the constraints in Σ (*i.e.*, $E \models \Sigma$) and $\bigvee \Delta \in E$. The calculus *CAEL* uses credulous autoepistemic sequents and extends LK and AC by the inference rules shown in Fig. 2. Bonatti and Olivetti [5] show soundness and completeness of *CAEL*.

$$(\text{cA1})\ \frac{\Gamma \vdash \Delta}{;\ \Gamma \hspace{-3pt}\sim\hspace{-3pt}\Delta}\ (\Gamma \cup \Delta \subseteq \mathcal{L})$$

$$(\text{cA2})\ \frac{\Gamma \vdash \alpha \qquad \Sigma;\ \Gamma \hspace{-3pt}\sim\hspace{-3pt}\Delta}{L\alpha,\ \Sigma;\ \Gamma \hspace{-3pt}\sim\hspace{-3pt}\Delta}\ (\alpha \in \mathcal{L})$$

$$(\text{cA3})\ \frac{\Gamma \nvdash \alpha \qquad \Sigma;\ \Gamma \hspace{-3pt}\sim\hspace{-3pt}\Delta}{\neg L\alpha,\ \Sigma;\ \Gamma \hspace{-3pt}\sim\hspace{-3pt}\Delta}\ (\Gamma \cup \{\alpha\} \subseteq \mathcal{L})$$

$$(\text{cA4})\ \frac{\neg L\alpha,\ \Sigma;\ \Gamma[L\alpha/\bot]\hspace{-3pt}\sim\hspace{-3pt}\Delta[L\alpha/\bot]}{\Sigma;\ \Gamma\hspace{-3pt}\sim\hspace{-3pt}\Delta} \qquad (\text{cA5})\ \frac{L\alpha,\ \Sigma;\ \Gamma[L\alpha/\top]\hspace{-3pt}\sim\hspace{-3pt}\Delta[L\alpha/\top]}{\Sigma;\ \Gamma\hspace{-3pt}\sim\hspace{-3pt}\Delta}$$

In rules (**cA4**) and (**cA5**) $L\alpha$ is a subformula of $\Gamma \cup \Delta$ and $\alpha \in \mathcal{L}$.

Fig. 2. Inference rules for the credulous autoepistemic calculus $CAEL$

Theorem 3 (Bonatti, Olivetti [5]). *A credulous autoepistemic sequent is true if and only if it is derivable in* $CAEL$.

We now investigate the complexity of proofs in $CAEL$, showing a very tight connection to proof size and length in the classical sequent calculus LK.

Theorem 4. $CAEL$ *obeys almost the same bounds on proof size and number of proof steps as* LK, *more precisely:* $s_{LK}(n) \leq s_{CAEL}(n) \leq n(s_{LK}(n) + n^2 + n)$ *and* $t_{LK}(n) \leq t_{CAEL}(n) \leq n(t_{LK}(n) + n + 1)$.

Proof. In the following we will explain all sequent proofs "backwards", i.e., we start the description with the rule that is immediately applied to derive the proven sequent and progress bottom up until we reach initial sequents or axioms. For the first inequality $s_{LK}(n) \leq s_{CAEL}(n)$ (and similarly $t_{LK}(n) \leq t_{CAEL}(n)$) it suffices to observe that each $CAEL$-proof of a sequent $\Gamma\hspace{-3pt}\sim\hspace{-3pt}\Delta$ with $\Gamma \cup \Delta \subseteq \mathcal{L}$ consists of one application of rule (**cA1**) followed by an LK-derivation of $\Gamma \vdash \Delta$. This holds as rules (**cA2**) to (**cA5**) are only applicable if $\Gamma\hspace{-3pt}\sim\hspace{-3pt}\Delta$ contain at least one occurrence of the L-operator.

We will now prove the remaining upper bounds, starting with $t_{CAEL}(n) \leq n(t_{LK}(n)+n+1)$. For $\alpha \in \mathcal{L}^{ae}$ we denote by $LC(\alpha)$ the number of occurrences of L in α. We extend this notation to $\Delta \subseteq \mathcal{L}^{ae}$ by defining $LC(\Delta) = \sum_{\alpha \in \Delta} LC(\alpha)$. Let $\Sigma; \Gamma\hspace{-3pt}\sim\hspace{-3pt}\Delta$ be a true credulous autoepistemic sequent of total size n (as a string). We will construct a $CAEL$-derivation of $\Sigma; \Gamma\hspace{-3pt}\sim\hspace{-3pt}\Delta$ starting from the bottom with the given sequent. We first claim that we can normalise the proof such that we start (always bottom-up) by eliminating all subformulas $L\alpha$ in $\Gamma \cup \Delta$ by using rules (**cA4**) and (**cA5**) and then use rules (**cA2**) and (**cA3**) to eliminate all provability constraints. Finally, one application of rule (**cA1**) follows. Thus the normalised proof will look as in Fig. 3. Let us argue that this normalisation is possible. By Theorem 3 there exists a proof Π of $\Sigma; \Gamma\hspace{-3pt}\sim\hspace{-3pt}\Delta$. At its top Π must contain exactly one application of (**cA1**). The rest of the proof are applications of (**cA2**) to (**cA5**). As (**cA2**) and (**cA3**) do not alter the part $\Gamma\hspace{-3pt}\sim\hspace{-3pt}\Delta$ of the

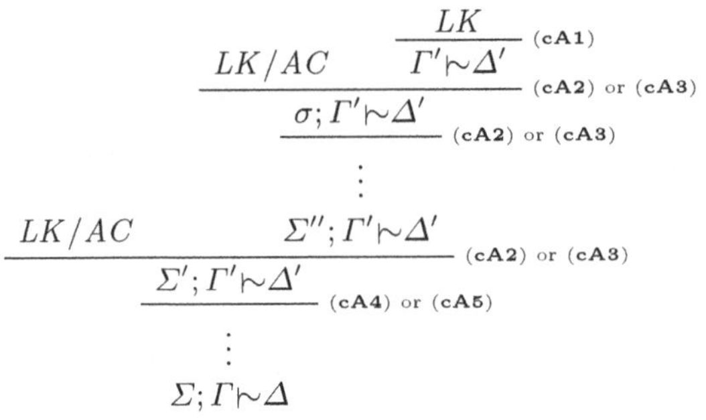

$$\cfrac{LK/AC \qquad \cfrac{\cfrac{LK}{\Gamma' \!\sim\! \Delta'} \; \text{(cA1)}}{\sigma; \Gamma' \!\sim\! \Delta'} \; \text{(cA2) or (cA3)}}{\text{(cA2) or (cA3)}}$$

$$\vdots$$

$$\cfrac{LK/AC \qquad \cfrac{\Sigma''; \Gamma' \!\sim\! \Delta'}{\text{(cA2) or (cA3)}}}{\cfrac{\Sigma'; \Gamma' \!\sim\! \Delta'}{\text{(cA4) or (cA5)}}}$$

$$\vdots$$

$$\Sigma; \Gamma \!\sim\! \Delta$$

Fig. 3. The structure of the $CAEL$-proof in Theorem 4. LK/AC denotes a proof in either LK or AC, LK denotes an LK-derivation, and σ is the last remaining constraint from Σ' after applications of (**cA2**) and (**cA3**).

sequent, they can be freely interchanged with applications of (**cA4**) and (**cA5**). This yields a normalised proof of the same size as Π.

We now estimate the length of this normalised proof. Eliminating all subformulas $L\alpha$ in $\Gamma \cup \Delta$ needs at most $LC(\Gamma \cup \Delta)$ applications of rules (**cA4**) and (**cA5**). The number of steps needed could be less than $LC(\Gamma \cup \Delta)$ as one step might delete several instances of $L\alpha$. After this process we obtain a sequent $\Sigma'; \Gamma' \!\sim\! \Delta'$ with $\Gamma' \cup \Delta' \subseteq \mathcal{L}$ and $|\Sigma'| \leq |\Sigma| + LC(\Gamma \cup \Delta) < n$. From this point on we use rules (**cA2**) and (**cA3**) until we have eliminated all constraints and then finally apply rule (**cA1**) once. This will result in $|\Sigma'| + 1 \leq n$ applications of rules (**cA1**) to (**cA3**). Each of these applications will invoke either an LK or an AC derivation of the left premise, but all these derived formulas are either from Σ or subformulas of Γ or Δ. Therefore all these LK and AC-derivations are used to prove formulas of size $\leq n$. To estimate the lengths of AC-proofs we use Proposition 2. In total this gives $\leq n(t_{LK}(n) + n + 1)$ steps to prove $\Sigma; \Gamma \!\sim\! \Delta$.

The bound for s_{CAEL} follows as each of the $< n$ applications of (**cA4**) and (**cA5**) leads to a sequent of size $\leq n$ and therefore this part of the proof is of size $\leq n^2$. Each of the (**cA2**) and (**cA3**) applications shortens the sequent $\Sigma'; \Gamma' \!\sim\! \Delta'$ which is of size $\leq n$ and incurs an LK or AC-derivation of a sequent of size $\leq n$. Using Proposition 2 and taking account of the final (**cA1**) application this contributes at most $n(s_{LK}(n) + n^2)$ to the size of the overall proof. $\square$

In the light of this result, proving either non-trivial lower or upper bounds to the proof size of $CAEL$ seems very difficult as such a result would directly imply a corresponding bound for LK which is know to be equivalent with respect to proof size to Frege systems. Showing any non-trivial lower bound for Frege is one of the hardest challenges in propositional proof complexity and this problem has been open for decades (cf. [4, 16]).

The connection between proof size in classical LK and credulous autoepistemic logic has further consequences. In particular, it allows to transfer intractability results from classical logic to autoepstemic reasoning. *Automatizability* asks whether proofs can be efficiently constructed, *i.e.*, whether a proof of φ in a proof system P can be found in polynomial time in the length of the shortest P-proof of φ [6]. Of course automatizability of a proof system is very desirable from a practical point of view. However, most known classical proof systems are not automatizable under cryptographic or complexity-theoretic assumptions. In particular, Bonet, Pitassi, and Raz [6] showed that Frege systems are not automatizable unless Blum integers can be factored in polynomial time (a Blum integer is the product of two primes which are both congruent 3 modulo 4). Frege systems are known to be equivalent to LK [8]. As credulous autoepistemic reasoning extends LK this result easily transfers to credulous autoepistemic reasoning:

Corollary 5. *CAEL is not automatizable unless factoring integers is possible in polynomial time.*

The same result also holds for the sceptical autoepistemic calculus analysed in the next section.

4   Lower Bounds for Sceptical Autoepistemic Reasoning

Bonatti and Olivetti [5] also introduce a calculus for sceptical autoepistemic reasoning. In contrast to the credulous calculus, sequents are simpler as they only consist of two components $\Gamma, \Delta \subseteq \mathcal{L}^{ae}$. An *SAEL* sequent is such a pair $\langle \Gamma, \Delta \rangle$, denoted by $\Gamma \hspace{-0.3em}\sim\hspace{-0.3em} \Delta$. Semantically, the *SAEL* sequent $\Gamma \hspace{-0.3em}\sim\hspace{-0.3em} \Delta$ is true, if $\bigvee \Delta$ holds in *all* expansions of Γ.

To give the definition of the *SAEL* calculus of Bonatti and Olivetti [5] we need some notation. An L-subformula of an $\mathcal{L}^{ae}$-formula φ is a subformula of φ of the form $L\theta$. By $LS(\varphi)$ we denote the set of all L-subformulas of φ. $ELS(\varphi)$ denotes the set of all *external* L-subformulas of φ, *i.e.*, all L-subformulas of φ that do not occur in the scope of another L-operator. The notation is extended to sets of formulas Φ by $LS(\Phi) = \bigcup_{\varphi \in \Phi} LS(\varphi)$ and $ELS(\Phi) = \bigcup_{\varphi \in \Phi} ELS(\varphi)$. We say that a set $\Gamma \subseteq \mathcal{L}^{ae}$ is *complete* with respect to $\Sigma \subseteq \mathcal{L}^{ae}$ if for all $\varphi \in \Sigma$, either $\varphi \in \Gamma$ or $\neg\varphi \in \Gamma$.

Bonatti and Olivetti's [5] calculus *SAEL* consists of the defining axioms and inference rules of LK and AC together with the rules shown in Fig. 4. Bonatti and Olivetti show soundness and completeness of this calculus for sceptical autoepistemic reasoning:

Theorem 6 (Bonatti, Olivetti [5]). *An SAEL sequent $\Gamma \hspace{-0.3em}\sim\hspace{-0.3em} \Delta$ is derivable in SAEL if and only if it is true.*

Let us comment a bit on the rules in Fig. 4. In rule (**sA2**), if $\neg L\alpha, \Gamma \hspace{-0.3em}\sim\hspace{-0.3em} \alpha$ is true, then $\neg L\alpha, \Gamma$ has no stable expansion and thus $\neg L\alpha, \Gamma \hspace{-0.3em}\sim\hspace{-0.3em} \Delta$ vacuously holds. The same applies to rule (**sA3**). If $L\alpha, \Gamma \not\vdash \alpha$ holds, then $L\alpha, \Gamma$ does not have any stable expansion and $L\alpha, \Gamma \hspace{-0.3em}\sim\hspace{-0.3em} \Delta$ is true (cf. [5, Theorem 5.14] for a detailed argument). Thus, to derive a sequent $\Gamma \hspace{-0.3em}\sim\hspace{-0.3em} \Delta$ where the antecedent Γ has a stable expansion, we can only use one of the rules (**sA1**) or (**sA4**) to immediately get $\Gamma \hspace{-0.3em}\sim\hspace{-0.3em} \Delta$. Note that rule (**sA1**) is quite powerful. Not only can it be used to

$$\textbf{(sA1)}\ \frac{\Gamma \vdash \Delta}{\Gamma \hspace{-0.3em}\sim\hspace{-0.3em} \Delta} \qquad \textbf{(sA2)}\ \frac{\neg L\alpha, \Gamma \hspace{-0.3em}\sim\hspace{-0.3em} \alpha}{\neg L\alpha, \Gamma \hspace{-0.3em}\sim\hspace{-0.3em} \Delta} \qquad \textbf{(sA3)}\ \frac{L\alpha, \Gamma \not\vdash \alpha}{L\alpha, \Gamma \hspace{-0.3em}\sim\hspace{-0.3em} \Delta}$$

where $\Gamma \cup \{L\alpha\}$ is complete wrt. $ELS(\Gamma \cup \{\alpha\})$ in rule **(sA3)**

$$\textbf{(sA4)}\ \frac{L\alpha, \Gamma \hspace{-0.3em}\sim\hspace{-0.3em} \Delta \qquad \neg L\alpha, \Gamma \hspace{-0.3em}\sim\hspace{-0.3em} \Delta}{\Gamma \hspace{-0.3em}\sim\hspace{-0.3em} \Delta}\ (L\alpha \in LS(\Gamma \cup \Delta))$$

Fig. 4. Inference rules for the sceptical autoepistemic calculus $SAEL$

derive sequents $\Gamma \hspace{-0.3em}\sim\hspace{-0.3em} \Delta$ with Γ and Δ comprising of only classical formulas, but it also applies to autoepistemic sequents $\Gamma \hspace{-0.3em}\sim\hspace{-0.3em} \Delta$ if L-subformulas are treated as propositional atoms. We give an example.

Example 7. Let Γ_n be the sequence $p_1 \leftrightarrow Lp_1, \ldots, p_n \leftrightarrow Lp_n, q$ and $\Delta_n = p_1 \vee Lq$. We obtain the derivation in Fig. 5. Neither Γ_n nor Δ_n consist of classical formulas and $\Gamma_n \vdash \Delta_n$ are no true classical sequents, but still **(sA1)** together with the omitted LK-derivations guarantee short proofs. Note that Γ_n has 2^n stable expansions (cf. also Example 1), but still the overall proofs of $\Gamma_n \hspace{-0.3em}\sim\hspace{-0.3em} \Delta_n$ are of linear length.

$$\frac{\dfrac{LK}{Lq, (p_i \leftrightarrow Lp_i)_{i \in [n]}, q \vdash p_1 \vee Lq}}{Lq, (p_i \leftrightarrow Lp_i)_{i \in [n]}, q \hspace{-0.3em}\sim\hspace{-0.3em} p_1 \vee Lq}\text{(sA1)} \qquad \frac{\dfrac{\dfrac{LK}{\neg Lq, (p_i \leftrightarrow Lp_i)_{i \in [n]}, q \vdash q}}{\neg Lq, (p_i \leftrightarrow Lp_i)_{i \in [n]}, q \hspace{-0.3em}\sim\hspace{-0.3em} q}\text{(sA1)}}{\dfrac{\neg Lq, (p_i \leftrightarrow Lp_i)_{i \in [n]}, q \hspace{-0.3em}\sim\hspace{-0.3em} p_1 \vee Lq}{}}\text{(sA2)}$$
$$\frac{}{(p_i \leftrightarrow Lp_i)_{i \in [n]}, q \hspace{-0.3em}\sim\hspace{-0.3em} p_1 \vee Lq}\text{(sA4)}$$

Fig. 5. Derivation of $\Gamma_n \hspace{-0.3em}\sim\hspace{-0.3em} \Delta_n$ in Example 7

In our next result we will show an exponential lower bound to the proof length (and therefore also to the proof size) in the sceptical calculus $SAEL$.

Theorem 8. *There exist sequents S_n of size $\mathcal{O}(n)$ such that every SAEL-proof of S_n has $2^{\Omega(n)}$ steps. Therefore, $s_{SAEL}(n), t_{SAEL}(n) \in 2^{\Omega(n)}$.*

Proof. Let Γ_n consist of the formulas $p_i \leftrightarrow Lp_i$, $p_i \leftrightarrow q_i$ with $i = 1, \ldots, n$ and $\Delta_n = \bigwedge_{i=1}^{n} (Lp_i \leftrightarrow Lq_i)$. We will prove that each $SAEL$-proof of $\Gamma_n \hspace{-0.3em}\sim\hspace{-0.3em} \Delta_n$ contains 2^n applications of rule **(sA4)**. Consider now sequents

$$(Lp_i : i \in I_p^+), (\neg Lp_i : i \in I_p^-), (Lq_i : i \in I_q^+), (\neg Lq_i : i \in I_q^-), \Gamma_n \hspace{-0.3em}\sim\hspace{-0.3em} \Delta_n \qquad (1)$$

where $I_p^+, I_p^-, I_q^+, I_q^- \subseteq [n]$ and $I_p^+ \cap I_p^- = I_q^+ \cap I_q^- = \emptyset$. If additionally $I_p^+ \cap I_q^- = I_p^- \cap I_q^+ = \emptyset$, we call a sequent of the form (1) a k-*sequent* for $k = |[n] \setminus (I_p^+ \cup I_p^- \cup I_q^+ \cup I_q^-)|$.

For a variable p let us denote by p^1 the variable p while p^{-1} stands for $\neg p$. We first note that each antecedent Γ of a k-sequent $\Gamma \hspace{-0.3em}\sim\hspace{-0.3em} \Delta$ has exactly 2^k stable expansions. Let $J = [n] \setminus (I_p^+ \cup I_p^- \cup I_q^+ \cup I_q^-)$ be the index set corresponding to the L-subformulas which are not already fixed by the antecedent. Then the stable expansions of Γ are generated by $p_j^{e_j}, q_j^{e_j}$ with $j \in J$ (together with $(p_i, q_i : i \in I_p^+ \cup I_q^+)$ and $(p_i^{-1}, q_i^{-1} : i \in I_p^- \cup I_q^-)$) where the variables $(e_j)_{j \in J}$ range over all 2^k elements of $\{-1, 1\}^k$.

We will now prove the following claim:

Claim. For all $k = 1, \ldots, n$, each *SAEL*-proof Π of $\Gamma_n \hspace{-0.3em}\sim\hspace{-0.3em} \Delta_n$ contains at least 2^k $(n-k)$-sequents. Moreover, all of these $(n-k)$-sequents appear as a premise of an application of (**sA4**) which has a $(n-k+1)$-sequent as its consequence.

For $k = n$ this claim yields the desired lower bound.

We prove the claim by induction on k. For the base case $k = 1$ observe that $\Gamma_n \hspace{-0.3em}\sim\hspace{-0.3em} \Delta_n$ is an n-sequent. We first determine which rule which was used in the proof Π to derive $\Gamma_n \hspace{-0.3em}\sim\hspace{-0.3em} \Delta_n$. The antecedent Γ_n has 2^n stable expansions. Therefore, $\Gamma_n \hspace{-0.3em}\sim\hspace{-0.3em} \Delta_n$ cannot have been derived by either rule (**sA2**) or (**sA3**) (cf. the discussion before Example 7). Likewise, $\Gamma_n \hspace{-0.3em}\sim\hspace{-0.3em} \Delta_n$ is not derivable by (**sA1**). This is so because even considering all subformulas Lp_i, Lq_i as propositional atoms, $\Gamma_n \hspace{-0.3em}\sim\hspace{-0.3em} \Delta_n$ is not a true propositional sequent. Therefore $\Gamma_n \hspace{-0.3em}\sim\hspace{-0.3em} \Delta_n$ is derived by an application of (**sA4**) by branching over some L-subformula Lp_i or Lq_i. This yields two distinct $(n-1)$-sequents.

For the inductive step let $\Gamma' \hspace{-0.3em}\sim\hspace{-0.3em} \Delta'$ be a $(n-k)$-sequent in Π which appears as a premise of an application of (**sA4**) and has a $(n-k+1)$-sequent as its consequence. Let us determine which rule which was used in the proof Π to derive $\Gamma' \hspace{-0.3em}\sim\hspace{-0.3em} \Delta'$. As $\Gamma' \hspace{-0.3em}\sim\hspace{-0.3em} \Delta'$ is a $(n-k)$-sequent, its antecedent Γ' has 2^{n-k} stable expansions (see above). Therefore, $\Gamma' \hspace{-0.3em}\sim\hspace{-0.3em} \Delta'$ cannot have been derived by either rule (**sA2**) or (**sA3**) (cf. the discussion before Example 7). Likewise, $\Gamma' \hspace{-0.3em}\sim\hspace{-0.3em} \Delta'$ is not derivable by (**sA1**). This is so because even considering all subformulas Lp_i, Lq_i as propositional atoms, $\Gamma' \hspace{-0.3em}\sim\hspace{-0.3em} \Delta'$ is not a true propositional sequent. Its succedent Δ' contains subformulas $Lp_i \leftrightarrow Lq_i$, $i \in [n] \setminus (I_p^+ \cup I_p^- \cup I_q^+ \cup I_q^-)$ which are not propositionally implied by the antecedent Γ'. Therefore $\Gamma' \hspace{-0.3em}\sim\hspace{-0.3em} \Delta'$ is derived by an application of (**sA4**) branching over an L-subformula Lx_i of $\Gamma' \cup \Delta'$ where x_i stands for either p_i or q_i. There are three cases according to the choice of variable x_i.

Case 1: $i \in I_p^+ \cap I_q^+$ or $i \in I_p^- \cap I_q^-$. In this case applying (**sA4**) yields two sequents, one of them a sequent with contradictory formulas in the antecedent, the other one again a $(n-k)$-sequent which deviates from $\Gamma' \hspace{-0.3em}\sim\hspace{-0.3em} \Delta'$ only in that Lx_i occurs repeatedly in Γ'. As this only increases the size of the overall proof, Case 1 does not occur in proofs of minimal size.

Case 2: $i \in I_p^+ \triangle I_q^+$ or $i \in I_p^- \triangle I_q^-$.[1] As both cases are symmetric let us assume $i \in I_p^+ \triangle I_q^+$. Then (**sA4**) yields the two sequents $Lx_i, \Gamma' \hspace{-0.3em}\sim\hspace{-0.3em} \Delta'$ and $\neg Lx_i, \Gamma' \hspace{-0.3em}\sim\hspace{-0.3em} \Delta'$. The latter sequent $\neg Lx_i, \Gamma' \hspace{-0.3em}\sim\hspace{-0.3em} \Delta'$ contains either both $\neg Lp_i$ and Lq_i (if $x_i = p_i$) or both of Lq_i and $\neg Lp_i$ (if $x_i = q_i$) in its antecedent. Therefore the antecedent

---

[1] Here $\triangle$ denotes symmetric difference, defined as $A \triangle B = (A \setminus B) \cup (B \setminus A)$.

is even propositionally unsatisfiable and hence the sequent $\neg Lx_i, \Gamma' \hspace{0.3em}\vdash\hspace{-0.75em}\sim\hspace{0.3em} \Delta'$ can be proven by an LK-derivation followed by (**sA1**).

The first sequent $Lx_i, \Gamma' \cup \Delta'$ is again a $(n - k)$-sequent (which, however, does not fulfil the second sentence of the inductive claim). We apply again our previous argument to this sequent: it must have been derived by (**sA4**). This application might fall again under Case 2, but this can only occur a constant number of times and eventually we will get an application of (**sA4**) to a $(n - k)$-sequent according to the only remaining Case 3.

Case 3: $i \in [n] \setminus (I_p^+ \cup I_p^- \cup I_q^+ \cup I_q^-)$. In this case (**sA4**) produces two ancestor sequents $Lx_i, \Gamma' \hspace{0.3em}\vdash\hspace{-0.75em}\sim\hspace{0.3em} \Delta'$ and $\neg Lx_i, \Gamma' \hspace{0.3em}\vdash\hspace{-0.75em}\sim\hspace{0.3em} \Delta'$. Both of these are $(n - k - 1)$-sequents and also fulfil the second condition of the inductive claim.

As we have seen, all three cases start with a $(n - k)$-sequent and lead to two $(n - k - 1)$-sequents, and all of these sequents fulfil the second condition of the inductive claim. By the induction hypothesis, Π contains 2^k many $(n - k)$-sequents. All of these are derived by one or more applications of (**sA4**) from prerequisite $(n - k - 1)$-sequents which are mutually distinct. Thus Π contains 2^{k+1} many $(n - k - 1)$-sequents, completing the argument. $\qquad\square$

We point out that our argument does not only work against tree-like proofs, but also rules out sub-exponential dag-like derivations for $\Gamma_n \hspace{0.3em}\vdash\hspace{-0.75em}\sim\hspace{0.3em} \Delta_n$. Thus, while dag-like derivations are typically shorter we also obtain an exponential lower bound in this stronger model.

5   Conclusion and Discussion

In this paper we have shown that with respect to lengths of proofs, proof systems for credulous autoepistemic reasoning and for propositional logic are very close to each other. On the other hand, we demonstrated exponential bounds for sceptical autoepistemic reasoning in the natural calculus of [5]. Such bounds are completely out of reach for the calculus LK in propositional logic. This situation closely resembles our findings for propositional default logic [2]. Credulous reasoning is Σ_2^p-complete for both default logic and autoepistemic logic while the sceptical reasoning tasks are both Π_2^p-complete as shown by Gottlob [12] (cf. also [3,9] for a refined analysis). Can this common underlying complexity of the decision problems serve as explanation for the similarities in proof complexity of these logics?

Let us dwell a bit on this theme. Although deciding credulous autoepistemic sequents is presumably harder than deciding tautologies (the former is Σ_2^p-complete [12], while the latter is complete for coNP), the difference disappears when we want to prove these objects. This becomes most apparent when we consider polynomially bounded proof systems: by the classical theorem of Cook and Reckhow [8], polynomially bounded propositional proof systems exist if and only if NP = coNP, while credulous autoepistemic reasoning (or any logic with a Σ_2^p-complete decision problem) has polynomially bounded proof systems if and only if NP = Σ_2^p. However, the assertions NP = coNP and NP = Σ_2^p are equivalent and this also extends to other proof lengths:

Proposition 9. *Let L be a language in Σ_2^p and let f be any monotone function. Then $\mathrm{TAUT} \in \mathrm{NTIME}(f(n))$ implies $L \in \mathrm{NTIME}(p(n)f(p(n)))$ for some polynomial p. In other words, for each propositional proof system P with $s_P(n) \le f(n)$ there exists a proof system P' for L with $s_{P'}(n) \le p(n)f(p(n))$.*

Proof. If $L \in \Sigma_2^p$, then there exists a polynomial-time nondeterministic oracle Turing machine M which decides L under oracle access to TAUT. Assume now $\mathrm{TAUT} \in \mathrm{NTIME}(f(n))$ via NTM N. We build an NTM N' for L by simulating M and replacing each oracle query θ to TAUT by the following nondeterministic procedure. Guess the answer to query θ. If the answer is yes, then simulate $N(\theta)$ and check that it accepts. Otherwise, if the answer is no, then guess an assignment α and verify that α satisfies $\neg\theta$. If p is the polynomial bounding the running time of M, then each oracle query is of size $\le p(n)$ and there can be at most $p(n)$ such queries. Therefore the running time of N' is bounded by $p(n)f(p(n))$. The second claim follows as each nondeterministic machine for L can be converted into a proof system for L (and vice versa).   □

This observation implies that from each propositional proof system P we can obtain a proof system for credulous autoepistemic logic which obeys almost the same bounds on the proof size. Theorem 4 tells us that the proof system for credulous autoepistemic reasoning constructed by this general method from *LK* is essentially the sequent calculus *CAEL* of Bonatti and Olivetti [5].

For *sceptical* autoepistemic (or default) reasoning—both of them Π_2^p-complete [12]—the situation is less clear. To the best of our knowledge it is not known whether a similar result as Proposition 9 holds for $L \in \Pi_2^p$. While sceptical autoepistemic reasoning has polynomially bounded proof systems if and only if this holds for TAUT (because $\mathrm{NP} = \mathrm{coNP}$ iff $\mathrm{NP} = \Pi_2^p$), we leave open whether this equivalence between extends to other bounds. Thus it is conceivable that lower bounds for sceptical reasoning are generally easier to obtain. This phenomenon particularly occurs with non-classical logics of even higher complexity as modal and intuitionistic logics which typically are PSPACE-complete and where exponential lower bounds are known for Frege and even extended Frege systems in these logics [14, 15].

In conclusion, the sequent calculi of Bonatti and Olivetti for credulous reasoning (both default and autoepistemic) are as good as one can hope for from a proof complexity perspective, whereas the calculi for sceptical reasoning call for stronger versions. This presents the double challenge of designing systems which are both natural and elegant and allow concise proofs. We remark that Krajíček and Pudlák [17] introduced very elegant sequent calculi G_i for quantified propositional logic, thus for logics with decision complexity ranging from Σ_2^p and Π_2^p through all the polynomial hierarchy up to PSPACE. However, no nontrivial lower bounds are known for these systems. As sceptical autoepistemic reasoning is Π_2^p-complete one could translate *SAEL*-sequents into propositional $\forall\exists$-formulas and use the sequent calculus G_2 from [17] (cf. also [7]).

Acknowledgements. I am grateful to the anonymous referees for useful comments on how to improve this article.

References

1. Beyersdorff, O., Kutz, O.: Proof complexity of non-classical logics. In: Bezhan-ishvili, N., Goranko, V. (eds.) ESSLLI 2010/2011. LNCS, vol. 7388, pp. 1–54. Springer, Heidelberg (2012)
2. Beyersdorff, O., Meier, A., Müller, S., Thomas, M., Vollmer, H.: Proof complexity of propositional default logic. Archive for Mathematical Logic 50(7), 727–742 (2011)
3. Beyersdorff, O., Meier, A., Thomas, M., Vollmer, H.: The complexity of reasoning for fragments of default logic. Journal of Logic and Computation 22(3), 587–604 (2012)
4. Bonatti, P.A.: A Gentzen system for non-theorems. Technical Report CD/TR 93/52, Christian Doppler Labor für Expertensysteme (1993)
5. Bonatti, P.A., Olivetti, N.: Sequent calculi for propositional nonmonotonic logics. ACM Transactions on Computational Logic 3(2), 226–278 (2002)
6. Bonet, M.L., Pitassi, T., Raz, R.: On interpolation and automatization for Frege systems. SIAM Journal on Computing 29(6), 1939–1967 (2000)
7. Cook, S.A., Nguyen, P.: Logical Foundations of Proof Complexity. Cambridge University Press (2010)
8. Cook, S.A., Reckhow, R.A.: The relative efficiency of propositional proof systems. The Journal of Symbolic Logic 44(1), 36–50 (1979)
9. Creignou, N., Meier, A., Vollmer, H., Thomas, M.: The complexity of reasoning for fragments of autoepistemic logic. ACM Transactions on Computational Logic 13(2) (2012)
10. Egly, U., Tompits, H.: Proof-complexity results for nonmonotonic reasoning. ACM Transactions on Computational Logic 2(3), 340–387 (2001)
11. Gentzen, G.: Untersuchungen über das logische Schließen. Mathematische Zeitschrift 39, 68–131 (1935)
12. Gottlob, G.: Complexity results for nonmonotonic logics. Journal of Logic and Computation 2(3), 397–425 (1992)
13. Gottlob, G.: Translating default logic into standard autoepistemic logic. J. ACM 42(4), 711–740 (1995)
14. Hrubeš, P.: On lengths of proofs in non-classical logics. Annals of Pure and Applied Logic 157(2-3), 194–205 (2009)
15. Jeřábek, E.: Substitution Frege and extended Frege proof systems in non-classical logics. Annals of Pure and Applied Logic 159(1-2), 1–48 (2009)
16. Krajíček, J.: Bounded Arithmetic, Propositional Logic, and Complexity Theory. Encyclopedia of Mathematics and Its Applications, vol. 60. Cambridge University Press (1995)
17. Krajíček, J., Pudlák, P.: Quantified propositional calculi and fragments of bounded arithmetic. Zeitschrift für Mathematische Logik und Grundlagen der Mathematik 36, 29–46 (1990)
18. Marek, V.W., Truszczyński, M.: Nonmonotonic Logics—Context-Dependent Reasoning. Springer, Heidelberg (1993)
19. Moore, R.C.: Semantical considerations on modal logic. Artificial Intelligence 25, 75–94 (1985)

Local Backbones

Ronald de Haan[1,*], Iyad Kanj[2], and Stefan Szeider[1,*]

[1] Institute of Information Systems, Vienna University of Technology, Vienna, Austria
[2] School of Computing, DePaul University, Chicago, IL

Abstract. A backbone of a propositional CNF formula is a variable whose truth value is the same in every truth assignment that satisfies the formula. The notion of backbones for CNF formulas has been studied in various contexts. In this paper, we introduce local variants of backbones, and study the computational complexity of detecting them. In particular, we consider k-backbones, which are backbones for sub-formulas consisting of at most k clauses, and iterative k-backbones, which are backbones that result after repeated instantiations of k-backbones. We determine the parameterized complexity of deciding whether a variable is a k-backbone or an iterative k-backbone for various restricted formula classes, including Horn, definite Horn, and Krom. We also present some first empirical results regarding backbones for CNF-Satisfiability (SAT). The empirical results we obtain show that a large fraction of the backbones of structured SAT instances are local, in contrast to random instances, which appear to have few local backbones.

1 Introduction

A *backbone* of a propositional formula φ is a variable whose truth value is the same for all satisfying assignments of φ. The term originates in computational physics [24], and the notion of backbones has been studied for SAT in various contexts. Backbones have also been considered in other contexts (e.g., knowledge compilation [5]) and for other combinatorial problems [25]. If a backbone and its truth value are known, then we can simplify the formula without changing its satisfiability, or the number of satisfying assignments. Therefore, it is desirable to have an efficient algorithm for detecting backbones. In general, however, the problem of identifying backbones is coNP-complete (this follows from the fact that a literal l is enforced by a formula φ if and only if $\varphi \wedge \neg l$ is unsatisfiable).

A variable can be a backbone because of *local properties* of the formula (such backbones we call *local backbones*). As an extreme example consider a CNF formula that contains a unit clause. In this case we know that the variable appearing in the unit clause is a backbone of the formula. More generally, we define the *order* of a backbone x of a CNF formula φ to be the cardinality of a smallest subset $\varphi' \subseteq \varphi$ such that x is a backbone of φ', and we refer to backbones of order $\leq k$ as *k-backbones*. Thus, unit clauses give rise to 1-backbones.

A natural generalization of k-backbones are variables whose truth values are enforced by repeatedly assigning k-backbones to their appropriate truth value and simplifying the formula according to this assignment. We call variables that are assigned

---

* Supported by the European Research Council (ERC), project COMPLEX REASON, 239962.

M. Järvisalo and A. Van Gelder (Eds.): SAT 2013, LNCS 7962, pp. 377–393, 2013.

by this iterative process *iterative k-backbones* (for a formal definition, see Section 2.1). For instance, iterative 1-backbones are exactly those variables whose truth values are enforced by unit propagation. The *iterative order* of a backbone x is the smallest k such that x is an iterative k-backbone.

Finding Local Backbones. For every constant k, we can clearly identify all k-backbones and iterative k-backbones of a CNF formula φ in polynomial time by simply going over all subsets of φ of size at most k (and iterating this process if necessary). However, if φ consists of m clauses, then this brute-force search requires us to consider at least m^k subsets, which is impractical already for small values of k. It would be desirable to have an algorithm that detects (iterative) k-backbones in time $f(k)\|\varphi\|^c$ where f is a function, $\|\varphi\|$ denotes the length of the formula, and c is a constant. An algorithm with such a running time would render the problem *fixed-parameter tractable* with respect to parameter k [7]. In this paper we study the question of whether the identification of (iterative) k-backbones of a CNF formula is fixed-parameter tractable or not, considering various restrictions on the CNF formula. We therefore define the following template for parameterized problems, where $\mathcal{C}$ is an arbitrary class of CNF formulas.

LOCAL-BACKBONE[$\mathcal{C}$]
Instance: a CNF formula $\varphi \in \mathcal{C}$, a variable x of φ, and an integer $k \geq 1$.
Parameter: The integer k.
Question: Is x a k-backbone of φ?

The problem ITERATIVE-LOCAL-BACKBONE is defined similarly. It is not hard to see that LOCAL-BACKBONE[$\mathcal{C}$] is closely related to the problem of finding a small unsatisfiable subset of a CNF formula (this is proven below in Lemmas 1 and 2). More precisely, for every class $\mathcal{C}$, the problem LOCAL-BACKBONE[$\mathcal{C}$] has the same parameterized complexity as the following problem, studied by Fellows et al. [10].

SMALL-UNSATISFIABLE-SUBSET[$\mathcal{C}$]
Instance: a CNF formula $\varphi \in \mathcal{C}$, and an integer $k \geq 1$.
Parameter: The integer k.
Question: Is there an unsatisfiable subset $\varphi' \subseteq \varphi$ consisting of at most k clauses?

This problem is of relevance also for classes $\mathcal{C}$ for which the satisfiability is decidable in polynomial time. For instance, given an inconsistent knowledge base in terms of an unsatisfiable set of Horn clauses, one might want to detect the cause for the inconsistency in terms of a small unsatisfiable subset.

Results. We draw a detailed parameterized complexity map of the considered problems LOCAL-BACKBONE[$\mathcal{C}$], ITERATIVE-LOCAL-BACKBONE[$\mathcal{C}$], and SMALL-UNSATISFIABLE-SUBSET[$\mathcal{C}$], for various classes $\mathcal{C}$. Table 1 provides an overview of our complexity results (FPT indicates that the problem is fixed-parameter tractable, W[1]-hardness indicates strong evidence that the problem is not fixed-parameter tractable; see Section 2.2 for details). It is interesting to observe that the non-iterative problems tend to be at least as hard as the iterative problems. Somewhat surprising is the W[1]-hardness of LOCAL-BACKBONE[KROM] and SMALL-UNSATISFIABLE-SUBSET[KROM] (which also implies the NP-hardness of the unparameterized versions of these problems). On the one hand, this seems to contrast with the fact that a shortest tree-like resolution

Table 1. Map of parameterized complexity results. (The classes $\mathcal{C}$ of formulas are defined in Section 2.1).

$\mathcal{C}$	Local-Backbone[$\mathcal{C}$]		Iterative-Local-Backbone[$\mathcal{C}$]	
CNF	W[1]-c	(Thm 2)	W[1]-h	(Cor 3)
DefHorn	W[1]-c	(Thm 2)	P	(Thm 7)
NuHorn	W[1]-c	(Thm 3)	W[1]-h	(Cor 3)
Krom	W[1]-c	(Thm 4)	P	(Thm 8)
VO_d	FPT	(Thm 5)	FPT	(Thm 6)

refutation of an unsatisfiable Krom formula can be found in polynomial time [4]. On the other hand, this is in line with the result that deciding whether a CNF formula can be refuted within k resolution steps (parameterized by k) is W[1]-complete [10]. The polynomial time solvability of finding iterative local backbones in Krom and definite Horn formulas is also interesting, especially in the light of the intractability of the corresponding problems of finding (non-iterative) local backbones.

We also provide some first empirical results on the distribution of local backbones in some benchmark SAT instances. We consider structured instances and random instances. For the structured instances that we consider we observe that a large fraction of the backbones are of relatively small iterative order. In contrast, the backbones of the random instances that we consider are of large iterative order. The results suggest that the distribution of the iterative order of backbones might be an indicator for a hidden structure in SAT instances.

Related Work. The notion of backbones has initially been studied in the context of optimization problems in computational physics [24]. The notion has later been applied to several combinatorial problems [25], including SAT. The relation between backbones and the difficulty of finding a solution for SAT has been studied by Kilby et al. [18], by Parkes [22] and by Slaney and Walsh [25]. The complexity of finding backbones has been studied theoretically by Kilby et al. [18]. The notion of backbones has also been used for improving SAT solving algorithms by Dubois and Dequen [8] and by Hertli et al. [14]. The problem of identifying unsatisfiable subsets of size at most k has been considered by Fellows et al. [10], who proved that this problem is W[1]-complete. Furthermore, they showed by the same reduction that finding a k-step resolution refutation for a given formula is W[1]-complete as well. Related notions of locally enforced literals have also been studied, including a notion of generalized unit-refutation [13,19].

Full Version. Because of space constraints some proofs have been omitted or shortened. Detailed proofs can be found in the full version, available at arxiv.org/abs/1304.5479.

2   Preliminaries

2.1   CNF Formulas, Unsatisfiable Subsets and Local Backbones

A *literal* is a propositional variable x or a negated variable $\neg x$. The *complement* $\overline{x}$ of a positive literal x is $\neg x$, and the complement $\overline{\neg x}$ of a negative literal $\neg x$ is x. A *clause* is a finite set of literals, not containing a complementary pair x, $\neg x$. A *unit clause* is a clause of size 1. We let $\bot$ denote the empty clause. A *formula* in conjunctive normal form (or CNF formula) is a finite set of clauses. We define the *length* $\|\varphi\|$ of a formula φ to be $\sum_{c \in \varphi} |c|$; the number of clauses of φ is denoted by $|\varphi|$. A formula φ is a k-CNF formula if the size of each of its clauses is at most k. A 2-CNF formula is also called a Krom formula. A clause is a *Horn clause* if it contains at most one positive literal. A Horn clause containing exactly one positive literal is a *definite Horn clause*. Formulas containing only Horn clauses are called *Horn formulas*. *Definite Horn formulas* are defined analogously. We denote the class of all Krom formulas by KROM, the class of all Horn formulas by HORN and the class of all definite Horn formulas by DEFHORN. We let NUHORN denote the class of Horn formulas not containing unit clauses (such formulas are always satisfiable). Let d be an integer. The class of CNF formulas such that each variable occurs at most d times is denoted by VO_d.

For a CNF-formula φ, the set $\mathrm{Var}(\varphi)$ denotes the set of all variables x such that some clause of φ contains x or $\neg x$; the set $\mathrm{Lit}(\varphi)$ denotes the set of all literals l such that some clause of φ contains l or $\bar{l}$. A formula φ is *satisfiable* if there exists an assignment $\tau : \mathrm{Var}(\varphi) \to \{0, 1\}$ such that every clause $c \in \varphi$ contains some variable x with $\tau(x) = 1$ or some negated variable $\neg x$ with $\tau(x) = 0$ (we say that such an assigment τ satisfies φ); otherwise, φ is *unsatisfiable*. φ is *minimally unsatisfiable* if φ is unsatisfiable and every proper subset of φ is satisfiable. It is well-known that any minimal unsatisfiable CNF formula has more clauses than variables (this is known as Tarsi's Lemma [1,20]). For two formulas φ, ψ, whenever all assignments satisfying φ also satisfy ψ, we write $\varphi \models \psi$. The reduct $\varphi|_L$ of a formula φ with respect to a set of literals $L \subseteq \mathrm{Lit}(\varphi)$ is the set of clauses of φ that do not contain any $l \in L$ with all occurrences of $\bar{l}$ for all $l \in L$ removed. For singletons $L = \{l\}$, we also write $\varphi|_l$. We say that a class $\mathcal{C}$ of formulas is *closed under variable instantiation* if for every $\varphi \in \mathcal{C}$ and every $l \in \mathrm{Lit}(\varphi)$ we have that $\varphi|_l \in \mathcal{C}$. For an integer k, a variable x is a k-*backbone* of φ, if there exists a $\varphi' \subseteq \varphi$ such that $|\varphi'| \leq k$ and either $\varphi' \models x$ or $\varphi' \models \neg x$. A variable x is a *backbone* of a formula φ if it is a $|\varphi|$-backbone. Note that the definition of the backbone of a formula φ that is used in some of the literature includes all literals $l \in \mathrm{Lit}(\varphi)$ such that $\varphi \models l$. For an integer k, a variable x is an *iterative k-backbone* of φ if either (i) x is a k-backbone of φ, or (ii) there exists $y \in \mathrm{Var}(\varphi)$ such that y is a k-backbone of φ, and for some $l \in \{y, \neg y\}$, $\varphi \models l$ and x is an iterative k-backbone of $\varphi|_l$.

For a Krom formula φ, we let $\mathrm{impl}(\varphi)$ be the *implication graph* (V, E) of φ, where $V = \{ x, \neg x : x \in \mathrm{Var}(\varphi) \}$ and $E = \{ (\bar{a}, b), (\bar{b}, a) : \{a, b\} \in \varphi \}$. We say that a path p in this graph *uses a clause* $\{a, b\}$ of φ if either one of the edges $(\bar{a}, b)$ and $(\bar{b}, a)$ occurs in p; we say that p *doubly uses* this clause if both edges occur in p.

2.2 Parameterized Complexity

Here we introduce the relevant concepts of parameterized complexity theory. For more details, we refer to text books on the topic [7,11,21]. An instance of a parameterized problem is a pair (I, k) where I is the main part of the instance, and k is the parameter. A parameterized problem is *fixed-parameter tractable* if instances (I, k) can be solved by a deterministic algorithm that runs in time $f(k)|I|^c$, where f is a computable function of k, and c is a constant (algorithms running within such time bounds are called *fpt-algorithms*). If $c = 1$, we say the problem is *fixed-parameter linear*. FPT denotes the class of all fixed-parameter tractable problems. Using fixed-parameter tractability, many problems that are classified as intractable in the classical setting can be shown to be tractable for small values of the parameter.

Parameterized complexity also offers a *completeness theory*, similar to the theory of NP-completeness. This allows the accumulation of strong theoretical evidence that a parameterized problem is not fixed-parameter tractable. Hardness for parameterized complexity classes is based on fpt-reductions, which are many-one reductions where the parameter of one problem maps into the parameter for the other. More specifically, a parameterized problem L is fpt-reducible to another parameterized problem L' (denoted $L \leq_{fpt} L'$) if there is a mapping R from instances of L to instances of L' such that (i) $(I, k) \in L$ if and only if $(I', k') = R(I, k) \in L'$, (ii) $k' \leq g(k)$ for a computable function g, and (iii) R can be computed in time $O(f(k)|I|^c)$ for a computable function f and a constant c.

Central to the completeness theory is the hierarchy $FPT \subseteq W[1] \subseteq W[2] \subseteq \cdots \subseteq$ para-NP. Each intractability class W[t] contains all parameterized problems that can be reduced to a certain parameterized satisfiability problem under fpt-reductions. The intractability class para-NP includes all parameterized problems that can be solved by a nondeterministic fpt-algorithm. Fixed-parameter tractability of any problem hard for any of these intractability classes would imply that the Exponential Time Hypothesis fails [11,16] (i.e., the existence of a $2^{o(n)}$ algorithm for n-variable 3SAT).

3   Local Backbones and Small Unsatisfiable Subsets

The straightforward reductions in the proofs of the following two lemmas, illustrate the close connection between LOCAL-BACKBONE and SMALL-UNSATISFIABLE-SUBSET.

Lemma 1. SMALL-UNSATISFIABLE-SUBSET $\leq_{fpt}$ LOCAL-BACKBONE.

Proof. Let (φ, k) be an instance of SMALL-UNSATISFIABLE-SUBSET. We construct an instance (φ', z, k) of LOCAL-BACKBONE, by letting $\varphi' = \{ c \cup \{z\} : c \in \varphi \}$ for some $z \notin Var(\varphi)$. We claim that $(\varphi, k) \in$ SMALL-UNSATISFIABLE-SUBSET if and only if $(\varphi', z, k) \in$ LOCAL-BACKBONE. A complete proof of this claim can be found in the full version of the paper.   □

Lemma 2. LOCAL-BACKBONE $\leq_{fpt}$ SMALL-UNSATISFIABLE-SUBSET.

Proof. Let (φ, z, k) be an instance of LOCAL-BACKBONE. We construct an instance (ψ, k) of SMALL-UNSATISFIABLE-SUBSET. For every variable $x \in Var(\varphi)$ we take two copies x_1, x_2. For $i \in \{1, 2\}$ we let φ_i be a copy of φ using the variables x_i. Now we define $\psi = \varphi_1|_{z_1} \cup \varphi_2|_{\neg z_2}$. In other words, ψ is the union of two disjoint copies of the reducts of φ with respect to z and $\neg z$. We claim that $(\varphi, z, k) \in$ LOCAL-BACK-BONE if and only if $(\psi, k) \in$ SMALL-UNSATISFIABLE-SUBSET. A complete proof of this claim can be found in the full version of the paper.   □

Theorem 1. LOCAL-BACKBONE *is* W[1]-*complete*.

Proof. Since SMALL-UNSATISFIABLE-SUBSET is W[1]-complete [10], the result follows from Lemmas 1 and 2.   □

4 Local Backbones of Horn Formulas

Restricting the problem of finding backbones in arbitrary formulas to Horn formulas reduces the classical complexity from co-NP-completeness to polynomial time solvability. It is a natural question whether the parameterized complexity of finding local backbones decreases in a similar way when the problem is restricted to Horn formulas. We will show that this is not the case. In order to do so, we define the parameterized problem SHORT-HYPERPATH, show that it is W[1]-hard, and then provide fpt-reductions from SHORT-HYPERPATH.

For a Horn formula φ and $s, t \in \mathrm{Var}(\varphi)$, we say that a subformula $\varphi' \subseteq \varphi$ is a *hyperpath* from s to t if (i) $t = s$ or (ii) $c = \{x_1, \ldots, x_n, t\} \in \varphi'$ and $\varphi' \backslash c$ is a hyperpath from s to x_i for each $1 \le i \le n$. If $|\varphi| \le k$ then φ is called a k-*hyperpath*. The parameterized problem SHORT-HYPERPATH takes as input a Horn formula φ, two variables $s, t \in \mathrm{Var}(\varphi)$ and an integer k. The problem is parameterized by k. The question is whether there exists a k-hyperpath from s to t. For a more detailed discussion on the relation between (backward) hyperpaths in hypergraphs and hyperpaths as defined above, we refer to a survey article by Gallo et al. [12].

For the hardness proof of SHORT-HYPERPATH, we reduce from the W[1]-complete problem MULTICOLORED-CLIQUE [9]. The MULTICOLORED-CLIQUE problem takes as input a graph G, some integer k, and a proper k-coloring c of the vertices of G. The problem is parameterized by k. The question is whether there is a properly colored k-clique in G.

Lemma 3. SHORT-HYPERPATH *is* W[1]-*hard, even for instances* (φ, s, t, k) *where* $\varphi \in 3\mathrm{CNF}$.

Proof. We give a reduction from MULTICOLORED-CLIQUE. Let (G, k, c) be an instance of MULTICOLORED-CLIQUE, where $G = (V, E)$ and $V_1, \ldots, V_k$ are the equivalence classes of V induced by the k-coloring c. We construct an instance (φ, s, t, k') of SHORT-HYPERPATH, where $k' = k + \binom{k}{2} + 1$ and

$$\mathrm{Var}(\varphi) = \{s, t\} \cup V \cup \{p_{i,j} : 1 \le i < j \le k\};$$
$$\varphi = \varphi_V \cup \varphi_p \cup \varphi_t;$$
$$\varphi_V = \{\{\neg s, v\} : v \in V\};$$
$$\varphi_p = \{\{\neg v_i, \neg v_j, p_{i,j}\} : 1 \le i < j \le k, v_i \in V_i, v_j \in V_j, \{v_i, v_j\} \in E\};$$
$$\varphi_t = \{\{\neg p_{i,j} : 1 \le i < j \le k\} \cup \{t\}\}.$$

This construction is illustrated for an example with $k = 3$ in Figure 1. We claim that $(G, k, c) \in$ MULTICOLORED-CLIQUE if and only if $(\varphi, s, t, k') \in$ SHORT-HYPERPATH. A complete proof of this claim can be found in the full version of the paper.

To see that clauses of size at most 3 in the hyperpath suffice, we slightly adapt the reduction. The only clause we need to change is the single clause $e \in \varphi_t$. This clause e is of the form $\{\neg p_1, \ldots, \neg p_m, t\}$, for $m = \binom{k}{2}$. We introduce new variables $v_1, \ldots, v_m$ and replace e by the $m+1$ many clauses $\{\neg p_1, v_1\}, \{\neg v_{i-1}, \neg p_i, v_i\}$ for all $1 < i \le m$ and $\{\neg v_m, t\}$. Clearly, the resulting Horn formula only has clauses of size at most 3. This adapted reduction works with the exact same line of reasoning as the reduction described above, with the only change that $k' = k + 2\binom{k}{2} + 1$.

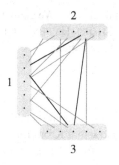

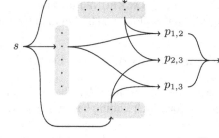

(a) A 3-partite graph G with a clique (in black)

(b) The B-hyperpath in H of size $k' = 3 + \binom{3}{2} + 1$ from s to t corresponding to the clique

Fig. 1. Illustration of the reduction in the proof of Lemma 3 for the case of a 3-colored clique

Note that even the slightly stronger claim holds that G has a properly colored k-clique if and only if there exists a (subset) minimal k'-B-hyperpath $\varphi' \subseteq \varphi$ for which we have $|\varphi'| = k'$. □

We are now in a position to prove the W[1]-hardness of LOCAL-BACKBONE[HORN]. In fact, we show that finding local backbones is already W[1]-hard for definite Horn formulas with a single unit clause. We also show that this hardness crucially depends on allowing unit clauses in the formula, since for definite Horn formulas without unit clauses the problem is trivial. In fact, the complexity jumps to W[1]-hardness already when allowing a single unit clause.

Theorem 2. LOCAL-BACKBONE[DEFHORN ∩ 3CNF] *is* W[1]*-hard, already for instances* (φ, x, k) *where* φ *has at most one unit clause.*

Proof. We show W[1]-hardness by reducing from SHORT-HYPERPATH. Let (φ, s, t, k) be an instance of SHORT-HYPERPATH. We can assume that $\varphi \in$ 3CNF. We construct an instance (ψ_φ, t, k') of LOCAL-BACKBONE. Here $k' = k + 1$. For each $\varphi' \subseteq \varphi$ we define a formula $\psi_{\varphi'}$, by letting $\text{Var}(\psi_{\varphi'}) = \text{Var}(\varphi')$ and $\psi_{\varphi'} = \{\{s\}\} \cup \varphi'$. Clearly $\psi_\varphi \in$ DEFHORN ∩ 3CNF and ψ_φ has only a single unit clause. We claim that $(\psi_\varphi, t, k') \in$ LOCAL-BACKBONE if and only if $(\varphi, s, t, k) \in$ SHORT-HYPERPATH. A complete proof of this claim can be found in the full version of the paper. □

Also, restricting the problem to Horn formulas without unit clauses unfortunately does not yield fixed-parameter tractability.

Theorem 3. LOCAL-BACKBONE[NUHORN ∩ 3CNF] *is* W[1]*-hard.*

Proof. We show the W[1]-hardness of LOCAL-BACKBONE[NUHORN ∩ 3CNF] by reducing from SHORT-HYPERPATH. Let (φ, s, t, k) be an instance of SHORT-HYPERPATH. We can assume without loss of generality that $\varphi \in$ 3CNF, and that each clause in which t occurs positively is of size 3. We construct an instance (ψ_φ, x_s, k) of LOCAL-BACKBONE. For each $\varphi' \subseteq \varphi$ we define a formula $\psi_{\varphi'}$.

$$\psi_{\varphi'} = \{ \{\neg x_a, \neg x_b, x_c\} : \{\neg a, \neg b, c\} \in \varphi', c \neq t \} \cup$$
$$\{ \{\neg x_a, \neg x_b\} : \{\neg a, \neg b, t\} \in \varphi' \}$$
$$\{ \{\neg x_a, x_b\} : \{\neg a, b\} \in \varphi' \}$$

Clearly we have that $\psi_\varphi \in \text{HORN} \cap 3\text{CNF}$ and that ψ_φ has no unit clauses. We claim that $(\psi_\varphi, x_s, k) \in \text{LOCAL-BACKBONE}$ if and only if $(\varphi, s, t, k) \in \text{SHORT-HYPERPATH}$. A complete proof of this claim can be found in the full version of the paper. $\square$

5   Local Backbones of Krom Formulas

We have seen that finding local backbones is already fixed-parameter intractable for Horn formulas, for which finding backbones is tractable. We show that Krom formulas have a similar property: even though finding backbones in Krom formulas is tractable, finding local backbones is fixed-parameter intractable.

Theorem 4. LOCAL-BACKBONE[KROM] *is* W[1]-*hard.*

Proof. We reduce from MULTICOLORED-CLIQUE. Let (G, k, c) be an instance of MULTICOLORED-CLIQUE, where $G = (V, E)$ and $V_1, \ldots, V_k$ are the equivalence classes of V induced by the k-coloring c. We construct an instance (φ, x, k') of LOCAL-BACKBONE[KROM]. Intuitively, we introduce a gadget for each V_i (see Figure 2a) and additionally a gadget for each pair (i, j) for $1 \leq i < j \leq k$ (see Figure 2b), and sequentially link these gadgets together (see Figure 2c). In the definition of (φ, x, k') for each $1 \leq i \leq k$ we define a formula φ_i^{guess} that corresponds to the gadget for V_i, and for each $1 \leq i < j \leq k$ we define a formula $\varphi_{i,j}^{\text{check}}$ that corresponds to the gadget for the pair (i, j). In the construction of φ we use variables $\sigma_{i,v}^j$ and $\tau_{i,v}^j$ that are used to encode the choice of vertex v in V_i for the clique, and that are used to verify whether v and the choice for V_j are connected. We let $x = g_1$ and $k' = k(2k+1) + 3\binom{k}{2} + 2$ and we define

$$\text{Var}(\varphi) = \{g_1, \ldots, g_{k+1}\} \cup \{c_{1,1}, c_{1,2} \ldots, c_{k-1,k}, c_{k,k+1}\} \cup$$
$$\{\sigma_{i,v}^j, \tau_{i,v}^j : 1 \leq i \leq k, 1 \leq j \leq k, v \in V_i\}, \text{ and}$$
$$\varphi = \bigcup_{1 \leq i \leq k} \varphi_i^{\text{guess}} \cup \bigcup_{1 \leq i < j \leq k} \varphi_{i,j}^{\text{check}} \cup \{\{\neg g_{k+1}, c_{1,1}\}, \{\neg c_{k,k+1}, \neg g_1\}\}.$$

For each $1 \leq i \leq k$, we define φ_i^{guess}, where $V_i = \{v_1, \ldots, v_n\}$, by letting

$$\varphi_i^{\text{guess}} = \{\{\neg g_i, \sigma_{i,v_l}^1\} : 1 \leq l \leq n\} \cup$$
$$\{\{\neg \sigma_{i,v_l}^j, \tau_{i,v_l}^j\} : 1 \leq j \leq k, 1 \leq l \leq n\} \cup$$
$$\{\{\neg \tau_{i,v_l}^j, \sigma_{i,v_l}^{j+1}\} : 1 \leq j < k, 1 \leq l \leq n\} \cup$$
$$\{\{\neg \tau_{i,v_l}^k, g_{i+1}\} : 1 \leq l \leq n\}.$$

Similarly, for each $1 \leq i < j \leq k$ we define the subformula $\varphi_{i,j}^{\text{check}}$ as follows. Here we let $E \cap (V_i \times V_j) = \{(v_1, v_1'), \ldots, (v_m, v_m')\}$. Also, we define the function next by letting $\text{next}(i, j) = (i, j+1)$ if $j \neq k$ and $\text{next}(i, j) = (i+1, i+2)$ if $j = k$.

$$\varphi_{i,j}^{\text{check}} = \{\{\neg c_{i,j}, \neg \tau_{i,v_l}^j\} : 1 \leq l \leq m\} \cup$$
$$\{\{\tau_{i,v_l}^j, \neg \sigma_{i,v_l}^j\}, \{\sigma_{i,v_l}^j, \neg \tau_{j,v_l'}^i\}, \{\tau_{j,v_l'}^i, \neg \sigma_{j,v_l'}^i\} : 1 \leq l \leq m\} \cup$$
$$\{\{\sigma_{j,v_l'}^i, c_{\text{next}(i,j)}\} : 1 \leq l \leq m\}.$$

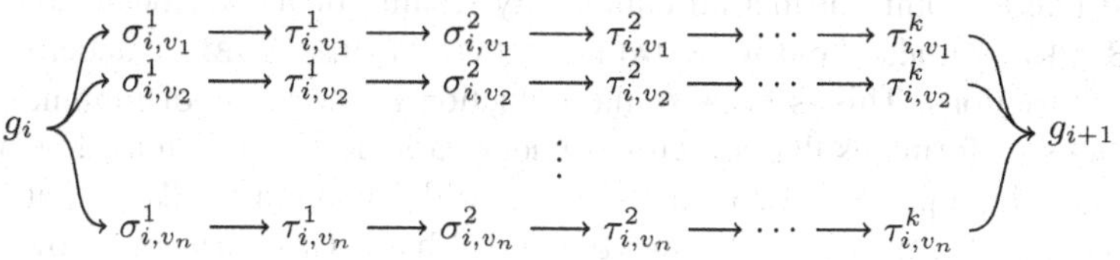

$$\gi \left\{ \begin{array}{ccccccccccc} \sigma_{i,v_1}^1 & \longrightarrow & \tau_{i,v_1}^1 & \longrightarrow & \sigma_{i,v_1}^2 & \longrightarrow & \tau_{i,v_1}^2 & \longrightarrow & \cdots & \longrightarrow & \tau_{i,v_1}^k \\ \sigma_{i,v_2}^1 & \longrightarrow & \tau_{i,v_2}^1 & \longrightarrow & \sigma_{i,v_2}^2 & \longrightarrow & \tau_{i,v_2}^2 & \longrightarrow & \cdots & \longrightarrow & \tau_{i,v_2}^k \\ & & & & \vdots & & & & & & \\ \sigma_{i,v_n}^1 & \longrightarrow & \tau_{i,v_n}^1 & \longrightarrow & \sigma_{i,v_n}^2 & \longrightarrow & \tau_{i,v_n}^2 & \longrightarrow & \cdots & \longrightarrow & \tau_{i,v_n}^k \end{array} \right\} \gi{i+1}$$

(a) Gadget $\varphi_i^{\mathrm{guess}}$ for partition $V_i = \{v_1, \ldots, v_n\}$

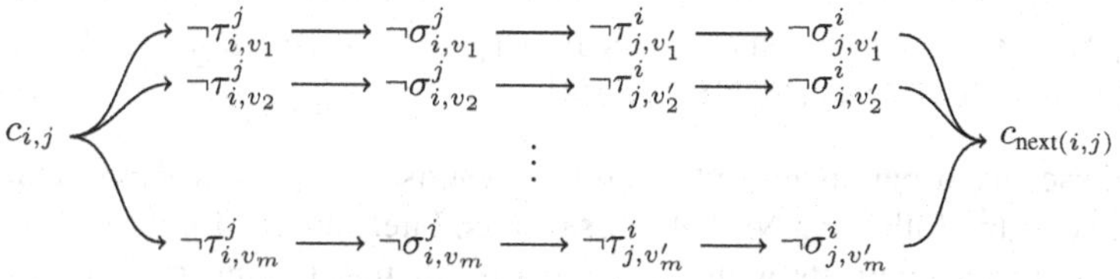

$$c_{i,j} \left\{ \begin{array}{ccccccc} \neg\tau_{i,v_1}^j & \longrightarrow & \neg\sigma_{i,v_1}^j & \longrightarrow & \neg\tau_{j,v_1'}^i & \longrightarrow & \neg\sigma_{j,v_1'}^i \\ \neg\tau_{i,v_2}^j & \longrightarrow & \neg\sigma_{i,v_2}^j & \longrightarrow & \neg\tau_{j,v_2'}^i & \longrightarrow & \neg\sigma_{j,v_2'}^i \\ & & & \vdots & & & \\ \neg\tau_{i,v_m}^j & \longrightarrow & \neg\sigma_{i,v_m}^j & \longrightarrow & \neg\tau_{j,v_m'}^i & \longrightarrow & \neg\sigma_{j,v_m'}^i \end{array} \right\} c_{\mathrm{next}(i,j)}$$

(b) Gadget $\varphi_{i,j}^{\mathrm{check}}$ for partitions i and j and
$E \cap (V_i \times V_j) = \{(v_1, v_1'), \ldots, (v_m, v_m')\}$

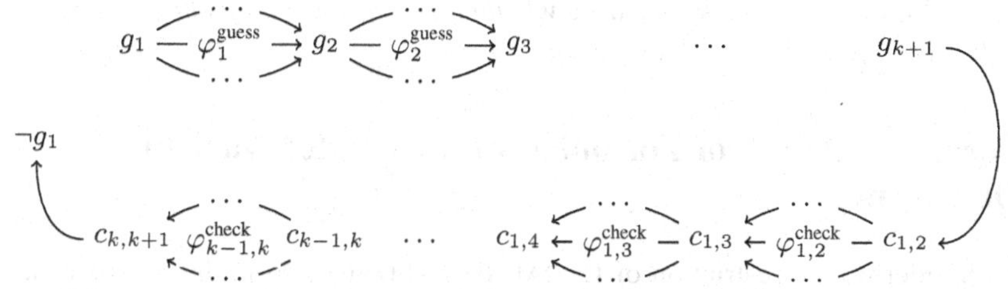

$$g_1 \overset{\cdots}{\underset{\cdots}{\rightarrow}} \varphi_1^{\mathrm{guess}} \rightarrow g_2 \overset{\cdots}{\underset{\cdots}{\rightarrow}} \varphi_2^{\mathrm{guess}} \rightarrow g_3 \qquad \cdots \qquad g_{k+1}$$

$$\neg g_1 \quad \uparrow$$

$$c_{k,k+1}\; \varphi_{k-1,k}^{\mathrm{check}}\; c_{k-1,k} \quad \cdots \quad c_{1,4} \leftarrow \varphi_{1,3}^{\mathrm{check}} - c_{1,3} \leftarrow \varphi_{1,2}^{\mathrm{check}} - c_{1,2} \leftarrow$$

(c) Linking the gadgets together in φ

Fig. 2. Gadgets for the reduction in the proof of Theorem 4

Intuitively, this reduction works as follows. Note that since g_1 occurs only negatively in φ we know that g_1 can only be a k'-backbone of φ if there exists a path from g_1 to $\neg g_1$ in $\mathrm{impl}(\varphi)$ that uses at most k' clauses. A path of length $2k + 1$ through $\mathrm{impl}(\varphi_i^{\mathrm{guess}})$ corresponds to guessing a vertex in the equivalence class V_i. Additionally, a path of length 5 through $\mathrm{impl}(\varphi_{i,j}^{\mathrm{check}})$ corresponds to verifying whether there is an edge in the graph that is in $V_i \times V_j$. So clearly, there exists a path of length $k(2k + 1) + 5\binom{k}{2} + 2$ in $\mathrm{impl}(\varphi)$ from g_1 to $\neg g_1$. However, many clauses in $\mathrm{impl}(\varphi_{i,j}^{\mathrm{check}})$ can be doubly used clauses, already used before in paths through $\mathrm{impl}(\varphi_i^{\mathrm{guess}})$. Concretely, there exists a path through $\mathrm{impl}(\varphi_{i,j}^{\mathrm{check}})$ that uses only 3 clauses that have not yet been used in paths through $\mathrm{impl}(\varphi_i^{\mathrm{guess}})$ if and only if the paths through $\mathrm{impl}(\varphi_i^{\mathrm{guess}})$ and $\mathrm{impl}(\varphi_j^{\mathrm{guess}})$ have selected vertices $v_i \in V_i$ and $v_j \in V_j$ such that $(v_i, v_j) \in E$. In other words, there exists a path in $\mathrm{impl}(\varphi)$ from g_1 to $\neg g_1$ that uses k' clauses if and only if G has a properly colored k-clique.

In the full version of the paper, we formally prove that G has a properly colored k-clique if and only if g_1 is a k'-backbone of φ. $\qquad \square$

We would like to point out that all complexity results for the various restrictions of LOCAL-BACKBONE also hold for SMALL-UNSATISFIABLE-SUBSET under the corresponding restrictions. This is because the reduction in the proof of Lemma 2 works for all classes of formulas that are closed under variable instantiations. For instance, the reduction in the proof of Lemma 2 together with Theorem 3 tells us that SMALL-UNSATISFIABLE-SUBSET[HORN ∩ 3CNF] is W[1]-hard. This does not follow from the reduction that Fellows et al. [10] use to prove the W[1]-hardness of SMALL-UNSATISFIABLE-SUBSET. In particular, the following previously unstated results hold.

Corollary 1. SMALL-UNSATISFIABLE-SUBSET[C] *is* W[1]-*hard for each* $C \in \{$DEF-HORN $\cap$ 3CNF, NUHORN $\cap$ 3CNF, KROM$\}$.

In fact, these fixed-parameter intractability results for SMALL-UNSATISFIABLE-SUBSET give us the following NP-hardness results. Interestingly, for the case of KROM formulas this result contrasts with the known result that finding minimal resolution refutations for KROM formulas can be done in polynomial time [3,4].

Corollary 2. *Let* $C \in \{$KROM, 3CNF $\cap$ DEFHORN, 3CNF $\cap$ NUHORN$\}$. *Given a formula* $\varphi \in C$ *and an integer* k, *deciding whether* φ *contains an unsatisfiable subset of size* $\leq k$ *is* NP-*hard.*

6 Local Backbones of Formulas with Bounded Variable Occurrence

When considering the restriction of LOCAL-BACKBONE to formulas where variables occur a bounded number of times, we get a fixed-parameter tractability result at last. This fixed-parameter tractability result is closely related to the result that SMALL-UNSATISFIABLE-SUBSET is fixed-parameter tractable for instances restricted to classes of formulas that have locally bounded treewidth [10]. Fellows et al. used a meta theorem to prove this. We give a direct algorithm to solve SMALL-UNSATISFIABLE-SUBSET[VO_d] in fixed-parameter linear time.

Let (φ, k) be an instance of SMALL-UNSATISFIABLE-SUBSET[VO_d]. The following procedure decides whether there exists an unsatisfiable subset $\varphi' \subseteq \varphi$ of size at most k, and computes such a subset if it exists. We let $\varphi^\star = \{c \in \varphi : |c| < k\}$. It suffices to consider subsets of $\varphi^\star$, since any unsatisfiable subset $\varphi' \subseteq \varphi$ contains a minimally unsatisfiable subset $\varphi'' \subseteq \varphi'$, and by Tarsi's Lemma we know that φ'' contains only clauses of size smaller than k.

Without loss of generality, we assume that the incidence graph of $\varphi^\star$ is connected. Otherwise, we can solve the problem by running the algorithm on each of the connected components. We guess a clause $c \in \varphi^\star$, we let $F_1 := \{c\}$, and we let all variables be unmarked initially. We compute F_{i+1} for $1 \leq i \leq k$ by means of the following (nondeterministic) rule:

1. take an unmarked variable $z \in \text{Var}(F_i)$;
2. guess a non-empty subset $F'_z \subseteq F_z$ for $F_z = \{c \in \varphi^\star : z \in \text{Var}(c)\}$;
3. let $F_{i+1} := F_i \cup F'_z$;
4. mark z.

If at any point all variables in F_i are marked, we stop computing F_{i+1}. For any F_i, if $|F_i| > k$ we fail. For each F_i, we check whether F_i is unsatisfiable. If it is unsatisfiable, we return with $\varphi' = F_i$. If it is satisfiable and if it contains no unmarked variables, we fail. It is easy to see that this algorithm is sound. If some $\varphi' \subseteq \varphi^\star$ is returned, then φ' is unsatisfiable and $|\varphi'| \leq k$. In order to see that the algorithm is complete, assume that there exists some unsatisfiable $\varphi' \subseteq \varphi^\star$ with $|\varphi'| \leq k$. Then, since we know that the incidence graph of F' is connected, we know that F' can be constructed as one of the F_i in the algorithm.

To see that this algorithm witnesses fixed-parameter linearity, we bound its running time. We have to execute the search process at most once for each clause of $\varphi^\star$. At each point in the execution of the algorithm, F_i contains at most k variables. Therefore, there are at most k choices to take an unmarked variable z. Since each variable occurs in at most d clauses, for each F_z used in the rule we know $|F_z| \leq d$. Thus, there are at most 2^d possible guesses for F'_z in each execution of the rule. Since we iterate the rule at most k times, we consider at most $(k2^d)^k$ sets F', each of size $O(k^2)$. Thus each (un)satisfiability check can be done in $O(2^k)$ time. Therefore, the total running time of the algorithm is $O(k^k 2^{dk} n)$, for n the size of the instance.

This algorithm also gives us a direct algorithm that shows that LOCAL-BACK-BONE[VO_d] is fixed-parameter linear.

Theorem 5. LOCAL-BACKBONE[VO_d] *is fixed-parameter linear.*

Proof. The result follows directly by using the reduction in the proof of Lemma 2 in combination with the above algorithm.   □

7   Iterative Local Backbones

We now consider the (parameterized) complexity of finding iterative local backbones. It is easy to see that ITERATIVE-LOCAL-BACKBONE is in para-NP. This is witnessed by a straightforward nondeterministic fpt-algorithm, that guesses a sequence of n witnesses (φ_i, l_i) with $|\varphi_i| \leq k$, and that verifies whether $\varphi_i \subseteq \varphi|_{\{l_1, \ldots, l_{i-1}\}}$ and whether $\varphi_i \models l_i$.

Some of the results we obtained for the problem of finding local backbones can be carried over.

Theorem 6. *Let $\mathcal{C}$ be a class of formulas such that* LOCAL-BACKBONE[$\mathcal{C}$] *is fixed-parameter tractable and $\mathcal{C}$ is closed under variable instantiation. Then* ITERATIVE-LOCAL-BACKBONE[$\mathcal{C}$] *is fixed-parameter tractable.*

Proof. We give an algorithm to solve ITERATIVE-LOCAL-BACKBONE[$\mathcal{C}$] that calls a subroutine to solve instances of SMALL-UNSATISFIABLE-SUBSET[$\mathcal{C}$]. This algorithm is given in the form of pseudo-code as Algorithm 1. By the fact that $\mathcal{C}$ is closed under variable instantiations we are able to apply the reduction in the proof of Lemma 2. Thus, we can assume that the question of whether some $\varphi \in \mathcal{C}$ contains an unsatisfiable subset of size at most k can be solved in $f(k)\|\varphi\|^c$ time, for some computable function f and some constant c. Then, the entire algorithm runs in $O(f(k)\|\varphi\|^{c+2})$ time. This proves the claim.   □

---

input : an instance (φ, x, k) of ITERATIVE-LOCAL-BACKBONE
output: yes iff $(\varphi, x, k) \in$ ITERATIVE-LOCAL-BACKBONE

$\psi \leftarrow \varphi$;   conseq $\leftarrow \emptyset$;
for $i \leftarrow 1$ **to** $|\mathrm{Lit}(\varphi)|$ **do**
    foreach *literal* $l \in \mathrm{Lit}(\psi)$ **do**
        if $(\psi|_{\overline{l}}, k) \in$ SMALL-UNSATISFIABLE-SUBSET **then**
            conseq $\leftarrow$ conseq $\cup \{l\}$;
    $\psi \leftarrow \psi|_{\text{conseq}}$;
return $\{x, \neg x\} \cap$ conseq $\neq \emptyset$

Algorithm 1. Deciding ITERATIVE-LOCAL-BACKBONE with a SMALL-UNSATISFIABLE-SUBSET oracle

Corollary 3. ITERATIVE-LOCAL-BACKBONE[NUHORN $\cap$ 3CNF] *is* W[1]-*hard.*

Proof. Observe that the proofs of Lemma 3 and Theorem 3 imply that it is already W[1]-hard to determine whether a formula $\varphi \in$ NUHORN $\cap$ 3CNF has a subset $\varphi' \subseteq \varphi$ of size exactly k witnessing that any $x \in \mathrm{Var}(\varphi)$ is a k-backbone. From this, it immediately follows that determining whether $(\varphi, x, k) \in$ ITERATIVE-LOCAL-BACKBONE is W[1]-hard as well. □

We identify several tractable cases for ITERATIVE-LOCAL-BACKBONE. The problem of finding iterative local backbones in definite Horn formulas is polynomial time solvable. Similarly, finding iterative local backbones in Krom formulas is solvable in polynomial time as well. Interestingly, for these restrictions the problem of finding (non-iterative) local backbones remains W[1]-hard. In order to show that finding iterative local backbones in definite Horn formulas is tractable, we will use the following observation.

Observation 1 *Let φ be any propositional formula, let l be any literal such that there exists a $\varphi' \subseteq \varphi$ with $|\varphi'| \leq k$ and $\varphi' \models l$, and let $\psi = \varphi|_l$. Then $x \in \mathrm{Var}(\psi)$ is an iterative k-backbone of ψ if and only if it is an iterative k-backbone of φ.*

Theorem 7. ITERATIVE-LOCAL-BACKBONE[DEFHORN] *is in* P.

Proof. We show that for any definite Horn formula φ and any $k \geq 1$ the set of iterative k-backbones of φ coincides with the set of variables $x \in \mathrm{Var}(\varphi)$ such that $\varphi \models x$. The claim then follows, since the entailment relation $\models$ can be decided in linear time for definite Horn formulas [6].

Fix an arbitrary integer $k \geq 1$ and an arbitrary definite Horn formula φ. Since definite Horn formulas cannot entail negative literals, we know that each iterative k-backbone x of φ is also a semantic consequence of φ. Now, let $x \in \mathrm{Var}(\varphi)$ be an arbitrary atom and assume that $\varphi \models x$. So there exist variables $x_1, \ldots, x_m \in \mathrm{Var}(\varphi)$ such that $x_m = x$ and for each x_i we have either (i) $\{x_i\} \in \varphi$ or (ii) $\{\neg x_{i_1}, \ldots, \neg x_{i_l}, x_i\} \in \varphi$ for some $i_1 < \cdots < i_l < i$. We prove by induction on m that each x_i is an iterative k-backbone. Take an arbitrary x_i. By the induction hypothesis, we can assume that every x_j for $j < i$ is an iterative k-backbone of φ. We proceed by case distinction for the justification of

x_i in the sequence. In case (i), we know that $\{x_i\} \in \varphi$. Therefore, it directly follows that x_i is a k-backbone of φ, and thus is an iterative k-backbone too. In case (ii), we know that $\{\neg x_{i_1}, \ldots, \neg x_{i_l}, x_i\} \in \varphi$ for some $i_1 < \cdots < i_l < i$. By the induction hypothesis, we know that each x_{i_j} is an iterative k-backbone of φ. By assumption, we have that $\varphi \models x_{i_j}$ for each x_{i_j}. By Observation 1, we get that x_i is an iterative k-backbone of φ if and only if it is an iterative k-backbone of $\varphi_{\{x_{i_1}, \ldots, x_{i_l}\}}$. It holds that $\{x_i\} \in \varphi_{\{x_{i_1}, \ldots, x_{i_l}\}}$. Thus, x_i is an iterative k-backbone of φ. $\qquad\square$

Theorem 8. ITERATIVE-LOCAL-BACKBONE[KROM] *is in* P.

Proof. We show that the iterative k-backbones of a Krom formula φ coincide with those backbones of φ that can be identified by iterated application of the following rule: if the implication graph of φ contains a path from a literal $l \in \{x, \neg x\}$ to its complement $\bar{l}$ of length at most k, conclude that x is a backbone and set $\varphi := \varphi|_{\bar{l}}$. Detection of such a path can be done in polynomial time. Also, at most $O(|\text{Var}(\varphi)|)$ iterated applications of this rule suffice to reach a fixpoint. All that remains is to show the correspondence.

The correspondence claim follows from the following property. Let $l \in \text{Lit}(\varphi)$. If $\text{impl}(\varphi)$ contains a path $\bar{l} \to^* l$ that uses at most k clauses and that doubly uses m of these clauses, then there exist literals $l_1, \ldots, l_{m+1} \in \text{Lit}(\varphi)$ such that (i) $l_{m+1} = l$ and (ii) for each $1 \leq i \leq m + 1$ the graph $\text{impl}(\varphi|_{\{l_1, \ldots, l_{i-1}\}})$ contains a path $\bar{l_i} \to^* l_i$ that uses at most k clauses and does not doubly use any clause. We prove this claim by induction on m. The case for $m = 0$ is trivial. Consider the case for $m \geq 1$. Since the path $\bar{l} \to^* l$ doubly uses some clause, we know that $\bar{l} \to^* a \to \bar{b} \to^* b \to \bar{a} \to^* l$, for some $a, b \in \text{Lit}(\varphi)$. We can assume without loss of generality that the path $\bar{b} \to^l b$ does not doubly use any clause. If this is not the case, the path $\bar{b} \to^l b$ contains a subpath $\bar{c} \to^* c$ that does not doubly uses any clauses, and we could select c instead of b. Also, we know that $l \leq k$. It is easy to see that $\text{impl}(\varphi|_b)$ contains the path $\bar{l} \to^* a \to \bar{a} \to^* l$, which uses at most k clauses and doubly uses $m - 1$ of these clauses. By the induction hypothesis, we obtain that there exist $l'_1, \ldots, l'_m$ such that $l'_m = l$ and for each $1 \leq i \leq m$ the graph $\text{impl}(\varphi|_{\{l'_1, \ldots, l'_{i-1}\}})$ contains a path $\bar{l'_i} \to^* l'_i$ that uses at most k clauses and does not doubly use any clause. Now let $l_1 = b$ and $l_i = l'_{i-1}$ for all $2 \leq i \leq m + 1$. It is straightforward to verify that $l_1, \ldots, l_{m+1}$ satisfy the required properties. $\qquad\square$

Somewhat related to our mechanism of computing enforced assignments via iterated k-backbones is the mechanism used to define *unit-refutation complete formulas of level k* [13,19]. This mechanism is based on mappings r_k from CNF formulas to CNF formulas. For a nonnegative integer k, the mapping r_k is defined inductively as follows. In the case for $k = 0$, we let $r_0(\varphi) = \{\bot\}$ if $\bot \in \varphi$, and $r_0(\varphi) = \varphi$ otherwise. In the case for $k > 0$, we let $r_k(\varphi) = r_k(\varphi|_l)$ if there exists a literal $l \in \text{Lit}(\varphi)$ such that $r_{k-1}(\varphi|_{\bar{l}}) = \{\bot\}$, and $r_k(\varphi) = \varphi$ otherwise. In particular, the mapping r_1 computes the result of applying unit propagation. Note that the result of $r_k(\varphi)$ is the application of a number of forced assignments to φ, i.e., $r_k(\varphi) = \varphi|_L$ for some $L \subseteq \text{Lit}(\varphi)$ such that for all $l \in L$ we have $\varphi \models l$. We let $L_k^{\text{UC}}(\varphi)$ denote the set of forced literals that are computed by r_k, i.e., $L_k^{\mathrm{UC}}(\varphi) = L \subseteq \mathrm{Lit}(\varphi)$ such that $r_k(\varphi) = \varphi|_L$. Similarly, we let $L_k^{\mathrm{ILB}}(\varphi)$ denote the set of forced literals that are found by computing iterative k-backbones.

The following observations relate the two mechanisms. Let φ be an arbitrary CNF formula. We have that $L_1^{\mathrm{UC}}(\varphi) = L_1^{\mathrm{ILB}}(\varphi)$. In fact, this set contains exactly those enforced literals that can be found by unit propagation. Also, for any $k \geq 2$ we have that $L_k^{\mathrm{ILB}}(\varphi) \subsetneq L_k^{\mathrm{UC}}(\varphi)$. The inclusion follows from the fact that each minimal subset φ' of size at most k that enforces a literal l has at most k literals (which is a direct result of Tarsi's Lemma). Whenever l is identified as an enforced literal in iterative k-backbone computation, it can then also be computed by r_k by first guessing $\bar{l}$, and subsequently obtaining a contradiction for each instantiation of the other variables in $\mathrm{Var}(\varphi')$. In order to see that the inclusion is strict, consider the family of formulas $(\varphi_n)_{n \in \mathbb{N}}$, where $\varphi_n = \{ \{\neg x_i, x_{i+1}\} : 1 \leq i < n \} \cup \{\neg x_n, \neg x_1\}$. For each φ_n, we know that $\varphi_n \models \neg x_1$. Furthermore, we have that $\neg x_1 \in L_2^{\mathrm{UC}}(\varphi_n)$, but x_1 is not an iterative k-backbone of φ_n for any $k < n$.

8   Experimental Results

In order to illustrate the relevance of the concept of local backbones and iterative local backbones, we provide some empirical evidence of the distribution of (iterative) local backbones in instances from different domains. We considered both randomly generated instances (3CNF instances with various variable-clause ratios around the phase transition) and instances originating from planning [15,17], circuit fault analysis [23], inductive inference [23], and bounded model checking [26]. We considered only satisfiable instances. For practical reasons, we used a method that gives us a lower bound on the number of k-backbone variables. By reducing the separate LOCAL-BACKBONE problems to SMALL-UNSATISFIABLE-SUBSET, we can use algorithms computing subset-minimal unsatisfiable subsets to approximate the number of iterative local backbones (we used MUSer2 [2]). In order to get the exact number, we would have to compute cardinality-minimal unsatisfiable subsets, which is difficult in practice.

The experimental results are shown in Figure 3. For each of the instances, we give the percentage of backbones that are of order k (dashed lines) and the percentage of backbones that are of iterative order k (solid lines), as well as the total number of backbones and the total number of clauses. There are instances with several backbones, most of which have relatively small order. This is the case for the instances from the domains of planning (*logistics*), circuit fault analysis (*ssa7552*) and bounded model checking (*bmc-ibm*). It is worth noting that already more than 75 percent of the backbones in all the considered *bmc-ibm* instances are of iterative order 2. We also found instances that have no backbones of small order or of small iterative order. This is the case for the instances from the domain of inductive inference (*ii32*) and the randomly generated instances. Some of these instances do have backbones, while others have no backbones at all. It would be interesting to confirm these findings by a more rigorous experimental investigation.

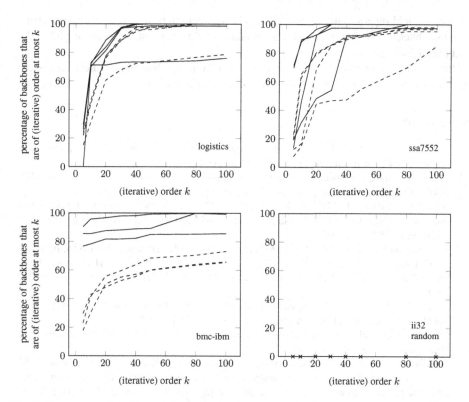

Fig. 3. Percentage of backbones that are of order at most k (dashed) and of iterative order at most k (solid), for SAT instances from planning (*logistics.[a–d]*, 828–4713 variables, 6718–21991 clauses, 437–838 backbones), circuit fault analysis (*ssa7552-[038,158–160]*, 1363–1501 variables, 3032–3575 clauses, 405–838 backbones), bounded model checking (*bmc-ibm-[2,5,7]*, 2810–9396 variables, 11683–41207 clauses, 405–557 backbones), inductive inference (*ii32[b–e][1–3]*, 222–824 variables, 1186–20862 clauses, 0–208 backbones) and random 3SAT instances (*random*, 200 variables, 820–900 clauses, 1–131 backbones).

9 Conclusions

We have drawn a detailed complexity map of the problem of finding local backbones and iterative local backbones, in general and for formulas from restricted classes. Additionally, we have provided some first empirical results on the distribution of (iterative) local backbones in some benchmark SAT instances. We found that in structured instances from different domains backbones are of quite low (iterative) order. This suggests that the notions of local backbones and iterative local backbones can be used to identify structure in SAT instances.

Some of our findings are somewhat surprising. (1) Finding local backbones in Horn and Krom formulas is fixed-parameter intractable, whereas backbones for these classes of formulas can be found in polynomial time. (2) In certain cases finding iterative local backbones is computationally easier than finding (non-iterative) local backbones.

(3) Local backbones and iterative local backbones seem to be a better indicator of structure than backbones. Random instances do have backbones, but these are of high order and iterative order.

Backbones and local backbones are implied unit clauses. It might be interesting to extend our investigation to implied clauses of larger fixed size, binary clauses in particular.

References

1. Aharoni, R., Linial, N.: Minimal non-two-colorable hypergraphs and minimal unsatisfiable formulas. J. Combin. Theory Ser. A 43, 196–204 (1986)
2. Belov, A., Marques-Silva, J.: MUSer2: An efficient MUS extractor. J. on Satisfiability, Boolean Modeling and Computation 8(1/2), 123–128 (2012)
3. Buresh-Oppenheim, J., Mitchell, D.: Minimum 2CNF resolution refutations in polynomial time. In: Marques-Silva, J., Sakallah, K.A. (eds.) SAT 2007. LNCS, vol. 4501, pp. 300–313. Springer, Heidelberg (2007)
4. Buresh-Oppenheim, J., Mitchell, D.: Minimum witnesses for unsatisfiable 2CNFs. In: Biere, A., Gomes, C.P. (eds.) SAT 2006. LNCS, vol. 4121, pp. 42–47. Springer, Heidelberg (2006)
5. Darwiche, A., Marquis, P.: A knowledge compilation map. J. Artif. Intell. Res. 17, 229–264 (2002)
6. Dowling, W.F., Gallier, J.H.: Linear-time algorithms for testing the satisfiability of propositional horn formulae. J. Logic Programming 1(3), 267–284 (1984)
7. Downey, R.G., Fellows, M.R.: Parameterized Complexity. Monographs in Computer Science. Springer, New York (1999)
8. Dubois, O., Dequen, G.: A backbone-search heuristic for efficient solving of hard 3-SAT formulae. In: Nebel, B. (ed.) Proceedings of the Seventeenth International Joint Conference on Artificial Intelligence, IJCAI 2001, Seattle, Washington, USA, August 4-10, pp. 248–253 (2001)
9. Fellows, M.R., Hermelin, D., Rosamond, F.A., Vialette, S.: On the parameterized complexity of multiple-interval graph problems. Theoretical Computer Science 410(1), 53–61 (2009)
10. Fellows, M.R., Szeider, S., Wrightson, G.: On finding short resolution refutations and small unsatisfiable subsets. Theoretical Computer Science 351(3), 351–359 (2006)
11. Flum, J., Grohe, M.: Parameterized Complexity Theory. Texts in Theoretical Computer Science. An EATCS Series, vol. XIV. Springer, Berlin (2006)
12. Gallo, G., Longo, G., Pallotino, S., Nguyen, S.: Directed hypergraphs and applications. Discrete Applied Mathematics 42, 177–201 (1993)
13. Gwynne, M., Kullmann, O.: Generalising and unifying SLUR and unit-refutation completeness. In: van Emde Boas, P., Groen, F.C.A., Italiano, G.F., Nawrocki, J., Sack, H. (eds.) SOFSEM 2013. LNCS, vol. 7741, pp. 220–232. Springer, Heidelberg (2013)
14. Hertli, T., Moser, R.A., Scheder, D.: Improving PPSZ for 3-SAT using critical variables. In: Schwentick, T., Dürr, C. (eds.) Symposium on Theoretical Aspects of Computer Science, vol. 9, pp. 237–248. Schloss Dagstuhl - Leibniz-Zentrum für Informatik (2011)
15. Hoos, H.H., Stützle, T.: SATLIB: An online resource for research on SAT. In: Gent, I., van Maaren, H., Walsh, T. (eds.) SAT 2000: Highlights of Satisfiability Research in the year 2000. Frontiers in Artificial Intelligence and Applications, pp. 283–292. Kluwer Academic (2000)
16. Impagliazzo, R., Paturi, R., Zane, F.: Which problems have strongly exponential complexity? J. of Computer and System Sciences 63(4), 512–530 (2001)
17. Kautz, H., Selman, B.: Pushing the envelope: planning, propositional logic, and stochastic search. In: Proceedings of the Thirteenth AAAI Conference on Artificial Intelligence, AAAI 1996, pp. 1194–1201. AAAI Press (1996)

18. Kilby, P., Slaney, J.K., Thiébaux, S., Walsh, T.: Backbones and backdoors in satisfiability. In: Proceedings of the Twentieth National Conference on Artificial Intelligence and the Seventeenth Innovative Applications of Artificial Intelligence Conference, AAAI 2005, Pittsburgh, Pennsylvania, USA, July 9-13, pp. 1368–1373 (2005)
19. Kullmann, O.: Investigating a general hierarchy of polynomially decidable classes of cnf's based on short tree-like resolution proofs. Electronic Colloquium on Computational Complexity (ECCC) 6(41) (1999)
20. Kullmann, O.: An application of matroid theory to the SAT problem. In: Fifteenth Annual IEEE Conference on Computational Complexity, pp. 116–124 (2000)
21. Niedermeier, R.: Invitation to Fixed-Parameter Algorithms. Oxford Lecture Series in Mathematics and its. Applications. Oxford University Press, Oxford (2006)
22. Parkes, A.J.: Clustering at the phase transition. In: Proceedings of the Fourteenth National Conference on Artificial Intelligence, AAAI 1997, pp. 340–345. AAAI Press (1997)
23. Prelotani, D.: Efficiency and stability of hypergraph SAT algorithms. In: Johnson, D.S., Trick, M.A. (eds.) Cliques, Coloring and Satisfiability, pp. 479–498. AMS (1996)
24. Schneider, J., Froschhammer, C., Morgenstern, I., Husslein, T., Singer, J.M.: Searching for backbones – an efficient parallel algorithm for the traveling salesman problem. Computer Physics Communications 96, 173–188 (1996)
25. Slaney, J.K., Walsh, T.: Backbones in optimization and approximation. In: Nebel, B. (ed.) Proceedings of the Seventeenth International Joint Conference on Artificial Intelligence, IJCAI 2001, Seattle, Washington, USA, August 4-10, pp. 254–259 (2001)
26. Strichman, O.: Tuning SAT checkers for bounded model checking. In: Emerson, E.A., Sistla, A.P. (eds.) CAV 2000. LNCS, vol. 1855, pp. 480–494. Springer, Heidelberg (2000)

Upper and Lower Bounds
for Weak Backdoor Set Detection*

Neeldhara Misra[1], Sebastian Ordyniak[2], Venkatesh Raman[3], and Stefan Szeider[4]

[1] Indian Institute of Science, Bangalore
[2] Masaryk University Brno
[3] Institute of Mathematical Sciences, Chennai
[4] Vienna University of Technology

Abstract. We obtain upper and lower bounds for running times of exponential time algorithms for the detection of weak backdoor sets of 3CNF formulas, considering various base classes. These results include (omitting polynomial factors), (i) a 4.54^k algorithm to detect whether there is a weak backdoor set of at most k variables into the class of Horn formulas; (ii) a 2.27^k algorithm to detect whether there is a weak backdoor set of at most k variables into the class of Krom formulas. These bounds improve an earlier known bound of 6^k. We also prove a 2^k lower bound for these problems, subject to the Strong Exponential Time Hypothesis.

1 Introduction

A backdoor set is a set of variables of a CNF formula such that fixing the truth values of the variables in the backdoor set moves the formula into some polynomial-time decidable class. Backdoor sets were independently introduced by Crama et al. [2] and by Williams et al. [16], the latter authors coined the term "backdoor." The existence of a small backdoor set in a CNF formula can be considered as an indication of "hidden structure" in the formula.

One distinguishes between various types of backdoor sets. Let $\mathcal{B}$ denote the base class of formulas under consideration. A *weak $\mathcal{B}$-backdoor set* of a CNF formula F is a set S of variables such that there is a truth assignment τ of the variables in S for which the formula $F[\tau]$, which is obtained from F by assigning the variables of S according to τ and applying the usual simplifications, is satisfiable and $F[\tau] \in \mathcal{B}$. A *strong $\mathcal{B}$-backdoor set* of F is a set S of variables such that for *each* truth assignment τ of the variables in S, the formula $F[\tau]$ is in $\mathcal{B}$.

The challenging problem is to find a weak or strong $\mathcal{B}$-backdoor set of size at most k if it exists. These problems are NP-hard for all reasonable base classes. However, if k is assumed to be small, an interesting complexity landscape evolves, which can be adequately analyzed in the context of *parameterized complexity*, where k is considered as the parameter (some basic notions of parameterized complexity will be reviewed in Section 2). This line of research was initiated by Nishimura et al. [14] who showed that

* All authors acknowledge support from the OeAD/DST (Austrian Indian collaboration grant, IN13/2011). Szeider acknowledges the support by the ERC, grant reference 239962.

M. Järvisalo and A. Van Gelder (Eds.): SAT 2013, LNCS 7962, pp. 394–402, 2013.

Table 1. Upper bounds (UB) and lower bounds (LB) for the time complexity of WB(3CNF, $\mathcal{B}$) for various base classes $\mathcal{B}$ (polynomial factors are omitted). The 2^k and $n^{\frac{k}{2}-\epsilon}$ lower bounds are subject to the Strong Exponential-Time Hypothesis, and the $2^{o(k)}$ lower bounds are subject to the Exponential-Time Hypothesis. Results marked [*] are obtained in this paper.

$\mathcal{B}$:	HORN	KROM	0-VAL	FOREST	RHORN	QHORN	MATCH
UB:	4.54^k [*]	2.27^k [*]	2.85^k [15]	$f(k)$ [7]	n^k [triv]	n^k [triv]	n^k [triv]
LB:	2^k [*]	2^k [*]	$2^{o(k)}$ [*]	2^k [*]	$n^{\frac{k}{2}-\epsilon}$ [8]	$n^{\frac{k}{2}-\epsilon}$ [6]	$n^{\frac{k}{2}-\epsilon}$ [*]

for the fundamental base classes HORN and KROM, the detection of strong backdoor sets is fixed-parameter tractable, whereas the detection of weak backdoor sets is not (under the complexity theoretic assumption FPT $\neq$ W[1]). However, if the width of the clauses of the input formula is bounded by a constant, then these hardness results do not hold any more and one achieves fixed-parameter tractability [8]. In order to discuss these results, we introduce the following problem template which is defined for any two classes $\mathcal{A}, \mathcal{B}$ of CNF formulas.

WB($\mathcal{A}, \mathcal{B}$)
Instance: A CNF formula $F \in \mathcal{A}$ with n variables, a non-negative integer k.
Parameter: The integer k.
Question: Does F have a weak $\mathcal{B}$-backdoor set of size at most k?

Thus, one could think of this problem as asking for a small weak backdoor "from $\mathcal{A}$ to $\mathcal{B}$." In this paper we focus on the special case where $\mathcal{A} = 3$CNF. In particular, we aim to draw a detailed complexity landscape of WB(3CNF, $\mathcal{B}$) for various base classes, providing improved lower and upper bounds. An overview of our results in the context of known results is provided in Table 1. The definitions of these classes appear in Section 2.

Gaspers and Szeider [8] showed that WB(3CNF, $\mathcal{B}$) is fixed-parameter tractable for every base class $\mathcal{B}$ which is defined by a property of individual clauses, such as the classes HORN, KROM, and 0-VAL. Their general algorithm provides a running time of 6^k (omitting polynomial factors). We improve this to 4.54^k for HORN and to 2.27^k for KROM. These results fit nicely with the recent 2.85^k algorithm for WB(3CNF, 0-VAL) by Raman and Shankar [15].

There are base classes for which the detection of weak backdoor sets remains fixed-parameter *intractable* (in terms of W[2]-hardness), even if the input is restricted to 3CNF. In particular, the W[2]-hardness of WB(3CNF, $\mathcal{B}$) is known for the base class RHORN [8] and for the base class of QHORN [6]: We extend this line of results with another example. We consider the class MATCH of *matched* CNF formulas [5], which are CNF formulas F where for each clause $C \in F$ one can select a unique variable x_C that appears in C positively or negatively, such that $x_C \neq x_D$ for $C \neq D$. Since all matched formulas are satisfiable, this class is particularly well suited as a base class for weak backdoor sets. It is known that WB(CNF, MATCH) is W[2]-hard, but the case WB(3CNF, MATCH) has been open. We show, that WB(3CNF, MATCH) is W[2]-hard as well.

We contrast the algorithmic upper bounds for the considered backdoor set detection problems by lower bounds. These lower bounds are either subject to the Exponential Time Hypothesis (ETH), or the Strong Exponential Time Hypothesis (SETH), see Section 4. Consequently, any algorithm that beats these lower bounds would provide an unexpected speedup for the exact solution of 3SAT or SAT, respectively. In particular, we explain how the W[2]-hardness proofs can be used to get lower bounds of the form $n^{\frac{k}{2}-\epsilon}$ under the SETH.

Full Version. Proofs of statements marked with ($\star$) are shortened or omitted due to space restrictions. Detailed proofs can be found in the full version, available at arxiv.org/abs/1304.5518.

2   Preliminaries

CNF Formulas and Assignments We consider propositional formulas in conjunctive normal form (CNF) as sets of clauses, where each clause is a set of literals, i.e., a literal is either a (positive) variable or a negated variable, not containing a pair of complementary literals. We say that a variable x is positive (negative) in a clause C if $x \in C$ ($\overline{x} \in C$), and we write var(C) for the set of variables that are positive or negative in C. A *truth assignment* τ is a mapping from a set of variables, denoted by var(τ), to $\{0, 1\}$. A truth assignment τ *satisfies* a clause C if it sets at least one positive variable of C to 1 or at least one negative variable of C to 0. A truth assignment τ satisfies a CNF formula F if it satisfies all clauses of F. Given a CNF formula F and a truth assignment τ, $F[\tau]$ denotes the *truth assignment reduct* of F under τ, which is the CNF formula obtained from F by first removing all clauses that are satisfied by τ and then removing from the remaining clauses all literals $x, \overline{x}$ with $x \in$ var(τ). Note that no assignment satisfies the empty clause. The *incidence graph* of a CNF formula F is the bipartite graph whose vertices are the variables and clauses of F, and where a variable x and a clause C are adjacent if and only if $x \in$ var(C).

We consider the following *classes of CNF formulas*.

- 3CNF: the class of CNF formulas where each clause contains at most 3 literals.
- KROM: the class of CNF formulas where each clause contains at most 2 literals (also called 2CNF).
- HORN: the class of *Horn* formulas, i.e., CNF formulas where each clause has at most 1 positive literal.
- RHORN: the class of *renameable* (or *disguised*) *Horn* formulas, i.e., formulas that can be made Horn by complementing variables.
- QHORN: the class of *q-Horn* formulas [1] (RHORN, KROM $\subseteq$ QHORN).
- 0-VAL: the class of 0-*valid* CNF formulas, i.e., formulas where each clause contains at least 1 negative literal.
- FOREST: the class of *acyclic* formulas (the undirected incidence graph is acyclic).
- MATCH: the class of *matched* formulas, formulas whose incidence graph has a matching such that each clause is matched to some unique variable.

All our results concerning the classes HORN and 0-VAL clearly hold also for the dual classes of *anti-Horn* formulas (i.e., CNF formulas where each clause has at most 1 positive literal), and 1-*valid* CNF formulas (i.e., formulas where each clause contains at least 1 positive literal), respectively.

Parameterized Complexity. Here we introduce the relevant concepts of parameterized complexity theory. For more details, we refer to text books on the topic [3,4,12]. An instance of a parameterized problem is a pair (I, k) where I is the main part of the instance, and k is the parameter. A parameterized problem is *fixed-parameter tractable* if instances (I, k) can be solved in time $f(k)|I|^c$, where f is a computable function of k, and c is a constant. FPT denotes the class of all fixed-parameter tractable problems. Hardness for parameterized complexity classes is based on *fpt-reductions*. A parameterized problem L is fpt-reducible to another parameterized problem L' if there is a mapping R from instances of L to instances of L' such that (i) $(I, k) \in L$ if and only if $(I', k') = R(I, k) \in L'$, (ii) $k' \leq g(k)$ for a computable function g, and (iii) R can be computed in time $O(f(k)|I|^c)$ for a computable function f and a constant c. Central to the completeness theory of parameterized complexity is the hierarchy FPT $\subseteq$ W[1] $\subseteq$ W[2] $\subseteq \ldots$. Each intractability class W[t] contains all parameterized problems that can be reduced to a certain parameterized satisfiability problem under fpt-reductions.

The problem HITTING SET is well-known to be W[2]-complete. This problem takes as input a family S of finite sets $S_1, \ldots, S_m$ and an integer $k > 0$, k is the parameter. The question is whether S has a hitting set of size at most k, i.e., a set $H \subseteq \bigcup_{1 \leq i \leq m} S_i$ such that $H \cap S_i \neq \emptyset$ for every $1 \leq i \leq m$ and $|H| \leq k$?

However, the restricted variant where all sets S_i are of size at most 3, is fixed-parameter tractable and can be solved in time 2.270^k, omitting polynomial factors [13].

3   Upper Bounds

Theorem 1. WB(3CNF, KROM) *can be solved in time* 2.270^k *(omitting polynomial factors).*

Proof. Let F and k be the given 3CNF formula and non-negative integer, respectively. Let S be the family of sets $\{\text{var}(C) : C \in F, |C| = 3\}$. We can find a weak KROM-backdoor set of size at most k by finding a hitting set H of S of size at most k and checking whether there is an assignment τ_H to the variables in H such that $F[\tau_H]$ is satisfiable. The correctness follows from the fact that if $F[\tau_H]$ is satisfiable for some τ_H, then we clearly have the desired backdoor set. On the other hand, if $F[\tau_H]$ is not satisfiable for *any* τ_H, then F was not satisfiable to begin with, and does not admit a weak backdoor set of any size. As $F[\tau_H] \in$ KROM, the satisfiability of $F[\tau_H]$ can be checked in polynomial-time. It follows that if we omit polynomial factors then the running time of this algorithm is the time required to find a hitting set of S of size at most k, i.e., 2.270^k [13], plus the time required to go over the at most 2^k assignments of the variables in the hitting set.   □

Theorem 2 (⋆)**.** WB(3CNF, HORN) *can be solved in time* $(\frac{1}{2}(1 + \sqrt{65}))^k < 4.54^k$ *(omitting polynomial factors).*

Proof. Let F and k be the given 3CNF formula and non-negative integer, respectively. If $F \in$ HORN then there is nothing to do. So suppose that $F \notin$ HORN and let NH(F) be the set of all clauses of F that are not horn. Then NH(F) can contain the following types of clauses: (C1) clauses that contain only positive literals (and at least two of them), and (C2) clauses that contain exactly two positive literals and one negative literal.

An assignment τ is *minimal* with respect to a formula F or to a clause C if $F[\tau] \in$ HORN or $\{C[\tau]\} \in$ HORN, respectively, but $F[\tau'] \notin$ HORN or $\{C[\tau']\} \notin$ HORN for every assignment τ' that agrees with τ but is defined on a strict subset of var(τ). Our algorithm uses the bounded search tree method to branch over all possible minimal assignments τ that set at most k variables of F such that $F[\tau] \in$ HORN. The algorithm then checks for each of these assignments whether $F[\tau]$ is satisfiable (because $F[\tau] \in$ HORN this can be done in polynomial-time). If there is at least one such assignment τ such that $F[\tau]$ is satisfiable, then the algorithm returns var(τ) as a weak HORN-backdoor set of F with witness τ. Otherwise, i.e., if there is no such assignment, the algorithm outputs that F does not have a weak HORN-backdoor set of size at most k.

At the root node of the search tree we set τ to be the empty assignment. Depending on the types and structure of the clauses of the formula $F[\tau]$, the algorithm then branches as follows: If $F[\tau]$ contains at least one clause of type (C1), then the algorithm branches on one of these clauses according to branching rule (R1). If $F[\tau]$ contains at least two clauses of type (C2) that are not variable-disjoint, then the algorithm branches on such a pair according to branching rule (R2). Otherwise, i.e., if NH($F[\tau]$) merely consists of clauses of type (C2) which are pairwise variable-disjoint the algorithm branches according to branching rule (R3).

We will now describe the branching rules (R1)–(R3) in detail. In the following let τ' be the assignment obtained before the current node in the search tree, and let x, x', y, y', z and z' be 6 pairwise distinct variables. Every branching rule will lead to a new assignment τ (extending the current assignment τ') where the parameter k decreases by $|\text{var}(\tau) \setminus \text{var}(\tau')|$.

Let $C \in F[\tau']$ be a clause of type (C1). Then branching rule (R1) is defined as follows. If $C = \{x, y, z\}$ then $\{C[\tau]\} \in$ HORN if and only if $\tau(x) = 1$ or $\tau(y) = 1$ or $\tau(z) = 1$ or $\tau(x) = 0 = \tau(y)$ or $\tau(x) = 0 = \tau(z)$, or $\tau(y) = 0 = \tau(z)$. Hence, there are 3 cases for which the parameter (the number of variables set in the backdoor) decreases by 1 and 3 cases for which the parameter decreases by 2. This leads to the recurrence relation $T(k) = 3T(k-1) + 3T(k-2) = (\frac{1}{2}(3 + \sqrt{21}))^k < 4.54^k$. Similarly, if $C = \{x, y\}$ then $\{C[\tau]\} \in$ HORN if and only if $\tau(x) = 0$ or $\tau(x) = 1$ or $\tau(y) = 0$ or $\tau(y) = 1$. Hence, there are 4 cases and in each of them the parameter decreases by 1. This leads to the recurrence function $T(k) = 4T(k-1) = 4^k < 4.54^k$.

Let $C \in F[\tau']$ and $C' \in F[\tau']$ be two distinct clauses of type (C2) that share at least one variable. Then branching rule (R2) is defined as follows.

We distinguish the following cases (due to space limitations we will only list the cases here; for a full description of the cases we refer to the full version of the paper). (1A) var(C) $\cap$ var(C') $= \{x\}$, $x \in C$ and $x \in C'$, (1B) var(C) $\cap$ var(C') $= \{x\}$, $\bar{x} \in C$ and $\bar{x} \in C'$, (1C) var(C) $\cap$ var(C') $= \{x\}$, $\bar{x} \in C$ and $x \in C'$, (2A) var(C) $\cap$ var(C') $= \{x, y\}$, $x, y \in C$ and $x, y \in C'$, (2B) var(C) $\cap$ var(C') $= \{x, y\}$, $x, \bar{y} \in C$

and $x, \bar{y} \in C'$, (2C) var$(C) \cap$ var$(C') = \{x, y\}$, $x, \bar{y} \in C$ and $\bar{x}, y \in C'$, and (3) var$(C) \cap$ var$(C') = \{x, y, z\}$.

Taking the maximum over the above cases we obtain the recurrence function $T(k) = T(k-1) + 16T(k-2) = (\frac{1}{2}(1 + \sqrt{65}))^k < 4.54^k$ for branching rule (R2).

Recall that after applying branching rule (R2) exhaustively all pairs of clauses of type (C2) are pairwise variable-disjoint. We will describe branching rule (R3), which makes use of this fact. Let $C = \{x, y, \bar{z}\}$. We set either $\tau(x) = 0$ or $\tau(x) = 1$. This leads to the recurrence function: $T(k) = 2T(k-1) = 2^k < 4.54^k$. Note that in contrast to the branching rules (R1) and (R2) the branching rule (R3) is not exhaustive. Indeed for every clause $C = \{x, y, \bar{z}\}$ of type (C2) there are 5 possible minimal assignments τ such that $\{C[\tau]\} \in$ HORN, i.e., the assignments $\tau(x) = 0$, $\tau(x) = 1$, $\tau(y) = 0$, $\tau(y) = 1$, and $\tau(z) = 0$. Because each of these assignments τ sets only 1 variable this would lead to a recurrence function $T(k) = 5T(k-1) = 5^k$ and hence $T(k) > 4.54^k$. It follows that in contrast to the branching rules R1 and R2 where we could exhaustively branch over all possible minimal assignments, this cannot be done for clauses of type (C2). However, because branching rule (R2) ensures that the remaining clauses of type (C2) are pairwise variable-disjoint it turns out that this is indeed not necessary (see Claim 1).

This concludes the description of our algorithm. The running time of the algorithm is the maximum branching factor over the cases described above, i.e., $(\frac{1}{2}(1+\sqrt{65}))^k < 4.54^k$ as required. To see that the algorithm is correct we need to show that it outputs an assignment τ if and only if the set var(τ) is a weak HORN-backdoor set of F of size at most k. Because the branching rules R1 and R2 branch exhaustively over all minimal assignments τ such that the corresponding clause(s) are reduced to Horn clauses, it only remains to show the correctness of branching rule R3. This is done by the following claim whose proof can be found in the full version of the paper.

Claim 1 $(\star)$. *Let F be a 3CNF formula, P be a set of pairwise variable-disjoint clauses of type (C2) such that $F \setminus P \in$ HORN. Furthermore, let L be a set of variables that consists of one positively occurring variable from each of the clauses in P. Then F has a weak HORN-backdoor set of size at most $|P|$ if and only if L is a weak HORN-backdoor set of F.* □

4   Lower Bounds

For our lower bounds we use the *Exponential Time Hypothesis* (ETH) and the *Strong Exponential Time Hypothesis* (SETH), introduced by Impagliazzo et al. [9,10], which state the following:

> **ETH**: There is no algorithm that decides the satisfiability of a 3CNF formula with n variables in time $2^{o(n)}$, omitting polynomial factors.
> **SETH**: There is no algorithm that decides the satisfiability of a CNF formula with n variables in time $(2 - \epsilon)^n$, omitting polynomial factors.

An *implication chain* is a CNF formula of the form $\{\{x_0\}, \{\bar{x}_0, x_1\}, \{\bar{x}_1, x_2\}, \dots, \{\bar{x}_{n-1}, x_n\}, \{\bar{x}_n\}\}$, $n \geq 1$ where the first $\{x_0\}$ and the last clause $\{\bar{x}_n\}$ can be missing. Let CHAINS denote the class of formulas that are variable-disjoint unions of implication chains.

Theorem 3. *Let $\mathcal{B}$ be a base class that contains* CHAINS. *Then* WB$(3\text{CNF}, \mathcal{B})$ *cannot be solved in time* $(2 - \epsilon)^k$ *(omitting polynomial factors) unless SETH fails.*

Proof. We show that an $(2-\epsilon)^k$ algorithm for WB$(3\text{CNF}, \text{CHAINS})$ implies an $(2-\epsilon)^n$ algorithm for SAT contradicting our assumption. Let F be a CNF formula with n variables. We will transform F into a 3CNF formula F_3 such that F is satisfiable if and only if F_3 has a weak CHAINS-backdoor set of size at most n. We obtain F_3 from F using a commonly known transformation that transforms an arbitrary CNF formula into a 3CNF formula that is satisfiability equivalent with the original formula. In particular, we obtain the formula F_3 from F by replacing every clause $C = \{x_1, \ldots, x_l\}$ where $l > 3$ with the clauses $\{x_1, x_2, y_1\}, \{\bar{y}_1, x_3, y_2\}, \ldots, \{\bar{y}_{l-3}, x_l\}$, where $y_1, \ldots, y_{l-3}$ are new variables. This completes the construction of F_3. Now, if F is satisfiable and τ is a satisfying assignment of F, then the variables of F form a weak CHAINS-backdoor set of size n of F_3 with witness τ. The reverse is immediate since F_3 is satisfiable, by virtue of having a weak backdoor set, and F is satisfiable if F_3 is satisfiable. □

As the classes HORN, KROM, and FOREST contain CHAINS, we have the following result.

Corollary 1. *Let $\mathcal{B} \in \{\text{HORN}, \text{KROM}, \text{FOREST}\}$. The problem* WB$(3\text{CNF}, \mathcal{B})$ *cannot be solved in time* $(2 - \epsilon)^k$ *(omitting polynomial factors) unless SETH fails.*

Interestingly, in the case of KROM the above result even holds if a hitting set for all clauses containing 3 literals is given with the input. The next lower bound is based on the observation that the VERTEX COVER problem can be considered as a special case of WB$(\text{KROM}, 0\text{-VAL})$, and on a corresponding lower bound for VERTEX COVER [11].

Theorem 4 $(\star)$. WB$(\text{KROM}, 0\text{-VAL})$, *and hence also* WB$(3\text{CNF}, 0\text{-VAL})$, *cannot be solved in time* $2^{o(k)}$ *(omitting polynomial factors) unless ETH fails.*

Let $\mathcal{B}$ be a base class. We say that a polynomial-time algorithm $\mathcal{A}$ is a *canonical HS reduction for* $\mathcal{B}$ if $\mathcal{A}$ takes as input an instance $(\mathcal{S}, k)$ of HITTING SET over n elements and m sets and outputs an instance (F, k) of WB$(3\text{CNF}, \mathcal{B})$ such that: (a) F has at most $O(nm)$ variables, and (b) $\mathcal{S}$ has a hitting set of size at most k if and only if F has a weak $\mathcal{B}$-backdoor set of size at most k .

Lemma 1 *Let $\mathcal{B}$ be a base class. If there is a canonical HS reduction for $\mathcal{B}$, then the following holds:*

1. WB$(3\text{CNF}, \mathcal{B})$ *is* W$[2]$-*hard, and*
2. *there is no algorithm that solves* WB$(3\text{CNF}, \mathcal{B})$ *in time* $O(n^{\frac{k}{2}-\epsilon})$ *unless SETH fails.*

Proof. Because HITTING SET is W$[2]$-complete and a canonical HS reduction is also an fpt-reduction, the first statement of the theorem follows. To see the second statement, we first note that it is is shown in [11, Theorem 5.8] that the DOMINATING SET problem cannot be solved in time $O(n^{k-\epsilon})$ for any $\epsilon > 0$ unless SETH fails (here n is the number of vertices of the input graph and k is the parameter). Using the standard reduction from DOMINATING SET to HITTING SET it follows that HITTING SET restricted to instances where the number of sets is at most the number of elements, cannot be solved in time $O(n^{k-\epsilon})$ for any $\epsilon > 0$, where n is the number of elements of the hitting set instance and k is the parameter. Now suppose that for some base class $\mathcal{B}$ it holds that WB(3CNF, $\mathcal{B}$) can be solved in time $n^{\frac{k}{2}-\epsilon}$ and $\mathcal{B}$ has a canonical HS reduction. Let $(\mathcal{S}, k)$ be an instance of HITTING SET with n_h elements and m_h sets. As stated above we can assume that $m_h \leq n_h$. We use the canonical HS reduction to obtain an instance (F, k) of WB(3CNF, $\mathcal{B}$) where F has at most $O(n_h m_h) \in O(n_h^2)$ variables. We now use the algorithm for WB(3CNF, $\mathcal{B}$) to solve HITTING SET in time $O((n_h^2)^{\frac{k}{2}-\epsilon}) \leq O(n_h^{k-\frac{\epsilon}{2}})$ which contradicts our assumption that there is no such algorithm for HITTING SET. $\square$

Lemma 2 ($\star$) *There is a canonical HS reduction for* MATCH.

Proof. Let $(\mathcal{S}, k)$ be an instance of HITTING SET with $\mathcal{S} = \{S_1, \ldots, S_m\}$ and $V = \bigcup_{i=1}^{m} S_i = \{x_1, \ldots, x_n\}$. We write $S_i = \{x_i^1, \ldots, x_i^{q_i}\}$, where $q_i = |S_i|$. We construct in linear-time a 3CNF formula F with $|V| + \sum_{1 \leq i \leq m}(q_i - 1) \leq n + nm \in O(nm)$ variables such that $\mathcal{S}$ has a hitting set of size at most k if and only if F has a weak MATCH-backdoor set of size at most k.

The variables of F consist of the elements of V and additional variables y_i^j for every $1 \leq i \leq m$ and $1 \leq j < q_i$. We let $F = \bigcup_{i=0}^{m} F_i$ where the formulas F_i are defined as follows. F_0 consists of n binary clauses $\{\bar{x}_1, x_2\}, \{\bar{x}_2, x_3\}, \ldots, \{\bar{x}_{n-1}, x_n\}$, $\{\bar{x}_n, x_1\}$. For $i > 0$, F_i consists of the clauses $\{y_i^1, x_i^1\}, \{\bar{y}_i^1, y_i^2, x_i^2\}, \{\bar{y}_i^2, y_i^3, x_i^3\}, \ldots, \{\bar{y}_i^{q_i-2}, y_i^{q_i-1}, x_i^{q_i-1}\}, \{\bar{y}_i^{q_i-1}, x_i^{q_i}\}$. This completes the construction of F.

We can show that $\mathcal{S}$ has a hitting set of size at most k if and only if F has a weak MATCH-backdoor set of size at most k. Hence the lemma follows. $\square$

We observe that the known W[2]-hardness proofs for WB(3CNF, RHORN) and WB(3CNF, QHORN) [8,6] are in fact canonical HS reductions. Hence, together with Lemmas 1 and 2 we arrive at the following result.

Theorem 5. *Let* $\mathcal{B} \in \{$MATCH, RHORN, QHORN$\}$. *Then* WB(3CNF, $\mathcal{B}$) *is* W[2]-*hard and cannot be solved in time* $O(n^{\frac{k}{2}-\epsilon})$ *for any* $\epsilon > 0$ *unless SETH fails.*

5   Conclusion

We have initiated a systematic study of determining the complexity of finding weak backdoor sets of small size of 3CNF formulas for various base classes. We have given improved algorithms for some of the base classes through the bounded search techniques.

Our lower bounds are among the very few known bounds based on the (Strong) Exponential-Time Hypotheses for parameterized problems where the parameter is the solution size (as opposed to some measure of structure in the input like treewidth).

Closing the gaps between upper and lower bounds of the problems we considered in this paper, and studying WB($\mathcal{A}, \mathcal{B}$) for classes $\mathcal{A}$ other than 3CNF are interesting directions for further research.

References

1. Boros, E., Hammer, P.L., Sun, X.: Recognition of q-Horn formulae in linear time. Discr. Appl. Math. 55(1), 1–13 (1994)
2. Crama, Y., Ekin, O., Hammer, P.L.: Variable and term removal from Boolean formulae. Discr. Appl. Math. 75(3), 217–230 (1997)
3. Downey, R.G., Fellows, M.R.: Parameterized Complexity. Monographs in Computer Science. Springer, New York (1999)
4. Flum, J., Grohe, M.: Parameterized Complexity Theory. Texts in Theoretical Computer Science. An EATCS Series, vol. XIV. Springer, Berlin (2006)
5. Franco, J., Van Gelder, A.: A perspective on certain polynomial time solvable classes of satisfiability. Discr. Appl. Math. 125, 177–214 (2003)
6. Gaspers, S., Ordyniak, S., Ramanujan, M.S., Saurabh, S., Szeider, S.: Backdoors to q-horn. In: Portier, N., Wilke, T. (eds.) Proceedings of the 30th International Symposium on Theoretical Aspects of Computer Science (STACS). Leibniz International Proceedings in Informatics (LIPIcs), vol. 20, pp. 67–79 (2013)
7. Gaspers, S., Szeider, S.: Backdoors to acyclic SAT. In: Czumaj, A., Mehlhorn, K., Pitts, A., Wattenhofer, R. (eds.) ICALP 2012, Part I. LNCS, vol. 7391, pp. 363–374. Springer, Heidelberg (2012)
8. Gaspers, S., Szeider, S.: Backdoors to satisfaction. In: Bodlaender, H.L., Downey, R., Fomin, F.V., Marx, D. (eds.) Fellows Festschrift 2012. LNCS, vol. 7370, pp. 287–317. Springer, Heidelberg (2012)
9. Impagliazzo, R., Paturi, R.: On the complexity of k-SAT. J. of Computer and System Sciences 62(2), 367–375 (2001)
10. Impagliazzo, R., Paturi, R., Zane, F.: Which problems have strongly exponential complexity? J. of Computer and System Sciences 63(4), 512–530 (2001)
11. Lokshtanov, D., Marx, D., Saurabh, S.: Lower bounds based on the exponential time hypothesis. Bulletin of the European Association for Theoretical Computer Science 105, 41–72 (2011)
12. Niedermeier, R.: Invitation to Fixed-Parameter Algorithms. Oxford Lecture Series in Mathematics and its. Applications. Oxford University Press, Oxford (2006)
13. Niedermeier, R., Rossmanith, P.: An efficient fixed-parameter algorithm for 3-hitting set. J. Discrete Algorithms 1(1), 89–102 (2003)
14. Nishimura, N., Ragde, P., Szeider, S.: Detecting backdoor sets with respect to Horn and binary clauses. In: Proceedings of SAT 2004 (Seventh International Conference on Theory and Applications of Satisfiability Testing), Vancouver, BC, Canada, May 10-13, pp. 96–103 (2004)
15. Raman, V., Shankar, B.S.: Improved fixed-parameter algorithm for the minimum weight 3-SAT problem. In: Ghosh, S.K., Tokuyama, T. (eds.) WALCOM 2013. LNCS, vol. 7748, pp. 265–273. Springer, Heidelberg (2013)
16. Williams, R., Gomes, C., Selman, B.: Backdoors to typical case complexity. In: Gottlob, G., Walsh, T. (eds.) Proceedings of the Eighteenth International Joint Conference on Artificial Intelligence, IJCAI 2003, pp. 1173–1178. Morgan Kaufmann (2003)

LearnSAT: A SAT Solver for Education

Mordechai (Moti) Ben-Ari

Department of Science Teaching
Weizmann Institute of Science
Rehovot 76100 Israel
http://www.weizmann.ac.il/sci-tea/benari/

Abstract. The extensive research on SAT solving and the development of software for applications have not been matched by the development of educational materials for introducing students to this field. LEARNSAT is a SAT solver designed for educational purposes. It implements the DPLL algorithm with CDCL and NCB. LEARNSAT produces detailed output of the execution of the algorithms. It generates assignment trees and the implication graphs of CDCL which are rendered by DOT. LEARNSAT is written in PROLOG so that the algorithms are concise and easy to read.

Keywords: education, CDCL SAT solver, Prolog.

1 Introduction

The literature on SAT solving is extensive: the *Handbook* [2] of almost one thousand pages covers theory, algorithms and applications. Since SAT solvers are widely used, it is essential that quality learning materials be available for students, even those who do not intend to become researchers, for example, undergraduate students taking a course in mathematical logic. Such learning materials will also be helpful for people using SAT solvers in applications.

Instructors should be enabled to create learning materials that demonstrate the central algorithms in detail. Furthermore, these demonstrations should use real problems in place of the artificial examples that appear in research papers.

LEARNSAT is a SAT solver designed for educational use. Design considerations include: (a) the student and instructor should be provided with maximum flexibility in specifying the trace output when running the SAT solver; (b) it should be simple to install and run on the vanilla computers used by students (Windows and Mac); (c) given the wide variety of programming languages taught to undergraduates, the program should be usable with only a superficial knowledge of a particular language; (c) the source code should be concise and very well documented so that advanced students can easily understand it.

2 The LEARNSAT SAT Solver

LEARNSAT implements the core algorithms of many modern SAT solvers: DPLL with conflict-driven clause learning (CDCL) and non-chronological backtracking (NCB).

M. Järvisalo and A. Van Gelder (Eds.): SAT 2013, LNCS 7962, pp. 403–407, 2013.

LEARNSAT can be run in three modes—plain DPLL, DPLL with CDCL, and DPLL with both CDCL and NCB—so that the student can examine the improvements obtained by each refinement. The user can specify the order in which literals are assigned. CDCL is implemented by backwards resolution from a conflict clause to a unique implication point (UIP). It is also possible to compute dominators in the implication graph although this computation is just displayed and not used.

3   The Output of LEARNSAT

The key to learning sophisticated algorithms like SAT solving is a trace of the step-by-step execution of the algorithm. The user of LEARNSAT can choose *any* subset of 25 display options in order to tailor the output to a specific learning context. The display options include elementary steps like decision assignments, unit propagations and identifying conflict clauses, as well as the advanced steps of CDCL: the resolution steps used to obtained a learned clause and the search for UIPs. The Appendix shows the (default) output for the example in [4] run in NCB mode.

LEARNSAT can generate two types of graphs that are rendered using the DOT tool (Figure 1): a tree showing the search through the assignments and the implication graphs that display the process for learning clauses from conflicts.[1] It is also possible to generate these graphs incrementally after each step in the algorithm that modifies the graphs.

4   Examples

The LEARNSAT archive includes the examples used in [3,4,5] to help advanced students read these articles. The archive also includes encodings of the following problems: (i) 4-queens,[2] (ii) Tseitin clauses associated with the graphs $K_{2,2}$ and $K_{3,3}$,[3] (iii) two and three-hole pigeonhole problems, and (iv) two- and three-level grid pebbling. With experience we will learn which of these problems is best for educational purposes.

The input to the program is a formula in clausal form written in a readable symbolic form; the 2-hole pigeonhole problem is:

```
hole2 :-
  dpll(
  [
    % Each pigeon in hole 1 or 2
  [p11, p12],   [p21, p22],   [p31, p32],
    % No pair is in hole 1
  [~p11, ~p21], [~p11, ~p31], [~p21, ~p31],
    % No pair is in hole 2
  [~p12, ~p22], [~p12, ~p32], [~p22, ~p32],
  ], _).
```

---

[1] In the Figure, the default color decoration for decision and conflict nodes has been changed to bold for black-and-white printing.

[2] The encoding and its solution by DPLL are explained in detail in [1, Section 6.4].

[3] See [1, Section 6.5].

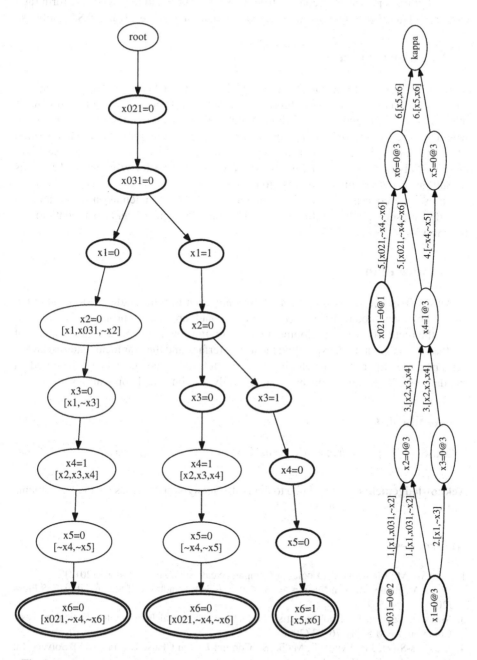

Fig. 1. Assignment tree for DPLL mode (left) and implication graph for NCB mode (right)

A program is provided to convert from DIMACS format to this symbolic form (and conversely) in order to faciliate the student's transition to more advanced SAT solvers.

5 Implementation

LEARNSAT is implemented in PROLOG which was chosen because PROLOG programs are extremely concise: the core algorithms take only 150 lines. The source code itself reads almost like pseudo-code (although making extensive modifications to the code would require mastery of the language). Furthermore, students are likely to know some PROLOG since it is often taught in a course on logic. Finally, the widely used high-quality SWI-PROLOG compiler is distributed with installers for Windows and Mac. Its default interface is minimal and easy to use.

The source code is extensively commented and the documentation in the archive includes: a user's guide, a tutorial using the example from [4], and documentation of the software.

6 Future Plans

LEARNSAT is not meant as a research tool nor even to train graduate students in the latest implementation techniques (MinSAT and Sat4j are more appropriate for this). Instead, the focus of future development will be on improving the pedagogical aspects of the tool. This will include expanding the user interface and the graphical features, and—perhaps more important—developing extensive tutorials. The tutorials will be based on standard puzzles like the 4-queens and hopefully also on actual applications.

7 Availability

LEARNSAT is open source and is available at http://code.google.com/p/mlcs/.

Acknowledgements. I would like to thank the anonymous referees for their comments and suggestions concerning LEARNSAT.

References

1. Ben-Ari, M.: Mathematical Logic for Computer Science, 3rd edn. Springer (2012)
2. Biere, A., Heule, M., Van Maaren, H., Walsh, T. (eds.): Handbook of Satisfiability. IOS Press (2009)
3. Malik, S., Zhang, L.: Boolean satisfiability from theoretical hardness to practical success. Commun. ACM 52(8), 76–82 (2009)
4. Marques-Silva, J.P., Lynce, I., Malik, S.: Conflict-Driven Clause Learning SAT Solvers. In: Biere, et al. (eds.) [2], ch. 4, pp. 131–153 (2009)
5. Marques-Silva, J.P., Sakallah, K.A.: GRASP—a new search algorithm for satisfiability. In: Proceedings of the 1996 IEEE/ACM International Conference on Computer-Aided Design, ICCAD 1996, pp. 220–227 (1996)

A   Output for the Example in [4]

```
LearnSAT v1.3.2. Copyright 2012-13 by Moti Ben-Ari. GNU GPL.
Decision assignment: x021=0@1
Decision assignment: x031=0@2
Decision assignment: x1=0@3
Propagate unit: ~x2 derived from: 1. [x1,x031,~x2]
Propagate unit: ~x3 derived from: 2. [x1,~x3]
Propagate unit: x4 derived from: 3. [x2,x3,x4]
Propagate unit: ~x5 derived from: 4. [~x4,~x5]
Propagate unit: ~x6 derived from: 5. [x021,~x4,~x6]
Conflict clause: 6. [x5,x6]
Not a UIP: two literals are assigned at level: 3
Clause: [x5,x6] unsatisfied
Complement of: x5 assigned true in the unit clause: [~x4,~x5]
Resolvent of the two clauses: [x6,~x4] is also unsatisfiable
Not a UIP: two literals are assigned at level: 3
Clause: [x6,~x4] unsatisfied
Complement of: x6 assigned true in the unit clause: [x021,~x4,~x6]
Resolvent of the two clauses: [x021,~x4] is also unsatisfiable
UIP: one literal is assigned at level: 3
Learned clause: [x021,~x4]
Non-chronological backtracking to level: 1
Skip decision assignment: x1=1@3
Skip decision assignment: x031=1@2
Decision assignment: x021=1@1
Decision assignment: x031=0@2
Decision assignment: x1=0@3
Propagate unit: ~x2 derived from: 1. [x1,x031,~x2]
Propagate unit: ~x3 derived from: 2. [x1,~x3]
Propagate unit: x4 derived from: 3. [x2,x3,x4]
Propagate unit: ~x5 derived from: 4. [~x4,~x5]
Propagate unit: x6 derived from: 6. [x5,x6]
Satisfying assignments:
[x021=1@1,x031=0@2,x1=0@3,x2=0@3,
 x3=0@3,x4=1@3,x5=0@3,x6=1@3]
Statistics: clauses=6,variables=8,units=10,decisions=6,conflicts=1
```

MUStICCa: MUS Extraction
with Interactive Choice of Candidates*

Johannes Dellert, Christian Zielke, and Michael Kaufmann

University of Tübingen, Germany

Abstract. Existing algorithms for minimal unsatisfiable subset (MUS) extraction are defined independently of any symbolic information, and in current implementations domain experts mostly do not have a chance to influence the extraction process based on their knowledge about the encoded problem. The MUStICCa tool introduces a novel graphical user interface for interactive deletion-based MUS finding, allowing the user to inspect and influence the structure of extracted MUSes.

The tool is centered around an explicit visualization of the explored part of the search space, representing unsatisfiable subsets (USes) as selectable states. While inspecting the contents of any US, the user can select candidate clauses to initiate deletion attempts. The reduction steps can be enhanced by a range of state-of-the-art techniques such as clause-set refinement, model rotation, and autarky reduction. MUStICCa compactly represents the criticality information derived for the different USes in a shared data structure, which leads to significant savings in the number of solver calls when multiple MUSes are explored. For automatization, our tool includes a reduction agent mechanism into which arbitrary user-implemented deletion heuristics can be plugged.

1 Introduction and Motivation

With a remarkable amount of recent work (e.g. [13,3,17]) on algorithms for computing minimal explanations of SAT formula unsatisfiability and a broad range of applications (e.g. [15,12]), the field of minimal unsatisfiable subset (MUS) extraction has become an emerging research field in the SAT community, leading to the introduction of a MUS track in the SAT competition 2011 [16].

The algorithms for MUS extraction can be characterized as *constructive*, *destructive* or *dichotomic* [6,14], but they all focus solely on the amount of clauses or clause sets (called *groups*) [2] present in the minimal explanation.

We introduce a completely new approach to guiding the destructive MUS extraction algorithm by Nadel [11]. The central idea of destructive MUS extraction, which was first proposed more than 20 years ago [4,1], is to perform a series of **reduction steps**, moving into smaller unsatisfiable subsets F' until

---

* This work was supported by DFG-SPP 1307, project "Structure-based Algorithm Engineering for SAT-Solving".

M. Järvisalo and A. Van Gelder (Eds.): SAT 2013, LNCS 7962, pp. 408–414, 2013.

all subsets $F'' \subset F'$ are satisfiable. The idea of **interactive MUS extraction** is to give a user full control over the individual reduction steps, providing an interface for interactively focusing on or eliminating parts of the search space. Within our MUStICCa tool, this is done by reverting to intermediate results, the non-minimal unsatisfiable subsets, and exploring new parts of the search space by choosing alternative deletion candidates. This feature will not only help domain experts (who usually have good intuitions about the clauses relevant for good explanations) to find meaningful explanations of unsatisfiability, but also researchers and students in analysing unsatisfiable instances and evaluating the effects of different heuristics for selecting deletion candidates.

By default, MUStICCa's graphical user interface (see Figure 1) consists of three main view components. The central view is a representation of the current knowledge about the search space. Explored USes are inspected in a separate US view which also provides the interface for starting reduction steps. The third view is responsible for administering automated reduction agents. Our tool automatically avoids unnecessary deletion tests by efficiently representing and reusing all the clause criticality information that was gained anywhere in the search space. The paper is organized as follows: Section 2 introduces the user interface, describes the main features, and gives some pointers to interesting implementation details. Section 3 describes some of our practical experiences with using the tool for an application. We conclude with a short summary in Section 4. The tool can be downloaded from `http://algo.inf.uni-tuebingen.de/?site=forschung/sat/MUStICCa`.

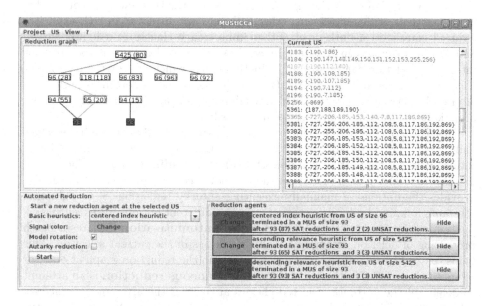

Fig. 1. Screenshot of MUStICCa's default user interface

2   The Application

2.1   The MUStICCa User Interface

The interface is based on the open-source Kahina framework [7] for graphical debugging, which was chosen because it already provided the needed view components and native support for managing a database of computation steps as nodes in a graph structure. Since deletion-based MUS extraction can be framed as a downward traversal of the powerset lattice for an unsatisfiable clause set F, the explored part of the search space is modeled by means of a **reduction graph**, which is a subsemilattice of the powerset lattice whose edges represent successful reduction attempts. In the visualization of this reduction graph, the number on each node gives the size of the corresponding US, followed in brackets by the number of clauses of unknown criticality. MUSes are marked in red, dark green marks non-MUSes where all reduction options have been explored, and light green color marks USes where all clauses are known to be either critical or unnecessary[1], but some unexplored reduction options remain.

Whenever a node in the reduction graph view is selected, MUStICCa derives which clauses in it are implied to be critical, and the US corresponding to the node is displayed in the US view. The default format for clauses in the US view consists of the clause ID (numbered according to the order in the input DIMACS file) and the set of integers representing its literals. If the file contains comment lines of the format c [variable] [symbol] between the header and the clause list, these are imported as a **symbol table**, and the respective symbols are displayed instead of the variable IDs. The clauses in the US view are colour-coded to reflect their criticality status: critical clauses are displayed in red, explicitly reduced clauses in a dark green, other unnecessary clauses in a lighter green, and clauses of unknown status in black. The colour codes make it easy to spot interesting deletion candidates for reduction steps, which are started by double-clicking on clauses in the US view. By default, clause set refinement is performed after each successful reduction, which results in much faster reduction at the cost of making parts of the search space unaccessible.

For advanced interactions, one or more clauses in the US view can be selected, and are then highlighted by a yellow background colour. The selection of interesting clauses is supported by a powerful **selection refinement** interface in the form of a hierarchy of submenus in the US view's context menu. This menu also provides the options of manually executing autarky reduction [8] or reduction attempts followed by model rotation, and two different options for executing reduction operations on sets of clauses. **Semi-automatization** initiates a batch processing of deletion attempts for all the clauses in the current selection. This helps to quickly open up several new branches in the reduction graph at once, or to speed up series of criticality checks. **Simultaneous reduction** is an attempt to delete all the currently selected clauses at once. If the attempt was successful, the new node (or link) will appear in the reduction graph, just like in the

---

[1] Note that not every unnecessary clauses is unusable. For an unnecessary clause C, we merely know that there is at least one MUS which does not contain C.

case of deleting a single clause. If a simultaneous reduction attempt fails, this only yields very weak criticality information, since it might have been possible to delete some of the clauses under the condition that others stay in.

Internally, the meta instance is compressed using an inferred **block structure** over the input clauses. In the development version, this block structure is exposed and used for visualizing the overlaps between encountered USes, often revealing interesting structural features of the input instance. These experimental features are described in Chapter 5 of the first author's master thesis [5].

2.2   Automatization through Reduction Agents

During a process of interactive extraction, a user will often want to quickly explore parts of the search space without having to manually execute hundreds of reduction attempts, especially in contexts where domain knowledge has not yet become relevant. For this purpose, MUStICCa includes an automated reduction mechanism in the form of **reduction agents** which in essence act like autonomous additional users who were given sets of simple instructions. In addition to predefined reduction agents which emulate standard deletion-based MUS extraction algorithms, user-defined reduction agents can be implemented as Java classes by inheriting from a plug-in interface for custom deletion heuristics. This interface is defined in Section 4.3 of the user's guide.

The most important option in the dialog for creating and starting new reduction agents serves to select one of the predefined heuristics from a drop-down menu. Each new agent is initialized with a random **signal colour** that can freely be redefined. Model rotation and autarky pruning can be activated or deactivated. The new reduction agent starts at the US that is currently selected in the reduction graph, and runs until it has determined a MUS. The downward path of an agent through the powerset lattice is visualized in the form of an **agent trace** highlighted in its signal colour.

2.3   Implementation details

Internally, MUStICCa stores its knowledge of the search space in a **reduction table** indexed by $V_i \in US$ and $C \in V_i$, where special values are used to represent that C was found to be critical in V_i, that C is known to be unnecessary, or that no criticality information about C is known yet. A positive integer value expresses that the deletion of the clause C in V_i has led to the US with that ID.

In the case of a successful reduction step that has led us from a subset V_1 to another unsatisfiable subset $V_2 \subset V_1$ by deletion of the clause C, we add a node v_2 and the edge (v_1, v_2) to the reduction graph. Since we typically use clause set refinement, we will often have $|V_1 \setminus V_2| > 1$. In this case, we additionally know that all clauses $C \in V_1 \setminus V_2$ are unnecessary. If the reduction step was unsuccessful, i.e. $V_1 \setminus \{C\}$ was found to be satisfiable, we store the information that C is critical in V_1. In the case of model rotation [17], we store this information for the entire set of clauses that was determined to be critical.

If the search space is explored in more than one direction, it becomes possible to reuse the criticality information derived for one US in other branches of the reduction graph. If a clause C is critical in some US V_1, it is also critical in all other USes $V_2 \subset V_1$. This suggests a simple downwards propagation of criticality information, but this would not be complete, since there can be subset relations between encountered USes that are not reflected in the reduction graph. For this reason, MUStICCa includes a mechanism for systematically sharing the derived criticality information across reduction graph nodes. The information is optimally exploited if we can quickly determine for any subset whether it is included in a set that is already known to be satisfiable as a result of failed reduction attempts. This leads to the concept of an upward wedge of satisfiable subsets or **sat wedge** through the powerset lattice which is created by every such attempt. The task of sharing criticality information now reduces to storing the sat wedges in a data structure that allows a quick check of sat wedge membership for any subset. The solution implemented in MUStICCa stores sat wedges as clauses over positive selector literals in an additional **meta instance**. Any subset can be expressed in terms of unit assumptions over selector literals. Solving the meta instance under these assumptions tells us whether the subset is implied to be satisfiable by sat wedge membership. The idea is very similar to the *Map* formula used for storing information about the search space in the recently proposed MARCO algorithm [10] for quick extraction of multiple MUSes. The meta instance allows us to quickly derive all the clauses implied to be critical in some subset V_i by merely propagating the corresponding selector assumptions. The resulting selector units correspond to the clauses which are implied to be critical in V_i. This allows us to quickly update the critical entries in the reduction table for any US. A formal description of this procedure, and a proof of its correctness, can be found in Chapter 4 of the mentioned thesis.

3   Experimental Results

Because of the general unavailability of MUS extraction instances that contain semantic information, we have only been able to evaluate MUStICCa as a tool for semantically guided MUS extraction in a single application from the first author's area of knowledge. Since the motivation and context of this application can only be outlined here due to space constraints, the interested reader is referred to the discussion in Chapter 6 of the thesis.

The application is based on a SAT encoding of context-free parsing, which is used to encode classical problems in the field of *symbolic grammar engineering*, a branch of computational linguistics which attempts to define formal models of natural languages. Unfeasibilities in the resulting SAT instances represent inadequacies and bugs in these models, and MUSes can be interpreted as containing instructions for minimal repairs. The number of different MUSes in the resulting instances turned out to be extremely high, making the enumeration of all MUSes by means of CAMUS [9] or similar tools entirely infeasible. Given the variable symbols generated by the SAT conversion tool as guidance, the manual selection of reduction candidates made it possible to narrow down bugs in the grammar to very small MUSes which were interpretable as minimal repairs, whereas the results returned by a simple MUS extractor were generally much larger and harder to interpret. Moreover, MUS extraction proved to be extremely useful for finding bugs already during the development of the SAT encoding, since it gave the developer direct access to problematic interactions in flexibly definable constraint subsets.

Beyond this application, we have used MUStICCa to explore the search spaces of other unsatisfiable benchmark instanced which are known to contain multiple MUSes, in particular the Daimler testset for automotive product configuration [15]. On these and similar test instances, MUStICCa proved useful for comparing the performance of different deletion candidate heuristics, and for analyzing properties of MUS extraction search spaces. For larger industrial instances, the large size of the resulting MUSes and the long duration of each reduction attempt makes interactive MUS extraction a less attractive option.

4 Summary

We present our tool MUStICCa that was designed to provide the user with the possibility to guide the well-known deletion-based MUS extraction algorithm [11] through the powerset lattice towards different MUSes. MUStICCa reuses criticality information from other parts of the search space to avoid unnecessary execution steps.

The main features of our tool can be summarized as follows:

- interactive execution of the deletion-based MUS extraction algorithm
- exploration of the search space starting at any already encountered US
- global reuse of clause criticality information
- well-structured graphical user interface
- automatized extraction of MUSes via reduction agents

We are confident that MUStICCa will be helpful to domain experts in analysing inconsistencies in SAT formulae. Moreover, our tool can be used to create and evaluate new deletion candidate selection heuristics. We are looking forward to receiving feedback from experts in different application domains about their experience with the tool, and are grateful to our anonymous reviewers for their many suggestions about interface improvements and feature requests. In the future, we plan to extend the concept to insertion-based and dichotomic MUS extraction by also representing satisfiable subsets in the reduction graph and allowing the user to work upwards in the powerset lattice. Furthermore, we are planning to provide support for the extraction of GMUSes as well as minimal unsatisfiable subformulae of propositional formulae in negation normal form.

References

1. Bakker, R.R., Dikker, F., Tempelman, F., Wognum, P.: Diagnosing and Solving Over-Determined Constraint Satisfaction Problems. In: Proceedings of IJCAI 1993, pp. 276–281. Morgan Kaufmann (1993)

2. Belov, A., Ivrii, A., Matsliah, A., Marques-Silva, J.: On Efficient Computation of Variable MUSes. In: Cimatti, A., Sebastiani, R. (eds.) SAT 2012. LNCS, vol. 7317, pp. 298–311. SAT, Heidelberg (2012)
3. Belov, A., Lynce, I., Marques-Silva, J.: Towards efficient MUS extraction. AI Commun. 25(2), 97–116 (2012)
4. Chinneck, J.W., Dravnieks, E.W.: Locating Minimal Infeasible Constraint Sets in Linear Programs. INFORMS Journal on Computing 3(2), 157–168 (1991)
5. Dellert, J.: Interactive Extraction of Minimal Unsatisfiable Cores Enhanced By Meta Learning. Diplomarbeit, Universität Tübingen (2013)
6. Desrosiers, C., Galinier, P., Hertz, A., Paroz, S.: Using heuristics to find minimal unsatisfiable subformulas in satisfiability problems. J. Comb. Optim. 18(2), 124–150 (2009)
7. Evang, K., Dellert, J.: Kahina - Trac. Web (2013), http://www.kahina.org/trac
8. Kullmann, O.: On the use of autarkies for satisfiability decision. Electronic Notes in Discrete Mathematics 9, 231–253 (2001)
9. Liffiton, M.: Mark Liffiton - CAMUS. Web (2013), http://sun.iwu.edu/~mliffito/camus (access date: January 22, 2013)
10. Liffiton, M.H., Malik, A.: Enumerating Infeasibility: Finding Multiple MUSes Quickly. In: Gomes, C., Sellmann, M. (eds.) CPAIOR 2013. LNCS, vol. 7874, pp. 160–175. Springer, Heidelberg (2013)
11. Nadel, A.: Boosting minimal unsatisfiable core extraction. In: FMCAD, pp. 221–229 (2010)
12. Oh, Y., Mneimneh, M.N., Andraus, Z.S., Sakallah, K.A., Markov, I.L.: AMUSE: a minimally-unsatisfiable subformula extractor. In: DAC, pp. 518–523 (2004)
13. Ryvchin, V., Strichman, O.: Faster Extraction of High-Level Minimal Unsatisfiable Cores. In: Sakallah, K.A., Simon, L. (eds.) SAT 2011. LNCS, vol. 6695, pp. 174–187. Springer, Heidelberg (2011)
14. Silva, J.P.M.: Minimal Unsatisfiability: Models, Algorithms and Applications (Invited Paper). In: ISMVL, pp. 9–14 (2010)
15. Sinz, C., Kaiser, A., Küchlin, W.: Formal Methods for the Validation of Automotive Product Configuration Data. Artificial Intelligence for Engineering Design, Analysis and Manufacturing 17(1), 75–97 (2003); special issue on configuration
16. The SAT association: The international SAT Competitions web page. Web (2011), http://www.satcompetition.org/
17. Wieringa, S.: Understanding, Improving and Parallelizing MUS Finding Using Model Rotation. In: Milano, M. (ed.) CP 2012. LNCS, vol. 7514, pp. 672–687. Springer, Heidelberg (2012)

SCSat: A Soft Constraint Guided SAT Solver

Hiroshi Fujita, Miyuki Koshimura, and Ryuzo Hasegawa

Dept. of Informatics, Kyushu University,
Fukuoka 819-0395, Japan
{fujita,koshi,hasegawa}@inf.kyushu-u.ac.jp

Abstract. SCSat is a SAT solver aimed at quickly finding a model for hard satisfiable instances using soft constraints. Soft constraints themselves are not necessarily maximally satisfied and may be relaxed if they are too strong to obtain a model. Appropriately given soft constraints can reduce search space drastically without losing many models, thus help find a model faster. In this way, we have succeeded to obtain several rare Ramsey graphs which contribute to raise the known best lower bound for the Ramsey number R(4,8) from 56 to 58.

Keywords: soft constraint, constraint relaxation, Ramsey number.

1   Introduction

We have been tackling hard combinatorial problems using theorem provers and SAT solvers [5,6,12]. In many cases we can hardly find a single model because the number of models, if any, is extremely small. For this, streamlining [10,11,18] and tunneling [14] appear to be promising. In both methods, providing good additional constraints, called *streamliners* and *tunnels* respectively, should be the key to a success.

Here we present yet another method based on *soft constraints,* that can easily be implemented on an off-the-shelf SAT solver such as MiniSat [1]. We prepare an additional set of constraints S which seemingly is suited for restricting search to preferable directions. The search with S may end with UNSAT very soon if it is too strong. Then, we will not immediately give up the constraints as a whole but take a relaxed S^r ($\subset S$), and restart a new search. A series of relaxations and restarts should be automated easily in a simple iterative procedure. In this setting, the initial constraints S can be arbitrarily strong, and is not required to be satisfied perfectly. So, we call it *soft constraints* (SC in short). Soft constraints are not required to be maximally satisfied as in MaxSAT [13]. We just require SC to be able to speed up finding a single model if relaxed adequately.

2   System Description

Outline of the SCSat Procedure. Given a CNF file (hard clauses) for the problem and a WCNF file [15] (soft clauses) for the soft constraints, which

M. Järvisalo and A. Van Gelder (Eds.): SAT 2013, LNCS 7962, pp. 415–421, 2013.

should be specified as a command line option "`-sc-file=<filename>`", repeat the MiniSat search enhanced with the following SC handling until a model is found or the set of SC-clauses becomes empty.

- *SC-UP:* Handle SC-clauses exactly in the same way as hard clauses for unit propagation.
- *SC-conflict:* If a SC-clause causes a conflict during a search, disregard the conflict, increase the *penalty score* for the SC-clause, and continue the search.
- *SC-restart:* When the search reaches UNSAT, pick up some portions of SC-clauses having higher penalty score, remove or tentatively deactivate them, and restart a new search with the relaxed SC-clauses.

Where penalty score for a SC-clause is quite similar to *activity* for a learnt clause in MiniSat. The higher penalty score a SC-clause has the earlier it would be deactivated or removed, while the higher activity a learnt clause has the longer it would be kept.

How to Provide Soft Constraints. In view of our experience, it seems sufficient to consider only binary clauses as soft constraints, which is often induced by equivalence of pairs of the variables appearing in the hard clauses, such as Z-SC described below. However, there is no reason to avoid non binary SC-clauses, and SCSat now properly handles any SC-clauses with arbitrary size except unit SC-clauses for which our design choices have not yet been fixed.

New variables, called *SC-variables,* that do not appear in the hard clauses but only in SC-clauses are allowed. The intention of SC-variables is described in the Ramsey Package section. SC-variables need not be assigned a value just as a SC-clause need not be satisfied. So, it is marked as "non decision variable." Also an SC-variable is currently inhibited to appear in a unit SC-clause.

How to Give Penalty Scores for Soft Constraints. To decide which portions of SC-clauses should be deactivated or removed at a SC-restart, we need to somehow discriminate bad SC-clauses from good ones. Here we take a *conflict driven clause degrading (CDCD)* approach, where a SC-clause is considered bad if it causes a conflict, directly or indirectly. Currently SCSat provides two options for the CDCD based SC-penalty scoring:

- **loose (default):** Only the conflict SC-clause, i.e. the direct reason of a conflict, will have a penalty. Take an arbitrary one if there exist more than one conflict SC-clauses at a single conflict.
- **severe:** Every SC-clause that is the reason of a literal in a learned clause created at each conflict, i.e. the indirect reason of the conflict, will have a penalty. Specify "`-sc-penalty=2`" in the command line to use this setting.

Generally speaking, taking severe may be better since more SC-clauses would be targeted, while taking loose may apt to concentrate very often on a very limited portions of SC-clauses. However, we consider that it would be better to take loose as default because it is a bit more efficient than severe.

The penalty score for a target SC-clause is calculated based on its weight and the decision level where a conflict occurred using the following formula:

$$penalty(SC) \mathrel{+}= (weight(SC) * (1 + decisionLevel()))^{-1},$$

where the value of $weight(SC)$ should be specified in the input WCNF file. Note that this weight value is used differently from MaxSAT.

Why this particular form? The intuition behind the formula is the following. On one hand, from the viewpoint of fairness, some kind of normalization has to be considered. As for $decisionLevel()$, for example, the exponent -1 seems better than 1. Because, the authors think, the more deeper level where a SC-clause is used on some search path, the more often it would be used at similar depth on other search paths as well. The $decisionLevel()^{-1}$ would cancel the effect of this tendency. On the other hand, from the viewpoint of symmetry breaking, such normalization might be unnecessary, or even harmful. There is no definite theory deciding which should be better.

How to Relax Soft Constraints. How much portions of SC-clauses should be deactivated or removed at each SC-restart? In some cases the higher rate may be better because of the fewer iterations. In other cases the things may be quite opposite. So, it is basically up to the user, who should specify preferable *reduction rate* for SC in a command line option like "-sc-rr=0.2". This value can have a large impact on the performance of the solver. In fact, 10 % was better than 20 % for obtaining $R(5,5,42)$, whereas 40 % was better than 20 % for $R(4,8,57)$. The current default 20 % seems a fairly good choice.

How to Handle Learned Clauses. Currently, SCSat clears all the leaned clauses at each SC-restart. Unit clauses accumulated in the previous SC-restart are to be canceled either. This is just for simplicity, and a future release should employ more sophisticated handling. See discussions below.

3   Ramsey Package

Ramsey Problem. The classical Ramsey number $R(s,t)$ is the smallest integer n such that in any two-coloring of the edges of K_n there is a monochromatic copy of K_s in the first color or a monochromatic copy of K_t in the second color [3]. A Ramsey graph $R(s,t,n)$ is a graph with n vertices, no clique of size s, and no independent set of size t [16]. Then, Ramsey theory tells us that there are only a finite number of Ramsey graphs for each s and t and for some n, but finding all such graphs, or even determining the largest n, i.e. the Ramsey number $R(s,t)$ minus one, for which they exist, is a famously difficult problem [2]. Some of the interesting instances can be found at several websites [4,8,16]. A recent summary of the state of the art for Ramsey numbers can be found in the Dynamic Survey [17].

Here we try to obtain a Ramsey graph denoted by $R(s, t, n)$ by encoding the condition for it to exist into a CNF, called *Ramsey clauses*, as follows:

$$C_{(s,t,n)} : \left(\bigwedge_{K_s \subset K_n} \bigvee_{e_{ij} \in K_s} \neg e_{ij} \right) \wedge \left(\bigwedge_{K_t \subset K_n} \bigvee_{e_{ij} \in K_t} e_{ij} \right),$$

where each e_{ij} is the propositional variable for the edge between vertices i and j, and assigning *true* to it means the edge is colored in the first color, otherwise the second. If $C_{(s,t,n)}$ is unsatisfiable, then no $R(s, t, n)$ exists. Otherwise a model representing a $R(s, t, n)$ can be obtained.

Z-SC. The most straightforward yet remarkably effective soft constraints, called *Zebra-SC* (Z-SC in short) is:

$$e_{ij} \equiv z_d \quad (0 \leq i < j < n, \; j - i = d),$$

where z_d $(1 \leq d \leq n - 1)$ are new propositional variables. The name zebra derives from the stripe like pattern observed in the adjacency matrix of the corresponding Ramsey graph. The intuition of Z-SC is the following. Since the general requirement for a Ramsey graph is highly symmetric in terms of vertex renaming, it seems natural to think that a special symmetry requirement like Z-SC, in which renaming every v_i to $v_{(i+d) \bmod n}$ becomes the identity mapping, might be met. The Z-SC with z_d variables is represented in CNF as follows:

$$Z_n : \bigwedge_{0 \leq i < j < n} \left((\neg e_{ij} \vee z_{j-i}) \wedge (e_{ij} \vee \neg z_{j-i}) \right).$$

We try to solve $C_{(s,t,n)} \wedge Z_n$. If it is satisfiable then $C_{(s,t,n)}$ is also satisfiable, and the obtained model should represent a Ramsey graph $R(s, t, n)$, which is called a Z-Ramsey graph and is denoted by $R^Z(s, t, n)$. Assignment for an e_{ij} quickly propagates to the corresponding e_{kl} $(l-k=j-i)$ via z_d $(d=j-i)$. That is, you only need to decide the coloring for $n - 1$ number of stripes.

U-SC. Some Ramsey graph $R(s', t', n')$ can be obtained easily using a smaller $R(s, t, n)$, such that $s \leq s'$, $t \leq t'$, $n < n'$, typically $s = s'$, $t = t' - 1$.

As shown in Figure 1, one of the author has succeeded to find a $R(4, 8, 57)$ using a $R(4, 7, 48)$ [7] where the edges e_{ij} $(0 \leq i < j < 48)$ in the $R(4, 8, 57)$ are colored exactly in the same color as in the $R(4, 7, 48)$. The rest of edges e_{ij} $(0 \leq i < 57, \; 48 \leq j < 56)$ were colored adequately by using SCSat with a simple Z-SC.

The constraints that will impose a fixed truth value for each propositional variable (a fixed color for each edge) is called a *Unit-SC* (U-SC in short). For instance, the coloring above can be represented as a set of unit clauses $\bigwedge l_{ij}$ $(0 \leq i < j < 48)$, where $l_{ij} = e_{ij}$ $(\neg e_{ij})$ if e_{ij} is true (false) for the $R(4, 7, 48)$.

Although SCSat is able to relax U-SCs as well as Z-SCs mechanically, the U-SC was not relaxed at all in the above case. Because the set of hard clauses is too large for SCSat input, we could not do without simplifying them with the exact U-SC in advance, thus making the input clauses small enough.

(a) Edges of the first color

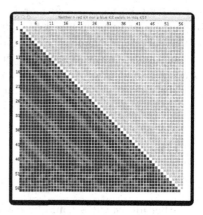

(b) Adjacency matrix

Fig. 1. A Ramsey graph $R(4, 8, 57)$

4 Experimental Results

As shown in Table 1, U-SCs work remarkably well compared to Z-SCs for smaller Ramsey graphs such as $R(3, 8, 27)$ and $R(3, 10, 39)$. The similar results on $R(3, 6, 17)$ and $R(3, 7, 22)$. It is quite the contrary for larger instances such as $R(4, 6, 33)$ and $R(5, 5, 42)$. The similar results on $R(4, 7, 47)$ and $R(4, 7, 48)$. As for $R(4, 8, 57)$, only a combination of a U-SC and a Z-SC was successful.

Table 1. Results on Ramsey instances

Inst.	C_n	$C_n \wedge Z_n$	$C_n \wedge U_{n'}$	$C_n \wedge U_{n'} \wedge Z_n$
$R(3, 8, 27)$	43.7	13022	2.4	11.1
$R(3, 10, 39)$	—	—	0.624	5.31
$R(4, 6, 33)$	—	27.8	4942	15574
$R(5, 5, 42)$	—	799	—	—

CPU time (sec.), — denotes for time out (>24 hrs).
Parameter settings: "best effort" (not optimal) for each entry.

Note that once a Ramsey graph $R(s, t, n)$ is obtained, any smaller $R(s, t, n')$, $(n' < n)$ becomes unimportant in the sense that it does not by itself improve lower bound of the Ramsey number $R(s, t)$. Nevertheless, it may be useful for searching greater $R(s, t', n'')$, $(t < t', n' < n'')$ if used as a U-SC.

5 Related Work

Streamlining [10,11,18] reforms the given problem by separating it into two disjoint subproblems with a streamliner constraint S and its complement $\overline{S}$

respectively so that the former becomes easy to solve. Our method does not employ $\overline{S}$ but just a relaxed S^r ($S^r \subset S$) instead.

The target problem of the tunnels method [14] as well as the essential idea of using additional constraints is indeed very close to ours. However, theirs is so deeply specialized to the specific problems on Van der Waerden Numbers that the method, as it is, seems hardly be applied even to the relative problems including those on Ramsey Numbers.

6   Discussions

There are many discussions about SCSat. We list only a few of them as follows.

- Other alternatives could be used for soft constraint removal. For example, if the formula is unsatisfiable then we can extract an unsatisfiable subformula that explains the reason of unsatisfiability.
- In practice we do not need to remove all learned clauses. The conflict analysis procedure can be extended to keep track of learned clauses that depend on soft constraints. Therefore, only soft learned clauses need to be removed. This may improve the overall performance of the solver.
- The relaxation scheme (governed by the reduction rate) is currently overlook the interaction between the soft constraints, as it simply deactivates the most conflicting soft clauses, and never puts them back again. In particular, when two sets of soft constraints are mostly incompatible, this approach would tend to remove both, while both of them (separately) could potentially speed up the search. It might become more problematic when one legitimately adds additional soft constraints, such as symmetry breaking constraints.

7   Conclusions

SCSat should be useful mainly for those who are interested in Ramsey problems. SCSat (ver.1 series) together with the companion tools would help find interesting Ramsey graphs using U-SC, Z-SC and its variations, and more. For those who wish to solve hard problems in general but other than Ramsey and those who are developing similar ideas or implementations, SCSat would provide some hints or data useful for comparative studies. In any case, the key to a real success depends essentially on how one can design a good recipe by making/selecting/combining appropriate SCs.

The tool can be found at SCSat homepage [9].

Acknowledgements. We would like to thank all the reviewers and meta-reviewers of the paper. The discussions listed above are contributed by them. Special thanks are due to Professor Bart Selman for his help and support in improving the paper. Their valuable suggestions also helped improve the tool.

References

1. Eén, N., Sörrenson, N.: The MiniSat Page, http://minisat.se
2. Erdös, P.: (Wikiquote), http://en.wikiquote.org/wiki/Paul_Erdos
3. Exoo, G.: On the Ramsey number R(4,6). Electron. J. Combin. 19 P66 (2012)
4. Exoo, G.: Ramsey numbers, http://ginger.indstate.edu/ge/RAMSEY
5. Fujita, M., Slaney, J., Bennett, F.: Automatic Generation of Some Results in Finite Algebra. In: IJCAI 1993 (1993)
6. Fujita, H.: Solving Hard Combinatorial Problems with SAT Solvers, Dagstuhl Seminar 09461 (2009)
7. Fujita, H.: A New Lower Bound for the Ramsey Number R(4,8), arXiv:1212.1328 [cs.DM] (2012)
8. Fujita, H.: RamseyGraphs, https://sites.google.com/site/ramseygraphs
9. Fujita, H.: SCSat, https://sites.google.com/site/scminisat
10. Gomes, C., Sellmann, M.: Streamlined Constraint Reasoning. In: Wallace, M. (ed.) CP 2004. LNCS, vol. 3258, pp. 274–289. Springer, Heidelberg (2004)
11. Gomes, C., Sabharwal, A., Selman, B.: Model Counting: A New Strategy for Obtaining Good Bounds. In: AAAI 2006 (2006)
12. Hasegawa, R., Fujita, H., Koshimura, M.: MGTP: A model generation theorem prover. In: Galmiche, D. (ed.) TABLEAUX 1997. LNCS, vol. 1227, pp. 1–15. Springer, Heidelberg (1997)
13. Koshimura, M., Zhang, T., Fujita, H., Hasegawa, R.: QMaxSAT: A Partial Max-SAT Solver. JSAT 8, 95–100 (2012)
14. Kouril, M., Franco, J.: Resolution tunnels for improved SAT solver performance. In: Bacchus, F., Walsh, T. (eds.) SAT 2005. LNCS, vol. 3569, pp. 143–157. Springer, Heidelberg (2005)
15. Max-SAT 2013, Benchmarks and Solver Requirements, http://maxsat.ia.udl.cat/requirements
16. McKay, B.D.: Ramsey Graphs, http://cs.anu.edu.au/~bdm/data/ramsey.html
17. Radziszowski, S.P.: Small Ramsey numbers. Electron. J. Combin., DS1 (2011), http://www.combinatorics.org/issue/view/Surveys
18. Smith, C., Gomes, C., Fernandez, C.: Streamlining Local Search for Spatially Balanced Latin Squares. In: IJCAI 2005 (2005)

Snappy: A Simple Algorithm Portfolio

Horst Samulowitz, Chandra Reddy, Ashish Sabharwal, and Meinolf Sellmann

IBM Watson Research Center, Yorktown Heights, NY 10598, USA
{samulowitz,creddy,ashish.sabharwal,meinolf}@us.ibm.com

Abstract. Algorithm portfolios try to combine the strength of individual algorithms to tackle a problem instance at hand with the most suitable technique. In the context of SAT the effectiveness of such approaches is often demonstrated at the SAT Competitions. In this paper we show that a competitive algorithm portfolio can be designed in an extremely simple fashion. In fact, the algorithm portfolio we present does not require any offline learning nor knowledge of any complex Machine Learning tools. We hope that the utter simplicity of our approach combined with its effectiveness will make algorithm portfolios accessible by a broader range of researchers including SAT and CSP solver developers.

1 Introduction

Algorithm portfolios [cf. 6, 8, 10] for combinatorial problems such as Boolean Satisfiability (SAT) and Constraint Satisfaction Problems (CSPs) have emerged as a highly successful approach for combining the strength and effectiveness of a variety of core solution techniques ("base solvers") that excel on various subsets of problem instances. By using Machine Learning based data-driven methods to select the most promising algorithm (or a schedule of several promising ones) based on past performance of each base solver on hundreds or even thousands of instances, such portfolios often achieve much more robust performance across a broad range of problem domains than any single base solver. In the context of SAT, a regression based portfolio called SATzilla2009 [16] showcased the benefits of algorithm portfolios by dominating several categories at annual SAT Competitions (www.satcompetition.org) in the past. Over the years, researchers working on portfolio algorithms have employed a number of techniques with their own benefits, including Scheduling based approaches [11, 15], Decision Forests [18], Nearest Neighbor classification [9], and Collaborative Filtering [14].

Unfortunately, despite the existence of such techniques for several years, portfolio algorithms have not truly been adopted by the SAT and CSP research communities. The users of these portfolios are often none other than their own creators, and sometimes a few other research groups working on competing portfolios. In other words, portfolios have not been embraced by the community as a generic tool. This is in sharp contrast with base solvers such as Minisat [13], Glucose [2], CryptoMinisat [12] and Lingeling [3], just to name a few, which are widely used by a large subset of the SAT community on a regular basis. Similarly, CSP solvers such as Gecode [5] and Choco [4] are commonly used.

M. Järvisalo and A. Van Gelder (Eds.): SAT 2013, LNCS 7962, pp. 422–428, 2013.
© Springer-Verlag Berlin Heidelberg 2013

Our hope is that the availability of an easy to use portfolio tool would encourage SAT and CSP researchers to explore new avenues. E.g., if there was a new restart strategy for Glucose that made it work better on some instances but worse on others, a simple 2-solver portfolio could be employed to try to achieve the best of both worlds.

The motivation for this work is the belief that a key reason for the lack of common adoption of portfolio solvers is usability. For example, the "portfolio builder" or trainer of some existing portfolios such as 3S is not publicly available due to proprietary reasons, while that of others such as SATzilla2012 requires a license to a relatively new version of MATLAB as well as enough familiarity with the offline training aspect of the code to be able to effectively adapt it to every new benchmark. Further, training can be expensive: it can take a few minutes to hours for SATzilla2009 and 3S, and much more for SATzilla2012 due to quadratic scaling in the number of base solvers [1]. The resulting effort and time investment often deters most researchers from benefiting from powerful portfolio technology.

The goal of this work is to develop an algorithm portfolio that (a) is simple to use and experiment with, (b) is data-driven and can thus exploit years of experience with various base solvers, (c) has *no offline training phase* or "portfolio builder", and (d) can improve its own performance through *online learning*.

To this end, we propose a training-less algorithm portfolio called snappy (Simple Neighborhood-based Algorithm Portfolio in PYthon). Packaged as a single file written in Python (www.python.org), it relies on a relatively simple prediction mechanism based on base solver performances on nearest neighbors. The availability of an extensive set of Python libraries allows the user to easily experiment with various aspects of interest such as distance measures, neighborhood size, weighting, cost function, feature reduction, etc. By skipping the traditional training phase altogether, snappy opens up the possibility of learning and improving itself on-the-fly when run on a number of test instances.

Our aim is not to create the best portfolio approach to date but to provide a tool that is effective and has a low barrier to being adopted by a wide range of algorithmic researchers. We provide empirical evidence that the simple, training-less approach embodied by snappy can be competitive with the current state of the art in portfolios for SAT. This is not to say that more sophisticated approaches to portfolios do not have value. For example, the SAT/UNSAT predictor in SATzilla is a powerful tool in itself, and so is the mixed integer programming based algorithm scheduler of 3S. Nonetheless, the sophisticated approaches also have a higher barrier to entry, which we hope to overcome through the simple yet competitive approach of snappy.

2   Background

We assume familiarity with the basics of SAT and CSP solvers, and briefly discuss algorithm portfolios in this context. The main concept behind any solver portfolio is to utilize data (collected offline) on the performance of various base solvers on a relatively large set of "training" instances in order to predict, given a new "test" instance, which single base solver or schedule of base solvers is most cost-effective for solving the test instance. The cost of interest is often the runtime, but may be other quantities such as solution quality in case of an optimization problem. The various portfolios referred to earlier differ in *how* they go about using training data to make this prediction. One aspect common to them is the abstract representation of instances in terms of a (small) set of "features", such as instance size, constraint density, etc.

Once the training data is collected, state-of-the-art portfolios such as 3S and SATzilla (winners of the past two SAT Competitions) go through an extensive offline training and internal cross-validation phase in order to learn the parameters of their prediction model. This phase typically requires understanding the portfolio builder well and can often be costly [1]. In the interest of space, we refer the reader to Xu et al. [16, 17] and Kadioglu et al. [9] for details.

3   Portfolio Tool Description

Our portfolio, snappy, is packaged as one Python file available at: http:// researcher.watson.ibm.com/researcher/files/us-samulowitz/snappy.zip Its basic usage is reported by running python snappy.py -h.

The tool can be run in two modes, analysis (ana) and execution (exe). In both modes, like all portfolio solvers, it expects "training data" which consists of a .times file specifying, in a header-less comma-separated format, the runtime of each base solver (columns) on a number of instances (rows), and a .features file specifying the features (columns) of each instance. Additionally, the desired timeout is specified for performance evaluation purposes. Options in both modes include what neighborhood size to consider, what penalty to use for performance evaluation when hitting the timeout, what distance measure to use (e.g., Euclidean or Minkowski), and whether to use distance-based weighting.

In the analysis (ana) mode, the tool expects a set of test instances and evaluates the performance of the portfolio on this test dataset in relation to the single best base solver (SBS) or the virtual best solver (VBS). The intent is that users can use the analysis mode to experiment with and tune the parameters of the portfolio for their benchmark of interest. One may optionally specify a given pre-schedule of base solvers, which is taken into account in all performance evaluations. Enabling the online "learning" mode makes snappy alter its behavior based on the runtimes it has observed on test instances.

In the execution (exe) mode, the tool expects a comma-separated list of features. Alternatively, the user may specify the name of a feature extraction tool and a test instance such that executing the tool will produce a comma-separated list of features of the instance.

The main components of snappy are:

Data Normalization: All feature values are cut off at a fixed maximum/minimum value and features are scaled to the same unit (e.g., [0..1]). This is motivated by the fact that we use distance to select nearest neighbors.

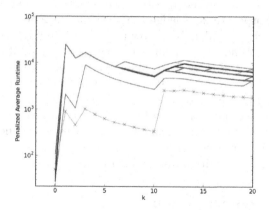

Fig. 1. A sample of the PAR score of various algorithms for varying k

Feature Space and Distance Measures: There exist various ways of computing the distance between feature vectors besides the Euclidean distance, such as standardized Euclidean, Minkowski, and Canberra, which are available through the `scipy.spatial.distance` Python package. More sophisticated measures such as the Mahalanobis distance impose requirements on the underlying data (e.g., being able to take the inverse of the Covariance matrix). To enable such measures we also support a PCA based Eigenvector representation of the data where features with low variance are eliminated. This allows the approach to stay somewhat more robust against uninformative and potentially misleading features.

Aggregation Schemes and Solver Selection: Since we do not fix a single neighborhood size k a priori but choose it dynamically (see e.g., [7]), we employ an aggregation scheme based on the performance of each base algorithm for each possible $k \leq k_{max}$. Essentially we have a $k_{max} \times \#$Algorithms matrix and each entry (k, A) contains a performance measure, namely the average penalized runtime, of algorithm A on the first k neighbors of the test instance. A sample visualization of such a matrix is shown in Figure 1 where we plot the penalized average runtime (PAR) per $k \in [0..20]$ and algorithm (1 to 18) including the (unknown) performance of each algorithm on the test instance ($k = 0$). The line corresponding to the selected algorithm is marked with crosses.

Based on this data one can deploy various aggregation schemes to select a solver. One simple scheme, which is the "default" setting and underlies all experiments reported in this paper, is one that selects the base algorithm that has the minimum PAR score for some $k \in [k_{min}, k_{max}]$:

$$\arg\min_{A \in \text{Algorithms}} \min_{k \in [k_{min}, k_{max}]} \text{PAR}(A, k \text{ nearest neighbors of the test instance})$$

One can also consider various other aggregation schemes such as distance-based weighted voting. Some of these are also available in **snappy**.

Online-Learning: Since we do not perform any offline learning, it gives us the liberty to easily incorporate knowledge that becomes available as the portfolio is run. To this end we consider the following ways of adding knowledge incrementally. First, every time a test instance is considered we add its feature vector to the current set of training instances and re-normalize the entire data again before selecting the next algorithm. We noticed that re-normalizing has a positive impact on performance. Second, after we select an algorithm for a given test instance, we add the actual runtime information for the selected algorithm on this instance to our data set. This means that, over time, the k-neighborhood of different algorithms might differ as some algorithms will be selected more often than others. This simple addition also seemed to have a positive effect.

4   Empirical Demonstration

We demonstrate the effectiveness of snappy by comparing it with two state-of-the-art algorithm portfolios for SAT that won in the last two competitions: 3S [9] and SATzilla2012 [18]. Our comparison does not rely on generating any new runtime or feature data, or running any base solver, but is based entirely on benchmarks used previously by the respective portfolio designers to showcase the merits of 3S and SATzilla, resp. Being competitive on these benchmarks thus speaks to the strength of our simple approach.

Our Python based tool is expected to work well across multiple platforms. All experiments reported here were conducted on a Linux machine with Python 2.6.6 installed with the following packages: Numpy 1.6.2, Scipy 0.11, and Matplotlib 1.2. All performance numbers of snappy are based on one, fixed, "default" setting of various parameters. We note that the performance of snappy varies with different settings of the command-line parameters and we expect the users of this tool to experiment with parameter settings that work best in their domain. Finally, all comparisons are performed on exactly the same datasets (in particular identical solver base) the results for 3S and SATzilla were obtained on.

We begin with a comparison with 3S using no scheduler[1] in Table 1. The first benchmark is the one with challenging training/test splits used in the original paper on 3S [9]. The other four benchmarks are based on an updated set of solvers and instances used to train the ISS solver (a information-sharing extension of 3S) for SAT Challenge 2012. These four benchmarks are divided into two categories based on the original 48 SATzilla features ("f1") and a new set of more efficiently computed 32 features ("f2") used by ISS. Within each category, there are cross-validation splits ("10-fold") and a competition-style split ("comp").

While we trained and evaluated 3S on all of these benchmarks, we were unsuccessful in doing so with SATzilla2012 as this appeared to require significant familiarity with the underlying MATLAB code and a need to adapt the code for every benchmark so as to give it a fair chance of success. Instead we compare the performance of snappy directly with the results reported by Xu et al. [17]. The

---

[1] When 3S and snappy use the same schedule, the relative difference in performance remains similar to what is reported in Table 1.

Table 1. 3S vs. snappy (with default settings): average percentage of instances solved and average PAR-1 score in seconds

Benchmark				3S		snappy	
Name	#Alg	#Feat.	Timeout	%	PAR-1	%	PAR-1
SAT-2011-splits	37	48	5000	91.23	772.8	94.52	512.5
SAT-2012-10fold-f1	72	48	2000	96.59	174.3	96.48	161.8
SAT-2012-comp-f1	72	48	2000	83.05	556.4	83.77	560.5
SAT-2012-10fold-f2	72	32	2000	97.23	146.1	96.17	167.5
SAT-2012-comp-f2	72	32	2000	85.42	499.1	85.42	526.3

Table 2. SATzilla vs. snappy (with default settings): average percentage of instances solved and average PAR-1 score in seconds

Benchmark				SATzilla		snappy	
Name	#Alg	#Feat.	Timeout	%	PAR-1	%	PAR-1
Industrial	18	125	5000	75.3	1685	72.6	1789
Crafted	15	125	5000	66.0	2096	63.3	2198
Random	9	125	5000	80.8	1172	80.3	1221

corresponding benchmark is available at the SATzilla webpage[2] and is comprised of three categories: application, crafted, random. The results are shown in Table 2. It is worth noting that this version of SATzilla employs a scheduler before selecting a long-running algorithm. While this data set comprises 125 features we only consider the first 48 features in snappy.

As the performance numbers in both tables demonstrate, snappy, despite its simplicity and ease of use, is quite competitive with the state of the art.

5   Conclusion

We presented a new algorithm portfolio approach that we hope will be easy to adopt by a broad range of researches, including those designing the base solvers that underlie any such portfolio. Our tool, snappy, does not only provide a strong baseline, but can also be easily extended by its users. For instance, if portfolio performance becomes an issue[3] one could use a priori k-means clustering to reduce the number of instances one needs to consider—which can be done by adding just few lines of code using the scipy library. Similarly, if one wants to automatically tune the high level parameters (e.g., distance measure), one can also add cross-validation using one line of Python. We believe this kind of experimentation flexibility can be immensely valuable from a research perspective.

---

[2] http://www.cs.ubc.ca/labs/beta/Projects/SATzilla. We thank Lin Xu for providing the cross-validation splits used in prior work [17] on this dataset.

[3] On the largest dataset used in this paper with about 5,000 instances, it takes around 1,5 seconds to select an algorithm.

428 H. Samulowitz et al.

References

[1] Amadini, R., Gabbrielli, M., Mauro, J.: An empirical evaluation of portfolios approaches for solving csps. In: Gomes, C., Sellmann, M. (eds.) CPAIOR 2013. LNCS, vol. 7874, pp. 316–324. Springer, Heidelberg (2013)

[2] Audemard, G., Simon, L.: Predicting learnt clauses quality in modern sat solver. In: IJCAI 2009 (July 2009)

[3] Biere, A.: Lingeling and friends at the sat competition 2011. Technical report, Johannes Kepler University, Altenbergerstr. 69, 4040 Linz, Austria (2011)

[4] Cnrs, L.: Choco: an open source java constraint programming library. In: White Paper 14th International Conference on Principles and Practice of Constraint Programming CPAI 2008 Competition, pp. 7–14 (2008), http://www.emn.fr/z-info/choco-solver/pdf/choco-presentation.pdf

[5] Gecode Team. Gecode: Generic constraint development environment (2006), http://www.gecode.org

[6] Gomes, C., Selman, B.: Algorithm portfolios. Artificial Intelligence Journal 126(1-2), 43–62 (2001)

[7] Huda, M.S., Alam, K.M.R., Mutsuddi, K., Rahman, M.K.S., Rahman, C.M.: A dynamic k-nearest neighbor algorithm for pattern analysis problem. In: 3rd International Conference on Electrical & Computer Engineering (2004)

[8] Rice, J.R.: The algorithm selection problem. Advances in Computers 15, 65–118 (1976)

[9] Kadioglu, S., Malitsky, Y., Sabharwal, A., Samulowitz, H., Sellmann, M.: Algorithm selection and scheduling. In: Lee, J. (ed.) CP 2011. LNCS, vol. 6876, pp. 454–469. Springer, Heidelberg (2011)

[10] Leyton-Brown, K., Nudelman, E., Andrew, G., McFadden, J., Shoham, Y.: A portfolio approach to algorithm selection. In: Proc. of the 15th Int. Joint Conference on Artificial Intelligence (IJCAI), pp. 1542–1543 (2003)

[11] O'Mahony, E., Hebrard, E., Holland, A., Nugent, C., O'Sullivan, B.: Using case-based reasoning in an algorithm portfolio for constraint solving. In: Irish Conference on Artificial Intelligence and Cognitive Science (2008)

[12] Soos, M.: CryptoMiniSat 3.1 (2013), http://www.msoos.org/cryptominisat2

[13] Sorensson, N., Een, N.: MiniSAT 2.2.0 (2010), http://minisat.se

[14] Stern, D., Samulowitz, H., Herbrich, R., Graepel, T., Pulina, L., Tacchella, A.: Collaborative expert portfolio management. In: AAAI (2010)

[15] Streeter, M., Smith, S.: Using decision procedures efficiently for optimization. In: ICAPS, pp. 312–319 (2007)

[16] Xu, L., Hutter, F., Hoos, H., Leyton-Brown, K.: Satzilla: Portfolio-based algorithm selection for sat. JAIR 32(1), 565–606 (2008)

[17] Xu, L., Hutter, F., Hoos, H., Leyton-Brown, K.: Evaluating component solver contributions to portfolio-based algorithm selectors. In: Cimatti, A., Sebastiani, R. (eds.) SAT 2012. LNCS, vol. 7317, pp. 228–241. Springer, Heidelberg (2012)

[18] Xu, L., Hutter, F., Shen, J., Hoos, H., Leyton-Brown, K.: Satzilla2012: Improved algorithm selection based on cost-sensitive classification models. solver description. In: SAT Challenge 2012 (2012b)

Scarab: A Rapid Prototyping Tool for SAT-Based Constraint Programming Systems

Takehide Soh, Naoyuki Tamura, and Mutsunori Banbara

Kobe University 1-1, Rokko-dai, Nada, Kobe, Hyogo 657-8501, Japan
{soh@lion.,tamura@,banbara@}kobe-u.ac.jp

Abstract. In this paper, we present the Scarab system which is a prototyping tool for developing SAT-based systems. It provides a rich constraint modeling language on Scala and enables a programmer to rapidly specify problems and to experiment with different modelings. Scarab also provides a simple way to realize incremental solving, solution enumeration, and dynamic addition and/or removal of constraints. In Scarab, we can use integer variables and arithmetic constraints, and all of them are encoded into SAT without the need of developing dedicated encoder. SAT solvers are then used for finding solutions.

1 Introduction

Remarkable improvements in the efficiency of SAT solvers have been made over the last decade [1]. Such improvements have enabled a programmer to develop SAT-based systems for planning, scheduling, and hardware/software verification. However, for a given problem, we usually need to develop a dedicated program that encodes it into SAT. We therefore cannot focus on problem modeling that plays an important role in the system development process.

In this paper, we present the Scarab system which is a prototyping tool for developing SAT-based systems. It provides a rich constraint modeling language on Scala [2, 3] and enables a programmer to rapidly specify problems and to experiment with different modelings. Scarab also provides a simple way to realize incremental solving, solution enumeration, and dynamic addition and/or removal of constraints. Scarab is implemented in Scala and consists of Constraint Programming Domain-Specific Language (DSL [4]), SAT encoding module, and interface to the back-end SAT solvers. The major design principle of Scarab is to provide an expressive, efficient, customizable, and portable workbench for SAT-based system developers.

Expressiveness: Scarab DSL can concisely write constraint modelings with the help of rich functionalities of Scala. The expressiveness of Scarab will be also shown by some prototyping examples of Square Packing, Latin Square, and an optimization version of Square Packing.

M. Järvisalo and A. Van Gelder (Eds.): SAT 2013, LNCS 7962, pp. 429–436, 2013.

```
1: import jp.kobe_u.scarab.csp._
2: import jp.kobe_u.scarab.solver._
3: import jp.kobe_u.scarab.sapp._
4:
5: val n = 15; val s =36
6:
7: for (i <- 1 to n)
8:   { int('x(i),0,s-i) ; int('y(i),0,s-i) }
9: for (i <- 1 to n; j <- i+1 to n)
10:    add((('x(i) + i <= 'x(j)) || ('x(j) + j <= 'x(i)) ||
11:        ('y(i) + i <= 'y(j)) || ('y(j) + j <= 'y(i)))
12:
13: if (find) println(solution)
```

Fig. 1. Scarab Program of $SP(15, 36)$ **Fig. 2.** Solution of $SP(15, 36)$

Efficiency: Scarab can be efficient in the sense that it uses an optimized version of the order encoding [5, 6] which an award-winning CSP solver Sugar [7] adopts. Scarab also can utilize current *state-of-the-art* SAT techniques.

Customizability: Scarab allows a programmer to customize his/her own constraints and to customize the search strategies. Scarab itself can be also customizable since it is 500 lines long without any comments. In particular, our core part of order encoding module is only 25 lines long.

Portability: The current version of Scarab adopts Sat4j [8] as the back-end SAT solver. The combination of Scarab and Sat4j makes it possible to develop portable SAT-based systems that run on any platform supporting Java.

2   Modeling in **Scarab**

Scarab DSL is implemented as an embedded DSL on Scala. Let T, V, C and B be the Scarab objects of `Term`, `Var`, `Constraint`, and `Bool`. Let Int, $String$, Seq, and Any be categories of integers, strings, sequences, and any objects of Scala respectively. The following shows the syntax of Scarab DSL. Note that Scarab programs are not restricted to this DSL, we can combine it with Scala program as is shown later by `alldiff` used in the Latin Square example.

$$T ::= V \mid \texttt{-} \; T \mid T \; \texttt{+} \; Int \mid T \; \texttt{+} \; T \mid T \; \texttt{-} \; Int \mid T \; \texttt{-} \; T \mid T \; \texttt{*} \; Int \mid$$
$$\quad \texttt{Sum}(V, \dots) \mid \texttt{Sum}(\texttt{Seq}(V, \dots))$$
$$V ::= \texttt{Var}(String, \; String, \dots) \mid V(Any, \dots)$$
$$C ::= B \mid T \; op \; T \mid \; \texttt{!} \; C \mid C \; \texttt{\&\&} \; C \mid C \; \texttt{||} \; C \mid \texttt{alldiff}(\texttt{Seq}(T, \dots)) \mid$$
$$\quad \texttt{And}(C, \dots) \mid \texttt{And}(\texttt{Seq}(C, \dots)) \mid \texttt{Or}(C, \dots) \mid \texttt{Or}(\texttt{Seq}(C, \dots))$$
$$op ::= \; \texttt{<=} \mid \texttt{<} \mid \texttt{>=} \mid \texttt{>} \mid \texttt{===} \mid \texttt{!==}$$
$$B ::= \texttt{Bool}(String, \; String, \dots) \mid B(Any, \dots)$$

Let us consider Square Packing $SP(n, s)$, which is a problem of packing a set of squares of sizes 1×1 to $n \times n$ into an enclosing square of size $s \times s$ without overlapping. For a given $SP(n, s)$, a direct model would be using integer variables $x_i, y_i \in \{0, \dots, s - i\}$ for each square i $(1 \leq i \leq n)$ such that each pair (x_i, y_i) represents the lower left coordinates of the square i. We then enforce the constraint $(x_i + i \leq x_j) \vee (x_j + j \leq x_i) \vee (y_i + i \leq y_j) \vee (y_j + j \leq y_i)$ to ensure that there is no overlapping for any distinct squares i and j $(1 \leq i < j \leq n)$.

```
1: var n: Int = 5
2: for (i <- 1 to n; j <- 1 to n)  int('x(i,j),1,n)
3: for (i <- 1 to n) {
4:   add(alldiff((1 to n).map(j => 'x(i,j))))
5:   add(alldiff((1 to n).map(j => 'x(j,i))))
6:   add(alldiff((1 to n).map(j => 'x(j,(i+j-1)%n+1))))
7:   add(alldiff((1 to n).map(j => 'x(j,(i+(j-1)*(n-1))%n+1))))}
8:
9: if (find)  println(solution)
```

2	3	5	1	4
5	1	4	2	3
4	2	3	5	1
3	5	1	4	2
1	4	2	3	5

Fig. 3. Scarab Program of $LS(5)$ **Fig. 4.** Solution of $LS(5)$

This modeling can be concisely written in Scarab (Fig. 1). In Scarab, we can use integer variables and arithmetic constraints, and all of them are encoded into SAT without the need of developing dedicated encoder. SAT solvers are then used for finding solutions. Fig. 2 shows a solution of $SP(15, 36)$.

Let us consider the Latin Square problem used in International CSP Solver Competition [9]. Latin Square $LS(n)$ is a problem of placing different n numbers into $n \times n$ matrix such that each number is occurring exactly once in each row, column, diagonally down right, and diagonally up right. For a given $LS(n)$, we use a $n \times n$ matrix of integer variables $x_{i,j} \in \{1, \ldots, n\}$ $(1 \leq i, j \leq n)$. The exact one constraints can be expressed by using *alldiff* constraints [10] that is one of the best known and most studied global constraints in Constraint Programming [11].

This modeling can be concisely written in Scarab with the help of `map` method of Scala (Fig. 3). For example, in line 4, `alldiff((1 to n).map(j => 'x(i,j)))` corresponds to *alldiff*$(x_{i,1}, x_{i,2}, \ldots, x_{i,n})$. Fig. 4 shows a solution of $LS(5)$.

3   SAT Encoding Module

Before encoding constraints to SAT, Scarab normalizes comparisons to be in the form of $\sum_{i=1}^{n} a_i x_i \leq c$, where a_i's are non-zero integers, c is an integer, and x_i's are integer variables. Constraints are then converted to be in CNF by introducing new Boolean variables as is known to Tseitin transformation [12, 13].

Order Encoding. In order encoding [6], we introduce a Boolean variable $p(x \leq c)$ for each integer variable x and each integer $c \in \{lb(x) - 1, \ldots, ub(x)\}$, where $lb(x)$ and $ub(x)$ are the lower and upper bounds of x, respectively. We also introduce $\bigwedge_{c=lb(x)+1}^{ub(x)-1}(\neg p(x \leq c - 1) \vee p(x \leq c))$ to encode an integer variable x.

To encode a comparison literal $\sum_{i=1}^{n} a_i x_i \leq c$, we use the translation (1). By recursively applying this translation, any normalized comparison can be converted to a CNF formula over Boolean variables.

```
1: def encodeLe(axs: Seq[(Int,Var)], c: Int):
2:       Seq[Seq[Literal]] = axs match {
3:
4:    case Seq((a,x)) =>
5:       if (a > 0) Seq(Seq(le(x, floorDiv(c, a))))
6:       else Seq(Seq(le(x, ceilDiv(c, a)-1).neg))
7:    case Seq((a,x), axs1 @ _*) => {
8:       if (a > 0) {
9:          for {
10:            b <- lb(x) to ub(x)
11:            clause <- encodeLe(axs1, c-a*b)
12:         } yield le(x, b-1) +: clause
13:      } else {
14:         for {
15:            b <- lb(x) to ub(x)
16:            clause <- encodeLe(axs1, c-a*b)
17:         } yield le(x, b).neg +: clause
18:   }}}
```

Fig. 5. Implementation of Order Encoding

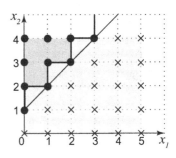

Fig. 6. Support and Conflict Region of $x_1 + 1 \leq x_2$

$$\sum_{i=1}^{n} a_i x_i \leq c = \begin{cases} x_1 \leq \lfloor c/a_1 \rfloor & (n = 1, a_1 > 0) \quad \text{(1a)} \\ \neg(x_1 \leq \lceil c/a_1 \rceil - 1) & (n = 1, a_1 < 0) \quad \text{(1b)} \\ \bigwedge_{b=lb(x_1)}^{ub(x_1)} \left((x_1 \leq b-1) \vee \sum_{i=2}^{n} a_i x_i \leq c - a_1 b \right) & (n \geq 2, a_1 > 0) \quad \text{(1c)} \\ \bigwedge_{b=lb(x_1)}^{ub(x_1)} \left(\neg(x_1 \leq b) \vee \sum_{i=2}^{n} a_i x_i \leq c - a_1 b \right) & (n \geq 2, a_1 < 0) \quad \text{(1d)} \end{cases}$$

Implementation of Order Encoding. Fig. 5 shows an implementation of the translation (1) in Scarab. The inputs are a sequence $\{(a_1, x_1), \ldots, (a_n, x_n)\}$, and an integer c (line 1). Lines 5 and 6 correspond to the translation (1a) and (1b), respectively. Lines 9 to 12 correspond to the translation (1c). In line 11, encodeLe is recursively called with $\{(a_2, x_2), \ldots, (a_n, x_n)\}$ and $c - a_1 b$. Similarly, lines 14 to 17 correspond to the translation (1d).

For instance, let us consider a part of non-overlapping constraint $x_1 + 1 \leq x_2$ in Square Packing $SP(2,6)$, where $x_1 \in \{0, \ldots, 5\}$ and $x_2 \in \{0, \ldots, 4\}$. At first, $x_1 + 1 \leq x_2$ is converted to a comparison literal $x_1 - x_2 \leq -1$. By calling encodeLe in Fig. 5 with arguments of $\{(1, x_1), (-1, x_2)\}$ and -1, we obtain the following CNF formula that represents the support and conflict region in Fig. 6: $\neg p(x_2 \leq 0) \wedge (p(x_1 \leq 0) \vee \neg p(x_2 \leq 1)) \wedge (p(x_1 \leq 1) \vee \neg p(x_2 \leq 2)) \wedge (p(x_1 \leq 2) \vee \neg p(x_2 \leq 3)) \wedge p(x_1 \leq 3) \wedge p(x_1 \leq 4)$.

Actual implementation of encodeLe is written in less than 25 lines. It is so compact that order encoding module can be modified to meet the developers' requirements. Developers also can implement a variety of encoding methods [14–19, 6, 20] by using the encoding interface of Scarab. Developers only need to implement methods of encoding integer variables, encoding clauses, and decoding.

```
1: def myalldiff(xs: Seq[Var]) = {
2:   And(And(for (Seq(x, y) <- xs.combinations(2)) yield x !== y))
3: }
```

(a) `myalldiff` version 1

```
1: def myalldiff(xs: Seq[Var]) = {
2:   var lb = for (x <- xs) yield csp.dom(x).lb
3:   var ub = for (x <- xs) yield csp.dom(x).ub
4:   And(And(for (Seq(x, y) <- xs.combinations(2)) yield x !== y),
5:     And(for (num <- lb.min to ub.max) yield Or(for (x <- xs) yield x === num)))
6: }
```

(b) `myalldiff` version 2

```
1: def myalldiff(xs: Seq[Var]) = {
2:   var lb = for (x <- xs) yield csp.dom(x).lb
3:   var ub = for (x <- xs) yield csp.dom(x).ub
4:   And(And(for (Seq(x, y) <- xs.combinations(2)) yield x !== y),
5:     And(for (num <- lb.min to ub.max) yield Or(for (x <- xs) yield x === num)),
6:     Or(for (x <- xs) yield !(x < lb.min+xs.size-1)),
7:     Or(for (x <- xs) yield !(x > ub.max-xs.size+1)))
8: }
```

(c) `myalldiff` version 3

Fig. 7. Customizing `alldiff` Constraints

4 System Development in **Scarab**

In Scarab, constraint models and search strategies are intended to be modeled and/or customized to meet developers' requirement. This section provides examples of customizing constraint models and search strategies.

Customizing Constraint Model. We present how to model and customize developers' own *alldiff* constraint by tracing a prototyping process in Scarab. Suppose that the first version of *alldiff* is given by Fig. 7 (a). Line 2 exactly corresponds to the original definition of *alldiff* constraint: $\bigwedge_{i<j}(x_i \neq x_j)$.

You can use this `myalldiff` for the Latin Square program (Fig. 3) by replacing `alldiff` with `myalldiff`. However, you probably encounter that your Latin Square program does not scale to $LS(8)$. Some improvement is then necessary.

Considering Latin Square, all integer variables $x_1, \ldots, x_n$ have the same domain $\{1, \ldots, n\}$. Then, permutation constraints would be helpful: $\bigwedge_{i=lb}^{ub} \bigvee_{j=1}^{n}(x_j = i)$, where lb and ub are the lower and upper bounds of $\{x_1, \ldots, x_n\}$. In fact, `myalldiff` version 2 in Fig. 7 (b) scales up to $LS(12)$ but $LS(13)$ cannot be solved within one hour. Thus, more improvements should be considered.

In the literature [7], Sugar adds extra pigeon hole constraints to *alldiff* constraints: $\neg \bigwedge(x_i < lb + n - 1)$ and $\neg \bigwedge(x_i > ub - n + 1)$. Then, we finally obtain `myalldiff` version 3 in Fig. 7 (c). In our environment, this version can solve $LS(13)$ within a second while other versions could not solve within an hour.

```
1: var lb = 15; var ub = s; int('m, lb, ub)
2:
3: for (i <- 1 to n)
4:    add(('x(i)+i <= 'm) && ('y(i)+i <= 'm))
5:
6: while(lb <= ub && find('m <= ub)) {
7:    add('m <= ub); ub -= 1
8: }
9:
10: while(find)
11:     println(solution)
```

```
1: var lb = 15; var ub = s; commit
2:
3: while(lb < ub) {
4:    var size = (lb + ub) / 2
5:    for (i <- 1 to n)
6:       add(('x(i)+i<=size)&&('y(i)+i<=size))
7:    if (find) {
8:       ub = size; commit
9:    } else {
10:      lb = size + 1; rollback
11:   }}
```

(a) Decremental Search   (b) Binary Search

Fig. 8. Search Strategies for Solving OSP

Customizing Search Strategy. Currently, Scarab supports the following solving techniques: (i) incremental SAT solving (in default), (ii) find method with assumption constraints, and (iii) commit and rollback of the addition of constraints. Also, solution enumeration is possible by multiple calls of find method. Using them, we show how to specify search strategies and customize them for an optimization version of Square Packing (OSP), which is a problem of finding the minimum size enclosing square capable of packing all squares of sizes 1×1 to $n \times n$ without overlapping.

Suppose that all constraints of Square Packing are defined. Then, the most direct model for OSP would be using an integer variable $m \in \{lb, \ldots, ub\}$ and the constraint $\bigwedge_{i=1}^{n}(x_i + i \leq m) \wedge (y_i + i \leq m)$, where lb and ub are the lower and upper bounds of the enclosing square size. Then, the objective is to minimize m such that encoded problems are satisfiable.

For solving OSP, a decremental search can be considered (Fig. 8 (a)). We can run this program by adding it to the bottom of the program of Fig. 1. In line 6, find method is called with an assumption $m \leq ub$. If a solution exists, a constraint $m \leq ub$ is permanently added and ub is decremented (line 7). If it is necessary, we can enumerate all optimal solutions by multiple calls of find method (line 10). In addition, Fig. 8 (b) shows a binary search by using commit/rollback methods of Scarab, which dynamically removes constraints added since the last commit point. As is shown in these programs, search strategies can be simply written, which allows us to focus on modeling those strategies.

Interface to Sat4j. Scarab provides an interface to the back-end SAT solver Sat4j [8]. Since Java libraries can be directly called from Scala, Scarab cooperates with Sat4j without generating CNF files, invoking outer processes, and parsing output logs. This interface module is also customizable and it is possible to introduce more advanced SAT technologies from Sat4j, which also provides Max-SAT, Pseudo-Boolean, and Minimal Unsatisfiable Subsets solvers.

5   Related Work and Future Work

Numberjack [21] is a modeling package written in Python for Constraint Programming. It is well designed and shares common features with Scarab. A difference is that Scarab adopts Scala, which is type-safe and has an advantage to Python in performance. There are other tools for Constraint Programming integrated in Scala [22–26]. In particular, Copris [22] is a Constraint Programming DSL embedded in Scala, which also adopts SAT solvers in its back-end. Differences from Copris are that Scarab is more compact and easier to customize.

An interesting extension is to introduce more advanced SAT technologies from Sat4j. It allows us to further SAT applications with the state-of-the-art technologies. The source code and information of Scarab is available in: http://kix.istc.kobe-u.ac.jp/~soh/scarab/.

References

1. Biere, A., Heule, M., van Maaren, H., Walsh, T.: Handbook of Satisfiability. IOS Press (2009)
2. Odersky, M., Altherr, P., Cremet, V., Emir, B., Maneth, S., Micheloud, S., Mihaylov, N., Schinz, M., Stenman, E., Zenger, M.: An overview of the scala programming language. Technical Report IC/2004/64, École Polytechnique Fédérale de Lausanne, Switzerland (2004)
3. Odersky, M., Spoon, L., Venners, B.: Programming in Scala, 2nd edn. Artima, Inc. (2010)
4. Mernik, M., Heering, J., Sloane, A.M.: When and how to develop domain-specific languages. ACM Computing Surveys 37(4), 316–344 (2005)
5. Tamura, N., Taga, A., Kitagawa, S., Banbara, M.: Compiling finite linear CSP into SAT. In: Benhamou, F. (ed.) CP 2006. LNCS, vol. 4204, pp. 590–603. Springer, Heidelberg (2006)
6. Tamura, N., Taga, A., Kitagawa, S., Banbara, M.: Compiling finite linear CSP into SAT. Constraints 14(2), 254–272 (2009)
7. Tamura, N., Tanjo, T., Banbara, M.: System description of a SAT-based CSP solver sugar. In: Proceedings of the 3rd International CSP Solver Competition, pp. 71–75 (2008)
8. Berre, D.L., Parrain, A.: The Sat4j library, release 2. 2. Journal on Satisfiability, Boolean Modeling and Computation 7, 59–64 (2010) system description
9. van Dongen, M.R.C., Lecoutre, C., Roussel, O. (eds.): Proceedings of the Second International CSP Solver Competition (2008)
10. Régin, J.C.: A filtering algorithm for constraints of difference in CSPs. In: AAAI 1994, pp. 362–367 (1994)
11. Rossi, F., van Beek, P., Walsh, T. (eds.): Handbook of Constraint Programming. Elsevier (2006)
12. Tseitin, G.S.: On the complexity of derivations in the propositional calculus. Studies in Mathematics and Mathematical Logic Part II, 115–125 (1968)
13. Prestwich, S.: 12: CNF encodings. In: Handbook of Satisfiability, pp. 75–97. IOS Press (2009)
14. Iwama, K., Miyazaki, S.: SAT-variable complexity of hard combinatorial problems. In: Proceedings of the World Computer Congress of the IFIP, pp. 253–258 (1994)

15. Walsh, T.: SAT v CSP. In: Dechter, R. (ed.) CP 2000. LNCS, vol. 1894, pp. 441–456. Springer, Heidelberg (2000)
16. Gent, I.P.: Arc consistency in SAT. In: Proceedings of the 15th Eureopean Conference on Artificial Intelligence, ECAI 2002, pp. 121–125 (2002)
17. Gent, I.P., Nightingale, P.: A new encoding of alldifferent into SAT. In: Proceedings of the 3rd International Workshop on Modelling and Reformulating Constraint Satisfaction Problems, pp. 95–110 (2004)
18. Ansótegui, C., Manyà, F.: Mapping problems with finite-domain variables to problems with boolean variables. In: Hoos, H.H., Mitchell, D.G. (eds.) SAT 2004. LNCS, vol. 3542, pp. 1–15. Springer, Heidelberg (2005)
19. Gavanelli, M.: The log-support encoding of CSP into SAT. In: Bessière, C. (ed.) CP 2007. LNCS, vol. 4741, pp. 815–822. Springer, Heidelberg (2007)
20. Tanjo, T., Tamura, N., Banbara, M.: Proposal of a compact and efficient SAT encoding using a numeral system of any base. In: Proceedings of the 1st International Workshop on the Cross-Fertilization Between CSP and SAT, p. 15 (2011)
21. Hebrard, E., O'Mahony, E., O'Sullivan, B.: Constraint programming and combinatorial optimisation in numberjack. In: Lodi, A., Milano, M., Toth, P. (eds.) CPAIOR 2010. LNCS, vol. 6140, pp. 181–185. Springer, Heidelberg (2010)
22. Copris: http://bach.istc.kobe-u.ac.jp/copris/
23. SCP: http://lara.epfl.ch/web2010/scp
24. scalasmt: http://code.google.com/p/scalasmt/
25. OscaR Team: Oscar: Scala in or (2012), https://bitbucket.org/oscarlib/oscar
26. JaCoP: web page, http://jacop.osolpro.com

Author Index